Slope

If $P_1(x_1, y_1)$ and $P_2(x_2, y_2)$ are two different points on a nonvertical line, then the slope m of the line is given by the formula

$$m = \frac{y_2 - y_1}{x_2 - x_1}.$$

Straight Line: Point-Slope Form

If a line contains the point $P(x_1, y_1)$ and has slope m, then the point slope form of the equation of a line is

$$y - y_1 = m(x - x_1).$$

Straight Line: Slope-Intercept Form

If a line has a y-intercept b and slope m, then the slope-intercept form of the equation of a line is $y = mx + b$.

Distance Formula

The distance between any two points $A(x_1, y_1)$ and $B(x_2, y_2)$ is given by the formula

$$d(A, B) = \sqrt{(x_2 - x_1)^2 + (y_2 - y_1)^2}.$$

Log/Exp Principle

For $b > 0$ and $b \neq 1$, $b^A = C$ is equivalent to $\log_b C = A$.

Rules of Logarithms

$\log_b 1 = 0$

$\log_b b = 1$

$\log_b(xy) = \log_b x + \log_b y$

$\log_b(x^t) = t \log_b x$

$\log_b\left(\dfrac{x}{y}\right) = \log_b x - \log_b y$

$\log_b\left(\dfrac{1}{x}\right) = -\log_b x$

Rules for Transformations

Transformation	Change in Equation	Effect on Graph
Horizontal translation	Replace x with $x - h$	If $h > 0$, graph moves $|h|$ units to the right. If $h < 0$, graph moves $|h|$ units to the left.
Vertical translation	Replace y with $y - k$	If $k > 0$, graph moves $|k|$ units up. If $k < 0$, graph moves $|k|$ units down.
Reflection across y-axis	Replace x with $-x$	Graph is reflected across y-axis.
Reflection across x-axis	Replace y with $-y$	Graph is reflected across x-axis.
Vertical change in shape	$y = f(x)$ becomes $y = b \cdot f(x)$	If $b > 1$, graph is stretched away from x-axis. If $0 < b < 1$, graph is shrunk down toward x-axis.
Horizontal change in shape	$y = f(x)$ becomes $y = f(ax)$	If $a > 1$, graph is shrunk toward y-axis. If $0 < a < 1$, graph is stretched away from y-axis.

College ALGEBRA and Trigonometry

College
ALGEBRA
and Trigonometry

TIMOTHY J. KELLY
Hamilton College

RICHARD H. BALOMENOS
University of New Hampshire

JOHN T. ANDERSON
Hamilton College

HOUGHTON MIFFLIN COMPANY   BOSTON
Dallas   Geneva, Ill.   Lawrenceville, N.J.   Palo Alto

Editorial Advisors
Donald J. Albers, Menlo College, Menlo Park, Calif.
Stephen P. L. Diliberto, University of California, Berkeley, Calif.
Chung C. Yang, U.S. Naval Research Laboratory, Washington, D.C.

Cover photograph by Michel Tcherevkoff
Chapter openers by Alan Davis/Omnigraphics

Copyright © 1987 by Houghton Mifflin Company.

All rights reserved. No part of the format or content of this work may be reproduced or transmitted in any form or any means, electronic or mechanical, including photocopying and recording, or by any information storage or retrieval system, except as may be expressly permitted by the 1976 Copyright Act or in writing by the publisher. Requests for permission should be addressed in writing to Houghton Mifflin Company, One Beacon Street, Boston, Massachusetts 02108.

Printed in the United States of America
Library of Congress Catalog Card Number: 86-81564

ISBN Numbers:
Text: 0-395-33287-7
Instructor's Edition: 0-395-33288-5
Solutions Manual: 0-395-33289-3
Alternate Testing Program: 0-395-33290-7
Study Guide: 0-395-41129-7
DEFGHIJ-H-89

Contents

1 Real Numbers and Algebraic Expressions — 1

1.1 Statements and Variables — 2
A. To review methods for writing equations that represent English sentences

1.2 The Real Numbers — 6
A. To review important subsets of the real numbers
B. To review the Properties of Real Numbers used in simplifying algebraic expressions

1.3 Absolute Value, Distance, and the Real Line — 11
A. To review and apply the definition of absolute value
B. To review the real number line and the ordering of real numbers
C. To review the notion of distance on the real line

1.4 The Integers as Exponents — 16
A. To review exponential expressions of the form a^n
B. To review the Rules of Exponents and to use these rules in simplifying expressions

1.5 Polynomials: Addition, Subtraction, and Multiplication — 22
A. To review the terminology used to describe polynomials
B. To review techniques for adding and subtracting polynomials
C. To review methods for multiplying polynomials and raising polynomials to powers

1.6 Polynomials: Factoring — 28
A. To find the greatest common factor of two or more monomials
B. To develop four strategies for factoring polynomials

1.7 Radicals and Rational Exponents — 36
A. To review the definition of the radical expression $\sqrt[n]{a}$ (nth root of a)
B. To review and apply the rules for simplifying radical expressions
C. To review methods for simplifying sums, differences, products, and quotients involving radicals
D. To review methods for simplifying expressions involving rational exponents ($a^{m/n}$)

1.8 Fractional Expressions — 45
A. To review methods for simplifying rational expressions (quotients of polynomials)
B. To review techniques for adding, subtracting, multiplying, and dividing rational expressions
C. To review a method of simplifying other fractional expressions

Chapter Review and Test — 51

2 Equations and Inequalities in One Variable — 55

2.1 Introduction to Equations — 56
A. To review methods for solving linear equations
B. To review methods for solving fractional equations
C. To review methods for solving absolute value equations

2.2 Quadratic Equations — 63
A. To solve quadratic equations by factoring
B. To solve quadratic equations by completing the square
C. To solve quadratic equations by using the quadratic formula

2.3 Nonlinear Equations — 70
A. To use the knowledge of quadratic equations to solve higher degree and fractional equations
B. To solve equations that are quadratic in form
C. To review methods for solving radical equations

2.4 Inequalities — 76
A. To review methods for solving linear inequalities
B. To solve absolute value inequalities
C. To solve other types of inequalities

2.5 Problem Solving — 86
A. To introduce three strategies for problem solving

Chapter Review and Test — 93

Cumulative Test: Chapters 1 and 2 — 97

3 Graphs and Functions — 99

3.1 Equations and Graphs — 100
A. To graph equations in a coordinate plane
B. To reflect a set of points across a line or point

3.2 Introduction to Functions — 108
A. To use a set of ordered pairs to describe a function
B. To evaluate functions defined by equations
C. To determine whether a graph is symmetric with respect to the x-axis, y-axis, or the origin, and to use symmetry to graph equations

3.3 Transformations and Graphing — 114
A. To graph functions by using horizontal or vertical translations
B. To graph functions by using reflections across the x-axis or y-axis
C. To graph functions by stretching or shrinking
D. To sketch the graph of a function requiring more than one transformation of a known graph

3.4 The Linear Function — 123
A. To determine the slope of a line and to sketch the line
B. To determine the equation of a line

3.5 The Quadratic Function — 130
A. To analyze the graphs of functions described by equations of the form $f(x) = a(x - h)^2 + k$
B. To graph functions of the form $f(x) = ax^2 + bx + c$

3.6 Composite Functions — 136
A. To evaluate composite functions
B. To find an equation for the composite function $f \circ g$ given equations for f and g
C. To describe a given function as a composition of other functions

Chapter Review and Test — 142

4 Polynomial Functions, Rational Functions, and Conic Sections — 147

4.1 Important Features of Polynomial Functions — 148
A. To study nth degree polynomial functions
B. To examine the graph of a polynomial function of the form $f(x) = ax^n$
C. To examine the behavior of a polynomial function for extreme values of x

4.2 Graphing Polynomial Functions — 157
A. To use x-intercepts to graph a polynomial function
B. To determine the zeros of a given polynomial function
C. To sketch the graph of a polynomial function

4.3 Rational Functions — 167
A. To determine the horizontal and vertical asymptotes of the graph of a rational function
B. To sketch the graph of a rational function

4.4 Conic Sections — 175
A. To sketch the graph of a parabola
B. To sketch the graph of a circle
C. To sketch the graph of an ellipse
D. To sketch the graph of a hyperbola

Chapter Review and Test — 189

5 Systems of Equations and Inequalities — 193

5.1 Systems of Equations in Two Variables — 194
A. To determine whether a given ordered pair is a solution of a system of equations
B. To solve a system of two linear equations in two variables by the addition method
C. To solve a system of two linear equations in two variables by the substitution method
D. To solve a system of nonlinear equations in two variables

5.2 Systems of Inequalities in Two Variables — 204
A. To graph the solution set of a single inequality in two variables
B. To graph the solution set of a system of inequalities in two variables

5.3 Systems of Linear Equations in Three Variables — 210
A. To plot ordered triples and graph planes
B. To solve a system of three linear equations in three variables

5.4 Matrices — 216
A. To represent a system of linear equations by means of a matrix and vice versa
B. To solve linear systems by using matrices

5.5 Determinants — 225
A. To use determinants to solve systems of two linear equations in two variables
B. To evaluate the determinant of an $n \times n$ matrix using signed minors
C. To use Cramer's Rule to solve a system of n linear equations in n variables

5.6 Linear Programming — 235
A. To solve linear programming problems in two variables

Chapter Review and Test — 243

Cumulative Test: Chapters 3–5 — 247

6 Exponential and Logarithmic Functions — 249

6.1 Exponential Functions — 250
A. To evaluate and simplify expressions containing exponents
B. To sketch the graph of an exponential function
C. To solve exponential equations

6.2 Logarithmic Functions — 255
A. To sketch the graph of an inverse function
B. To rewrite an exponential equation as a logarithmic equation and vice versa
C. To graph logarithmic functions
D. To use the rules of logarithms to simplify and evaluate logarithmic expressions

6.3 Common Logarithms — 263
A. To use a table of values to estimate common logarithms
B. To use a table to estimate common antilogarithms

6.4 Linear Interpolation — 267
A. To use linear interpolation to approximate function values
B. To use linear interpolation to estimate common logarithms and antilogarithms

6.5 Applications and the Natural Logarithm — 271
A. To use common logarithms to solve applied problems
B. To use natural logarithms to solve applied problems

Chapter Review and Test — 277

7 Trigonometry: An Introduction — 281

7.1 Triangles — 282
A. To determine sides and angles of the special types of triangles
B. To solve problems involving similar triangles

7.2 Trigonometric Ratios of Acute Angles 289
A. To determine the trigonometric ratios of an acute angle in a right triangle
B. To determine the trigonometric ratios of special angles (angles with measure 30°, 45°, or 60°)
C. To determine trigonometric ratios of an angle given one ratio

7.3 Angles of Rotation 296
A. To sketch an arbitrary angle as a rotation
B. To determine trigonometric function values of arbitrary angles

7.4 Identities and Tables 302
A. To use basic trigonometric identities
B. To use the tables of trigonometric function values

7.5 Reference Angles 309
A. To determine the sign of a trigonometric function value
B. To determine the reference angle for an angle of any measure
C. To use reference angles to determine trigonometric function values

Chapter Review and Test 313

8 Trigonometric Functions 317

8.1 Radian Measure and the Unit Circle 318
A. To determine points and standard arcs on the unit circle
B. To determine the radian measure of an angle
C. To solve problems using radian measure

8.2 Trigonometric Functions of Real Numbers 325
A. To evaluate trigonometric functions whose domains are real numbers

8.3 The Trigonometric Functions: Basic Graphs and Properties 333
A. To graph the sine and cosine functions
B. To graph the other trigonometric functions

8.4 Transformations of the Trigonometric Functions 344
A. To graph functions of the form $y = a \sin bx$ and $y = a \cos bx$
B. To graph functions of the form $y = k + a \sin b(x - h)$, and $y = k + a \cos b(x - h)$
C. To graph by the addition of ordinates

Contents   xi

8.5 The Inverse Trigonometric Functions   351
A. To use special angles to evaluate the inverse trigonometric functions
B. To use right triangles to evaluate trigonometric and inverse trigonometric functions.

Chapter Review and Test   359

Cumulative Test: Chapters 6–8   363

9 Trigonometric Identities and Equations   365

9.1 Basic Trigonometric Identities   366
A. To simplify an expression by using identities
B. To verify simple identities

9.2 Sum and Difference Identities   371
A. To apply the sum and difference identities

9.3 The Double-Angle and Half-Angle Identities   378
A. To apply the double-angle and half-angle identities

9.4 Identities Revisited   385
A. To simplify a trigonometric expression using identities
B. To verify trigonometric identities
C. To use conversion identities

9.5 Trigonometric Equations   391
A. To solve trigonometric equations using knowledge of the special angles
B. To solve trigonometric equations by using tables or a calculator
C. To solve equations involving inverse trigonometric functions

Chapter Review and Test   399

10 Applications in Trigonometry   403

10.1 Applications Involving Right Triangles   404
A. To solve a right triangle
B. To solve applied problems by using right triangles

10.2 Law of Sines   410
A. To solve a triangle, given one side and two angles
B. To solve a triangle given two sides and an angle opposite one of them
C. To solve applied problems by using the Law of Sines

10.3 Law of Cosines — 416
A. To solve a triangle given two sides and the included angle, or three sides
B. To solve problems involving the area of a triangle

10.4 Vectors in the Plane — 421
A. To determine the magnitude and direction of a vector
B. To describe vectors as ordered pairs

10.5 Vector Applications and the Dot Product — 428
A. To solve applied problems by using vector methods
B. To compute the dot product of two vectors

10.6 Simple Harmonic Motion — 434
A. To solve problems involving simple harmonic motion

10.7 Polar Coordinates — 440
A. To plot points given their polar coordinates
B. To translate ordered pairs and equations from rectangular form to polar form and vice versa
C. To graph a polar equation

Chapter Review and Test — 447

Cumulative Test: Chapters 9 and 10 — 451

11 Complex Numbers — 453

11.1 Introduction to Complex Numbers — 454
A. To identify and simplify complex numbers
B. To add, subtract, and multiply complex numbers

11.2 Properties of Complex Numbers — 464
A. To write the quotient of two complex numbers in the form $a + bi$
B. To use the symbol z to represent the conjugate of a complex number
C. To represent a complex number as a point in the complex plane

11.3 Zeros of Polynomial Functions — 472
A. To solve quadratic equations over the set of complex numbers
B. To determine the complex zeros of a quadratic function
C. To solve higher degree polynomial equations

11.4 Trigonometric Form of Complex Numbers — 480
A. To express a complex number in trigonometric form
B. To determine the product and quotient of two complex numbers
C. To determine powers and roots of a complex number

Chapter Review and Test — 487

12 Sequences, Series, and Limits — 491

12.1 Arithmetic Sequences and Sums — 492
A. To determine the terms of a sequence
B. To determine the terms of an arithmetic sequence
C. To find the sum of the first n terms of an arithmetic sequence

12.2 Geometric Sequences and Sums — 499
A. To determine the terms of a geometric sequence
B. To find the sum of the first n terms of a geometric sequence

12.3 Series — 503
A. To use sigma notation to describe a sum
B. To compute partial sums of a series
C. To compute the sum of a series, if the sum exists

12.4 Limits — 510
A. To evaluate the limit of a function described by a graph
B. To determine the limit of a sequence

12.5 Mathematical Induction — 515
A. To use mathematical induction to prove that a statement is true for all positive integers

12.6 The Binomial Theorem — 519
A. To use the Binomial Theorem to expand expressions of the form $(a + b)^n$

Chapter Review and Test — 526

Cumulative Test: Chapters 11 and 12 — 531

Appendix — A1
Calculator Exercises — A1

Tables — A23

Answers to Exercises — A35

Index — A109

Preface

In view of the large number of college algebra and trigonometry texts that are currently available, one might justifiably question the need for writing this particular COLLEGE ALGEBRA AND TRIGONOMETRY series. The answer is simple. The books in this series have been written *to be read!* Students consistently report that precalculus textbooks are unreadable. In developing the manuscript for the Kelly/Balomenos/Anderson COLLEGE ALGEBRA AND TRIGONOMETRY books, the authors have been guided by specific student comments indicating places where more explanation, illustration, or review is needed in order to assure understanding and maintain the flow in reading. It is hoped that both the student and instructor will come to rely on these books for their sound, complete, yet careful explanations.

Pedagogical Features of the Text

Readability As indicated above, this text represents a very serious effort to produce exposition that is accessible to the precalculus student. At each stage of manuscript development, student comments were seriously considered in making explanations more comprehensible. Ultimately, a Field Test Edition was used in actual classroom trials across the country. Final revisions to the manuscript reflect both student and instructor comments from these trials.

Organization Each chapter of this book is divided into sections, and within each section, the material is organized around specific goal statements. These goals clearly identify for the student the purpose of the discussion that is to follow. Furthermore, exercise sets, detailed chapter reviews, and comprehensive chapter tests are all organized by this same goal-referencing system.

Examples As a further aid to student understanding, worked examples in the text have been enhanced by annotations that explain how the various steps in the solution were obtained. The intent is to simulate, as closely as possible, the verbal explanation that an instructor would supply in presenting problems at the blackboard. Both students and instructors involved in field testing the text found this feature to be of enormous value.

Exercise Sets Exercises following each section are classified according to the goals stated in the section. Under each specific goal, problems are quite simple at the outset, then gradually increase in difficulty. Problems in any odd-even pairing (as exercises 5 and 6, for example) are always at the same level of difficulty. At the end of each exercise set, there are "Superset" problems, which are intended to extend and challenge the student's understanding.

Content Features of the Text

Content Emphases Decisions with respect to topical inclusion, exclusion, and sequencing reflect the authors' beliefs regarding what is most crucial and most beneficial to a mathematics student in a precalculus course. Moreover, some highly useful information came as a result of two national surveys, one in 1982, and the other in 1985. Finally, the judgements of the mathematical advisers and the field-testers of the program served as valuable input in forming the present version of the program.

Functions The text presents functions as sets of ordered pairs as well as mappings (or rules). Both approaches seem necessary if students are to be prepared for both discrete and continuous follow-ups to this course. In this text, graphs have been fully exploited as a way of exploring a function's properties. It is hoped that students will leave this course with a significantly enlarged set of visualization skills.

Transformations Translations, stretchings, and shrinkings have been used extensively in unifying the algebra and geometry of functions. Students are frequently reminded of how simple changes in the algebraic form of a function correspond to changes in its graphical form.

Problem Solving One of the authors, Professor Kelly, wishes especially to acknowledge the insight in this regard that was offered to him by his teacher, George Polya. Above all, Professor Polya emphasized the development of a "repertoire" of problem solving strategies that a student could bring to bear on a problem. It is hoped that such an approach is fostered, in some modest way, in the treatment of Problem Solving strategies in the text.

Calculators Calculator exercises are referenced in the Exercise Sets throughout the text, by means of the symbol $\boxed{=}$. In many instances these exercises provide students with the opportunity to discover, through computation, some patterns that are of significance in mathematics. In other cases, they present students with some computational complexities that are sometimes unavoidable in applied problems.

Acknowledgements

An expression of gratitude is due to the following field testers for their contribution to the shaping of this program:

John Graves, Auburn University at Montgomery
Wayne Mackey, Johnson County Community College
Rosario Diprizio, Oakton Community College
Philip Farmer, College of Marin
M. R. Childers, University of Texas at Arlington
Phyllis Cox, Shelby State Community College
Enrico Serpone, Phoenix College
Arthur Dull, Diablo Valley College

In addition, the following individuals have assisted greatly in making valuable mathematical and pedagogical suggestions at various times in the development of the manuscript.

Larry Knop, Hamilton College
Anne Ludington, Loyola College (Baltimore)
Ellen O'Keefe, Wheaton College (Northampton)
Michael Schramm, Hamilton College

Supplements

A number of supplements have been created as aids for both students and instructors. There are: an *Instructor's Edition*, which contains answers to all problems; a *Study Guide* for the student which contains a guide to each chapter and complete solutions to the odd-numbered problems; a *Solutions Manual*, which contains complete solutions to all problems; and an *Alternate Testing Program*, which contains a complete battery of goal-referenced chapter tests, cumulative tests, and a final exam, available to the instructor.

T. J. K.
R. H. B.
J. T. A.

1 Real Numbers and Algebraic Expressions

Real Numbers

Irrational Numbers

Rational Numbers

Integers

0

Positive Integers

Negative Integers

1.1 Statements and Variables

Goal A *to review methods for writing equations that represent English sentences*

In mathematics, as in everyday language, we use statements to communicate ideas. In algebra, our statements are about numbers. But don't be mistaken! Algebra can be used to solve a wide variety of "real-life" problems. The skill in using algebra lies in translating such problems into statements about numbers.

We begin our review of algebra with two kinds of problems which may be familiar to you. They represent two different types of translation problems: describing a specific unknown value, and making generalizations about many values.

Problem 1. Five less than twice a certain number is one more than the number. What is the number?

Here, we are given information about a "certain number." Even though we do not yet know what the value of the number is, we can give it a name: x. We call x a **variable.** We use a variable to talk about a number, whose value is not yet known.

English Sentence	Five less than twice a certain number	is	one more than the number
Algebraic Equation	$2x - 5$	$=$	$x + 1$

Recall that when two **expressions,** such as $2x - 5$ and $x + 1$ represent the same value, they can be written on opposite sides of an equals sign. The resulting statement is called an **equation.** The equation

$$2x - 5 = x + 1$$

summarizes, in algebraic form, the information about the unknown number x. The equation is then "solved for x." For now, don't worry about solving equations—that is reviewed in Chapter 2. Our purpose in stating Problem 1 is to suggest one important use of algebra:

> *To write an equation that summarizes the information about a specific number whose value is not yet known. By solving the equation, we discover the value of this number.*

Now consider a slightly different type of problem.

Problem 2. Think of a number. Double it. Add ten. Now take half of that result. Subtract your original number. Your answer is five.

To help understand the reason this "number trick" works, let us see what happens to a number that undergoes the process. Here we let the variable x stand for any number.

- Think of a number............................ x
- Double it...................................... $2x$
- Add 10.. $2x + 10$
- Take half..................................... $\frac{1}{2}(2x + 10)$
- Subtract your original number .... $\frac{1}{2}(2x + 10) - x$
- Your answer is 5.............................. $\frac{1}{2}(2x + 10) - x = 5$

The reason the trick works is that the equation

$$\frac{1}{2}(2x + 10) - x = 5$$

is always true: no matter what number is chosen as the value of x, the expression $\frac{1}{2}(2x + 10) - x$ is always equal to 5. (Try it for some values.) Problem 2 suggests another important use of algebra:

> *To state an* **identity,** *i.e. a statement that is true for all numbers.*

Let us now summarize some important terminology.

Word	Meaning	Examples
Variable	A symbol (usually a letter) used to represent a number	x is the variable in $3x - 1$; x and y are the variables in $3x + 7y$.
Expression	A combination of numbers, variables, and operation symbols $(+, -, \times, \div, \sqrt{\ })$	$4 + 2$, $5x - 13$, $\frac{x + y}{5}$, $5x^2 - xyz$
Equation	A statement that two expressions are equal	$2x + 1 = 11$, $x + y = 1$

Note that if two expressions, such as $7 + 2$ and $8 + 10$, do not represent the same number, then we can write $7 + 2 \neq 8 + 10$, where the symbol "$\neq$" is read "is not equal to."

Example 1 Translate each of the following sentences into equations.

(a) Three times a certain number is eight less than the number.
(b) Three times a certain number is eight less the number.
(c) The product of two consecutive integers is four more than the smaller integer.
(d) The product of two numbers is five more than their average.
(e) The sum of two numbers is the same, regardless of the order in which they are added.
(f) One hundred more than a certain number is the same as one hundred less than the number.

Solution

(a) $3x = x - 8$

(b) $3x = 8 - x$

(c) $n(n + 1) = n + 4$ ■ Since the two integers are consecutive, we can call the smaller one n and the larger one $n + 1$, thereby using only one variable.

(d) $xy = \frac{1}{2}(x + y) + 5$ ■ Since there is only one relationship between the two numbers, we must use two variables.

(e) $x + y = y + x$ ■ This is an identity; it is true for any two numbers represented by x and y.

(f) $x + 100 = x - 100$ ■ The equation faithfully represents the English sentence given. Of course, there can be no such number x. If there were, then a pay raise of $100 and a pay cut of $100 would mean the same thing.

Exercise Set 1.1

Goal A

In exercises 1–8, translate the given statement into an equation with one variable.

1. Eight times a certain number is forty-seven.
2. Seven less than a number is fourteen.
3. Five more than a number is twice the number.
4. Ten more than a number is five times the number.
5. One third added to three-fifths of a given number is twice the number.
6. Seventy-five percent of a number is twelve less than the number.
7. The sum of two consecutive integers is three times the smaller integer.
8. The product of two consecutive integers is seven more than the square of the smaller integer.

In exercises 9–16, translate the given statement into an equation with two variables.

9. The sum of two numbers is eleven.
10. The product of two numbers is fifty-three.
11. Twice a number is fifteen less than another number.
12. Five times a number is ten more than another number.
13. Six more than one number is eight less than twice another number.
14. Fifteen more than one number is two more than twice another number.
15. The average of two numbers is nineteen more than their product.
16. The average of two numbers is eleven less their product.

In exercises 17–22, translate the given phrase into an algebraic expression.

17. The value in cents of n dimes.
18. The number of feet in N inches.
19. The cost of an item selling for d dollars per dozen.
20. The cost of S square yards of carpeting selling for $19.95 per square yard.
21. The distance travelled if you walk n hours at 3 mph and run m hours at 7 mph.
22. The total interest on two accounts, one containing n dollars at 5% interest and the other containing m dollars at 9%.

Superset

In exercises 23–26, write an equation that represents the problem.

23. Twenty-two people were riding on a bus. At one stop no one got off, but several people got on the bus. Then there were thirty-six people on the bus. How many got on? (Let x represent the number of people who got on.)
24. A student sold 18 paperbacks to a used book dealer and had 32 left. How many did the student have originally? (Let p represent the number of paperbacks that the student had originally.)
25. Your cost for renting a car last week was $198.00. The charge quoted at the time of rental was a flat rate of $99.00 for the week, plus 20 cents per mile. How many miles did you drive the rental car? (Let m represent the number of miles that the rental car was driven.)
26. The length and width of a rectangular patio are each doubled, and as a result, the area of the patio is four times larger. If the final area is now 82.9 square meters, what was the original area? (Let a represent the original area.)

1.2 The Real Numbers

Goal A   *to review important subsets of the real numbers*

Suppose you sold subscriptions for a newspaper and had been promised a commission of 50¢ for each subscription sold. Your paycheck stub lists your total commission as $87.30. If we let n represent the number of subscriptions sold, then the equation

$$87.30 = .50n$$

models the situation. The solution of this equation is 174.6, since replacing the variable n with 174.6 produces the true statement

$$87.30 = .50(174.6).$$

But this solution makes no sense for the given problem. You could not have sold 174.6 subscriptions, since the number of subscriptions must be a whole number. Therefore, we conclude that 174.6 is not a meaningful replacement for the variable n, and the amount listed ($87.30) is in error.

When writing an equation to solve a problem, you should ask yourself, "What kinds of numbers make sense for the variable in this problem?" We now review terminology that may help you answer this question.

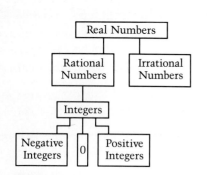

Figure 1.1

Set	Description
Integers	$\{\ldots, -3, -2, -1, 0, 1, 2, 3, \ldots\}$ The set of positive integers is $\{1, 2, 3, \ldots\}$ The set of negative integers is $\{-1, -2, -3, \ldots\}$
Rational Numbers	A rational number is any number that can be represented as a fraction $\frac{p}{q}$, where p and q are both integers, and $q \neq 0$ (for example, $\frac{17}{3}$, $-\frac{1}{3}$). Every integer p is a rational number since it can be written as the fraction $\frac{p}{1}$.
Irrational Numbers	Numbers such as $\sqrt{2}$, $-\sqrt[3]{5}$, π, and $\frac{2}{\sqrt[3]{3}}$, which are not rational numbers, are called irrational.
Real Numbers	All rational and irrational numbers, taken together, form the set of real numbers.

Every real number can be written in decimal form. One way of distinguishing a rational number from an irrational number is by looking at the number's decimal expansion.

In decimal form, *a rational number will either terminate or repeat blocks of digits.*

Terminating: $\dfrac{3}{8} = 0.375, \quad -\dfrac{7}{2} = -3.5, \quad \dfrac{1013}{500} = 2.026$

Repeating: $1\dfrac{7}{11} = 1.63636363\ldots = 1.\overline{63}$ ■ The bar over 63 means that it repeats indefinitely.

In decimal form, *an irrational number will neither terminate nor repeat.*

$$\sqrt{7} = 2.6457513\ldots$$ ■ The decimal expansion continues, but without any pattern.

Careful! When using a calculator, don't forget that certain displayed values are not exact—they are only approximations. For example, the calculator will display 2.6457513 for $\sqrt{7}$ and 1.6363636 for $1\tfrac{7}{11}$. But neither $\sqrt{7}$ nor $1\tfrac{7}{11}$ can be expressed as a terminating decimal.

Example 1 Classify each of the numbers below as one or more of the following: integer, rational number, irrational number, real number.

(a) -8 (b) 0 (c) $-\sqrt{11}$ (d) $0.958\overline{37}$ (e) $0.21356928\ldots$

Solution (a) integer, rational, real   (b) integer, rational, real   (c) irrational, real

(d) rational, real   (e) irrational, real

Not every expression represents a real number. For example, $\sqrt{-1}$ is not a real number, because -1 has no real-valued square root; that is, there is no real number whose product with itself is equal to -1. We define a special symbol for $\sqrt{-1}$:

$$\sqrt{-1} = i.$$

With this definition we can write the nonreal number $\sqrt{-4}$ as $2i$, since $\sqrt{-4} = \sqrt{4} \cdot \sqrt{-1} = 2i$. Numbers involving i are called **complex numbers.** We shall discuss complex numbers in a later chapter.

Goal B *to review the Properties of Real Numbers used in simplifying algebraic expressions*

We now list several important generalizations about the behavior of real numbers. We call these generalizations the Properties of Real Numbers. Unless otherwise noted, each variable used in the following table may represent any real number.

PROPERTIES OF REAL NUMBERS
Associative Property of Addition $(a + b) + c = a + (b + c)$
Associative Property of Multiplication $(ab)c = a(bc)$
Commutative Property of Addition $a + b = b + a$
Commutative Property of Multiplication $ab = ba$
Additive Identity Property   There exists a unique number 0 such that $a + 0 = 0 + a = a$
Multiplicative Identity Property   There exists a unique number 1 such that $a \cdot 1 = 1 \cdot a = a$
Additive Inverse Property   For each real number a, there exists a unique real number $-a$ such that $a + (-a) = (-a) + a = 0$
Multiplicative Inverse Property   For each real number a not equal to 0, there exists a unique real number $\frac{1}{a}$ such that $a \cdot \frac{1}{a} = \frac{1}{a} \cdot a = 1$
Distributive Property of Multiplication over Addition $a(b + c) = ab + ac$ $(b + c)a = ba + ca$

Example 2 Determine whether each statement is true or false.

(a) $3(8 + 4) = 24 + 12 = 36$

(b) $3(8 \cdot 4) = 24 \cdot 12 = 288$

(c) $16x + x = (16 + 1)x = 17x$

(d) $(4 + x) \cdot 3 = 4 + 3x$

(e) The additive inverse of -5 is 5.

(f) The multiplicative inverse of -5 is $-\frac{1}{5}$.

Solution (a) True. ■ This is an application of the Distributive Property.
(b) False. ■ Don't be tricked into seeing this problem as an application of the Distributive Property. The parentheses simply tell you what to multiply first: $3(8 \cdot 4) = 3(32) = 96$.
(c) True. ■ Remember that x and $1x$ mean the same thing.
(d) False. ■ The Distributive Property requires that each term within the parentheses be multiplied by 3: $(4 + x) \cdot 3 = 12 + 3x$.
(e) True. ■ This fact is sometimes represented as $-(-5) = 5$.
(f) True. ■ There are three ways to write a negative fraction: $\frac{-1}{5}$, $\frac{1}{-5}$, or $-\frac{1}{5}$; the most common is $-\frac{1}{5}$.

Careful! Don't assume that a variable without a negative sign in front of it represents a positive number. A variable, such as x, might stand for a positive number, a negative number, or zero. Likewise, don't assume that an expression such as $-2x$ must be negative. For example, if x represents -3, then $-2x$ represents 6 (i.e., $-2x = -2(-3) = +6$).

We conclude this section with some number substitution problems. Pay close attention to the signs as you practice the arithmetic of real numbers.

Example 3 Evaluate each expression, given that $x = 5$, $y = -8$, and $z = 10$.

(a) $-y$ (b) $-(-z)$ (c) $y - x$ (d) $-[x + (-y)]$ (e) $xz + x(-y)$

Solution (a) $-y = -(-8) = 8$ (b) $-(-z) = -(-10) = 10$

(c) $y - x = -8 - 5 = -13$ (d) $-[x + (-y)] = -[5 + 8] = -13$

(e) $xz + x(-y) = 5(10) + 5(8) = 90$

Exercise Set 1.2 □= Calculator Exercises in the Appendix

Goal A

In exercises 1–8, classify each number as one or more of the following: integer, rational number, irrational number, real number.

1. $\dfrac{3}{5}$
2. -5
3. 0
4. $\sqrt{7}$
5. $\dfrac{22}{7}$
6. 3.715
7. $14.\overline{36}$
8. $\sqrt[3]{8}$

Goal B

In exercises 9–18, determine whether each statement is true or false.

9. $\sqrt{3} + 2 = 2 + \sqrt{3}$
10. $7\pi = \pi \cdot 7$
11. $3 \cdot 0 \cdot 8 = 24$
12. $3(6 - 6) = 0$
13. $(2 + 7)8 = 72$
14. $8(7 - 2) = 54$
15. $9x - x = 8$
16. $3x - 7x = 4x$
17. $-(3 \cdot 5) = (-3)(-5) = 15$
18. $2(3 \cdot 4) = (2 \cdot 3)(2 \cdot 4) = 48$

In exercises 19–30, determine the additive and multiplicative inverses.

19. 5
20. -3
21. $-\dfrac{2}{3}$
22. -1
23. 0
24. 1
25. π
26. $\sqrt{2}$
27. x
28. $-x$
29. $\dfrac{1}{x}$
30. $-xy$

In exercises 31–38, evaluate each expression, given that $a = 2$, $b = -10$, and $c = 5$.

31. $-a$
32. $-b$
33. $-(a)$
34. $-(-b)$
35. $-(c - a)$
36. $-(8 - c)$
37. $ac + \dfrac{a(-b)}{c}$
38. $ca + \dfrac{c(-b)}{-a}$

Superset

In exercises 39–44, translate each statement into an equation.

39. One-fourth of a certain number is five times its multiplicative inverse.
40. The sum of a number and ten is twenty-four times its multiplicative inverse.
41. Three times the sum of a number and its reciprocal is four more than the number.
42. Twice the sum of a number and its reciprocal is one less than the number.
43. The additive inverse of the sum of two numbers is five less than their product.
44. The additive inverse of the product of two numbers is six more than twice their sum.

Specify whether each of the statements in exercises 45–50 is true or false. If false, illustrate this with an example.

45. $(2x)(2y) = 2xy$
46. $3(x + 7) = 3x + 7$
47. $-(3 - x) = x - 3$
48. $5 - (x + 2) = 5 - x + 2$
49. $\dfrac{1}{2x} = \dfrac{1}{2}x$
50. $\dfrac{1}{\left(\dfrac{x}{y}\right)} = \dfrac{y}{x}$

In exercises 51–54, add parentheses and brackets to each expression so that the value of the resulting expression is 12.

51. $3 \cdot 7 + 2 - 5$
52. $3 \cdot 5 - 10 - 7$
53. $12 - 3 \cdot 8 \div 6$
54. $12 + 3 \cdot 8 \div 3$

1.3 Absolute Value, Distance, and the Real Line

Goal A *to review and apply the definition of absolute value*

Suppose the variable a represents some nonzero real number.

If a represents a positive number, then $-a$ is negative.
(For example, if $a = 7$, then $-a = -7$.)

If a represents a negative number, then $-a$ is positive.
(For example, if $a = -10$, then $-a = -(-10) = 10$.)

Thus, for any nonzero real number a, one of the numbers, a or $-a$, must be positive. The positive one is referred to as the *absolute value* of a.

Definition
The **absolute value** of a nonzero real number a is the positive number in the set $\{a, -a\}$. The absolute value of 0 is 0. The absolute value of any real number a is denoted $|a|$. Notice that $|a| = |-a|$.

Example 1 Evaluate each of the following.

(a) $|5|$ (b) $|-10|$ (c) $-\left|-\frac{1}{2}\right|$ (d) $|-4| - |-6|$ (e) $|0|$

Solution (a) $|5| = 5$ (b) $|-10| = 10$ (c) $-\left|-\frac{1}{2}\right| = -\left(\frac{1}{2}\right) = -\frac{1}{2}$

(d) $|-4| - |-6| = (4) - (6) = -2$ (e) $|0| = 0$

Notice that the absolute value of a nonnegative number is the number itself, whereas the absolute value of a negative number is the additive inverse (opposite) of the number. We can use these two facts to write an alternate definition for absolute value:

$$|x| = \begin{cases} x, & \text{if } x \text{ represents a positive number or } 0, \\ -x, & \text{if } x \text{ represents a negative number.} \end{cases}$$

Goal B *to review the real number line and the ordering of real numbers*

It is often helpful to visualize the set of real numbers as labels for the points on a line. Suppose we agree to choose one point on the line and label it 0. We call this point the **origin.** Then, suppose we let each real number a correspond to a point lying $|a|$ units from zero. If a is positive, the point lies to the right of 0; if a is negative, the point lies to the left of 0.

The real number corresponding to a particular point on the line is called the **coordinate** of the point. Points A, B, C, and D have coordinates $-\frac{7}{2}$, $-\sqrt{2}$, $\frac{3}{4}$, and π, respectively.

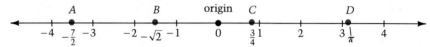

Figure 1.2 The Real Number Line.

To each real number there corresponds exactly one point on the real line, and to each point on the real line, there corresponds exactly one real number. Such a correspondence between the points on a line and the set of real numbers is called a **one-to-one correspondence.**

The relative position of two different points on the real line suggests a "greater than" or "less than" relationship between the two coordinates. For example, since 2 is to the right of -3, we say that 2 is greater than -3 and write $2 > -3$. Furthermore, since -3 is to the left of 2, we say that -3 is less than 2 and write $-3 < 2$. Relationships such as these are called **order relationships** or **inequalities.**

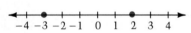

Figure 1.3 $-3 < 2$ or $2 > -3$.

Example 2 In each case, use the given symbol and numbers to write an inequality.

(a) $<$: 5, -10 (b) $>$: -1, -8 (c) $<$: $\frac{1}{3}$, 0.33   (d) $>$: π, 3.14

Solution

(a) $-10 < 5$ (b) $-1 > -8$ ■ Since -1 is to the right of -8, it is greater than -8.

(c) $0.33 < \frac{1}{3}$ (d) $\pi > 3.14$ ■ Since π can be approximated by 3.1416, we conclude π is greater than 3.14.

In Example 3, we use inequalities to translate word problems into algebraic statements.

Example 3 Translate each of the following sentences into an inequality.

(a) Three times a certain number is less than seven.
(b) Eight less than a certain number is greater than twice the number.
(c) The product of two consecutive integers is less than their sum.
(d) The quarters in my pocket amount to more than five dollars.
(e) The sum of two consecutive integers is greater than the absolute value of the sum of the smaller integer and five.

Solution

(a) $3x < 7$ ▪ Here, x represents the given number.

(b) $x - 8 > 2x$ ▪ Here, x represents the "certain" number.

(c) $n(n + 1) < 2n + 1$ ▪ Here, n represents the smaller integer, and $n + 1$ the larger. The sum is $n + (n + 1)$ or $2n + 1$.

(d) $.25q > 5$ ▪ Here, q represents the number of quarters; thus, $.25q$ represents the value of those quarters in dollars.

(e) $2n + 1 > |n + 5|$ ▪ Here, n represents the smaller integer, and $n + 1$ the larger.

Example 4 Graph each of the following sets of numbers on the real line.

(a) The set of integers between -5 and 1.
(b) The set of real numbers between -5 and 1.
(c) The set of real numbers less than or equal to 2.
(d) The set of real numbers whose absolute values are greater than 2.

Solution

(a)

(b)

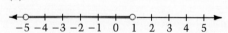

When the endpoint of a set of real numbers is excluded, the dot is left open.

(c)

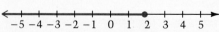

(d)

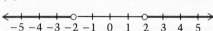

When the endpoint of a set of real numbers is included, the dot is filled in.

Goal C *to review the notion of distance on the real line*

When someone asks you to estimate the distance between two cities, your answer is certainly not a negative number. In "real-life," as well as in mathematics, distance is a nonnegative quantity.

In Figure 1.4, the distance between points A and B is 7. To determine this, we find the absolute value of the difference between the coordinates 2 and 9.

$$|2 - 9| = |-7| = 7$$

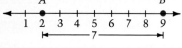

Figure 1.4 The distance between A and B is 7.

Definition

Suppose points A and B lie on the real number line and have coordinates a and b, respectively. The **distance** between A and B, denoted $d(A,B)$, is given by the equation

$$d(A,B) = |a - b|.$$

Figure 1.5 $d(A,B) = |a - b|$.

When you calculate the distance between two points, it does not matter which coordinate you consider to be a and which you consider to be b. For the example above, the distance between A and B can be found from either of the following equations:

$$d(A,B) = |2 - 9| = |-7| = 7, \text{ or}$$
$$d(B,A) = |9 - 2| = |7| = 7.$$

In general, $d(A,B) = d(B,A)$, for any points A and B.

Example 5 Using the information in the figure below, find each distance.

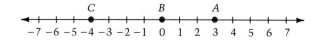

(a) $d(C,B)$ (b) $d(A,C)$ (c) $d(C,A)$ (d) $d(A,A)$

Solution

(a) $d(C,B) = |(-4) - (0)| = |-4| = 4$ ■ The distance between the origin and a point is just the absolute value of its coordinate.

(b) $d(A,C) = |(3) - (-4)| = |7| = 7$

(c) $d(C,A) = 7$ ■ Here, we use the result of part (b), together with the fact that $d(C,A) = d(A,C)$.

(d) $d(A,A) = |(3) - (3)| = |0| = 0$ ■ The distance between a point and itself is 0.

Exercise Set 1.3 = Calculator Exercises in the Appendix

Goal A

In exercises 1–10, evaluate the expression.

1. $|-4|$
2. $\left|\dfrac{5}{9}\right|$
3. $-|0|$
4. $-|-\sqrt{3}|$
5. $-\left|-\dfrac{1}{5}\right|$
6. $-|3 + (-5)|$
7. $|10| + |-3|$
8. $|-5| - |6|$
9. $|-3||-10|$
10. $\dfrac{|7 - (-2)|}{-5}$

Goal B

In exercises 11–16, use the given symbol and numbers to write an inequality.

11. $>$: $0, -11$
12. $<$: $-20, -30$
13. $>$: $\dfrac{3}{5}, \dfrac{4}{7}$
14. $<$: $-0.05, -0.5$
15. $>$: $-\dfrac{5}{8}, -0.624$
16. $<$: $\dfrac{\pi}{2}, 1.577$

In exercises 17–20, translate each sentence into an inequality.

17. Seven less than twice a certain number is greater than 20.
18. Fifteen more than a certain number is less than -10.
19. Six more than a certain number is positive.
20. Five less than a certain number is nonnegative.

In exercises 21–26, graph each set of numbers on the real line.

21. $\left\{-1, -\dfrac{1}{2}, \sqrt{2}, 2\right\}$
22. $\{10, -30, 60\}$
23. The set of integers between -3 and 2.
24. The set of real numbers greater than or equal to 0.
25. The set of real numbers less than -2.
26. The set of integers greater than $-\tfrac{1}{2}$ and less than $2\tfrac{1}{2}$.

Goal C

In exercises 27–32, use the following graph to determine each distance.

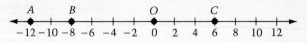

27. $d(O,A)$
28. $d(O,B)$
29. $d(C,A)$
30. $d(A,B)$
31. $d(A,C)$
32. $d(B,B)$

Superset

33. For the following graph, $d(A,B) = 7$, $d(B,C) = 10$, and $d(O,C) = 5$. Determine the values of a, b, and c.

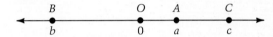

34. If a point A has coordinate a_1 and a point B has coordinate b_1, show that $d(A,B) = d(B,A)$.

In exercises 35–42, determine whether each statement is true or false. If false, illustrate this with an example.

35. $-x$ represents a negative number.
36. The absolute value of a number is always positive.
37. The sum of two negative numbers is always negative.
38. The product of two negative numbers is always negative.
39. Distance is never negative.
40. Distance is always positive.
41. $|x| = x$
42. $|x| > -x$

1.4 The Integers as Exponents

Goal A *to review exponential expressions of the form a^n*

The product $3 \cdot 3 \cdot 3 \cdot 3 \cdot 3$ is said to be in **expanded form.** It can be rewritten in **exponential form** as 3^5. In the expression 3^5, 3 is called the **base,** and 5 is called the **exponent.** In general, for any real number a and any positive integer n, the symbol a^n represents the product of n factors, each equal to a.

$$\underbrace{a \cdot a \cdot a \cdot \cdots \cdot a \cdot a}_{n \text{ factors}} = a^n$$

- n is the exponent.
- a is the base.

The symbol a^n is called "the nth power of a", "a (raised) to the nth power," or simply "a to the n." If the exponent n is 1, it is usually omitted; thus a^1 is written a.

You probably learned your first rule of exponents by noticing a pattern like the one shown below.

$$(5 \cdot 5 \cdot 5) \cdot (5 \cdot 5 \cdot 5 \cdot 5) = 5 \cdot 5 \cdot 5 \cdot 5 \cdot 5 \cdot 5 \cdot 5$$
$$5^3 \quad \cdot \quad 5^4 \quad = \quad 5^7$$

The rule is stated as follows.

Rule

If a is a real number, and if m and n are positive integers, then

$$a^m \cdot a^n = a^{m+n}.$$

If, in the equation above, we do not wish to restrict ourselves to positive exponents only, then we must explore the meaning of the symbol a^n when the exponent is 0 or a negative integer.

We begin with the case of a^0. The rule $a^m \cdot a^n = a^{m+n}$ requires that

$$a^0 \cdot a^n = a^{0+n} = a^n.$$

That is,

$$a^0 \cdot a^n = a^n,$$

and thus

$$a^0 = 1.$$

We require that $a \neq 0$ to avoid the undefined expression 0^0.

Now suppose that n is a positive integer. Then a^{-n} has a negative exponent. The rule $a^m \cdot a^n = a^{m+n}$ requires that the following be true: $a^{-n} \cdot a^n = a^{-n+n} = a^0 = 1$. That is, $a^{-n} \cdot a^n = 1$, and so a^{-n} must be the reciprocal of a^n.

$$a^{-n} = \frac{1}{a^n}$$

We require that $a \neq 0$ to avoid the undefined expression $\frac{1}{0}$.

Example 1 Evaluate.

(a) $(-2)^4$ (b) 5^{-3} (c) $\frac{1}{5^{-3}}$ (d) 0^{-4} (e) $(-3)(-3)^4$

Solution (a) $(-2)^4 = (-2)(-2)(-2)(-2) = 16$

■ Don't confuse $(-2)^4$ with -2^4.
$-2^4 = -(2^4) = -(16) = -16$.

(b) $5^{-3} = \frac{1}{5^3} = \frac{1}{125}$

(c) $\frac{1}{5^{-3}} = \frac{1}{\frac{1}{5^3}} = 5^3 = 125$

■ In (b) and (c) we have "moved" the exponential factor 5^{-3} from the numerator to the denominator or vice versa. Doing so requires that we change the sign of the exponent.
$\left(5^{-3} = \frac{1}{5^3} \text{ and } \frac{1}{5^{-3}} = 5^3.\right)$

(d) 0^{-4} is not defined.

(e) $(-3)(-3)^4 = (-3)^1(-3)^4$
$= (-3)^5 = -243$

■ Remember: a negative number to an odd power is negative, and to an even power is positive.

Goal B to review the Rules of Exponents and to use these rules in simplifying expressions

Agreement

In the rules, examples, and exercises that follow, we will be using variables for bases and exponents. Let us agree that these variables have been restricted so as to avoid undefined terms of the form $\frac{1}{0}$, 0^0, and 0 to a negative power.

We now list the rules for integer exponents that you studied in your first course in algebra.

RULES OF EXPONENTS	
Rule 1 $a^m \cdot a^n = a^{m+n}$	$a^3 \cdot a^4 = a \cdot a \cdot a \cdot a \cdot a \cdot a \cdot a = a^7$
Rule 2 $(ab)^n = a^n b^n$	$(ab)^3 = (ab)(ab)(ab) = (aaa)(bbb) = a^3 b^3$
Rule 3 $(a^m)^n = a^{mn}$	$(a^2)^3 = a^2 \cdot a^2 \cdot a^2 = (aa)(aa)(aa) = a^6$
Rule 4 $\dfrac{a^m}{a^n} = a^{m-n}$	$\dfrac{a^5}{a^2} = \dfrac{a \cdot a \cdot a \cdot a \cdot a}{a \cdot a} = a^3$
Rule 5 $\left(\dfrac{a}{b}\right)^n = \dfrac{a^n}{b^n}$	$\left(\dfrac{a}{b}\right)^3 = \dfrac{a}{b} \cdot \dfrac{a}{b} \cdot \dfrac{a}{b} = \dfrac{a \cdot a \cdot a}{b \cdot b \cdot b} = \dfrac{a^3}{b^3}$
Remember also the three important facts:	
$a^0 = 1 \qquad\qquad a = a^1 \qquad\qquad a^{-n} = \dfrac{1}{a^n}$	

Example 2 Write each expression in the form a^n.

(a) $(2^3)^{-2}$ (b) $6^5 \left(\dfrac{1}{6}\right)^5$ (c) $\dfrac{(3^2 \cdot 3^3)^4}{3^7}$ (d) $2^3 \cdot 4^5$

Solution

(a) $(2^3)^{-2} = 2^{3(-2)} = 2^{-6}$ ■ Rule 3 is used.

(b) $6^5 \left(\dfrac{1}{6}\right)^5 = \left(6 \cdot \dfrac{1}{6}\right)^5 = 1^5 = 1$ ■ Rule 2 is used.

(c) $\dfrac{(3^2 \cdot 3^3)^4}{3^7} = \dfrac{(3^5)^4}{3^7}$ ■ $3^2 \cdot 3^3 = 3^5$ by Rule 1.

$= \dfrac{3^{20}}{3^7}$ ■ $(3^5)^4 = 3^{20}$ by Rule 3.

$= 3^{13}$ ■ $\dfrac{3^{20}}{3^7} = 3^{13}$ by Rule 4.

(d) $2^3 \cdot 4^5 = 2^3 (2^2)^5$ ■ 4 is written as a power of 2.

$= 2^3 \cdot 2^{10}$ ■ We can use Rule 1 since both bases are 2.

$= 2^{13}$

Example 3 Simplify each expression so that x, y, and z appear at most once, with no powers of powers, or negative exponents.

(a) $(3x^2y)(2xy)(4x^3y^2)$ (b) $(6xy^2)^3$ (c) $\left(\dfrac{4y}{x^2}\right)^3 x^6$

Solution

(a) $(3x^2y)(2xy)(4x^3y^2) = (3 \cdot 2 \cdot 4)(x^2 \cdot x \cdot x^3)(y \cdot y \cdot y^2)$
$= 24x^6y^4$

■ Constants and like variables are grouped together before applying Rule 1.

(b) $(6xy^2)^3 = 6^3x^3(y^2)^3 = 216x^3y^6$

■ When using Rule 2 here, remember to raise each factor, including the constant, to the third power.
■ Rule 5 is used.

(c) $\left(\dfrac{4y}{x^2}\right)^3 x^6 = \dfrac{(4y)^3}{(x^2)^3} x^6$

$= (4y)^3 x^{-6} x^6$

■ $\dfrac{1}{(x^2)^3} = \dfrac{1}{x^6} = x^{-6}$.

$= (4y)^3 x^0$

■ Rule 1 is used.

$= (4y)^3$

■ $x^0 = 1$.

$= 64y^3$

■ Rule 3 is used.

One important result of the fact that $a^{-n} = \dfrac{1}{a^n}$ is that a fraction to a negative power $-n$ is equal to the reciprocal to the positive power n. For example,

$$\left(\dfrac{2}{3}\right)^{-5} = \dfrac{1}{\left(\dfrac{2}{3}\right)^5} = \dfrac{1^5}{\left(\dfrac{2}{3}\right)^5} = \left(\dfrac{1}{\dfrac{2}{3}}\right)^5 = \left(\dfrac{3}{2}\right)^5.$$

Careful! Remember that you can only use Rule 1 if the bases are equal. Thus,

$$3^4 \cdot 9^5 = 3^4 \cdot (3^2)^5 = 3^4 \cdot 3^{10} = 3^{14},$$

and

$$3^4 \cdot 6^5 = 3^4 \cdot (3 \cdot 2)^5 = 3^4 \cdot 3^5 \cdot 2^5 = 3^9 \cdot 2^5.$$

In the second case, we cannot further rewrite $3^9 \cdot 2^5$ because the bases are not equal.

Exercise Set 1.4 ▭ Calculator Exercises in the Appendix

Goal A

In exercises 1–20, evaluate the given expression.

1. $(-5)^4$
2. $(-3)^4$
3. -5^4
4. -3^4
5. 2^{-3}
6. 3^{-2}
7. 0^8
8. 8^0
9. 0^{-7}
10. 0^0
11. 9^{-1}
12. 9^{-2}
13. $\left(\dfrac{2}{3}\right)^{-1}$
14. $\left(\dfrac{5}{8}\right)^{-1}$
15. $\left(\dfrac{1}{2}\right)^{-3}$
16. $\left(\dfrac{1}{3}\right)^{-4}$
17. $\dfrac{1}{3^{-2}}$
18. $\dfrac{2}{4^{-3}}$
19. $\dfrac{3^2}{7^{-5}}$
20. $\dfrac{5^{-4}}{6^{-2}}$

In exercises 21–32, rewrite the given expression with a positive exponent.

21. 6^{-2}
22. $(-8)^{-3}$
23. $\left(\dfrac{3}{7}\right)^{-5}$
24. $\left(\dfrac{5}{8}\right)^{-3}$
25. a^{-2}
26. b^{-10}
27. $\dfrac{1}{x^{-3}}$
28. $\dfrac{1}{y^{-7}}$
29. $\dfrac{1}{-3^{-4}}$
30. -2^{-3}
31. x^{-1}
32. $-y^{-1}$

In exercises 33–44, rewrite the given expression with a negative exponent.

33. $\dfrac{1}{8^3}$
34. $\dfrac{1}{(-10)^4}$
35. 4^5
36. 3^9
37. $\dfrac{1}{x^5}$
38. $\dfrac{1}{(-y)^7}$
39. $-\dfrac{1}{3^8}$
40. $-\dfrac{3}{4^3}$
41. $\left(\dfrac{2}{3}\right)^5$
42. $\left(\dfrac{3}{5}\right)^3$
43. $\dfrac{1}{y}$
44. $\dfrac{2}{-x}$

In exercises 45–52, evaluate each expression, given that $x = 3$, $y = -5$, and $z = -4$.

45. $x^2 + y^2$
46. $x^2 - y^2$
47. $xy - z^2$
48. $x(y - z)^2$
49. y^x
50. x^0
51. $x^{-2} + y^{-1}$
52. $(x + y)^{-2}$

Goal B

In exercises 53–60, write each expression in the form 3^n.

53. $3^3 \cdot 3^4$
54. $3^2 \cdot 3^5$
55. $3^2 \cdot 3^5 \cdot 3^{-7}$
56. $3^4 \cdot 3^{-3} \cdot 3^{-2}$
57. $(9^3)^5$
58. $[(27)^2]^{-3}$
59. $3^6 \cdot 9^6$
60. $3^5 \cdot \left(\dfrac{1}{9}\right)^2$

In exercises 61–84, simplify each expression so that x, y, and z appear at most once, and so that there are no negative exponents.

61. $(2x^2)(-2xy^3)$
62. $(3z^3)(4xyz)$
63. $(2x)(7yx)(x^2)$
64. $(5x^2)(xyz)$
65. $(3x^2)(5x^{-4}y)$
66. $(2xyz)(x^{-1}yz)$
67. $(6xy)(2z^{-5})$
68. $(xy)(xz)(yz)$
69. $(2x)^3$
70. $(3y)^2$
71. $(6xy^2z)^2$
72. $(4x^3yz)^2$
73. $(-5z^4)^3$
74. $(-6x^3y)^4$
75. $\dfrac{2xz^{-3}}{3yz^{-2}}$
76. $\dfrac{8y^{-2}z}{11yz^{-3}}$
77. $\dfrac{7x^{-2}y^{-5}}{10x^{-5}z^{-2}}$
78. $\dfrac{5y^{-3}z^2}{6x^{-4}z^{-8}}$
79. $\left(\dfrac{x}{y}\right)^3\left(\dfrac{x^2y}{z}\right)^2$
80. $\left(\dfrac{xy}{z}\right)\left(\dfrac{x}{yz^2}\right)^3$

81. $\left(\dfrac{2xy}{3z}\right)^{-3}$ 82. $\left(\dfrac{5x}{7yz}\right)^{-2}$

83. $\left(\dfrac{x^2}{3z^3}\right)^{-5}$ 84. $\left(\dfrac{y^3}{2x^2}\right)^{-4}$

Superset

In exercises 85–94, determine whether each statement is true or false.

85. $3^2 \cdot 3^4 = 3^8$ 86. $(5^2)^2 = 5^4$

87. $(5^2)^3 = 5^5$ 88. $(-1)^0 = -1$

89. $(-6)^{-2} = \dfrac{1}{6^2}$ 90. $\dfrac{1}{(-3)^2} = (3)^{-2}$

91. $3^4 + 3^5 = 3^9$ 92. $2^3 \cdot 5^4 = 10^7$

93. $(-3)^4 = -3^4$ 94. $\dfrac{3^8}{3^4} = 3^2$

In exercises 95–102, simplify each expression so that x, y, and z appear at most once, and so that there are no negative exponents.

95. $\dfrac{(2x^2y)^2}{xy^3z^2}$ 96. $\dfrac{3x^2y^2z}{(7x^3z^4)^3}$

97. $\dfrac{(4x^3z^{-1})^2}{(x^2yz^{-3})^{-1}}$ 98. $\dfrac{(y^{-5}z^2)^{-3}}{(3x^3y^4)^{-2}}$

99. $\dfrac{(xz^5)^{-4}(y^2z^{-2})^3}{(x^{-1})(x^2y^4)^3}$ 100. $\dfrac{(x^7y^3)^2(yz^{-1})^{-5}}{(xy^{-3})^{-1}(y^4z)^{-3}}$

101. $\dfrac{(x+y)^3(x+y)^{-2}}{(x+y)^4}$ 102. $\dfrac{(y-z)^{-5}(y-z)^{-6}}{(y-z)^{-7}}$

For exercises 103–110, suppose that all you know is that $x^3 = 3$, $y^2 = 2$, and $z^4 = 4$. Use the Rules of Exponents to determine whether each inequality is true or false. (Do not use radicals.)

103. $\dfrac{1}{x^6} < \dfrac{1}{y^6}$ 104. $z^8 > x^9y^{-2}$

105. $\left(\dfrac{xy}{z^2}\right)^6 > 1$

106. $2x^{-3} - y^{-2} > z^{-4}$

107. $\left(\dfrac{2}{x^2}\right)^3 + \dfrac{1}{z^4+1} > 1$

108. $2^{-4} + z^{-4} < \dfrac{2}{x^3y^2}$

109. $\dfrac{1}{x^{-6}} > \dfrac{1}{y^{-4}} + \dfrac{1}{z^{-4}}$

110. $(x^{-1}yz)(xyz)^{-5} > 0.01$

In exercises 111–118, determine the desired area or volume, rounded to the nearest tenth. The necessary formulas from geometry are listed inside the front cover.

111. Determine the area of a circle whose radius is $2x^3$ cm.

112. Determine the volume of a sphere of radius $3a^2$ ft.

113. Determine the area of a square whose side measures $5p^{-1}$ in.

114. Determine the volume of a cube whose side measures $0.1y^3$ ft.

115. Determine the volume of a cylinder whose radius is n^2 cm and whose height is 1 m.

116. Determine the volume of a cylinder whose radius is $.04s^{-1}$ ft and whose height is 12 in.

117. Compare the volume of a sphere having radius 2^{-3} m, with the volume of a cylinder having radius 3^{-2} m and height $\tfrac{3}{2}$ m.

118. Which is greater, the area of a circle of radius $\left(\tfrac{3}{2}\right)^{-2}$ m or the surface area of a cube having side $\tfrac{1}{3}$ m?

1.5 Polynomials: Addition, Subtraction, and Multiplication

Goal A *to review the terminology used to describe polynomials*

A jet, beginning its takeoff, is at one end of a runway. If *a* represents its acceleration (in feet per second per second), and if it takes *t* seconds to take-off, then the length of runway needed for takeoff is given by the expression

$$\frac{1}{2}at^2.$$

This expression is called a monomial. In general, a **monomial** is either a nonzero constant or the product of a nonzero constant and one or more variables, each of which is raised to a positive power. The constant is usually referred to as the **numerical coefficient** or simply the **coefficient** of the monomial. The **degree of the monomial** is the sum of the exponents of its variables.

-3
$4x$
$16a^2bc^3$

Examples of Monomials

$$-8x^2y^3 \quad \blacksquare \text{ The coefficient is } -8; \text{ the degree is 5.}$$

The word **term** is often used to mean monomial. A **polynomial** is then defined as a sum of a finite number of terms. The **degree of a polynomial** is the highest degree of the terms in the polynomial.

$3x - 2$
$8x^2 - 5xy + z^3$

Examples of Polynomials

$$\underbrace{8x^2y}_{\substack{\text{degree}\\3}} - \underbrace{9x^3y^2}_{\substack{\text{degree}\\5}} - \underbrace{x^4y^2z}_{\substack{\text{degree}\\7}} \quad \overset{\text{3 terms}}{}$$

■ The degree of the polynomial is 7.

Some polynomials have special names; a polynomial with two terms, such as $5x - 3$, is called a **binomial,** and a polynomial with three terms, such as

$$x^2 - 5x + 7,$$

is called a **trinomial.** A nonzero number, such as 16, is called a **constant polynomial,** and has degree zero. The number 0, called the **zero polynomial,** is said to have no degree.

Terms having exactly the same variables with exactly the same exponents are called **like terms.**

$5x^2y \quad \sqrt{2}x^2y \quad -x^2y$ ■ like terms

$7xy^2z \quad 7x^2yz$ ■ not like terms

Example 1 Select the appropriate expression to complete each sentence.

(a) The polynomial $5x^4 - 2x + 4y$ has _____ terms. [two/three/four]
(b) In the polynomial $3x^3 - 4x^2 - 7$, the coefficient of the x^2-term is _____. [−4/4]
(c) The degree of the monomial $5x^4y^2z$ is _____. [4/6/7/8]
(d) The degree of the polynomial $2xy - 3x^2y$ is _____. [2/3/5]
(e) The degree of the monomial 4 is _____. [0/1/not defined]
(f) The degree of the zero polynomial 0 is _____. [0/1/not defined]

Solution (a) three (b) −4 (c) 7 (d) 3 (e) 0 (f) not defined

Goal B *to review techniques for adding and subtracting polynomials*

Polynomials can be simplified by combining like terms. This is accomplished by using the Commutative and Distributive Properties.

Example 2 Simplify the expression $5xy^3 - 6y^3 - 8xy^3$.

Solution
$5xy^3 - 6y^3 - 8xy^3 = (5xy^3 - 8xy^3) - 6y^3$ ▪ Like terms are grouped together.
$= (5 - 8)xy^3 - 6y^3$ ▪ The Distributive Property is used: $(ac - bc) = (a - b)c$.
$= -3xy^3 - 6y^3$

When adding two or more polynomials you simply collect like terms. To subtract a polynomial, you add its additive inverse. Be careful when finding the additive inverse of a polynomial.

$$-(3x^2 - 7xy + 18y^2) = -3x^2 + 7xy - 18y^2$$

The negative sign in front of the parentheses changes the sign of each term inside the parentheses.

Example 3   Subtract $5x - 7y$ from $8x + 4y$.

Solution
$$(8x + 4y) - (5x - 7y) = 8x + 4y - 5x + 7y$$
$$= (8x - 5x) + (4y + 7y)$$
$$= 3x + 11y$$

- Remember: $-(A - B) = -A + B$.
- Like terms are grouped together.

Goal C   *to review methods for multiplying polynomials and raising polynomials to powers*

The simplest type of polynomial multiplication involves multiplying a polynomial by a monomial.

Example 4   Multiply $3x^2$ by $5xy - 2x^3y^2$.

Solution
$$3x^2(5xy - 2x^3y^2) = (3x^2)(5xy) - (3x^2)(2x^3y^2)$$
$$= 15x^3y - 6x^5y^2$$

- The Distributive Property is used.
- Rule 1 of Exponents is used.

Example 5   Multiply $(5x + 2y)(3x^2 - 7y^2 + 4xy)$.

Solution

$(A + B) \cdot (\ \ \ C\ \ \ ) = A \cdot (\ \ \ C\ \ \ ) + B \cdot (\ \ \ C\ \ \ )$

$(5x + 2y) \cdot (3x^2 - 7y^2 + 4xy) = 5x(3x^2 - 7y^2 + 4xy) + 2y(3x^2 - 7y^2 + 4xy)$

- The Distributive Property is used.

$$= 15x^3 - 35xy^2 + 20x^2y + 6x^2y - 14y^3 + 8xy^2$$

- The Distributive Property is used again.

$$= 15x^3 + (20x^2y + 6x^2y) + (-35xy^2 + 8xy^2) - 14y^3$$

- Like terms are grouped together.

$$= 15x^3 + 26x^2y - 27xy^2 - 14y^3$$

Since products of binomials occur frequently in our work, they merit special attention. Suppose we wish to multiply two binomials, $(a + b)$ and $(c + d)$. Think of a and c as the first terms in the binomials, a and d as the outer terms, b and c as the inner terms, and b and d as the last terms.

$$(a + b) \cdot (c + d) = a \cdot (c + d) + b \cdot (c + d)$$

$$= \underset{\text{First terms}}{ac} + \underset{\text{Outer terms}}{ad} + \underset{\text{Inner terms}}{bc} + \underset{\text{Last terms}}{bd}$$

This procedure for finding the product of two binomials is sometimes called the **FOIL** method.

Example 6 Multiply $(3x - 7)(4x + 3y)$.

Solution
$$(3x - 7)(4x + 3y) = \underset{\text{First}}{(3x)(4x)} + \underset{\text{Outer}}{(3x)(3y)} + \underset{\text{Inner}}{(-7)(4x)} + \underset{\text{Last}}{(-7)(3y)}$$
$$= 12x^2 + 9xy - 28x - 21y$$

Special Products

$(\boldsymbol{a + b})^2 = a^2 + 2ab + b^2$

$(\boldsymbol{a - b})^2 = a^2 - 2ab + b^2$

$(\boldsymbol{a + b})(\boldsymbol{a - b}) = a^2 - b^2$

Three special products of binomials are displayed at the left. Their derivations are shown below (notice the use of the FOIL rule).

$(a + b)^2 = (a + b)(a + b) = a^2 + ab + ba + b^2 = a^2 + 2ab + b^2$
$(a - b)^2 = (a - b)(a - b) = a^2 - ab - ba + b^2 = a^2 - 2ab + b^2$
$(a + b)(a - b) = a^2 - ab + ba - b^2 = a^2 - b^2$

Example 7 Determine $(6y + 5x)^2$.

Solution
$$(6y + 5x)^2 = (6y)^2 + 2(6y)(5x) + (5x)^2$$
$$= 36y^2 + 60xy + 25x^2$$

■ Square the first term; double the product of the two terms; square the last term.

Example 8 Multiply $(2x - 7y)(2x + 7y)$.

Solution
$$(2x - 7y)(2x + 7y) = (2x)^2 - (7y)^2$$
$$= 4x^2 - 49y^2$$

■ Here we used the special product: $(a - b)(a + b) = a^2 - b^2$.

To raise a polynomial to a power, first write it in expanded form (as a product) and then multiply.

Example 9 Determine $(2x + 3)^3$.

Solution
$$(2x + 3)^3 = (2x + 3)^2(2x + 3)$$
$$= (4x^2 + 12x + 9)(2x + 3)$$
$$= 4x^2(2x + 3) + 12x(2x + 3) + 9(2x + 3)$$
$$= 8x^3 + 12x^2 + 24x^2 + 36x + 18x + 27$$
$$= 8x^3 + 36x^2 + 54x + 27$$

■ Rewrite as a product of a square and a first power.
■ Use the Distributive Property.
■ Use the Distributive Property again.

Exercise Set 1.5

Goal A

In exercises 1–10, determine the number of terms and degree of each polynomial.

1. $6x^2y^5z$
2. $10a^4b^4c$
3. $5x^2 - 2x - 7$
4. $x^4 - x^3 + 5xy$
5. $17xy^2 - 5x^2y + 3xy - 10$
6. $6x^2y - 17xyz + 4y^4$
7. $x^5 - 5x^2y^2 + 2xz + 1$
8. $5abc^3 + 7a^2b^2 - 5ac - 8b$
9. 10
10. 0

Goal B

In exercises 11–18, add or subtract as instructed.

11. Add $2x^2 + 5x$ to $3x^3 - 7x^2 + 5x - 4$.
12. Add $-x^3 - 9x^2 + 5$ to $x^2 - 3x + 14$.
13. Subtract $u^2 - u + 1$ from $4u^4 - 2u^2 + 1$.
14. Subtract $x^3 + x^2 - 1$ from $4x^4 - 2x^2 + 1$.
15. $(x^4 - x^2 + 1) + (2x^2 - 8x + 5x^4 + 10)$

16. $(14a^2 - 6a - 5) + (7a^2 + 5a - 18)$
17. $(3v^2 - 8v + 14) - (5v^2 - 2v - 6)$
18. $-2(x + y) - (x^2 - y^2 + 2x + 2y)$

Goal C

In exercises 19–44, perform the indicated operations.

19. $2x(5x^2 - 7x + 1)$
20. $-3a(a^2 + 7ab - b^2)$
21. $(-1)(a^2 - b^2)$
22. $(-1)(uv - vu)$
23. $(3x - 2)(7x + 3)$
24. $(5x - 6)(2x + 7)$
25. $(4m - 3)(5m - \frac{1}{2})$
26. $(9n - 4)(2n - \frac{2}{3})$
27. $(5x)(6x + 5y)(3x + 2y)$
28. $(4uv)(7u + 2v)(3u + 4v)$
29. $(3x + \frac{2}{5})(x^2 + 3x - 4)$
30. $(2y - \frac{1}{3})(y^2 - 5y + 2)$
31. $(4m + 3)(2m^3 - 3m^2 + 5m - 4)$
32. $(5x - 2)(x^4 - 2x^2 + x - 10)$
33. $(x^2 + x - 1)(3x^2 - 5x + 2)$
34. $(u - v + 7)(u + 3v - 2)$
35. $(x^2 + y^2)(x^2 - y^2)$
36. $(2m^2 + 3n^2)(2m^2 - 3n^2)$
37. $(3a + 7)^2$
38. $(4m - 3)^2$
39. $(5x + 3y)^2$
40. $(7a - 4b)^2$
41. $(x + 2)^3$
42. $(x - 2)^3$
43. $(x - 2)^2 - (x + 2)^2$
44. $(p + 5)^2 - (p - 5)^2$

Superset

In exercises 45–51, each answer should be written as a polynomial in simplified form.

45. The square at the left has a side with length $x + 2$. Find a polynomial that describes its area.

$x + 2$

46. The square at the left has a side with length $2x + 1$. Find a polynomial that describes its area.

$2x + 1$

47. The triangle at the left has an altitude of length $2x$ and a base of length $x - 2$. Find a polynomial that describes its area.

48. The circle at the left has a radius of length $x + 3$. Find a polynomial that describes its area.

49. The sphere at the left has a radius of length $3x + 2$. Find a polynomial that describes its volume.

$3x + 2$

50. The cube at the left has an edge of length $x + 3$. Find a polynomial that describes its volume.

$x + 3$

51. The height of the box at the left is $2x - 1$, and the area of the shaded top is $x^2 + 4$. Find a polynomial that describes the volume of the box.

$2x - 1$

1.6 Polynomials: Factoring

Goal A *to find the greatest common factor of two or more monomials*

Suppose a number a is the product of two numbers, b and c. We write

$$a = bc \text{ and say that } b \text{ and } c \text{ are } \textbf{factors} \text{ of } a.$$

$6 = 2 \cdot 3$
2 and 3 are factors of 6

We could also say that a is **factored** into the product bc. We use the same terminology when we talk about monomials and polynomials.

$$3x^2yz = (3x^2)(yz)$$
■ $3x^2$ and yz are factors of $3x^2yz$.

$$x^2 - 5x + 6 = (x - 2)(x - 3)$$
■ $(x - 2)$ and $(x - 3)$ are factors of $x^2 - 5x + 6$; or, $x^2 - 5x + 6$ is factored into the product $(x - 2)(x - 3)$.

A **prime number** is a positive integer greater than 1 whose only positive factors are 1 and itself. The first five prime numbers are

$$2, 3, 5, 7 \text{ and } 11.$$

The **greatest common factor (GCF)** of two integers is the largest factor that divides both integers. The GCF is found by first listing the prime factors of the numbers.

$$360 = 2^3 \cdot 3^2 \cdot 5$$
$$84 = 2^2 \cdot 3 \cdot 7$$
■ Since 360 and 84 have 2^2 and 3 but no other numbers as common factors, the GCF is $2^2 \cdot 3 = 12$.

To determine the GCF of two or more monomials, we begin by looking at the factors of each monomial. We then form the product of those factors common to all the monomials.

Example 1   Find the GCF of $-16x^2yz^3$, $12xyz^3$ and $20xz^2$.

Solution
$$-16x^2yz^3 = (-1)2^4 \quad \cdot x^2 \cdot y \cdot z^3$$
$$12xyz^2 = \quad 2^2 \cdot 3 \cdot x \,\cdot y \cdot z^2$$
$$20xy^2 = \quad 2^2 \cdot 5 \cdot x \quad\quad \cdot z^2$$
$$\quad\quad\quad\quad\quad\blacktriangle \quad\quad\, \blacktriangle \quad\quad\quad \blacktriangle$$
$$\quad\quad\quad\quad\quad 2^2, \quad x \text{ and } z^2 \text{ are the common factors.}$$

The GCF is $2^2 \cdot x \cdot z^2 = 4xz^2$.

Goal B *to develop four strategies for factoring polynomials*

Factoring Strategy 1. Factor out a common monomial.
The simplest type of polynomial factoring requires that you "factor out" the GCF from each term of the polynomial. Consider the polynomial,

$$12x^2yz^2 - 18xy^2z^3.$$

Its terms are

$$12x^2yz^2 \quad \text{and} \quad -18xy^2z^3,$$

and the GCF of these terms is $6xyz^2$. In factored form, the polynomial will look like

$$6xyz^2(\quad - \quad).$$

To find the expression inside the parentheses, we must determine "what's left" in each term of the polynomial after we factor out the GCF. In each case, "what's left" is the original term divided by the GCF.

Term	GCF	$\dfrac{\text{Term}}{\text{GCF}}$	=	What's left
$12x^2yz^2$	$6xyz^2$	$\dfrac{12x^2yz^2}{6xyz^2}$	=	$2x$
$-18xy^2z^3$	$6xyz^2$	$\dfrac{-18xy^2z^3}{6xyz^2}$	=	$-3yz$

Therefore,

$$12x^2yz^2 - 18xy^2z^3 = 6xyz^2(2x - 3yz),$$

where the expression on the right is called the **factored form** of the given polynomial.

Example 2   Factor (a) $16x^5 + 8x^3 - 6x^2$ (b) $14a^2bc^2 - 42ab^3c + 21ab^2c^4$

Solution   (a) $16x^5 + 8x^3 - 6x^2 = 2x^2(8x^3 + 4x - 3)$ ■ Check your work by multiplying the factors in your answer. The result should be the original polynomial.

(b) $14a^2bc^2 - 42ab^3c + 21ab^2c^4 = 7abc(2ac - 6b^2 + 3bc^3)$

Factoring Strategy 2. Factor by observing a pattern.

In each line of the table below, the polynomial on the left side of the equals sign is found by multiplying the factors on the right side. You should do the multiplication to verify the four identities.

	Polynomial		Factored Form
Difference of squares	$A^2 - B^2$	=	$(A + B)(A - B)$
Perfect trinomial square	$A^2 + 2AB + B^2$ $A^2 - 2AB + B^2$	= =	$(A + B)^2$ $(A - B)^2$
Difference of cubes	$A^3 - B^3$	=	$(A - B)(A^2 + AB + B^2)$
Sum of cubes	$A^3 + B^3$	=	$(A + B)(A^2 - AB + B^2)$

We will use these patterns to factor these four types of polynomials. In practice, A and B may represent any real number or algebraic expression.

Example 3   Factor $x^2 + 6x + 9$.

Solution
$$x^2 + 6x + 9$$
$$A^2 + 2AB + B^2 = (A + B)^2$$
$$(x + 3)^2$$

- This is a perfect trinomial square with $A = x$ and $B = 3$. In a perfect trinomial square, two of the terms are perfect squares of some expressions A and B, and the other term is twice the product of A and B.

Thus, $x^2 + 6x + 9 = (x + 3)^2$.

Example 4   Factor $16x^2 - 81y^2$.

Solution
$$16x^2 - 81y^2 = (4x)^2 - (9y)^2$$
$$= (4x + 9y)(4x - 9y)$$

- The polynomial has the form $A^2 - B^2$, where $A = 4x$ and $B = 9y$.
- The pattern $A^2 - B^2 = (A + B)(A - B)$ is used.

Example 5   Factor $27x^3 + 64y^3$.

Solution

$$27x^3 + 64y^3 = (3x)^3 + (4y)^3$$
$$= (3x + 4y)(9x^2 - 12xy + 16y^2)$$

- This expression is of the form $A^3 + B^3$, where $A = 3x$ and $B = 4y$.
- The following pattern is used: $A^3 + B^3 = (A + B)(A^2 - AB + B^2)$. **Careful!** The term $12xy$ in the second factor is AB, not $2AB$.

Example 6   Factor $(2x + y)^3 - (3y)^3$.

Solution

$$(\underset{\blacktriangledown}{A})^3 - (\underset{\blacktriangledown}{B})^3 = (\underset{\blacktriangledown}{A} - \underset{\blacktriangledown}{B})[(\underset{\blacktriangledown}{A})^2 + (\underset{\blacktriangledown}{A})(\underset{\blacktriangledown}{B}) + (\underset{\blacktriangledown}{B})^2]$$

$$(2x + y)^3 - (3y)^3 = (2x + y - 3y)[(2x + y)^2 + (2x + y)(3y) + (3y)^2]$$
$$= (2x - 2y)[(4x^2 + 4xy + y^2) + (6xy + 3y^2) + (9y^2)]$$
$$= 2(x - y)(4x^2 + 10xy + 13y^2)$$

- In order to factor completely, we have factored out the constant 2.

Factoring Strategy 3. Factor by grouping.

Sometimes a polynomial containing four or more terms can be simplified by grouping pairs (or triples) of terms, and then factoring each grouping. This technique is demonstrated in the following example.

Example 7   Factor $x^3 - 2x^2 + 4x - 8$.

Solution

$$x^3 - 2x^2 + 4x - 8 = (x^3 - 2x^2) + (4x - 8)$$
$$= x^2(x - 2) + 4(x - 2)$$
$$= (x^2 + 4)(x - 2)$$

- Two groupings are formed.
- A monomial is factored out of each grouping to produce the common factor $(x - 2)$ in each term.
- $(x - 2)$ is factored out of each term.

In Example 7, we could have grouped the terms as $(x^3 + 4x) - (2x^2 + 8)$. In general, there is always more than one way to form the groupings.

Factoring Strategy 4. Factor a trinomial by trial and error.

We already know that multiplying the factors $3x + 5$ and $2x + 3$ produces the trinomial $6x^2 + 19x + 15$. Suppose instead, we begin with the trinomial and need to find its two factors. How do we go about finding them? We must factor $6x^2 + 19x + 15$ into a product of the form $(ax + b)(cx + d)$, which equals $acx^2 + adx + bcx + bd$, or $acx^2 + (ad + bc)x + bd$ by the FOIL rule.

$$acx^2 + (ad + bc)x + bd = (ax + b)(cx + d)$$
$$6x^2 + 19x + 15 = (\ ?\ )(\ ?\ )$$

First clue: $ac = 6$. Thus, the factorization has one of the following forms:

$$(3x + b)(2x + d) \quad \text{or} \quad (x + b)(6x + d).$$

$\underline{ac = 6}$
$3 \times 2 = 6$
$1 \times 6 = 6$

Second clue: $bd = 15$, a positive number. Thus, b and d have the same sign. The coefficient of the x-term is 19; therefore, we conclude b and d are positive. The possible factorizations are:

$\underline{bd = 15}$
$3 \times 5 = 15$
$1 \times 15 = 15$

$$(3x + 15)(2x + 1) \qquad (x + 15)(6x + 1)$$
$$(3x + 1)(2x + 15) \qquad (x + 1)(6x + 15)$$
$$(3x + 5)(2x + 3) \qquad (x + 5)(6x + 3)$$
$$(3x + 3)(2x + 5) \qquad (x + 3)(6x + 5)$$

Third clue: the product of the "inner terms" is $19x$. Only $(3x + 5)(2x + 3)$ satisfies this condition.

We conclude that $6x^2 + 19x + 15 = (3x + 5)(2x + 3)$.

Example 8   Factor $2y^2 + 7y - 4$.

Solution   Trial and error: we want the form $(ay + b)(cy + d)$.
Since $ac = 2$, the form is $(2y + b)(y + d)$.
Since $bd = -4$, b and d have different signs. The possible factorizations are:

$(2y + 2)(y - 2) \quad (2y + 4)(y - 1) \quad (2y + 1)(y - 4)$
$(2y - 2)(y + 2) \quad (2y - 4)(y + 1) \quad (2y - 1)(y + 4)$ ■ Only $(2y - 1)(y + 4)$ produces the middle term $7y$.

Thus, $2y^2 + 7y - 4 = (2y - 1)(y + 4)$.

Example 9 Factor $x^2 + x + 4$.

Solution Trial and error: The possible factorizations are $(x + 4)(x + 1)$ or $(x + 2)(x + 2)$. But neither produces the correct middle term. Thus, $x^2 + x + 4$ cannot be factored.

When a polynomial cannot be factored into a product of nonconstant polynomials with coefficients in some specified set, the polynomial is said to be **irreducible over the set.** Thus, $x^2 + x + 4$ is irreducible over the set of integers.

Example 10 Factor $15y^3 + 27y^2 - 6y$ completely.

Solution
$$15y^3 + 27y^2 - 6y = 3y(5y^2 + 9y - 2)$$
$$= 3y(5y - 1)(y + 2)$$

■ To begin, factor out a common monomial. Then factor the remaining binomial by trial and error.

Example 11 Factor $3x^3 + 2x^2 - 27x - 18$ completely.

Solution
$$3x^3 + 2x^2 - 27x - 18 = (3x^3 + 2x^2) - (27x + 18)$$
$$= x^2(3x + 2) - 9(3x + 2)$$
$$= (x^2 - 9)(3x + 2)$$
$$= (x + 3)(x - 3)(3x + 2)$$

■ To begin, factor by grouping.

■ Now factor the difference of squares.

Example 12 Factor $10x^2z^2 - 105z^2 + 5xz^2$ completely.

Solution
$$10x^2z^2 - 105z^2 + 5xz^2 = 5z^2(2x^2 - 21 + x)$$
$$= 5z^2(2x^2 + x - 21)$$
$$= 5z^2(2x + 7)(x - 3)$$

■ Factor out a common monomial.

■ The terms of the trinomial have been reordered; now factor by trial and error.

Exercise Set 1.6

Goal A

In exercises 1–6, find the greatest common factor of the given monomials.

1. $5a^2b, 10ab, 15ab^2$
2. $6uv^2, 12u^2v, 18uv^2$
3. $4xy^2, 6x^2y^2, 8x^2y^2z$
4. $10m^2n^2, 20mn, 35m^3$
5. $30pqr, 60p^2, 75q^2, 30r^2$
6. $18x^2y^2z^3, 24xy^2, 12x^2z^2, 78xyz$

Goal B

In exercises 7–14, fill in the blanks with the correct polynomial.

7. $2x^3 + 6x^2 + 10x = 2x(\underline{})$
8. $6y^2 - 9y + 12y^3 = 3y(\underline{})$
9. $6a - 7b = -1(\underline{})$
10. $-x + y - z = -1(\underline{})$
11. $2m - 2n + 4 = -2(\underline{})$
12. $6x^2 - 9xy + 15y^2 = -3(\underline{})$
13. $5m^2n - 10mn^2 - 25m^2n^2 = -mn(\underline{})$
14. $2a^2b + 4ab^2 - 6ab = -ab(\underline{})$

In exercises 15–76, factor completely. If a polynomial cannot be factored, then state this fact.

15. $5m - m^2$
16. $3x^2 - 15x$
17. $4v^3 - 5v^2 - 6v$
18. $2p^2q - 8pq - 14pq^2$
19. $x^2 - 100$
20. $v^2 - 64$
21. $4m^2 - 0.25$
22. $0.16p^2 - 49$
23. $x^2 - y^2$
24. $x^2 - 4y^2$
25. $4m^2 + 9n^2$
26. $18x^2 - 50y^2$
27. $25v^2 + 49u^2$
28. $8v^2 - 18u^2$
29. $x^3 - 8$
30. $27 - z^3$
31. $1 - 64m^3$
32. $125v^3 - 8$
33. $8x^3 + 27y^3$
34. $8x^3 - 27y^3$
35. $0.001a^3b + b^4$
36. $x^5 - 0.008x^2y^3$
37. $x^2 + 14x + 49$
38. $n^2 - 14n + 49$
39. $x^2 + 10x + 25$
40. $m^2 - 10m + 25$
41. $4v^2 + 12v + 9$
42. $16p^2 - 40p + 25$
43. $3a^2 - 18a + 27$
44. $8a^2 - 8a + 2$
45. $2x^2 + xy - 2xy - y^2$
46. $3v^2 - vw + 3vw - w^2$
47. $m^3 - 2m^2 + m - 2$
48. $a^3 - 3a^2 + 2a - 6$
49. $x^2 + 7x + 12$
50. $m^2 + 7m + 10$
51. $x^2 + x + 1$
52. $y^2 + 2y - 15$
53. $z^2 + 5z - 14$
54. $x^2 - 2x + 2$
55. $p^2 + p - 6$
56. $q^2 - q - 12$
57. $2x^2 - 5x - 3$
58. $4v^2 + 4v - 3$
59. $d^2 + 15d + 36$
60. $s^2 + 2s - 35$
61. $15t^2 - 37t - 8$
62. $12t^2 + 8t - 15$
63. $6x^2 + 5xy - 4y^2$
64. $4x^2 - 4xy - 35y^2$
65. $x^4 - 2x^2 - 8$
66. $x^4 - 2x^2 + 1$
67. $9x^2y - 16y^3$
68. $8a^6 - 27b^6$
69. $x^5 + 5x^4 - 14x^3$
70. $2x^4 - 14x^3 - 36x^2$
71. $x^4 - 25y^2$
72. $(a + b)^2 - c^2$
73. $u^4 + v^4$
74. $u^6 + v^6$
75. $x^2 - 4x + 4 - y^2$
76. $a^2 - (9 - 6b + b^2)$

Superset

In exercises 77–88, factor completely. *Assume all variable exponents represent positive integers.*

77. $x^n - 8x^{n+1}$
78. $5y^m + y^{2m}$
79. $x^{2n} + 2x^n y^n + y^{2n}$
80. $m^{2x} - 2m^x n^x + n^{2x}$
81. $a^{2x} - b^{2x}$
82. $x^{2r} - y^{2s}$
83. $x^{2n+1} - xy^{2n}$
84. $x^{4r} - y^{4s}$
85. $x^{3n} - y^{3n}$
86. $x^{3n} + y^{3n}$
87. $(a + b)^2 + 2(a + b)c + c^2$
88. $(x + y)^2 - 2(x + y)z + z^2$

89. The area of a square is given by the expression $x^2 + 8x + 16$. Describe the length of each side of the square in terms of x.

90. The area of a square is given by the expression $4x^2 - 12x + 9$. Describe the length of each side of the square in terms of x.

91. The area of the rectangle below is given by the expression $2c^2 - c - 15$. Describe the length of the rectangle in terms of c.

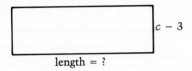

92. The area of the triangle below is given by the expression $t^2 - 0.5t - 3$. Express the length of the indicated base in terms of t.

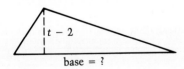

93. The volume of the box below is given by the expression $8a^3 - 27$. Find the area of the (shaded) top in terms of a.

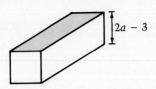

94. The area of the rectangle below is given by the expression $x^2 + 7x + 12$. Determine the area of the shaded region.

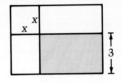

95. In the rectangle below, describe the area of the shaded region in terms of x.

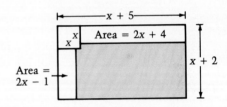

96. Draw a square with sides equal to $x + y$. Use the square to show that
$$(x + y)(x + y) \neq x^2 + y^2.$$

In exercises 97–102, use the factoring patterns of this section to verify the given statement, without raising any number to a power.

97. $10^2 - 9^2 = 19$
98. $100^2 - 99^2 = 199$
99. $1000^2 - 999^2 = 1999$
100. $1000^2 - 990^2 = 19{,}990$
101. $100^2 - 90^2 = 1900$
102. $1000^2 - 900^2 = 190{,}000$

1.7 Radicals and Rational Exponents

Goal A *to review the definition of the radical expression $\sqrt[n]{a}$ (nth root of a)*

The equation $3^4 = 81$ tells us that "81 is 3 to the fourth power," but what does it tell us about 3? We say that 3 is a "fourth root of 81." This can be written as

$$3 = \sqrt[4]{81}.$$

$\sqrt[n]{b}$

*n is called the **index**, b is called the **radicand**, and $\sqrt{}$ is called the **radical sign**.*

In general, if n is a positive integer, and a and b are real numbers, then the equation

$$a^n = b$$

tells us that a is an **nth root of b.** Using a radical, we might describe a as

$$a = \sqrt[n]{b}.$$

The index n must be a positive integer. When the index is 2, it is not written. Thus, $\sqrt{7}$ means $\sqrt[2]{7}$.

When n is even, a number b will have two (real) nth roots. (For example, since

$$5^2 = 25 \quad \text{and} \quad (-5)^2 = 25,$$

both 5 and -5 are second (or square) roots of 25.) In this case, the positive root is called the principal nth root. (5 is the principal square root of 25.) The symbol $\sqrt[n]{}$ denotes the principal (positive) nth root when n is even.

	Real Number Values of $\sqrt[n]{b}$	
b	n is even	n is odd
$b = 0$	$\sqrt[n]{0} = 0$	$\sqrt[n]{0} = 0$
$b > 0$	Two nth roots, but $\sqrt[n]{b}$ denotes only the positive one. $\sqrt{9} = 3$; $\sqrt[4]{16} = 2$; $\sqrt[6]{1} = 1$	One nth root. $\sqrt[3]{8} = 2$; $\sqrt[5]{\dfrac{1}{32}} = \dfrac{1}{2}$
$b < 0$	$\sqrt[n]{b}$ is not defined. $\sqrt{-9}$ and $\sqrt[4]{-16}$ are not real numbers.	One nth root. $\sqrt[3]{-8} = -2$; $\sqrt[5]{-\dfrac{1}{32}} = -\dfrac{1}{2}$

Example 1 Evaluate, if possible.

(a) $\sqrt[5]{-32}$ (b) $\sqrt[3]{\frac{27}{8}}$ (c) $\sqrt{0}$ (d) $\sqrt{-25}$ (e) $-\sqrt{25}$

(f) $-\sqrt[3]{-8}$ (g) The principal square root of $\frac{9}{4}$.

Solution (a) $\sqrt[5]{-32} = -2$ (b) $\sqrt[3]{\frac{27}{8}} = \frac{3}{2}$ (c) $\sqrt{0} = 0$

(d) $\sqrt{-25}$ is not a real number. (e) $-\sqrt{25} = -5$ (f) $-\sqrt[3]{-8} = 2$

(g) The principal square root of $\frac{9}{4}$ is $\frac{3}{2}$.

Goal B *to review and apply the rules for simplifying radical expressions*

Study the following statements:

$\sqrt{3^2} = 3$ ■ True, since $\sqrt{3^2} = \sqrt{9} = 3$.
$\sqrt{(-3)^2} = 3$ ■ True, since $\sqrt{(-3)^2} = \sqrt{9} = 3$.
$\sqrt[5]{3^5} = 3$ ■ True, since $\sqrt[5]{3^5} = \sqrt[5]{243} = 3$.
$\sqrt[5]{(-3)^5} = -3$ ■ True, since $\sqrt[5]{(-3)^5} = \sqrt[5]{-243} = -3$.

The first two statements suggest that when n is even, $\sqrt[n]{a^n}$ is never negative, no matter what a is. Thus, we use $|a|$ to define $\sqrt[n]{a^n}$ for even n. The last two statements imply that when n is odd, the value of $\sqrt[n]{a^n}$ is just a.

Rule 1 of Radicals

If a is any real number, then

$\sqrt[n]{a^n} = |a|$, when n is an even positive integer;
$\sqrt[n]{a^n} = a$, when n is an odd positive integer.

In the next example, we use Rule 1 of Radicals in a variety of situations involving numbers and variables.

Example 2 Evaluate.

(a) $\sqrt[6]{(-4)^6}$ (b) $\sqrt[3]{2^3}$ (c) $\sqrt[3]{(-2)^3}$ (d) $\sqrt[4]{x^4}$ (e) $\sqrt[5]{x^5}$

Solution (a) $\sqrt[6]{(-4)^6} = |-4| = 4$ (b) $\sqrt[3]{2^3} = 2$ (c) $\sqrt[3]{(-2)^3} = -2$

(d) $\sqrt[4]{x^4} = |x|$ (e) $\sqrt[5]{x^5} = x$

From this point on we shall assume that the variables in radical expressions have been restricted so that the expressions represent real numbers. For example, when you see $\sqrt{y}$, you may assume that the radicand y is not negative.

Other Rules of Radicals

For all real numbers a and b and positive integers n and m such that the radical expressions represent real numbers, the following rules apply.

Rule 2 $\sqrt[n]{a^m} = (\sqrt[n]{a})^m$

Rule 3 $\sqrt[n]{a} \cdot \sqrt[n]{b} = \sqrt[n]{ab}$

Rule 4 $\dfrac{\sqrt[n]{a}}{\sqrt[n]{b}} = \sqrt[n]{\dfrac{a}{b}}, \quad b \neq 0$

When simplifying a radical expression you should have two goals in mind: the final radicand should contain no powers greater than or equal to the index, and there should be no fraction under the radical sign or radical in the denominator. (For example, $\sqrt[3]{4^5}$ and $\frac{2}{\sqrt{5}}$ are not simplified.)

In the next example, we shall practice how to simplify radical expressions using the Rules of Radicals.

Example 3 Simplify each of the following.

(a) $\sqrt[3]{48}$ (b) $\sqrt{8x^2y^5}$ (c) $\sqrt[4]{64x^8y^3}$ (d) $\sqrt[3]{\dfrac{7}{100}}$ (e) $\sqrt{\dfrac{3x^3y}{2z^5}}$

Solution

(a) $\sqrt[3]{48} = \sqrt[3]{2^3 \cdot 6}$ ■ Since the index is 3, we look for perfect cubes in the radicand.
$= \sqrt[3]{2^3} \cdot \sqrt[3]{6}$ ■ Rule 3 of Radicals is used.
$= 2\sqrt[3]{6}$ ■ Rule 1 of Radicals is used.

(b) $\sqrt{8x^2y^5} = \sqrt{2^2 \cdot 2 \cdot x^2 \cdot (y^2)^2 \cdot y}$ ■ Since the index is 2, we look for perfect squares in the radicand.
$= \sqrt{2^2} \cdot \sqrt{2} \cdot \sqrt{x^2} \cdot \sqrt{(y^2)^2} \cdot \sqrt{y}$ ■ Rule 3 of Radicals is used.
$= 2 \cdot \sqrt{2} \cdot |x| \cdot y^2 \cdot \sqrt{y}$ ■ Rule 1 of Radicals is used.
$= 2|x|y^2\sqrt{2y}$ ■ Rule 3 of Radicals is used.

(c) $\sqrt[4]{64x^8y^3} = \sqrt[4]{2^4 \cdot 2^2 \cdot (x^2)^4 \cdot y^3}$ ■ Since the index is 4, we look for perfect fourth powers.
$= \sqrt[4]{2^4} \cdot \sqrt[4]{2^2} \cdot \sqrt[4]{(x^2)^4} \cdot \sqrt[4]{y^3}$ ■ Rule 3 of Radicals is used.
$= 2 \cdot \sqrt[4]{4} \cdot x^2 \cdot \sqrt[4]{y^3}$ ■ Rule 1 of Radicals is used.
$= 2x^2 \cdot \sqrt[4]{4y^3}$ ■ Rule 3 of Radicals is used.

(d) $\sqrt[3]{\dfrac{7}{100}} = \sqrt[3]{\dfrac{7}{100} \cdot \dfrac{10}{10}}$ ■ The index is 3. To simplify, we want a perfect cube in the denominator. Multiply by $\dfrac{10}{10}$ under the radical sign to get a denominator of $1000 = 10^3$.
$= \dfrac{\sqrt[3]{70}}{\sqrt[3]{10^3}}$
$= \dfrac{\sqrt[3]{70}}{10}$

(e) $\sqrt{\dfrac{3x^3y}{2z^5}} = \sqrt{\dfrac{3x^3y}{2z^5} \cdot \dfrac{2z}{2z}}$ ■ The index is 2. We want perfect squares in the denominator: $2z^5 \cdot 2z = 2^2 \cdot z^6 = 2^2 \cdot (z^3)^2$.
$= \sqrt{\dfrac{6x^2 \cdot xyz}{2^2 \cdot (z^3)^2}}$
$= \dfrac{|x|}{2|z^3|}\sqrt{6xyz}$ ■ x and z could both be negative.

Goal C *to review methods for simplifying sums, differences, products, and quotients involving radicals*

Let us agree that two radical terms will be called **like radicals,** provided they have the same index and same radicand. ($5\sqrt[3]{x}$ and $-2\sqrt[3]{x}$ are like radicals; $6\sqrt{y}$ and $6\sqrt[3]{y}$ are not.) To simplify sums and differences of radical expressions, you simplify individual terms first, and then combine like radicals.

Example 4 Simplify each of the following.

(a) $4\sqrt[3]{16} - 5\sqrt[3]{2} + 4\sqrt[3]{6}$ (b) $\sqrt{50x^2y^3} - 3\sqrt{8y}$

Solution (a) $4\sqrt[3]{16} - 5\sqrt[3]{2} + 4\sqrt[3]{6} = 4\sqrt[3]{2^3 \cdot 2} - 5\sqrt[3]{2} + 4\sqrt[3]{6}$ ■ Simplify individual terms.
$= 4 \cdot 2\sqrt[3]{2} - 5\sqrt[3]{2} + 4\sqrt[3]{6}$
$= (8\sqrt[3]{2} - 5\sqrt[3]{2}) + 4\sqrt[3]{6}$ ■ Like terms are combined.
$= 3\sqrt[3]{2} + 4\sqrt[3]{6}$

(b) $\sqrt{50x^2y^3} - 3\sqrt{8y} = \sqrt{5^2 \cdot 2 \cdot x^2 \cdot y^2 \cdot y} - 3\sqrt{2^2 \cdot 2y}$ ■ y cannot be negative; otherwise $\sqrt{50x^2y^3}$ and $3\sqrt{8y}$ would not be defined. Thus, $|y|$ is not necessary.
$= 5|x|y\sqrt{2y} - 3 \cdot 2\sqrt{2y}$
$= (5|x|y - 6)\sqrt{2y}$

Careful! An expression such as $\sqrt{7} - \sqrt{2}$ cannot be simplified any further. Don't make the mistake of writing:

$$\sqrt{7} - \sqrt{2} = \sqrt{5}.$$

Example 5 Multiply each of the following.

(a) $\sqrt{5x} \cdot \sqrt{10xy}$ (b) $(3 - \sqrt{6})^2$ (c) $(5 + \sqrt{6x})(5 - \sqrt{6x})$

Solution (a) $\sqrt{5x} \cdot \sqrt{10xy} = \sqrt{5^2 \cdot 2 \cdot x^2 \cdot y}$ ■ Rule 3 of Radicals is used.
$= 5x\sqrt{2y}$ ■ x cannot be negative; thus $|x|$ is unnecessary.

(b) $(3 - \sqrt{6})^2 = 3^2 - 2(3)(\sqrt{6}) + (\sqrt{6})^2$ ■ $(a - b)^2 = a^2 - 2ab + b^2$
$= 9 - 6\sqrt{6} + 6$
$= 15 - 6\sqrt{6}$

(c) $(5 + \sqrt{6x})(5 - \sqrt{6x}) = 5^2 - (\sqrt{6x})^2$ ■ $(a + b)(a - b) = a^2 - b^2$
$= 25 - 6x$

In Example 5(c), the expressions $5 + \sqrt{6x}$ and $5 - \sqrt{6x}$ are called **conjugates.** When conjugates contain square roots, the product of the conjugates will not contain any radicals. In the following example we use the conjugate of $3 - \sqrt{6}$ to remove the radical from the denominator. This process is known as **rationalizing the denominator.**

Example 6 Simplify $\dfrac{5}{3 - \sqrt{6}}$. (Rationalize the denominator.)

Solution
$$\dfrac{5}{3 - \sqrt{6}} = \dfrac{5}{3 - \sqrt{6}} \cdot \dfrac{3 + \sqrt{6}}{3 + \sqrt{6}}$$
■ Using the conjugate of the denominator, multiply by 1 in the form $\dfrac{3 + \sqrt{6}}{3 + \sqrt{6}}$.

$$= \dfrac{15 + 5\sqrt{6}}{3^2 - (\sqrt{6})^2} = \dfrac{15 + 5\sqrt{6}}{3} = 5 + \dfrac{5}{3}\sqrt{6}.$$

Goal D *to review methods for simplifying expressions involving rational exponents* ($a^{m/n}$)

For any nonnegative real number a and for any positive integer n, we can use a rational exponent to write the nth root of a: $a^{1/n} = \sqrt[n]{a}$.

For example, $9^{1/2} = 3$, and $8^{1/3} = 2$. It turns out that the standard Rules of Exponents hold when the exponents are rational numbers. But what does an expression like $3^{4/5}$ mean? We can use Rule 3 of Exponents to write

$$3^{4/5} = (3^{1/5})^4 = (\sqrt[5]{3})^4 = \sqrt[5]{3^4}.$$

Thus, $3^{4/5}$ and $\sqrt[5]{3^4}$ mean the same thing.

Definition
For any nonnegative real number a, and for any positive integers m and n,

$$a^{m/n} = \sqrt[n]{a^m} = (\sqrt[n]{a})^m.$$

In addition,

$$a^{-m/n} = \dfrac{1}{a^{m/n}}, \quad \text{provided } a \neq 0.$$

Example 7 Evaluate each of the following.

(a) $4^{3/2}$ (b) $\left(\dfrac{8}{27}\right)^{5/3}$ (c) $7^{4/5} \cdot 7^{2/5}$

(d) $3^{2/3} \cdot 3^{-2/3}$ (e) $\sqrt[5]{x} \cdot \sqrt{x}$

Solution

(a) $4^{3/2} = (\sqrt{4})^3 = 2^3 = 8$

(b) $\left(\dfrac{8}{27}\right)^{5/3} = \left(\sqrt[3]{\dfrac{8}{27}}\right)^5 = \left(\dfrac{2}{3}\right)^5 = \dfrac{32}{243}$

(c) $7^{4/5} \cdot 7^{2/5} = 7^{4/5+2/5} = 7^{6/5} = \sqrt[5]{7^6} = \sqrt[5]{7^5 \cdot 7} = 7\sqrt[5]{7}$

(d) $3^{2/3} \cdot 3^{-2/3} = 3^{2/3+(-2/3)} = 3^0 = 1$

(e) $\sqrt[5]{x} \cdot \sqrt{x} = x^{1/5} \cdot x^{1/2}$ ■ Since the bases are equal, we can use Rule 1 of Exponents. To do so, we must write the exponents with a common denominator.
$= x^{2/10} \cdot x^{5/10}$
$= x^{7/10}$ or $\sqrt[10]{x^7}$

Example 8 Write each of the following in the form $cx^r y^q$, where c is a real number, and r and q are rational numbers.

(a) $(3x^{-1/5} y^{1/4})^2$ (b) $\dfrac{4x^{2/3}}{7x^{1/6} y^{-1/2}}$ (c) $\dfrac{5x^{1/2} y^{3/4}}{x^{-1/3} y^{2/3}}$

Solution

(a) $(3x^{-1/5} y^{1/4})^2 = (3)^2 (x^{-1/5})^2 (y^{1/4})^2$ ■ Rule 2 of Exponents is used.
$= 9x^{-2/5} y^{1/2}$ ■ Rule 3 of Exponents is used.

(b) $\dfrac{4x^{2/3}}{7x^{1/6} y^{-1/2}} = \dfrac{4}{7} x^{(2/3-1/6)} y^{1/2}$ ■ $\dfrac{1}{a^{-r}} = a^r$ and Rule 4 of Exponents.
$= \dfrac{4}{7} x^{1/2} y^{1/2}$

(c) $\dfrac{5x^{1/2} y^{3/4}}{x^{-1/3} y^{2/3}} = 5x^{(1/2+1/3)} y^{(3/4-2/3)}$ ■ Rule 4 of Exponents is used.
$= 5x^{5/6} y^{1/12}$

Exercise Set 1.7   = Calculator Exercises in the Appendix

Goal A

In exercises 1–12, evaluate each expression if possible. If the expression does not represent a real number, then state this fact.

1. $\sqrt[3]{1000}$
2. $\sqrt[4]{16}$
3. $\sqrt{\dfrac{36}{49}}$
4. $-\sqrt{\dfrac{81}{25}}$
5. $\sqrt{0.0016}$
6. $\sqrt[3]{0.008}$
7. $\sqrt{-1}$
8. $\sqrt[3]{-64}$
9. $-\sqrt[3]{-125}$
10. $\sqrt[4]{-81}$
11. $\sqrt{-0.0001}$
12. $\sqrt[3]{-0.027}$

Goal B

In exercises 13–26, simplify each expression.

13. $\sqrt[3]{40}$
14. $\sqrt[5]{96}$
15. $-\sqrt{27}$
16. $\sqrt{50}$
17. $\sqrt{\dfrac{9}{10}}$
18. $-\sqrt{\dfrac{4}{7}}$
19. $\sqrt[3]{\dfrac{-25}{9}}$
20. $\sqrt[3]{\dfrac{1}{16}}$
21. $\dfrac{2}{\sqrt{3}}$
22. $\dfrac{1}{\sqrt[3]{4}}$
23. $\sqrt[3]{8x^4}$
24. $\sqrt[4]{10v^4}$
25. $\sqrt{\dfrac{x}{y}}$
26. $\sqrt[3]{\dfrac{u}{v^2}}$

In exercises 27–38, simplify each expression if possible. Assume that all variables represent positive real numbers.

27. $\sqrt{8x^2y^3}$
28. $\sqrt[3]{15u^4v}$
29. $\sqrt[4]{32a^2b^5c}$
30. $\sqrt[4]{x^7y^2z^5}$
31. $\sqrt[5]{(8m^6n^4)^2}$
32. $\sqrt[5]{(-3x^2y^4)^3}$
33. $\sqrt{\dfrac{x^3}{y^4}}$
34. $\sqrt[3]{\dfrac{-u}{v^6}}$
35. $\sqrt[3]{\dfrac{10x^2}{2y^2}}$
36. $\sqrt{\dfrac{8z^3}{5w}}$
37. $\sqrt{(-5xyz)^2}$
38. $\sqrt[3]{(-6vw)^3}$

Goal C

In exercises 39–64, perform the indicated operations and simplify. *Assume that all variables represent positive real numbers.*

39. $\sqrt{12} - \sqrt{3}$
40. $8\sqrt{48} - \sqrt{63}$
41. $2\sqrt{5} - 5\sqrt{45}$
42. $\sqrt{2} + \dfrac{\sqrt{2}}{2}$
43. $\sqrt{8} \cdot \sqrt{10}$
44. $\sqrt{6} \cdot \sqrt{12}$
45. $\sqrt[3]{5} \cdot \sqrt[3]{30}$
46. $\sqrt[3]{7} \cdot \sqrt[3]{54}$
47. $\dfrac{\sqrt{12}}{\sqrt{18}}$
48. $\dfrac{\sqrt{10}}{\sqrt{20}}$
49. $\sqrt{3}(\sqrt{6} - 2)$
50. $\sqrt{2}(5 - \sqrt{6})$
51. $(2 + \sqrt{5})(2 - \sqrt{5})$
52. $(3 - \sqrt{7})(4 + \sqrt{2})$
53. $(1 - \sqrt{7})^2$
54. $(\sqrt{2} + \sqrt{3})^2$
55. $\sqrt[3]{6pq^6}\,\sqrt[3]{12p^4}$
56. $\sqrt{8ab^5}\,\sqrt{10a^3b}$
57. $\dfrac{\sqrt{8a^3}}{\sqrt{12b^4}}$
58. $\dfrac{\sqrt[3]{81u^5}}{\sqrt[3]{64v}}$
59. $3\sqrt{x} - 2\sqrt{x^3} + \sqrt{x^2}$
60. $2\sqrt[3]{a^2b^4} + 7\sqrt[3]{a^2b} - \sqrt[3]{a^5b}$
61. $\sqrt{x}(\sqrt{x} + \sqrt{y})$
62. $\sqrt[3]{a^2}(\sqrt[3]{a} - \sqrt[3]{b^4})$
63. $(\sqrt{x} + \sqrt{y})(\sqrt{x} - \sqrt{y})$
64. $(2\sqrt{x} - \sqrt{3})(2\sqrt{x} + \sqrt{3})$

Goal D

In exercises 65–74, evaluate each expression.

65. $100^{3/2}$
66. $25^{5/2}$
67. $8^{2/3}$
68. $81^{1/4}$
69. $\left(\dfrac{25}{30}\right)^{-3/2}$
70. $\left(\dfrac{9}{16}\right)^{-3/2}$

71. $3^{2/3} \cdot 3^{4/3}$

72. $9^{3/5} \cdot 9^{4/5}$

73. $2^{4/5} \cdot 4^{1/5}$

74. $3^{5/2} \cdot 27^{-1/2}$

In exercises 75–88, simplify the given expression. Write your answer in the form $cx^r y^q$ where c is a real number, and r and q are rational numbers.

75. $(5x^{2/3})(4x^{2/3})$

76. $(-8y^{2/5})(4y^{4/5})$

77. $(2a^{1/2})(3a^{1/3})$

78. $(4m^{1/5})(5m^{1/4})$

79. $(10xy^{1/2})(-2x^{1/3}y^{3/2})$

80. $(6u^2 v^{1/4})(-7uv^{5/4})$

81. $(8x^{-1/2}y^{1/4}z)^3$

82. $(-2m^{3/5}np^8)^2$

83. $(16x^2 y)^{3/4}$

84. $(8u^{1/4}v^{3/8})^{1/3}$

85. $\dfrac{3x^{3/5}}{6x^{1/10}y^{-2}}$

86. $\dfrac{14m^{3/4}n^{1/2}}{18m^{-1/2}n^{3/8}}$

87. $\left(\dfrac{1}{3x}\right)^{1/2}$

88. $\left(\dfrac{2}{3x^2}\right)^{1/3}$

Superset

In exercises 89–98, rewrite each expression in exponential form; then simplify if possible. Answer with a single radical of smallest possible index. *Assume all variables represent positive real numbers.*

89. $\sqrt[8]{x^4}$

90. $\sqrt[4]{a^2}$

91. $\sqrt[3]{m} \cdot \sqrt[4]{v}$

92. $\sqrt{4v} \cdot \sqrt[4]{v}$

93. $\sqrt[4]{\sqrt{z}}$

94. $\sqrt{\sqrt[3]{v}}$

95. $\sqrt[6]{x\sqrt{y}}$

96. $\sqrt[5]{a^2 \sqrt[3]{c}}$

97. $\dfrac{\sqrt[15]{14}}{\sqrt{a}}$

98. $\dfrac{\sqrt[4]{a^3}}{\sqrt[3]{a}}$

Specify whether each of the statements in exercises 99–110 is true or false. If false, illustrate this with an example.

99. $\sqrt{x^2} = x$

100. $\sqrt[3]{x^3} = x$

101. $\sqrt{x+y} = \sqrt{x} + \sqrt{y}$

102. $\sqrt{x^2 + y^2} = x + y$

103. $\dfrac{1}{\sqrt{x} + \sqrt{y}} = \dfrac{1}{\sqrt{x}} + \dfrac{1}{\sqrt{y}}$

104. $\dfrac{1}{\sqrt{xy}} = \dfrac{1}{\sqrt{x}} \cdot \dfrac{1}{\sqrt{y}}$

105. $x^{1/n} = \dfrac{1}{x^n}$

106. $\sqrt[3]{x} \cdot \sqrt{x} = \sqrt[6]{x}$

107. $\sqrt[3]{x} \cdot \sqrt{x} = \sqrt[6]{x^2}$

108. $\sqrt[3]{x} \cdot \sqrt{x} = \sqrt[6]{x^5}$

109. $\sqrt{x^2 - 1} = x - 1$

110. $\sqrt{x^2 + 2x + 1} = x + 1$

111. How does doubling a number affect its square root?

112. How does tripling a number affect its square root?

113. What would you do to a number in order to double its cube root?

114. What would you do to a number in order to triple its cube root?

115. If a number is increased ten-fold, is its square root more than doubled? More than tripled? More than quadrupled?

116. Verify that $\sqrt{8 + 2\sqrt{15}} = \sqrt{5} + \sqrt{3}$.

117. Is $\sqrt{5 - 2\sqrt{6}} = \sqrt{2} - \sqrt{3}$? Explain.

118. Is $\sqrt{2} \cdot \sqrt{10 - 2\sqrt{21}} = \sqrt{14} - \sqrt{6}$? Explain.

119. Verify that $\sqrt{\pi} \cdot \sqrt{3 + 2\sqrt{2}} = \sqrt{\pi} + \sqrt{2\pi}$.

120. Determine the area of the shaded region in the figure below, given that figure *ABCD* is a square with area 2 square feet, and figure *EFGA* is a square with area 3 square feet. Explain your solution carefully.

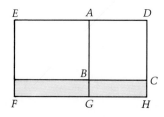

1.8 Fractional Expressions

Goal A *to review methods for simplifying rational expressions (quotients of polynomials)*

Quotients of polynomials are commonly referred to as **rational expressions.** The rules for adding, subtracting, multiplying and dividing rational expressions are patterned after the rules for the arithmetic of fractions.

Recall the basic rule for reducing fractions: for any real number a, and any nonzero real numbers b and c,

$$\frac{ac}{bc} = \frac{a}{b} \cdot \frac{c}{c} = \frac{a}{b}$$

$\frac{14}{15} \qquad \frac{3xy^2 - 6x^3}{xy^4}$

$\frac{8x^2 - 3x + 40}{x - 1}$

Examples of Rational Expressions

$\frac{8}{12} = \frac{2 \cdot 4}{3 \cdot 4} = \frac{2}{3}$

Example of the method of reducing fractions

The number c is called a **common factor.** If c is the greatest common factor, then $\frac{a}{b}$ is said to be in **lowest terms.**

$$\frac{x^2 + 4x + 3}{x^2 - 9} = \frac{(x + 1)(x + 3)}{(x - 3)(x + 3)} = \frac{(x + 1)}{(x - 3)}$$

factor and cancel

The original expression is not defined when

$$x = 3 \quad \text{or} \quad x = -3,$$

since both of these x-values produce a zero denominator. Thus, we should write

$$\frac{x^2 + 4x + 3}{x^2 - 9} = \frac{x + 1}{x - 3}, \quad \text{provided } x \neq 3 \text{ and } x \neq -3.$$

In the examples and exercises that follow, we assume that numbers producing zero denominators have been excluded as values of the variables.

Example 1 Simplify $\dfrac{6x^2 - 2x - 4}{x^2 - 1}$.

Solution $\dfrac{6x^2 - 2x - 4}{x^2 - 1} = \dfrac{2(3x + 2)(x - 1)}{(x + 1)(x - 1)} = \dfrac{2(3x + 2)(x - 1)}{(x + 1)(x - 1)} = \dfrac{2(3x + 2)}{x + 1}$

Goal B *to review techniques for adding, subtracting, multiplying and dividing rational expressions*

Recall the rules for multiplying and dividing fractions.

$$\frac{a}{b} \cdot \frac{c}{d} = \frac{ac}{bd} \quad (b \text{ and } d \text{ are nonzero})$$

$$\frac{a}{b} \div \frac{c}{d} = \frac{a}{b} \cdot \frac{d}{c} = \frac{ad}{bc} \quad (b, c, \text{ and } d \text{ are nonzero})$$

These rules are also true when a, b, c, and d are replaced with polynomials.

Example 2 Perform the indicated operations and then simplify.

(a) $\dfrac{x}{2x+5} \cdot \dfrac{2x^2 + 3x - 5}{x^2 + 3x}$ (b) $\dfrac{xy - 3x}{x+2} \div \dfrac{y^2 - 9}{3x^2 - x - 14}$

Solution (a) $\dfrac{x}{2x+5} \cdot \dfrac{2x^2 + 3x - 5}{x^2 + 3x} = \dfrac{\cancel{x} \cdot \cancel{(2x+5)}(x-1)}{\cancel{(2x+5)} \cdot \cancel{x} \cdot (x+3)} = \dfrac{x-1}{x+3}$

(b) $\dfrac{xy - 3x}{x+2} \div \dfrac{y^2 - 9}{3x^2 - x - 14} = \dfrac{xy - 3x}{x+2} \cdot \dfrac{3x^2 - x - 14}{y^2 - 9}$ ■ The second expression is inverted.

$= \dfrac{x\cancel{(y-3)}(3x-7)\cancel{(x+2)}}{\cancel{(x+2)}(y+3)\cancel{(y-3)}}$

$= \dfrac{x(3x-7)}{y+3}$

The crucial step to remember when adding or subtracting fractions is that you must first make the denominators the same. Recall the basic rules:

$$\frac{a}{c} + \frac{b}{c} = \frac{a+b}{c} \quad (c \neq 0),$$

$$\frac{a}{c} - \frac{b}{c} = \frac{a-b}{c} \quad (c \neq 0).$$

We call c a common denominator.

Example 3   Combine $\dfrac{2}{x} - \dfrac{3}{x-1} + \dfrac{x+3}{x^2-1}$.

Solution

$$\begin{array}{ll} x\text{:} & x \\ x-1\text{:} & x-1 \\ x^2-1\text{:} & x-1 \quad x+1 \\ \text{LCD}\text{:} & x \cdot x - 1 \cdot x + 1 \end{array}$$

■ Begin by factoring each of the denominators to determine the factors needed for the Least Common Denominator (LCD). The LCD is the product of the greatest power of each factor present in any of the factorizations.

$$\dfrac{2}{x} - \dfrac{3}{x-1} + \dfrac{x+3}{x^2-1} = \dfrac{2}{x}\cdot\left[\dfrac{(x-1)(x+1)}{(x-1)(x+1)}\right] - \dfrac{3}{x-1}\left[\dfrac{x(x+1)}{x(x+1)}\right] + \dfrac{x+3}{x^2-1}\left[\dfrac{x}{x}\right]$$

■ Each term has been multiplied by a form of 1 so that the denominator is the LCD.

$$= \dfrac{2(x^2-1)}{x(x^2-1)} - \dfrac{3x^2+3x}{x(x^2-1)} + \dfrac{x^2+3x}{x(x^2-1)}$$

$$= \dfrac{2x^2 - 2 - 3x^2 - 3x + x^2 + 3x}{x(x^2-1)}$$

$$= \dfrac{-2}{x(x^2-1)}$$

Goal C   *to review a method of simplifying other fractional expressions*

Up to this point we have been concerned with rational expressions, that is, quotients of polynomials. A **radical fractional expression** contains radicals in its numerator or denominator. Simplifying these expressions requires that denominators be rationalized. We sometimes use the conjugate (see Section 1.7) to do this.

Example 4   Simplify $\dfrac{1}{\sqrt{x}+\sqrt{2}}$.

Solution $\dfrac{1}{\sqrt{x}+\sqrt{2}} = \dfrac{1}{\sqrt{x}+\sqrt{2}} \cdot \dfrac{\sqrt{x}-\sqrt{2}}{\sqrt{x}-\sqrt{2}} = \dfrac{\sqrt{x}-\sqrt{2}}{(\sqrt{x})^2-(\sqrt{2})^2} = \dfrac{\sqrt{x}-\sqrt{2}}{x-2}$

The second type of fractional expression to be considered is called a **complex fractional expression.** In such an expression, the numerator or denominator itself contains a fraction.

Example 5 Simplify the following expression.

$$\frac{1 - \frac{2}{x}}{1 - \frac{3}{x} + \frac{2}{x^2}}$$

Solution

$$\frac{1 - \frac{2}{x}}{1 - \frac{3}{x} + \frac{2}{x^2}} = \frac{1\left(\frac{x}{x}\right) - \frac{2}{x}}{1\left(\frac{x^2}{x^2}\right) - \frac{3}{x}\left(\frac{x}{x}\right) + \frac{2}{x^2}}$$

■ Begin by combining terms in the numerator and denominator.

$$= \frac{\frac{x}{x} - \frac{2}{x}}{\frac{x^2}{x^2} - \frac{3x}{x^2} + \frac{2}{x^2}}$$

$$= \frac{\frac{x - 2}{x}}{\frac{x^2 - 3x + 2}{x^2}}$$

■ The problem is now a fraction divided by a fraction; we shall invert and multiply.

$$= \frac{x - 2}{x} \cdot \frac{x^2}{x^2 - 3x + 2}$$

$$= \frac{\cancel{x - 2}}{\cancel{x}} \cdot \frac{\cancel{x^2}^{x}}{\cancel{(x - 2)}(x - 1)}$$

$$= \frac{x}{x - 1}$$

Exercise Set 1.8   = Calculator Exercises in the Appendix

Goal A

In exercises 1–16, simplify each expression.

1. $\dfrac{(3-x)^2}{x-3}$
2. $\dfrac{2a-2b}{b-a}$
3. $\dfrac{3m^2-27}{m+3}$
4. $\dfrac{m^3-27}{m-3}$
5. $\dfrac{v^3-4v}{v^2-v-6}$
6. $\dfrac{u^4-9u^2}{u^3+7u^2+12u}$
7. $\dfrac{x^3-27}{x^2-9}$
8. $\dfrac{m^3+64}{m^2-16}$
9. $\dfrac{2a^2+8a+8}{3a^2+4a-4}$
10. $\dfrac{a^2+6a-27}{3a^2-27a+54}$
11. $\dfrac{2x+4-x^3-2x^2}{3x^2-12}$
12. $\dfrac{5v^2+10v+5}{v^3+v^2-v-1}$
13. $\dfrac{-v^2+30v-225}{4v^2-58v-30}$
14. $\dfrac{6-x-x^2}{x^2-15x-54}$
15. $(m^2n^2+2m^2n-24m^2)(n+6)^{-1}$
16. $-(x-y)(y^2+4xy-5x^2)^{-1}$

Goal B

In exercises 17–34, perform the indicated operation.

17. $\left(-\dfrac{5a}{b^2}\right)\left(\dfrac{3b}{a^3}\right)$
18. $\left(\dfrac{9xy}{z^3}\right)\left(-\dfrac{xz^2}{y^3}\right)$
19. $\left(\dfrac{x-1}{x}\right)\left(\dfrac{x+1}{x}\right)$
20. $\left(\dfrac{3+y}{3-y}\right)\left(\dfrac{3+y}{y}\right)$
21. $\dfrac{5xyz^2}{10x^2}\left(\dfrac{yz}{x}\right)^{-2}$
22. $\dfrac{7a^2bc}{12c^3}\left(\dfrac{ab}{2c}\right)^{-3}$
23. $(3m^2)(2n)^{-2}(8n)(2m)^{-3}$
24. $x^3(y^2)^{-1}(3x)^{-2}(2y)^{-3}$
25. $\left(\dfrac{3v^2-12}{5v-15}\right)\left(\dfrac{2v^2-18}{v^2+5v+6}\right)$
26. $\left(\dfrac{x^2+10x+25}{x^2+7x+10}\right)\left(\dfrac{5-x}{2x^2-50}\right)$
27. $\dfrac{a}{a-b} \div \dfrac{b}{b-a}$
28. $\dfrac{x}{x^2+2x} \div \dfrac{x}{x+2}$
29. $\dfrac{2v^2+20v}{v^2-9} \div \left(\dfrac{v}{v+3}\right)^2$
30. $\dfrac{9m^2-18m-16}{6m+4} \div \dfrac{6m+16}{9m^2-64}$
31. $\dfrac{3}{v+2} - \dfrac{4}{v-2}$
32. $\dfrac{6}{m} - \dfrac{8m}{m^2-3m}$
33. $\dfrac{x^2-10}{x^2-2x-15} - \dfrac{5x}{2x^2-3x-35}$
34. $\dfrac{a+3}{2a^2+5a-12} - \dfrac{a-1}{2a^2+7a-4}$

Goal C

In exercises 35–48, simplify the given expression.

35. $\dfrac{1}{3\sqrt{v}+\sqrt{v}}$
36. $\dfrac{\sqrt{w}-4}{\sqrt{w}+3}$
37. $\dfrac{5-\sqrt{t}}{1-\sqrt{t}}$
38. $\dfrac{\sqrt{m}-\sqrt{2}}{\sqrt{m}+\sqrt{2}}$
39. $\dfrac{x}{\sqrt{x}-2\sqrt{y}}$
40. $\dfrac{a-2b}{2\sqrt{a}-\sqrt{8b}}$
41. $(x+2+\sqrt{y})^{-1}$
42. $(\sqrt{a}+\sqrt{b})^{-2}$
43. $\dfrac{u-\dfrac{4}{u}}{1-\dfrac{2}{u}}$
44. $\dfrac{y-y^{-2}}{y-y^{-1}}$
45. $\dfrac{1-\dfrac{1}{v+1}}{1+\dfrac{1}{v-1}}$
46. $\dfrac{1+\dfrac{1}{w+1}}{1+\dfrac{1}{w-1}}$

47. $\dfrac{\dfrac{1}{m+2} - \dfrac{m+1}{m}}{\dfrac{m}{m+2} + \dfrac{1}{m}}$

48. $\dfrac{\dfrac{n}{1+n} + \dfrac{1-n}{n}}{\dfrac{n}{1+n} - \dfrac{1-n}{n}}$

58.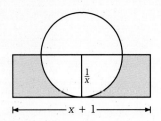

Superset

Specify whether each of the statements in exercises 49–54 is true or false. If false, illustrate this with an example.

49. $\dfrac{1}{x+y} = \dfrac{1}{x} + \dfrac{1}{y}$

50. $\dfrac{3}{4} + \dfrac{x}{y} = \dfrac{3+x}{4+y}$

51. $(x+y)^{-2} = x^{-2} + y^{-2}$

52. $\left(\dfrac{x}{y} + \dfrac{2}{3}\right)^{-1} = \dfrac{y}{x} + \dfrac{3}{2}$

53. $x^{-2} - y^{-2} = (x^{-1} - y^{-1})(x^{-1} + y^{-1})$

54. $(x \div y) \div z = x \div (y \div z)$

In exercises 55–58, determine the area of each shaded region. Write your answers as a single rational expression in x.

55.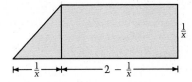

56.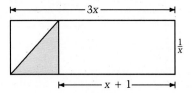

57.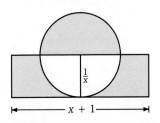

The **weighted mean** of a set of values, $x_1, x_2, \ldots, x_m$, takes into account the number of times each value occurs, $n_1, n_2, \ldots, n_m$, respectively, and is given by the expression

$$\dfrac{x_1 n_1 + x_2 n_2 + \cdots + x_m n_m}{n_1 + n_2 + \cdots + n_m}$$

In exercises 59–61, you will need to determine a weighted mean.

59. The seniors at a certain high school took the Advanced Placement Exam in Computer Science. Possible scores are 1, 2, 3, 4, and 5. The chart below shows how many students received each grade.

Grade	1	2	3	4	5
Students	8	10	24	19	5

Determine the average score of the students at that school. (Round to the nearest tenth.)

60. A person has invested $5000 at 5.5%, $12,000 at 7%, $6000 at 9.5% and $3000 at 12.5%. What is the average return (rate of profit) on these investments? (Round to the nearest tenth of a percent.)

61. A corporation has four levels of middle management positions. The mean salary is
$21,500 for its 40 Level I managers,
$26,800 for its 65 Level II managers,
$33,000 for its 28 Level III managers, and
$42,600 for its 15 Level IV managers.
What is the average salary paid to the corporation's middle management employees? (Round to the nearest dollar.)

Chapter Review & Test

Chapter Review

1.1 Statements and Variables (pp. 2–5)

A *variable* is a symbol used to represent a number. (p. 2)   An *expression* is a combination of numbers and variables (e.g., $3x^2 + 2yz$). An *equation* is a statement that two expressions have the same value. (p. 3)

1.2 The Real Numbers (pp. 6–10)

The set of *real numbers* is composed of *rational numbers* (e.g., $\frac{2}{3}$ and 8) and *irrational numbers* (e.g., $\sqrt{2}$ and π). The set of rational numbers is composed of *integers and noninteger fractions.* (p. 6)

1.3 Absolute Value, Distance, and the Real Number Line (pp. 11–15)

The *absolute value* of a nonzero real number a, denoted $|a|$, is the positive number in the set $\{a, -a\}$. The absolute value of zero is zero. (p. 11)

The set of real numbers can be placed in a one-to-one correspondence with the points on a line, called the *real line.* The number associated with a given point is the *coordinate* of the point. (p. 12)   If points A and B have coordinates a and b respectively, then the *distance* between A and B, denoted $d(A,B)$, is equal to $|a - b|$. (p. 14)

1.4 The Integers as Exponents (pp. 16–21)

$$a^m \cdot a^n = a^{m+n}, \quad (ab)^n = a^n b^n, \quad (a^m)^n = a^{mn}, \quad \frac{a^m}{a^n} = a^{m-n}, \quad \left(\frac{a}{b}\right)^n = \frac{a^n}{b^n}.$$

In addition, if $a \neq 0$, then $a^0 = 1$ and $\frac{1}{a^n} = a^{-n}$. (p. 18)

1.5 Polynomials: Addition, Subtraction, and Multiplication (pp. 22–27)

A *term*, or *monomial*, is either a constant or a product of constants and variables. Its *degree* is the sum of the exponents of its variables. The constant multiplier is called the *coefficient. Like terms* have the same variables with the same exponents. (p. 22)

A *polynomial* is a sum of terms, and can be simplified by combining like terms. (p. 23)   A polynomial with two terms is called a *binomial,* and one with three terms, a *trinomial.* The FOIL method is used to find the product of two binomials. (p. 25)

1.6 Polynomials: Factoring (pp. 28–35)

To factor a polynomial, use one or more of the following strategies:

1. Factor out a common factor.
2. Observe a pattern: $A^2 + 2AB + B^2 = (A + B)^2$
$$A^2 - B^2 = (A + B)(A - B)$$
$$A^3 + B^3 = (A + B)(A^2 - AB + B^2)$$
$$A^3 - B^3 = (A - B)(A^2 + AB + B^2)$$
3. Grouping
4. Trial and Error (p. 29)

1.7 Radicals and Rational Exponents (pp. 36–44)

If n is even and $b > 0$, then $\sqrt[n]{b}$ refers to the positive *(principal) nth root.*
If n is even and $b < 0$, then $\sqrt[n]{b}$ is not a real number.
If n is odd, then $\sqrt[n]{b}$ refers to the one real nth root, regardless of the sign of b. (p. 36)
If a is nonnegative, the principal nth root of a can be written $a^{1/n}$. (p. 41)

1.8 Fractional Expressions (pp. 45–50)

The arithmetic of fractional expressions containing variables follows the same rules as the arithmetic of fractions. Reducing fractional expressions requires factoring numerator and denominator first. Adding fractional expressions requires common denominators. (p. 45)

Chapter Test

1.1A Translate each of the following sentences into an equation:

  1. Thirteen more than twice a certain number is fifty-six.

  2. The average of three numbers is twenty-seven.

1.2A Classify each number as one or more of the following: integer, rational number, irrational number, real number.

  3. -16 4. 0 5. $\sqrt[3]{11}$

1.2B Determine whether each statement is true or false. If false, find the error.

  6. $5(3 + 4) = 5 \cdot 3 + 4 = 19$ 7. $6 - (8 + 2) = (6 - 8) + 2 = 0$

1.3A Evaluate **8.** $|52|$ **9.** $|-7| - |5|$ **10.** $-|-3| - |3|$

1.3B Graph each set on the real line. **11.** $\{-50, 10, 35\}$ **12.** $\left\{\frac{1}{8}, 2, \pi\right\}$

1.3C Use the graph at the right to determine the following distances.

$$\begin{array}{c} A \qquad\quad O \qquad\quad B \\ \leftarrow\!\!+\!\!+\!\!\bullet\!\!+\!\!+\!\!\bullet\!\!+\!\!+\!\!\bullet\!\!+\!\!+\!\!\rightarrow \\ -12\;-9\;-6\;-3\;\;0\;\;3\;\;6\;\;9\;\;12 \end{array}$$

13. $d(A,B)$ **14.** $d(B,O)$ **15.** $d(O,B)$

1.4A Rewrite with a positive exponent: **16.** x^{-5} **17.** $\frac{1}{x^{-7}}$ **18.** a^{-1}

1.4B Simplify each expression so that x, y, and z appear at most once, with no powers of powers or negative exponents.

19. $(5x^2)(y^2)(3x^{-5})$ **20.** $(3xy^2z)^4$ **21.** $\left(\frac{2x}{5y^2}\right)^{-3}$

1.5A Complete the following. Select your answer from the choices offered in brackets.

22. The degree of $3x^2 + 5xy^2z^2 + 6x^4$ is _____. [3/4/5]

23. The terms of $8w^2 - 7$ are $8w^2$ and _____. [7w/7/-7]

1.5B Perform the indicated operation.

24. $(u^2 + 3u - 7) + (u^3 - 15u + 8)$

25. Subtract $6n^2 - 5n - 7$ from $3n^3 + 12n - 4$.

1.5C Perform the indicated operation.

26. $3x(x^2 - 7x + 4)$ **27.** $(8m - 2)(3m + 5)$ **28.** $(9z - 2)^2$

1.6A Find the greatest common factor of the following terms.

29. 42, 140, 154   **30.** $12x^2yz$, $8x^2z^3$, $2x^5yz^2$

1.6B Factor completely. If a polynomial cannot be factored, state this fact.

31. $6x^2y - 2xy^2 + 7x^2y^2$ **32.** $6t^2 - 5t - 4$ **33.** $8z^3 - 27$

1.7A Evaluate. If the expression is not a real number, state this fact.

34. $\sqrt[3]{-216}$ **35.** $\sqrt{0.0004}$ **36.** $\sqrt[4]{-16}$

1.7B Simplify. Assume that all variables represent positive real numbers.

37. $\sqrt{24x^2y^2z^2}$ 38. $\sqrt[4]{\dfrac{u^3}{32v^5}}$

1.7C Simplify. Assume that all variables represent positive real numbers.

39. $\sqrt{12} - \sqrt{27}$ 40. $4\sqrt[3]{x}(6 - \sqrt[3]{x^2y})$ 41. $(x - \sqrt{y})(x + \sqrt{y})$

1.7D Simplify. If your answer involves variables, write it in the form cx^ry^q.

42. $25x^{1/2}(y^{1/2})^{2/3}$ 43. $(2x^{1/3}y^{2/5})^2$ 44. $\dfrac{3x^{2/3}}{8x^{1/9}y^{-1/2}}$

1.8A Simplify: 45. $\dfrac{m^3 - 9m}{3m^2 + 8m - 3}$ 46. $\dfrac{n^3 + n^2 + n + 1}{n^2 - 1}$

1.8B Perform the indicated operations.

47. $\left(\dfrac{3a^2bc}{7c^3}\right)\left(\dfrac{15b^2c}{9a^2c^3}\right)^{-1}$ 48. $\left(\dfrac{10u + 10v}{u^3 + 8v^3}\right)\left(\dfrac{u^2 + 3uv + 2v^2}{u^2 - 5uv - 6v^2}\right)$

1.8C Simplify the following: 49. $\dfrac{\dfrac{1}{x} - 1}{1 - \dfrac{1}{x}}$ 50. $\dfrac{2 + \dfrac{1}{u - 5}}{\dfrac{19}{u - 5} - 12}$

Superset

51. Determine whether each of the following statements is true or false. If false, illustrate this with an example.

(a) The sum of a positive and negative number is never positive.
(b) For any real numbers x and y, $|x + y| = |x| + |y|$.
(c) For any real number x, $\sqrt[4]{x^4} = x$.

52. For the graph at the right the following statements are true:

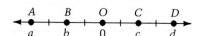

$d(A,C) = 7$ $d(A,C) = d(B,D)$
$d(B,C) = 3$ $d(B,O) = d(O,C)$

Determine $d(O,D)$.

53. In the figure at the right a circle of radius $x - 1$ has been inscribed in a square. Find a polynomial to describe the shaded area.

2

Equations and Inequalities in One Variable

2.1 Introduction to Equations

Goal A *to review methods for solving linear equations*

An **equation** is a statement that two expressions represent the same number. The equation $\frac{1}{2}(5 + 7) = (-2)(-3)$ is true, since both expressions represent the same number, 6. An equation containing variables, such as $2x + 5 = 4x - 1$, may be true for some values of the variables, and false for others. Those values of the variables that make an equation true are called **solutions** or **roots** of the equation, and are said to "satisfy" the equation. Thus, 3 is a solution of the above equation, but 7 is not.

Example 1 Determine whether $\frac{1}{3}$ is a solution of $7x - 3 = x + 1$.

Solution
$$7x - 3 = x + 1$$ ■ Replace x with $\frac{1}{3}$.
$$7\left(\frac{1}{3}\right) - 3 = \frac{1}{3} + 1$$
$$-\frac{2}{3} = \frac{4}{3}$$ ■ False.

Since the last equation is false, $\frac{1}{3}$ is not a solution.

Example 2 Determine whether -2 and 3 are solutions of $x^2 - x - 6 = 0$.

Solution

$$x^2 - x - 6 = 0$$
$$(-2)^2 - (-2) - 6 = 0$$
$$4 + 2 - 6 = 0$$
$$0 = 0 \quad \blacksquare \text{ True}$$

$$x^2 - x - 6 = 0$$
$$(3)^2 - (3) - 6 = 0$$
$$9 - 3 - 6 = 0$$
$$0 = 0 \quad \blacksquare \text{ True}$$

Both -2 and 3 are solutions of the equation $x^2 - x - 6 = 0$.

To "solve" an equation means to find *all* its solutions. The set of all solutions is called the **solution set.** Two equations with the same solution set are called **equivalent equations.** The strategy used to solve an equation is to transform it into simpler equivalent equations, until the solution set is obvious. The following properties will help you accomplish this.

Addition Property for Equivalent Equations
Adding the same real number (or variable expression) to both sides of an equation produces an equivalent equation.

Multiplication Property for Equivalent Equations
Multiplying both sides of an equation by the same nonzero real number (or variable expression) produces an equivalent equation.

Example 3   Solve $2(x - 30) = 30 - (x - 9)$.

Solution

$2(x - 30) = 30 - (x - 9)$ ■ Begin by simplifying. Rewrite the equation without parentheses and collect like terms.
$2x - 60 = 30 - x + 9$
$2x - 60 = 39 - x$
$2x + x - 60 = 39 - x + x$ ■ The Addition Property is used to get a single x-term on one side and a constant term on the other.
$3x - 60 = 39$
$3x - 60 + 60 = 39 + 60$
$3x = 99$
$\dfrac{1}{3} \cdot 3x = \dfrac{1}{3} \cdot 99$ ■ The Multiplication Property is used to get the coefficient of x equal to 1.
$x = 33$ ■ $x = 33$ is equivalent to the original equation.

Replacing x with 33 in the original equation produces a true statement. Thus, 33 is a solution.

Definition

A **linear equation** is an equation containing a first degree monomial, but no monomial of higher degree. A linear equation written in the form $ax + b = 0$, where a and b represent constants and $a \neq 0$, is called the **standard form of a linear equation in x.**

Any linear equation that can be written in the standard form $ax + b = 0$ has $-\frac{b}{a}$ as its solution. Therefore, we conclude that such an equation has exactly one solution. Certain linear equations, however, cannot be written in standard form. Two types of such equations are represented in Example 4 below. One type has no solutions, and the other has infinitely many solutions.

Example 4 Solve for x: (a) $2x + 4 = 2(5 + x)$ (b) $2(3 - 6x) = 3(2 - 4x)$

Solution (a)
$$2x + 4 = 2(5 + x)$$
$$2x + 4 = 10 + 2x$$
$$2x + (-2x) + 4 = 10 + 2x + (-2x)$$
$$4 = 10$$

Since the final equation is never true, no matter what the value of x is, the original equation has no solutions.

(b)
$$2(3 - 6x) = 3(2 - 4x)$$
$$6 - 12x = 6 - 12x$$
$$6 - 12x + 12x = 6 - 12x + 12x$$
$$6 = 6$$

Since the final equation is always true, no matter what the value of x is, the original equation has the set of all real numbers as its solution set.

One final note regarding linear equations: an equation such as

$$3x^2 + 2x = 3x^2 + 10$$

is not linear, but when we add $-3x^2$ to both sides, the linear equation $2x = 10$ is produced. The original equation is said to be equivalent to a linear equation.

Goal B *to review methods for solving fractional equations*

An equation containing fractional expressions involving variables is called a **fractional equation.** Usually, the first step in solving a fractional equation is to multiply both sides of the equation by the least common denominator (LCD) of all fractional expressions in the problem.

$$\frac{5}{x - 3} - 1 = 4$$

An example of a fractional equation.

Careful! Multiplying both sides of an equation by a variable expression might produce a nonequivalent equation that has more solutions than the original equation. Thus, you must check solutions after multiplying by a variable expression.

Example 5 Solve $\dfrac{x}{x+1} - 1 = \dfrac{2}{x^2 - 1}$.

Solution

$$\dfrac{x}{x+1} - 1 = \dfrac{2}{x^2 - 1}$$

$$\left(\dfrac{x}{x+1} - 1\right)(x^2 - 1) = \left(\dfrac{2}{x^2 - 1}\right)(x^2 - 1)$$

■ Both sides are multiplied by the LCD $(x^2 - 1)$.

$$\dfrac{x(x^2-1)}{x+1} - 1(x^2 - 1) = \dfrac{2(x^2-1)}{x^2-1}$$

■ Think of $x^2 - 1$ as $(x + 1)(x - 1)$ and cancel.

$$x(x - 1) - 1(x^2 - 1) = 2$$

$$x^2 - x - x^2 + 1 = 2$$

$$-x = 1$$

$$x = -1$$

■ This equation may not be equivalent to the given equation. The value *must* be checked.

Must check:

$$\dfrac{x}{x+1} - 1 = \dfrac{2}{x^2 - 1}$$

$$\dfrac{-1}{-1+1} - 1 = \dfrac{2}{(-1)^2 - 1}$$

$$\dfrac{-1}{0} - 1 = \dfrac{2}{0}$$

■ Since -1 produces a zero denominator, it cannot be a solution.

The given equation has no solution.

Goal C *to review methods for solving absolute value equations*

Suppose $|A| = 19$. Since the only real numbers having an absolute value of 19 are 19 and -19, then either

$$A = 19 \quad \text{or} \quad A = -19.$$

If A represents a variable expression, then we can draw the same conclusions about A.

Suppose $|1 - 4x| = 19$. Then either

$$1 - 4x = 19 \quad \text{or} \quad 1 - 4x = -19$$
$$-4x = 18 \qquad\qquad -4x = -20$$
$$x = -\frac{9}{2} \qquad\qquad x = 5$$

Thus, $-\frac{9}{2}$ and 5 are solutions of the equation $|1 - 4x| = 19$.

The technique used to solve the previous equation can be used even when the variable appears on both sides of the equation. For instance, if $|3x - 2| = 7x + 5$, we solve for x by first writing the following:

$$3x - 2 = 7x + 5 \quad \text{or} \quad 3x - 2 = -(7x + 5).$$

Rule

To solve the absolute value equation $|A| = B$, you must solve the two equations in the statement "$A = B$ or $A = -B$." You *must* check solutions.

Example 6 Solve $|2x - 3| = 5 - 8x$.

Solution

$$|2x - 3| = 5 - 8x$$

■ To solve, begin by writing two equations.

$$2x - 3 = 5 - 8x \quad \text{or} \quad 2x - 3 = -(5 - 8x)$$
$$10x - 3 = 5 \qquad\qquad 2x - 3 = -5 + 8x$$
$$10x = 8 \qquad\qquad -3 = -5 + 6x$$
$$x = \frac{4}{5} \qquad\qquad 2 = 6x$$
$$\qquad\qquad \frac{1}{3} = x$$

■ Both values, $\frac{4}{5}$ and $\frac{1}{3}$, must be checked.

Must Check:

$$|2x - 3| = 5 - 8x \qquad\qquad |2x - 3| = 5 - 8x$$
$$\left|2\left(\frac{4}{5}\right) - 3\right| = 5 - 8\left(\frac{4}{5}\right) \qquad \left|2\left(\frac{1}{3}\right) - 3\right| = 5 - 8\left(\frac{1}{3}\right)$$
$$\left|-\frac{7}{5}\right| = -\frac{7}{5} \quad \blacksquare \text{ False.} \qquad \left|-\frac{7}{3}\right| = \frac{7}{3} \quad \blacksquare \text{ True.}$$

The only solution is $\frac{1}{3}$.

Exercise Set 2.1 □ Calculator Exercises in the Appendix

Goal A

In exercises 1–8, determine whether the values 2 and/or -2 are solutions.

1. $5x + 7 = 3(x + 1)$
2. $5 - y = 6y - 4$
3. $2u^2 - 3 = u$
4. $5t + 3 = t^2 - 2$
5. $|3s - 4| = s$
6. $|7x + 10| = 2x$
7. $\dfrac{1}{v - 2} = 2 - v$
8. $\dfrac{u + 2}{u - 2} = 2 + u$

In exercises 9–24, solve the given equation.

9. $5y + 7 = -13$
10. $8 - 3z = -1$
11. $3 = 5 - 7u$
12. $16 + 3v = 5$
13. $9m - 4 = 15 - 2m$
14. $1 - 8n = 3 + 2n$
15. $5(s + 2) = 6s - 4$
16. $9t - 1 = 9(1 - t)$
17. $9t - 1 = 9(t - 1)$
18. $1 - 9(t - 1) = 9t$
19. $8s + 3(s - 4) = 10s$
20. $3(s - 4) - 5s = 2(6 - s)$
21. $p(p - 4) = p^2 + 3p + 2$
22. $x(2x - 1) = (5 + x)(2x) + 1$
23. $(4x - 3)(x + 4) = (2x - 3)^2$
24. $(6y - 5)(y + 1) = (3y + 1)(2y) - 1$

In exercises 25–38, solve for the variable x.

25. $3x - b = 7$
26. $a - 2x = 9$
27. $2 - abx = c$
28. $a^2 + b^2x = c$
29. $5x + a = mx$
30. $x = b - mx$
31. $2(a - x) = -2(x - a)$
32. $a(x + 2) = 2b(x + 1)$
33. $ax + 2 = 2x(b - 1)$
34. $ax + 2(x - b) = x(2 + 3a)$
35. $a(x - 1) - x(b + 2) = 0$
36. $2ax(b + 1) - ab(x - 2) = 0$
37. $x(x - 1) + x^2 - 1 = 2x^2$
38. $1 - (x^2 - 2x) + 3x^2 = 2x^2 - 1$

Goal B

In exercises 39–60, solve the given equation.

39. $\dfrac{8}{x} = \dfrac{4}{5}$
40. $\dfrac{2}{3} = \dfrac{t}{12}$
41. $\dfrac{2}{s + 3} = \dfrac{9}{10}$
42. $\dfrac{6}{1 - v} = \dfrac{3}{4}$
43. $\dfrac{w}{w - 1} = \dfrac{3}{5}$
44. $\dfrac{2}{5m + 1} = \dfrac{3}{m}$
45. $\dfrac{3}{n} + \dfrac{3}{4n} = 6$
46. $\dfrac{1}{u} = 7 - \dfrac{3}{2u}$
47. $\dfrac{1}{5x} + \dfrac{3}{2x} = \dfrac{17}{10x}$
48. $\dfrac{1}{5x} - \dfrac{3}{2x} = \dfrac{17}{10x}$
49. $\dfrac{1}{x} - \dfrac{3}{2x} + \dfrac{4}{3x} = \dfrac{1}{2}$
50. $\dfrac{1}{n} + \dfrac{1}{2n} + \dfrac{1}{4n} = \dfrac{2}{n}$
51. $\dfrac{x}{x - 2} = \dfrac{x + 2}{x + 4}$
52. $\dfrac{w + 1}{w + 2} = \dfrac{w - 3}{w - 1}$
53. $\dfrac{y}{y + 2} = 1 - \dfrac{1}{y + 3}$
54. $\dfrac{2x}{x + 3} = 3 - \dfrac{x}{x + 9}$
55. $\dfrac{x}{x - 4} - 2 = \dfrac{4}{x - 4}$
56. $\dfrac{v}{v - 3} = 2 + \dfrac{3}{v - 3}$
57. $\dfrac{6t}{t + 1} - \dfrac{10}{t^2 + 4t + 3} = \dfrac{6t + 1}{t + 3}$
58. $\dfrac{5x}{x + 2} = \dfrac{30}{x^2 + 7x + 10} + \dfrac{5x}{x + 5}$
59. $\dfrac{2p}{p - 3} = \dfrac{p}{p + 4} + \dfrac{24 + p^2}{p^2 + p - 12}$
60. $\dfrac{5m^2 + 1}{2m^2 + m - 1} - \dfrac{m}{2m - 1} = \dfrac{2m}{m + 1}$

Goal C

In exercises 61–70, solve the given equation.

61. $|3 + 2x| = 8$
62. $|4x - 5| = 15$
63. $|8 - 3y| = 7$
64. $|12 - 7p| = 31$
65. $|6 - 5x| = 11$
66. $|-4s + 1| = 11$
67. $|2 - 3v| = 7v - 3$
68. $|3t - 4| = 2t + 5$
69. $|2n + 5| = 4n - 5$
70. $|9u - 2| = 3u - 8$

Superset

In exercises 71–82, solve for the indicated variable in terms of the other variables.

71. Solve $C = \frac{5}{9}(F - 32)$ for F.
72. Solve $A = \frac{x + y + z}{3}$ for z.
73. Solve $P = 2(L + W)$ for W.
74. Solve $a = P(1 + rt)$ for t.
75. Solve $s = \frac{1}{2}at^2$ for a.
76. Solve $V = \frac{1}{3}\pi r^2 h$ for h.
77. Solve $P = 2wh + 2lh + 2lw$ for h.
78. Solve $\frac{1}{R} = \frac{1}{R_1} + \frac{1}{R_2}$ for R_1.
79. Solve $s = s_0 + v_0 t + \frac{1}{2}at^2$ for v_0.
80. Solve $S = \frac{a}{1 - r}$ for r.
81. Solve $t = a + (n - 1)d$ for n.
82. Solve $x = ta^2 + tb^2 + a^2 b^2$ for t.

In exercises 83–86, determine the value of a so that the solution set is the set of all real numbers.

83. $3(x + 4) = x - 2(a - x)$
84. $5x - 3(x - a) = x - (a - x)$
85. $\frac{1}{a} + \frac{4ax + 1}{2a} + \frac{1}{3a} = \frac{60x + 11}{30}$
86. $\frac{3 - 2x}{a} - \frac{3 + 4x}{2a} + 6 = \frac{27 - 8x}{4}$

In each of the figures in exercises 87–88, the shaded area is 10 square inches. Write an equation describing this fact in terms of the given variables.

87.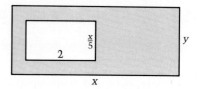

88.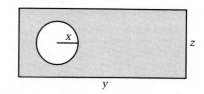

89. In the figure below, the area of the shaded region is 60 square yards. Find the dimensions of the larger rectangle. (Find x and $x + 2$.)

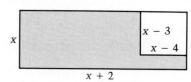

90. In the figure below, the area of the shaded region is 18 square inches. What is the height of the parallelogram (i.e., what is the value of x)?

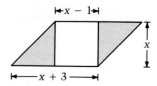

2.2 Quadratic Equations

Goal A *to solve quadratic equations by factoring*

All the problems in Section 2.1 ultimately involved solving equations of the form $ax + b = 0$. We now consider methods for solving equations of the form $ax^2 + bx + c = 0$.

Definition
An equation containing a second degree monomial, but no monomial of higher degree, is called a **quadratic equation.** The form $ax^2 + bx + c = 0$, where a, b, and c are constants, and $a \neq 0$, is called the **standard form of a quadratic equation in x.**

The key to solving quadratic equations is the Property of Zero Products. This property says that if the product of two or more factors is zero, then at least one of the factors must be zero.

Property of Zero Products
If A and B are real numbers (or expressions representing real numbers) and if $A \cdot B = 0$, then $A = 0$, or $B = 0$, or both.

For example, to solve $(x - 2)(3x - 7) = 0$, we solve the two equations $x - 2 = 0$ and $3x - 7 = 0$. Note that in order to apply the Property of Zero Products, one side of the given equation must be factored and the other side must be 0.

Example 1 Solve $2x^2 + x = 15$.

Solution

$$2x^2 + x = 15$$
$$2x^2 + x - 15 = 0 \quad \blacksquare \text{ All nonzero terms are on the left side.}$$
$$(2x - 5)(x + 3) = 0 \quad \blacksquare \text{ The nonzero side is factored.}$$

$2x - 5 = 0 \quad | \quad x + 3 = 0 \quad \blacksquare$ The Property of Zero Products is applied.

$$x = \frac{5}{2} \quad \quad \quad x = -3$$

The solutions are $\frac{5}{2}$ and -3.

Example 2 Solve $3x^2 - 30x + 75 = 0$.

Solution

$$3x^2 - 30x + 75 = 0$$
$$3(x^2 - 10x + 25) = 0 \quad \blacksquare \text{ 3 is a common factor.}$$
$$(x - 5)(x - 5) = 0 \quad \blacksquare \text{ Both sides are divided by 3, and } x^2 - 10x + 25 \text{ is factored.}$$
$$x - 5 = 0$$
$$x = 5 \quad \blacksquare \text{ Since the factors are the same, we only need to solve } x - 5 = 0 \text{ once.}$$

The solution is 5.

In Example 2, the solution 5 is called a *root of multiplicity* 2, since the factor that produced it appeared twice in the factorization. In general, if a linear factor $ax + b$ appears exactly n times in the factorization, then $-\frac{b}{a}$ (the solution of $ax + b = 0$) is called a **root of multiplicity n.**

Example 3 Solve $x^2 = c$ for x. (Treat c as a nonnegative constant.)

Solution

$$x^2 = c$$
$$x^2 - c = 0$$
$$x^2 - (\sqrt{c})^2 = 0 \quad \blacksquare \text{ Since } c \geq 0, c \text{ is rewritten as } (\sqrt{c})^2. \text{ The resulting difference of squares can be factored.}$$
$$(x - \sqrt{c})(x + \sqrt{c}) = 0$$
$$x - \sqrt{c} = 0 \quad | \quad x + \sqrt{c} = 0 \quad \blacksquare \text{ The Property of Zero Products is applied.}$$
$$x = \sqrt{c} \quad | \quad x = -\sqrt{c}$$

The solutions are $\sqrt{c}$ and $-\sqrt{c}$. This can also be written $\pm\sqrt{c}$.

Property of Square Roots
If $A^2 = c$, where A represents a variable expression, and c is a nonnegative real number, than $A = \pm\sqrt{c}$.

Example 4 Solve $(x - 4)^2 = 7$.

Solution
$(x - 4)^2 = 7$ ■ The equation has the form $A^2 = 7$, where A is $x - 4$.
$x - 4 = \pm\sqrt{7}$ ■ By the Property of Square Roots, $A = \pm\sqrt{7}$.
$x = 4 \pm \sqrt{7}$

The solutions are $4 + \sqrt{7}$ and $4 - \sqrt{7}$, which we write as $4 \pm \sqrt{7}$.

Goal B *to solve quadratic equations by completing the square*

Solving $x^2 + 4x - 9 = 0$ presents a problem since the binomial cannot be factored easily. Example 5 demonstrates a method called *completing the square* which allows us to rewrite the equation in the form $(x + a)^2 = c$. An equation of this form can then be solved by the Property of Square Roots.

Example 5 Solve $x^2 + 4x - 9 = 0$.

Solution
$x^2 + 4x - 9 = 0$ ■ We want an expression of the form $(x + a)^2$ on the left.
$x^2 + 4x + \square = 9 + \square$ ■ Rewrite so that only terms in x are on the left. The symbol $\square$ refers to the number to be added to each side.
$x^2 + 4x + 4 = 9 + 4$ ■ Take half the coefficient of x and square it. $(\frac{1}{2} \cdot 4)^2 = (2)^2 = 4$. Add 4 to both sides.
$(x + 2)^2 = 13$ ■ The left side is now a perfect trinomial square. Thus, we can use the Property of Square Roots.
$x + 2 = \pm\sqrt{13}$
$x = -2 \pm \sqrt{13}$

The solutions are $-2 \pm \sqrt{13}$.

Method for Completing the Square

An expression of the form $x^2 + bx$ becomes a perfect trinomial square when $\left(\frac{b}{2}\right)^2$ is added: $x^2 + bx + \left(\frac{b}{2}\right)^2 = \left(x + \frac{b}{2}\right)^2$.

To apply the method for completing the square to the equation $3x^2 + 2x - 4 = 0$, an equation whose x^2-coefficient is not 1, we must first multiply both sides of the equation by $\frac{1}{3}$ so that the coefficient of x^2 is 1.

Example 6 Solve for x: $ax^2 + bx + c = 0$. Assume that $a > 0$.

Solution

$$ax^2 + bx + c = 0$$

$$x^2 + \frac{b}{a}x + \frac{c}{a} = 0$$

■ Multiplying both sides by $\frac{1}{a}$ makes the coefficient of x^2 equal to 1. Then complete the square.

$$x^2 + \frac{b}{a}x + \square = -\frac{c}{a} + \square$$

■ The coefficient of x is $\frac{b}{a}$. Thus, $(\frac{1}{2} \cdot \frac{b}{a})^2$ must be added to both sides to complete the square.

$$x^2 + \frac{b}{a}x + \left(\frac{b}{2a}\right)^2 = -\frac{c}{a} + \left(\frac{b}{2a}\right)^2$$

$$\left(x + \frac{b}{2a}\right)^2 = -\frac{c}{a} \cdot \frac{4a}{4a} + \frac{b^2}{4a^2}$$

■ Combine fractions.

$$\left(x + \frac{b}{2a}\right)^2 = \frac{b^2 - 4ac}{4a^2}$$

■ We assume $b^2 - 4ac \geq 0$, and apply the Property of Square Roots.

$$x + \frac{b}{2a} = \pm\sqrt{\frac{b^2 - 4ac}{4a^2}}$$

■ We assumed $a > 0$; thus $\sqrt{4a^2} = 2a$. (The solutions will have the same form even if $a < 0$.)

$$x = -\frac{b}{2a} \pm \frac{\sqrt{b^2 - 4ac}}{2a}$$

$$x = \frac{-b \pm \sqrt{b^2 - 4ac}}{2a}$$

The solutions are $\dfrac{-b \pm \sqrt{b^2 - 4ac}}{2a}$

Goal C to solve quadratic equations by using the quadratic formula

The last example gives us the following important result.

The Quadratic Formula

Given an equation $ax^2 + bx + c = 0$, with $a \neq 0$, its solutions can be determined by the formula

$$x = \frac{-b \pm \sqrt{b^2 - 4ac}}{2a}.$$

Example 7 Use the quadratic formula to solve $3x^2 + 2x - 4 = 0$.

Solution
$$3x^2 + 2x - 4 = 0$$
$$x = \frac{-b \pm \sqrt{b^2 - 4ac}}{2a}$$
■ Write the quadratic formula. Here $a = 3$, $b = 2$, and $c = -4$.
$$x = \frac{-2 \pm \sqrt{2^2 - 4(3)(-4)}}{2(3)}$$
$$x = \frac{-2 \pm \sqrt{52}}{6}$$
$$x = \frac{-2 \pm 2\sqrt{13}}{6} = \frac{-1 \pm \sqrt{13}}{3}$$
■ $\frac{-2 \pm 2\sqrt{13}}{6}$ reduces to $\frac{-1 \pm \sqrt{13}}{3}$.

The solutions are $\frac{-1 \pm \sqrt{13}}{3}$.

Example 8 Use the quadratic formula to solve $4 = 12x - 9x^2$.

Solution
$$4 = 12x - 9x^2$$
$$9x^2 - 12x + 4 = 0$$
$$x = \frac{-b \pm \sqrt{b^2 - 4ac}}{2a}$$
■ In this problem $a = 9$, $b = -12$, and $c = 4$.
$$x = \frac{-(-12) \pm \sqrt{(-12)^2 - 4(9)(4)}}{2(9)}$$
$$x = \frac{12 \pm \sqrt{0}}{18} = \frac{12}{18} = \frac{2}{3}$$
■ There is only one root, a root of multiplicity 2.

The solution is $\frac{2}{3}$.

Example 9 Use the quadratic formula to solve $y^2 + y = -2$.

Solution
$$y^2 + y + 2 = 0$$
■ Write the equation in standard form.
$$y = \frac{-1 \pm \sqrt{1^2 - 4(1)(2)}}{2(1)}$$
■ In this problem $a = 1$, $b = 1$, and $c = 2$.
$$y = \frac{-1 \pm \sqrt{-7}}{2}$$
■ The radicand is negative; thus there are no real number solutions.

The equation has no real number solutions.

The last three examples demonstrate the three different situations which can arise when we solve a quadratic equation. For real numbers a, b, and c, the expression $b^2 - 4ac$, called the **discriminant**, can be used to determine the nature of the solutions.

Discriminant	Nature of Solutions
$b^2 - 4ac > 0$	two different real number solutions
$b^2 - 4ac = 0$	one real number solution (of multiplicity 2)
$b^2 - 4ac < 0$	no real number solutions

Exercise Set 2.2 = Calculator Exercises in the Appendix

Goal A

In exercises 1–32, solve each equation by factoring.

1. $x^2 + 9x + 8 = 0$
2. $y^2 - 5y + 6 = 0$
3. $s^2 - 3s - 10 = 0$
4. $t^2 + 4t - 21 = 0$
5. $u^2 - 8u + 16 = 0$
6. $m^2 + 12m + 36 = 0$
7. $n^2 - 64 = 0$
8. $p^2 - 100 = 0$
9. $3v^2 - 18v + 27 = 0$
10. $2q^2 + 16q - 18 = 0$
11. $5r^2 + 18r - 8 = 0$
12. $25z^2 - 1 = 0$
13. $16u^2 - 49 = 0$
14. $9x^2 - 6x + 1 = 0$
15. $y^2 + 10y = 0$
16. $2m^2 - m = 0$
17. $4s^2 + 4s - 3 = 0$
18. $7t^2 - 13t + 6 = 0$
19. $15n^2 - 37n - 8 = 0$
20. $6x^2 + 29x + 20 = 0$
21. $4z^2 = 16$
22. $y^2 = y$
23. $w^2 + 10w = 24$
24. $u^2 + 14u = -49$
25. $x^2 - 4 = 3x$
26. $4p^2 - 9 = 27$
27. $3q - q^2 = -54$
28. $-60 - s^2 = 19s$
29. $n(12n - 35) = 3$
30. $r(4r - 23) = -15$
31. $(3w - 1)^2 = 6w - 3$
32. $27v(1 - 3v) = 27v - 1$

Goal B

In exercises 33–42, solve each equation by completing the square.

33. $x^2 + 4x - 21 = 0$
34. $u^2 + 10u = 24$
35. $s^2 + 14s = -49$
36. $v^2 + 16v = 0$
37. $t^2 - 5t = -6$
38. $w^2 + 11w + 30 = 0$
39. $4z^2 + 16z + 15 = 0$

40. $3p^2 - 20p = 32$

41. $5m^2 - 4m - 3 = 0$

42. $3n^2 + 6n = 1$

Goal C

In exercises 43–70, use the quadratic formula to solve each equation.

43. $x^2 + 5x + 4 = 0$
44. $y^2 + 11y + 18 = 0$
45. $v^2 - 8v + 16 = 0$
46. $w^2 + 4w - 60 = 0$
47. $z^2 + z + 10 = 0$
48. $p^2 + p - 10 = 0$
49. $q^2 - 64 = 0$
50. $s^2 + 64 = 0$
51. $4t^2 + 4t + 1 = 0$
52. $v^2 - 8v + 13 = 0$
53. $3x^2 - 7x - 6 = 0$
54. $3y^2 - 7y + 6 = 0$
55. $u^2 - 11 = 0$
56. $p^2 + 11p = 0$
57. $s^2 + 3s - 5 = 0$
58. $5t^2 - 11t + 2 = 0$
59. $36q^2 + 12q + 1 = 0$
60. $2m^2 + 12m + 6 = 0$
61. $-x^2 + 2x - 2 = 0$
62. $-x^2 - 2x + 2 = 0$
63. $8r - r^2 - 14 = 0$
64. $s^2 - 2s = 24$
65. $2n^2 = 9n$
66. $7 - 14t + 5t^2 = 0$
67. $3q^2 - 8q = 2$
68. $9 - 16z^2 = 0$
69. $x(4x - 3) = -1$
70. $(u + 8)^2 = 57 - u^2$

Superset

In exercises 71–78, use the discriminant to determine the nature of the roots without solving.

71. $x^2 - 5x + 3 = 0$
72. $3x^2 - 5x + 2 = 0$
73. $x^2 + x + 1 = 0$
74. $x^2 + x - 1 = 0$
75. $3x^2 - 6x + 5 = 8$
76. $2x^2 - 7x + 7 = 3$
77. $8 - 5x^2 = 4x$
78. $7x - 3 = 9x^2$

For what real number value(s) of a will each of the equations in exercises 79–84 have one real root of multiplicity 2. (If no such value exists, then state this.)

79. $2x^2 - ax - 1 = 0$
80. $x^2 - ax - 7 = 11$
81. $x^2 + ax + 1 = 0$
82. $2x^2 + ax + a = 0$
83. $x^2 + (a - 3)x + a^2 = 0$
84. $ax^2 + 2ax + a = 0$

In each of the figures in exercises 85–86, the shaded region is 10 square feet. Write an equation to describe this fact.

85.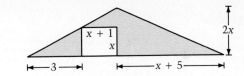

86.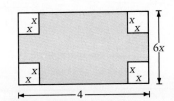

87. When an object is propelled upwards from a height of h_0 feet above the ground, with a velocity of v_0 ft/sec, its height h (in feet) after t seconds is given by the formula
$$h = h_0 + v_0 t - 16t^2.$$
If an object is shot upwards with a velocity of 88 ft/sec from the top of a 500 ft tall building, after how many seconds will it be 620 ft above the ground? Explain the reason you get two answers.

88. Approximately how many seconds after it is shot upward will the object in exercise 87 hit the ground?

2.3 Nonlinear Equations

Goal A *to use the knowledge of quadratic equations to solve higher degree and fractional equations*

We now consider a variety of situations where we must use the Property of Zero Products to solve equations.

Example 1   Solve $x^5 - 5x^4 = 4x^3$.

Solution

$$x^5 - 5x^4 = 4x^3$$
$$x^3(x^2 - 5x - 4) = 0$$

$x^3 = 0$ | $x^2 - 5x - 4 = 0$
$x = 0$ | $x = \dfrac{5 \pm \sqrt{41}}{2}$

- Begin by rewriting with all nonzero terms on one side. Then apply the Property of Zero Products.

- Zero is a root of multiplicity 3 since zero is the solution of the equation $x^3 = 0$.

The solutions are 0, $\dfrac{5 + \sqrt{41}}{2}$, and $\dfrac{5 - \sqrt{41}}{2}$.

The equation given in Example 1 is called a **polynomial equation** since each side of the equation is a polynomial. We say that it is a fifth degree polynomial equation since the highest degree of any monomial is 5.

Example 2   Solve $3x^3 + 2x^2 - 12x - 8 = 0$.

Solution

$$3x^3 + 2x^2 - 12x - 8 = 0$$
$$x^2(3x + 2) - 4(3x + 2) = 0$$
$$(x^2 - 4)(3x + 2) = 0$$

$x^2 - 4 = 0$ | $3x + 2 = 0$
$x = \pm 2$ | $x = -\dfrac{2}{3}$

- There is no common factor. The four terms suggest factoring by grouping.

- The Property of Zero Products is now applied.

The solutions are 2, -2, and $-\frac{2}{3}$.

2 / Equations and Inequalities in One Variable

Example 3 demonstrates that when solving fractional equations, you always must check solutions.

Example 3 Solve $\dfrac{x+7}{x-1} + \dfrac{x-1}{x+1} = \dfrac{4}{x^2-1}$.

Solution

$\left(\dfrac{x+7}{x-1} + \dfrac{x-1}{x+1}\right)(x^2-1) = \left(\dfrac{4}{x^2-1}\right)(x^2-1)$ ■ Begin by multiplying both sides by the LCD (x^2-1).

$\dfrac{x+7}{\cancel{x-1}} \cdot \cancel{(x^2-1)}^{(x+1)} + \dfrac{x-1}{\cancel{x+1}} \cdot \cancel{(x^2-1)}^{(x-1)} = \dfrac{4}{\cancel{x^2-1}} \cdot \cancel{(x^2-1)}$

$(x+7)(x+1) + (x-1)(x-1) = 4$

$x^2 + 8x + 7 + x^2 - 2x + 1 = 4$

$2x^2 + 6x + 4 = 0$

$2(x+2)(x+1) = 0$ ■ The Property of Zero Products is applied.

$2(x+2) = 0 \quad | \quad x+1 = 0$

$x = -2 \quad | \quad x = -1$

Must Check:

$x = -2$:

$\dfrac{-2+7}{-2-1} + \dfrac{-2-1}{-2+1} = \dfrac{4}{(-2)^2-1}$

$\dfrac{-5}{3} + 3 = \dfrac{4}{3}$ ■ True.

$x = -1$:

$\dfrac{-1+7}{-1-1} + \dfrac{-1-1}{-1+1} = \dfrac{4}{(-1)^2-1}$

$\dfrac{6}{-2} + \dfrac{-2}{0} = \dfrac{4}{-3}$ ■ $\dfrac{-2}{0}$ is not defined.

The only solution is -2.

Goal B *to solve equations that are quadratic in form*

An equation is said to be **quadratic in form** if substituting an expression with a single variable makes the equation quadratic in the new variable. Consider the equation $x^6 + 2x^3 - 3 = 0$.

$x^6 + 2x^3 - 3 = 0$

$(x^3)^2 + 2(x^3) - 3 = 0$ ■ Replace x^3 with u.

$u^2 + 2u - 3 = 0$ ■ The result is quadratic in u.

As another example, consider $y^{2/3} - y^{1/3} + 10 = 0$. It can be rewritten as $(y^{1/3})^2 - (y^{1/3}) + 10 = 0$. Replacing $y^{1/3}$ with u produces $u^2 - u + 10 = 0$ which is quadratic in u.

Example 4   Solve $x^4 + 3x^2 - 10 = 0$.

Solution

$$x^4 + 3x^2 - 10 = 0$$
$$(x^2)^2 + 3x^2 - 10 = 0$$
$$u^2 + 3u - 10 = 0 \quad \blacksquare \ x^2 \text{ is replaced by } u.$$
$$(u + 5)(u - 2) = 0 \quad \blacksquare \ \text{The Property of Zero Products is now applied.}$$

$u + 5 = 0$	$u - 2 = 0$
$u = -5$	$u = 2$
$x^2 = -5$	$x^2 = 2$
$x = \pm\sqrt{-5}$	$x = \pm\sqrt{2}$

■ Remember that u represents x^2.
■ $\sqrt{-5}$ is not a real number.

The only solutions are $\pm\sqrt{2}$.

Example 5   Solve $5p^{-2} + 4p^{-1} - 1 = 0$.

Solution

$$5p^{-2} + 4p^{-1} - 1 = 0$$
$$5(p^{-1})^2 + 4(p^{-1}) - 1 = 0$$
$$5u^2 + 4u - 1 = 0 \quad \blacksquare \ p^{-1} \text{ is replaced by } u.$$
$$(5u - 1)(u + 1) = 0 \quad \blacksquare \ \text{The Property of Zero Products is applied.}$$

$5u - 1 = 0$	$u + 1 = 0$
$u = \dfrac{1}{5}$	$u = -1$
$p^{-1} = \dfrac{1}{5}$	$p^{-1} = -1$
$\dfrac{1}{p} = \dfrac{1}{5}$	$\dfrac{1}{p} = -1$
$p = 5$	$p = -1$

■ Remember that u represents p^{-1}.
■ Recall that $p^{-1} = \dfrac{1}{p}$.

The solutions are -1 and 5.

Goal C   *to review methods for solving radical equations*

In order to solve a radical equation such as $\sqrt{x - 1} = x - 3$, it is necessary to rewrite it in a form that does not contain radicals. We can accomplish this by squaring both sides of the equation, but this does not necessarily produce an equivalent equation. For that reason you *must* check solutions.

We now apply this technique to $\sqrt{x-1} + 3 = x$.

$$\sqrt{x-1} + 3 = x$$
$$\sqrt{x-1} = x - 3 \qquad \blacksquare \text{ The radical term is isolated.}$$
$$x - 1 = x^2 - 6x + 9 \qquad \blacksquare \text{ Squaring eliminates the radical.}$$
$$0 = x^2 - 7x + 10 \qquad \blacksquare \text{ One side must be zero to use the Property of Zero Products.}$$
$$0 = (x - 5)(x - 2)$$

$x - 5 = 0$	$x - 2 = 0$
$x = 5$	$x = 2$

When you check the answers 5 and 2, you find that 2 does not satisfy the original equation but 5 does.

If both sides of an equation are raised to any positive integral power, then the solution set of the resulting equation will contain the solutions of the original equation. However, it may also contain "extra" solutions called **extraneous roots.** In the above example, 2 is an extraneous root because it does not satisfy the original equation, $\sqrt{x-1} + 3 = x$. Thus you *must* always check solutions found after raising the sides of an equation to a power.

The Power Property for Equations

If $A = B$ and if n is a positive integer, then the solution set of the equation $A^n = B^n$ contains all the solutions of the equation $A = B$. However, $A = B$ and $A^n = B^n$ are not necessarily equivalent. Solutions *must* be checked.

Example 6   Solve $\sqrt[3]{x^2 - 1} = 2$.

Solution

$$\sqrt[3]{x^2 - 1} = 2 \qquad \blacksquare \text{ Cube both sides to eliminate the radical term.}$$
$$(\sqrt[3]{x^2 - 1})^3 = 2^3$$
$$x^2 - 1 = 8$$
$$x^2 = 9$$
$$x = \pm 3$$

Must Check:

$\sqrt[3]{x^2 - 1} = 2$		$\sqrt[3]{x^2 - 1} = 2$	
$\sqrt[3]{3^2 - 1} = 2$		$\sqrt[3]{(-3)^2 - 1} = 2$	
$\sqrt[3]{8} = 2$	$\blacksquare$ True.	$\sqrt[3]{8} = 2$	$\blacksquare$ True.

The solutions are ± 3.

Example 7 Solve $1 + \sqrt{y + 4} = \sqrt{3y + 1}$.

Solution

$$1 + \sqrt{y + 4} = \sqrt{3y + 1}$$
$$(1 + \sqrt{y + 4})^2 = (\sqrt{3y + 1})^2$$ ■ Use the Power Property.
$$1 + 2\sqrt{y + 4} + y + 4 = 3y + 1$$
$$2\sqrt{y + 4} = 2y - 4$$ ■ Multiply both sides by $\frac{1}{2}$.
$$\sqrt{y + 4} = y - 2$$
$$(\sqrt{y + 4})^2 = (y - 2)^2$$ ■ Use the Power Property again to eliminate the remaining radical term.
$$(y + 4) = y^2 - 4y + 4$$
$$0 = y^2 - 5y$$
$$0 = y(y - 5)$$ ■ The Property of Zero Products is applied.

$y = 0 \quad | \quad y - 5 = 0$
$ y = 5$

Must Check:
$1 + \sqrt{y + 4} = \sqrt{3y + 1} \qquad 1 + \sqrt{y + 4} = \sqrt{3y + 1}$
$1 + \sqrt{0 + 4} = \sqrt{3 \cdot 0 + 1} \qquad 1 + \sqrt{5 + 4} = \sqrt{3 \cdot 5 + 1}$
$1 + 2 = 1 \qquad\qquad\qquad 1 + 3 = 4$
$3 = 1$ ■ False. $\qquad\qquad 4 = 4$ ■ True.

The only solution is 5.

Exercise Set 2.3

Goal A

In exercises 1–26, solve the given equation.

1. $3s^3 + 8s = 14s^2$
2. $5m^6 + 4m^5 = 12m^4$
3. $v^4 + v^3 + v^2 = 0$
4. $w^4 = w^3 + w^2$
5. $2t^5 + 4t^3 = 7t^4$
6. $n^8 = 8n^6$
7. $2p^3 - p^2 - 18p + 9 = 0$
8. $s^3 + 3s^2 - 25s - 75 = 0$
9. $3t^3 - 2t^2 + 48t - 32 = 0$
10. $8v^3 - 3v^2 + 32v - 12 = 0$
11. $27r^4 - 27r^3 + r - 1 = 0$
12. $x^4 + 7x^3 - 8x - 56 = 0$
13. $5u^5 - 20u^4 - 4u^3 + 16u^2 = 0$
14. $7z^4 - 6z^3 + 28z^2 - 24z = 0$
15. $\dfrac{x + 5}{x^2 - 2x + 1} = 2$
16. $\dfrac{2x^2 - 2}{x^2 - 4x + 11} = 3$
17. $\dfrac{1}{x} - \dfrac{2}{3} = \dfrac{1}{3x^2}$
18. $\dfrac{7}{x^2} - 1 = \dfrac{9}{4x}$
19. $\dfrac{3}{x - 1} + \dfrac{4}{x + 1} = 1$
20. $\dfrac{3}{x + 5} = 1 - \dfrac{4}{x - 5}$

21. $\dfrac{1}{x-2} = 1 + \dfrac{2}{x^2 - 2x}$

22. $\dfrac{5}{x+2} + \dfrac{10}{x^2 + 2x} = -2$

23. $\dfrac{3x}{x+2} - \dfrac{x}{x-1} = \dfrac{18}{x^2 + x - 2}$

24. $\dfrac{2x}{x+3} = \dfrac{x}{x-2} + \dfrac{1}{x^2 + x - 6}$

25. $\dfrac{6x}{2x-3} - \dfrac{2}{x} = 2$

26. $\dfrac{3x+2}{x-1} + \dfrac{2x}{x+1} = \dfrac{7x+3}{x^2 - 1}$

Goal B

In exercises 27–40, solve the given equation.

27. $x^4 - 5x^2 - 14 = 0$
28. $x^4 - 11x^2 + 24 = 0$
29. $x^6 - 3x^3 = 10$
30. $2x^6 - 7x^3 - 18 = 0$
31. $x^4 + x^2 = 1$
32. $x^4 - 5x^2 + 3 = 0$
33. $x^4 = 2$
34. $x^4 = -2$
35. $3p^{-2} - 4p^{-1} = 0$
36. $\dfrac{6}{p^2} - \dfrac{5}{p} = -1$
37. $(p-1)^2 + 4(p-1) + 3 = 0$
38. $(t+4)^2 - 10(t+4) + 22 = 0$
39. $5x^{2/3} - 11x^{1/3} + 2 = 0$
40. $x^{1/2} - 8x^{1/4} + 15 = 0$

Goal C

In exercises 41–50, solve the given equation.

41. $\sqrt{5x+7} = 3$
42. $\sqrt[3]{3x-1} = 8$
43. $x = 5 + \sqrt{x+7}$
44. $3 + \sqrt{5-x} = x$
45. $\sqrt{x} + \sqrt{x-3} = 3$
46. $\sqrt{2x} + \sqrt{x-3} = 1$
47. $\sqrt{2x-5} - \sqrt{x-3} = 1$
48. $\sqrt{3x+1} - \sqrt{x-1} = 2$
49. $\sqrt{2x-2} = 1 + \sqrt{x+6}$
50. $\sqrt{3x+1} = 2 + \sqrt{1-x}$

Superset

In exercises 51–56, solve each equation.

51. $\sqrt{1 - \sqrt{x}} = 1 - \sqrt{x}$
52. $\sqrt{3 - \sqrt{3-x}} = \sqrt{x}$
53. $1 + \dfrac{1}{\sqrt{x}+1} = \dfrac{1}{\sqrt{x}-1}$
54. $\dfrac{4}{3} - \dfrac{1}{\sqrt{x}+1} = \dfrac{1}{\sqrt{x}-1}$
55. $\dfrac{12}{\sqrt{x}-9} - \sqrt{x} = \dfrac{\sqrt{x}-9}{-2}$
56. $\dfrac{2}{1-x} = \dfrac{2}{1-\sqrt{x}} - \dfrac{1}{1+\sqrt{x}}$

57. In the equation

$$T = 2\pi\sqrt{\dfrac{x}{32}},$$

T is the time in seconds that it takes for one complete swing of a pendulum x ft long. Find the approximate length of a certain pendulum, if a complete swing takes 3.1 seconds. (Round to the nearest tenth and use 3.14 for π.)

58. The equation $V = \sqrt{64h}$ gives the velocity, V (in ft/sec), of an object when it strikes the ground, after it has been dropped from a height of h ft. From what height must an object have been dropped if its velocity is 30 mph when it strikes the ground?

In exercises 59–62, solve each equation by first making a u-substitution.

59. $x - 3\sqrt{x} + 2 = 0$
60. $2\sqrt{x} - 7\sqrt[4]{x} + 6 = 0$
61. $\sqrt[6]{x} + 2\sqrt[3]{x} - 1 = 0$
62. $(\sqrt{x} + 1)^2 + \sqrt{x} - 1 = 0$
(Hint: Let $u = \sqrt{x} + 1$. You will need to rewrite the equation before making the substitution.)

2.4 Inequalities

Goal A *to review methods for solving linear inequalities*

In Chapter 1 we saw that the relative position of two different points on the real number line can be used to write an inequality involving the coordinates of the points. In this section we will consider inequalities that contain variables, such as

$$3x - 2 < 7 \quad \text{or} \quad |x - 7| > 3.$$

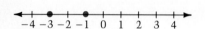

Figure 2.1 Recall that $-3 < -1$ and $-1 > -3$ represent the same order relationship.

To *solve* an inequality means to determine its **solution set,** the set of numbers that make the inequality a true statement.

Consider the linear inequality $3x - 2 < 7$. We test -5 and 4 to determine whether either is a solution.

$$\begin{array}{cc} 3x - 2 < 7 & 3x - 2 < 7 \\ 3(-5) - 2 < 7 & 3(4) - 2 < 7 \\ -17 < 7 \quad \blacksquare \text{ True.} & 10 < 7 \quad \blacksquare \text{ False.} \end{array}$$

The number -5 is a solution since replacing x with -5 produces a true statement. However, 4 is not a solution.

Two inequalities having the same solution set are called **equivalent inequalities.** The following properties suggest ways to produce equivalent inequalities.

$x + 3 < 4$
is equivalent to
$x + 3 + (-3) < 4 + (-3)$

Addition Property for Equivalent Inequalities

Adding the same real number (or variable expression) to both sides of an inequality produces an equivalent inequality.

$2x > 7$
is equivalent to
$3 \cdot 2x > 3 \cdot 7$

Positive Multiplication Property for Equivalent Inequalities

Multiplying both sides of an inequality by the same positive real number (or variable expression) produces an equivalent inequality.

$8y > 5$
is equivalent to
$-6 \cdot 8y < -6 \cdot 5$

Negative Multiplication Property for Equivalent Inequalities

Multiplying both sides of an inequality by the same negative real number (or variable expression) and reversing the direction of the inequality produces an equivalent inequality.

There are four types of inequalities corresponding to the symbols

$>$ (meaning: "is greater than"),
$<$ (meaning: "is less than"),
$\geq$ (meaning: "is greater than or equal to"),
$\leq$ (meaning: "is less than or equal to").

The strategy for solving an inequality is to transform it into an equivalent inequality whose solution set is obvious.

Example 1 Solve $13x - 17 > 4x + 1$ and graph the solution set.

Solution

$$13x - 17 > 4x + 1$$
$$13x + (-4x) - 17 > 4x + (-4x) + 1$$
$$9x - 17 > 1$$
$$9x - 17 + 17 > 1 + 17$$
$$9x > 18$$
$$\frac{1}{9} \cdot 9x > \frac{1}{9} \cdot 18$$
$$x > 2$$

■ Use the Addition Property to get a single x-term on one side and a constant on the other.

■ Now use the Positive Multiplication Property to make the x-coefficient 1.

The solution set is the set of all real numbers greater than 2.

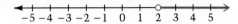

Note: A good way to check your work is to select one value in your solution set and verify, by substitution, that it is a solution. Then select one value not in your solution set and verify, by substitution, that it is not a solution.

Example 2 Solve $2 - 3x \geq 17$, and graph the solution set.

Solution

$$2 - 3x \geq 17$$
$$-3x \geq 15$$
$$-\frac{1}{3}(-3x) \leq -\frac{1}{3}(15)$$
$$x \leq -5$$

■ Use the Addition Property.

■ Use the Negative Multiplication Property. Remember to reverse the direction of the inequality.

The solution set is the set of all real numbers less than or equal to -5.

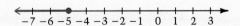

In addition to graphing, *set notation* and *interval notation* provide two more ways of describing the solution set of an inequality.

In **set notation** the solution sets for Examples 1 and 2 are written:

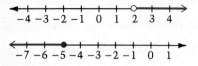

Figure 2.2

$\{x \mid x > 2\}$ ■ This is read "The set of all real numbers x such that x is greater than 2."

$\{x \mid x \leq -5\}$ ■ This is read "The set of all real numbers x such that x is less than or equal to -5."

In **interval notation** the solution set in Example 1 is written $(2, +\infty)$. The solution in Example 2 is written $(-\infty, -5]$. The symbols $-\infty$ and $+\infty$ are not real numbers. The $+\infty$ in $(a, +\infty)$ denotes that the interval contains all real numbers greater than a; the $-\infty$ in $(-\infty, a)$ denotes that the interval contains all real numbers less than a.

The eight different types of intervals are shown below.

Interval Notation	Set Notation	Graph
(a, b)	$\{x \mid a < x < b\}$	○———○ a b
$[a, b)$	$\{x \mid a \leq x < b\}$	●———○ a b
$(a, b]$	$\{x \mid a < x \leq b\}$	○———● a b
$[a, b]$	$\{x \mid a \leq x \leq b\}$	●———● a b
$(a, +\infty)$	$\{x \mid x > a\}$	○——→ a
$[a, +\infty)$	$\{x \mid x \geq a\}$	●——→ a
$(-\infty, a)$	$\{x \mid x < a\}$	←——○ a
$(-\infty, a]$	$\{x \mid x \leq a\}$	←——● a

In the intervals above, a and b are called **endpoints**. If endpoints are included, as in $[a, b]$, the interval is called a **closed interval**. If the endpoints are not included, as in (a, b) or $(a, +\infty)$, the interval is called an **open interval**. Intervals such as $[a, b)$ or $(-\infty, a]$, which contain only one endpoint, are called **half-open intervals**.

Example 3 Describe each of the following intervals in set notation, and sketch the graph.

(a) $(3, +\infty)$ (b) $[-3, \sqrt{7})$ (c) $(-\infty, 0]$

Solution (a) $\{x \mid x > 3\}$

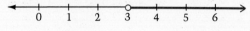

(b) $\{x \mid -3 \leq x < \sqrt{7}\}$

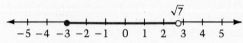

(c) $\{x \mid x \leq 0\}$

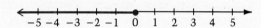

Goal B to solve absolute value inequalities

Two simple inequalities, such as

$$x > 5 \quad \text{and} \quad x \leq -4,$$

can be used to produce a **compound inequality.** There are two types of compound inequalities:

$\{x \mid x > 7 \text{ or } x < -3\}$ ■ This set is formed by *combining all numbers* greater than 7 or less than -3 into a single set.

$\{x \mid x \geq 2 \text{ and } x < 9\}$ ■ This set is formed by taking those *numbers common to both* the sets $\{x \mid x \geq 2\}$ and $\{x \mid x < 9\}$.

Example 4 Graph each of the following sets.

(a) $\{x \mid x < 3 \text{ and } x \geq -5\}$ (b) $\{x \mid x > 2 \text{ or } x < -2\}$ (c) $\{x \mid -10 < x < 5\}$

Solution (a)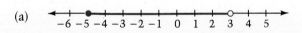

■ "All numbers greater than or equal to -5 **and** less than 3."

(b)

■ "All numbers greater than 2 **or** less than -2."

(c)

■ "All numbers greater than -10 **and** less than 5."

We can use compound inequalities to describe the solution set of an absolute value inequality. Suppose you must solve the inequality $|x| > 2$. Only numbers greater than 2 or less than -2 have absolute values that are greater than 2. Thus the sketch of $|x| > 2$ is as follows.

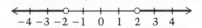

- The graph represents the solution set of $|x| > 2$. It also represents the compound inequality $x > 2$ **or** $x < -2$.

On the other hand, consider the inequality $|x| < 2$. Only numbers between -2 and 2 have absolute values less than 2.

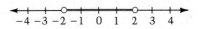

- The graph represents the solution set of $|x| < 2$. It also represents the compound inequality $x > -2$ **and** $x < 2$.

Rule

If A is a variable expression and c is a positive real number, then the following statements are true:
1. The solution set of the inequality $|A| > c$ is described by the compound inequality $A > c$ **or** $A < -c$.
2. The solution set of the inequality $|A| < c$ is described by the compound inequality $A < c$ **and** $A > -c$ which can also be written $-c < A < c$.

In the above rule the symbols $>$ and $<$ can be replaced by $\geq$ and $\leq$, respectively.

Example 5 Solve $|2x - 3| \geq 7$. Describe the solution set in set notation and sketch the graph.

Solution

$$|2x - 3| \geq 7$$

$2x - 3 \geq 7$ **or** $2x - 3 \leq -7$

$2x \geq 10 \qquad\qquad 2x \leq -4$

$x \geq 5 \qquad\qquad x \leq -2$

- Since the inequality symbol is $\geq$, part 1 of the above rule applies. Thus, we use the word "or".

The solution set is $\{x \mid x \geq 5 \text{ or } x \leq -2\}$.

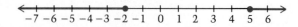

Example 6 Solve $|5 - 3x| < 8$ and describe the solution set in interval notation.

Solution

$$|5 - 3x| < 8$$

$5 - 3x < 8$	**and**	$5 - 3x > -8$
$-3x < 3$		$-3x > -13$
$x > -1$		$x < \dfrac{13}{3}$

■ Since the inequality symbol is $<$, part 2 of the above rule applies. Thus we use the word "and".

The solution set is $\left(-1, \dfrac{13}{3}\right)$.

Goal C *to solve other types of inequalities*

In applied mathematics we frequently need to solve inequalities involving a quadratic expression. An inequality that contains a quadratic expression is called a **quadratic inequality.** To solve such an inequality, we shall use the techniques we learned for factoring polynomials as well as the simple rules about multiplying positive and negative numbers.

To solve the inequality $x^2 - 4x < 12$, we must first collect all the terms on one side of the inequality and then factor.

$$x^2 - 4x < 12$$
$$x^2 - 4x - 12 < 0$$
$$(x + 2)(x - 6) < 0$$

This last inequality tells us that the product of the two factors $(x - 6)$ and $(x + 2)$ is negative. We know that this can happen only if one factor is positive and the other factor is negative. Thus, there are two cases.

Case 1: $x + 2 > 0$ and $x - 6 < 0$
$x > -2$ $x < 6$

Case 2: $x + 2 < 0$ and $x - 6 > 0$
$x < -2$ $x > 6$

In Case 1, values of x such that $x > -2$ and $x < 6$, that is, $-2 < x < 6$, satisfy the condition $(x + 2)(x - 6) < 0$. In Case 2, there are no values of x which can be less than -2 and greater than 6 at the same time. Thus, the values of x such that $-2 < x < 6$ (Case 1) are the only solutions to the original inequality. In interval notation, the solution set of the inequality $x^2 - 4x < 12$ is the interval $(-2, 6)$.

To understand fully what is happening to the quadratic expression $x^2 - 4x - 12$ as the variable x takes on different values, let us observe when the factors $(x + 2)$ and $(x - 6)$ are positive and when they are negative (see Figure 2.3).

The factors $(x + 2)$ and $(x - 6)$ are positive on the interval $(6, +\infty)$. Thus, their product is positive on the interval $(6, +\infty)$.

The factors $(x + 2)$ and $(x - 6)$ are negative on the interval $(-\infty, -2)$. Thus, their product is also positive on the interval $(-\infty, -2)$.

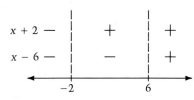

Figure 2.3

Note in Fig. 2.3 that their product is negative on the interval $(-2, 6)$; i.e., $x^2 - 4x - 12$ is less than zero on the interval $(-2, 6)$. Thus, the interval $(-2, 6)$ is the solution set of the inequality $x^2 - 4x < 12$.

An easier method for solving a quadratic inequality involves first solving the associated equality. For example, in order to solve the inequality

$$x^2 - 4x < 12,$$

we look at $x^2 - 4x = 12$.

> **Rule**
>
> For a given quadratic inequality, the solutions of the associated equality break the real number line into intervals. Within each interval, the quadratic expression is either always positive or always negative.

Using the above rule, we need only test one value in each interval to determine whether the quadratic expression is positive or negative in that interval.

Example 7 Solve the inequality $x^2 + 4x > 21$. Describe the solution set in interval notation.

Solution To solve the inequality $x^2 + 4x > 21$, we first consider the associated equality $x^2 + 4x = 21$.

$$x^2 + 4x = 21 \quad \blacksquare \text{ Collect terms on one side of the equation.}$$
$$x^2 + 4x - 21 = 0 \quad \blacksquare \text{ Factor the polynomial.}$$
$$(x - 3)(x + 7) = 0 \quad \blacksquare \text{ Now apply the Property of Zero Products.}$$
$$\begin{array}{c|c} x - 3 = 0 & x + 7 = 0 \\ x = 3 & x = -7 \end{array}$$

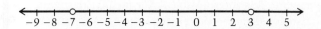
■ The solutions -7 and 3 break the real number line into three intervals.

Now we choose test values for each of the three intervals.

Interval	Test Value	Value of the Polynomial	Sign of the Polynomial
$(-\infty, -7)$	-9	$(-9)^2 + 4(-9) - 21 = 24$	$+$
$(-7, 3)$	0	$(0)^2 + 4(0) - 21 = -21$	$-$
$(3, +\infty)$	5	$(5)^2 + 4(5) - 21 = 24$	$+$

Thus, the quadratic expression $x^2 + 4x - 21$ is positive on the intervals $(-\infty, -7)$ and $(3, +\infty)$; that is, the solution to the inequality $x^2 + 4x > 21$ in interval notation is $(-\infty, -7)$ or $(3, +\infty)$.

This rule applies to other types of inequalities such as inequalities containing rational expressions. We can use the same methods to solve such inequalities.

Example 8 Solve the inequality $\dfrac{6}{2x - 5} \leq 2$. Describe the solution set in interval notation.

Solution We first consider the associated equality $\dfrac{6}{2x - 5} = 2$.

$$\dfrac{6}{2x - 5} = 2 \qquad \text{■ Collect terms on one side of the equation.}$$

$$\dfrac{6}{2x - 5} - 2 = 0$$

$$\dfrac{6}{2x - 5} - \dfrac{2(2x - 5)}{2x - 5} = 0$$

$$\dfrac{16 - 4x}{2x - 5} = 0 \qquad \text{■ Multiply each side of the equation by } 2x - 5 \text{ and solve for } x.$$

$$x = 4 \qquad \text{■ You } must \text{ check that this result is a solution of the original equation.}$$

The associated equality has the solution 4. For inequalities containing rational expressions, we must also consider the cases where the denominator is 0. Setting the denominator $2x - 5$ equal to 0, we obtain the result $x = \frac{5}{2}$. Thus, the real number line is divided into the intervals $(-\infty, \frac{5}{2})$, $(\frac{5}{2}, 4)$ and $(4, +\infty)$. Now we select test values from these three intervals.

Interval	Test Value	Value of the Expression	Sign of the Expression
$(-\infty, \frac{5}{2})$	-6	$\frac{6}{2(-6) - 5} - 2 = -\frac{40}{17}$	$-$
$(\frac{5}{2}, 4)$	3	$\frac{6}{2(3) - 5} - 2 = 4$	$+$
$(4, +\infty)$	7	$\frac{6}{2(7) - 5} - 2 = -\frac{4}{3}$	$-$

The solution of the inequality $\frac{6}{2x - 5} \leq 2$ in interval notation is $(-\infty, \frac{5}{2})$ or $[4, +\infty)$.

Exercise Set 2.4   ▣ Calculator Exercises in the Appendix

Goal A

In exercises 1–8, describe the interval in two ways: use set notation, and sketch the graph.

1. $(3, 8)$
2. $[3, 10]$
3. $(-2.5, 6]$
4. $[-4, 5.5)$
5. $(-\infty, -2)$
6. $(-2, +\infty)$
7. $[-\sqrt{3}, \sqrt{3}]$
8. $(-\infty, \pi)$

In exercises 9–20, solve the inequality. Indicate the solution set (a) in set notation, (b) in interval notation, and (c) with a graph. (If there are no solutions, then state this fact.)

9. $3x - 8 > 4$
10. $9m + 2 < 7$
11. $2 \geq 3v + 5$
12. $9 > 2w - 6$
13. $8x - 7 > 5 - 2x$
14. $4u + 1 \leq 6u + 4$
15. $\frac{1}{4} - \frac{3}{4}x \geq 5x - 3$
16. $7 - \frac{10}{3}m > \frac{14}{5}m$
17. $5(2s + 1) < s + 3(3s + 1)$
18. $1 - 3(w + 4) \geq 2w - 3(w + 7)$
19. $6y - 2(y + 3) > 5(y - 2) - y$
20. $5z - 2(z - 1) > 3z - 1$

Goal B

In exercises 21–26, graph each set.

21. $\{x \mid x \leq 10 \text{ and } x \geq -7\}$
22. $\{x \mid x > 3 \text{ and } x < 12\}$
23. $\{x \mid x \geq \frac{7}{2} \text{ or } x \leq -\frac{8}{5}\}$
24. $\{x \mid x < 3 \text{ or } x > \pi\}$
25. $\{x \mid x \geq \frac{\sqrt{3}}{2} \text{ and } x < \sqrt{2}\}$
26. $\{x \mid x \geq -\frac{\sqrt{3}}{2} \text{ or } x < -\sqrt{2}\}$

In exercises 27–40, solve the inequality. Indicate the solution set (a) in set notation, and (b) with a graph. (If there are no solutions, then state this fact.)

27. $|x| > 1$
28. $|s| < 5$
29. $2|v| \leq 7$
30. $3|w| \geq 10$
31. $|y| > -1$
32. $|z| < -1$
33. $|3x - 5| > 10$
34. $|5v - 7| \leq 18$
35. $|8 - 3u| \leq 13$
36. $|3 - 10s| > 12$
37. $|8r - 1| \leq -1$
38. $|7 - 9t| \geq 22$
39. $|s - 7| \leq 0$
40. $|s - 7| \geq 0$

Goal C

In exercises 41–58, solve the inequality. State your answer in interval notation.

41. $x^2 + 3x + 15 > 15$
42. $x^2 + 11x + 2 < 2$
43. $y^2 - 6y - 5 \geq -5$
44. $z^2 - 9z - 20 < -20$
45. $x^2 - 7x + 4 \geq 12$
46. $v^2 - 13v > -36$
47. $2m^2 - 4m \leq -1$
48. $t^2 + 6t + 3 < 0$
49. $x^3 - 9x < 0$
50. $y^3 - 36y \geq 0$
51. $\dfrac{10}{x - 1} - 3 < 2$
52. $\dfrac{6}{x + 4} - 2 \geq 1$
53. $\dfrac{2}{3 - x} + 5 > 4$
54. $\dfrac{6}{6 - x} - 1 < -3$
55. $\dfrac{8}{3x - 1} + 3 \leq 7$
56. $\dfrac{40}{4x + 3} - 7 \geq 1$
57. $\dfrac{-3}{2x - 9} < \dfrac{1}{3}$
58. $\dfrac{-2}{15 - \frac{1}{2}x} > \dfrac{1}{10}$

Superset

In exercises 59–62, determine whether each statement is true or false. If false, illustrate this with an example.

59. $a < b$ always implies that $ac < bc$.
60. $a < b$ always implies that $\frac{1}{a} < \frac{1}{b}$, provided that neither a nor b is zero.
61. $a < b$ always implies that $\frac{a}{b} < 1$, provided $b \neq 0$.
62. $a < b$ always implies that $a^2 < b^2$.

For each figure in exercises 63–66, determine an interval for x, so that the perimeter of each figure will be between 16.9 and 17.1 inches.

63.

64.

65.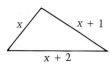

66.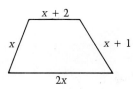

When written between two intervals, the symbol $\cup$ signifies the set of all the elements from either set, and the symbol $\cap$ signifies the set of only those elements common to both sets. In exercises 67–72, graph the resulting set.

67. $[0, 4) \cup (5, 8]$
68. $(-1, 1) \cap [0, 2]$
69. $[-10, -3] \cap [-2, 0]$
70. $[-1, 0] \cap [0, 1]$
71. $(-\infty, 0) \cap (-1, +\infty)$
72. $(-\infty, 5] \cup (5, +\infty)$

2.5 Problem Solving

Goal A *to introduce three strategies for problem solving*

Solving problems is very much like playing a game. In order to play the game, you must first learn the moves. Only after the moves have become second nature can you hope to be proficient in selecting the moves that will win the game.

The goal in problem solving is, generally, to determine some unknown quantity. The moves that lead you toward this goal are referred to as **strategies.** We are mainly concerned with problems that can be represented by **models,** that is, equations or inequalities. We will discuss some strategies that will guide you in producing such models. Just like the moves in a game, this list of strategies should become second nature to you. They should serve as a mental checklist each time you approach a problem-solving situation.

Although the goal in solving a specific problem may be to find a certain value, the heart of problem solving is the *search for a useful strategy.* At this point we restrict ourselves to three problem-solving strategies.

> **Three Problem-Solving Strategies**
> 1. **Translate** the problem directly into an equation or an inequality.
> 2. Draw a **sketch** showing how the various details of the problem are related.
> 3. Make a **chart** that relates the known and unknown quantities in the problem.

The checklist of these three strategies below will serve as a guide in the examples in this section. Notice that the strategies can be used singly or in combinations to solve a problem.

> - **Translate**
> - **Sketch**
> - **Chart**

Translation

It is often possible to translate a problem directly into an algebraic equation or an inequality which then can be solved. Consider the following problem.

> A certain number is 15 more than twice its additive inverse. What is the number?

- **Translate**
- ○ Sketch
- ○ Chart

In *analyzing* this problem you should ask yourself what is it that you must find. A careful reading of the problem shows that you must determine the value of an unknown number. Direct translation produces an equation that is used to find this number.

Now you need to *organize* your work by introducing a variable so that you can create a model:

Let x = the unknown number.

The *model* for the problem is the following equation:

$$x = 15 + 2(-x).$$

The *answer*, 5, is found by solving the equation.

Sketch

The purpose of a sketch is to help you visualize the information in a problem. The sketch may suggest an equation or an inequality that models the problem. This strategy is particularly useful in problems involving geometric figures, where the sketch may remind you of an appropriate formula.

> The length and width of a rectangular patio have been increased by 3 ft and 6 ft respectively. The enlarged patio is a square. If the area of the old patio was 130 ft² (square feet), what is the area of the new patio?

In *analyzing* this problem you should recognize that you must find the dimensions of the new patio, which depend on the dimensions of the old patio. Your first impulse might be to draw a sketch. If so, your impulse is a good one.

To *organize* your thinking about the problem, define the variables. Let l = length of the old patio and let w = width of the old patio.

- ○ Translate
- **Sketch**
- ○ Chart

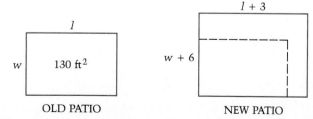

Figure 2.4

From the sketches we can conclude that *models* for the problem are the following two equations.

$$l \cdot w = 130$$ ■ Recall $A = lw$, where $A = 130$ ft^2.

$$l + 3 = w + 6$$ ■ Since the new patio is a square, the sides are equal.

We now combine the two equations to get a *simpler model* of one equation in one variable. Since the second equation can be rewritten as

$$w = l - 3,$$

we can substitute $l - 3$ for w in the first equation to get a simpler model in one variable:

$$l(l - 3) = 130.$$

Solving this equation produces the result $l = 13$ or $l = -10$. The latter value is discarded since length cannot be negative. Since $l = 13$, the new patio has sides equal to $l + 3 = 16$ or 16 ft. The *answer*, then, is 256 ft^2.

Chart

Very often the data in a problem is too complicated for simple translation, or is not easily represented in a sketch. A chart may help organize the information in a way that suggests the equation or inequality. That is the case in the next problem.

> At 5 P.M. a jet leaves an airport and flies due west at an average speed of 520 mph. At 6 P.M. a second jet leaves the same airport and flies due west at 650 mph. When does the second jet catch the first?

○ Translate
○ Sketch
● Chart

In *analyzing* this problem you should recognize that the two jets must cover the same distance. Since we must find the time it takes to do this, we use the formula $d = rt$, where d is the distance traveled, r is the rate of speed, and t is the time.

Often when a formula applies, a chart is useful to *organize* information. Let $t =$ the time in hours for Jet 1 to cover the distance. Then $t - 1 =$ the time in hours for Jet 2 to cover the same distance *in one less hour*.

	distance	=	rate	×	time
Jet 1	$520t$	=	520	×	t
Jet 2	$650(t - 1)$	=	650	×	$t - 1$

At the moment when Jet 2 catches up, both planes have covered the same distance, so it makes sense to equate the two different expressions for the same distance. This produces the *model:*

$$520t = 650(t - 1).$$

Solving the equation reveals that $t = 5$, and so the *answer* is that Jet 2 catches Jet 1 at 10 P.M., five hours after Jet 1 has taken off.

This kind of problem serves to remind us that the equation that models a given problem is often found by equating two different expressions for the same quantity. This useful fact will be used many times in this section.

You might have noticed that in each problem we emphasized the words ANALYZE, ORGANIZE, MODEL, and ANSWER. These terms represent *four important stages* in problem solving. Notice how these stages help to guide us through the problem-solving process.

Example 1 The sum of two numbers is 15, and their product is 44. Find the two numbers.

Solution

ANALYZE We must find two numbers that satisfy two conditions. Let us try translating the information in the problem directly into an equation.

ORGANIZE Let the two numbers be represented by x and y. We ultimately want to find an equation in one unknown. The problem presents two pieces of information which translate into the following equations.

- **Translate**
- ○ Sketch
- ○ Chart

$x + y = 15$ ■ "The sum of two numbers is 15 . . ."
$xy = 44$ ■ ". . . their product is 44."

Solving the first equation for y shows that $y = 15 - x$. Now replace y with $15 - x$ in the second equation $xy = 44$.

MODEL
$$x(15 - x) = 44$$
$$15x - x^2 - 44 = 0$$
$$x^2 - 15x + 44 = 0$$
$$(x - 11)(x - 4) = 0$$
$$x - 11 = 0 \quad | \quad x - 4 = 0$$
$$x = 11 \quad | \quad x = 4$$

■ The expressions $x(15 - x)$ and 44 both represent the product.

■ Both sides are multiplied by -1 to make the x^2-term positive.

■ The Property of Zero Products is applied.

ANSWER If $x = 11$, then since $x + y = 15$, we have $y = 4$. Thus, the numbers are 11 and 4. (If $x = 4$, then $y = 11$, which is the same answer.)

Example 2 Two bikers have a pair of walkie-talkies that work up to a distance of 40 mi. One biker leaves from the town line traveling at an average rate of 16 mph. One hour later the other biker leaves from the same place traveling 21 mph in the opposite direction. How long will the second biker ride before the walkie-talkies fail?

Solution

ANALYZE The unknown is the *time* it takes the two bikers to be 40 mi apart. Complicated data, such as in this problem, is best organized in a chart.

ORGANIZE Let t = the time that the *second* biker rides before the walkie-talkies fail.
Let $t + 1$ = the time that the *first* biker rides before the walkie-talkies fail.

○ Translate
● Sketch
● Chart

A chart helps us find a model for the distances each biker travels.

	d	=	r × t
first biker	$16(t + 1)$	=	$16 \times (t + 1)$
second biker	$21t$	=	$21 \times t$

A sketch helps us visualize this information.

```
        1st Biker          2nd Biker
    ←──────────────┼──────────────→
        16(t + 1)         21t
    ├──────────────── 40 ────────────────┤
```

The distance at which the walkie-talkies stop working is 40 mi. From the sketch we see the total distance between the bikers is described by $16(t + 1) + 21t$. Thus, we want to equate these two quantities.

MODEL
$$16(t + 1) + 21t = 40$$
$$t = \frac{24}{37}$$

ANSWER The second biker travels for $\frac{24}{37}$ hr, or approximately 39 min before the walkie-talkies stop working.

Example 3 A man 6 ft tall is walking toward a street light that is 21 ft tall. What is the length of the man's shadow on the ground when he is 10 ft away from the light?

Solution

ANALYZE You are to find the length of the man's shadow when he is 10 ft away from the light. Draw a sketch to help visualize the situation.

ORGANIZE

○ Translate
● Sketch
○ Chart

■ Let x = the length of the man's shadow.

The sketch suggests two similar triangles, one whose height is the man's and another whose height is the street light's.

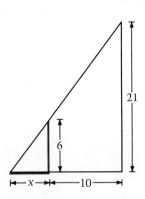 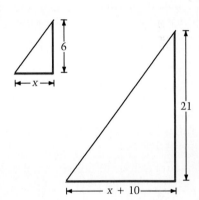

Use the ratios of the corresponding sides to set up a proportion.

MODEL

$$\frac{6}{x} = \frac{21}{x + 10}$$

$6(x + 10) = 21x$ ■ Two steps have been combined by cross multiplying.

$4 = x$

ANSWER   The man's shadow is 4 ft long when he is 10 ft away from the light.

Exercise Set 2.5   ▦ Calculator Exercises in the Appendix

Goal A

1. A number is increased by 5, and the result is twice the number. Find the original number.

2. Six more than a certain positive number is five more than twice the reciprocal of the number. What is the number?

3. The sum of three consecutive integers is 36. What are the three integers?

4. The sum of three consecutive even integers is twice the value of the smallest. What are the three integers?

5. A used car dealer recently purchased 3 cars. The station wagon costs $3000 less than the sports car, but is twice the cost of the subcompact. The total cost of the three cars is $32,760. Find the cost of each car.

6. Five years ago a tree was planted and measured 4 ft less than its present height. It is predicted that seven years from now it will be 16.5 ft high, triple its height at planting. How tall is the tree now?

7. One hundred and twenty feet of fencing are used to enclose a rectangular (or square) garden. If the garden must have an area of 900 ft^2, what are the dimensions of the garden?

8. A square piece of aluminum is used to manufacture a baking pan by cutting off from each corner a square 2 in on a side. The flaps are then folded up to form the sides of the pan. How large a piece of aluminum should be used to produce a pan whose volume is 162 in^3?

9. An investor wishes to invest a total of $10,000 in two accounts for one year. The first account offers an 8% annual percentage rate, and the second offers 12%. How much should be invested in each account so that the total interest is $920?

10. The entire stock of designer jeans at Hutton's Department Store would produce revenues of $4000, since there are 100 pairs and the price of each pair is $40. The manager decides to offer a 10% discount and later increases the discount to 25%. How many pairs should be sold at each discount to assure a total income of $3,375?

11. When a small plane flew against a 40 mph wind from Portland to Watertown, the trip took $2\frac{1}{2}$ hr. On the return trip the plane flew with the wind, and the trip took $1\frac{1}{2}$ hr. On both legs of the trip the pilot flew at the maximum speed. What was the plane's speed in calm air?

12. A train leaves Detroit heading for Chicago on the westbound track at an average speed of 60 mph. One hour later a train leaves Chicago on the eastbound track for Detroit at a speed of 45 mph. If Detroit and Chicago are 270 mi apart, how long has the second train been traveling when the trains meet? How far from Chicago will they be?

Superset

13. A camper's motorboat has a top speed of 6 mph. The camper sets out downstream traveling at full speed. The motor breaks down, and the camper rows back to the dock at a rate of 2.5 mph. If the return trip takes the camper three times as long as the trip downstream, how fast is the current flowing?

14. A farmer prepares feed for the pigs by mixing two brands of feed. Feed A provides 140 grams of a certain nutrient per ounce, and Feed B provides 110 grams per ounce. Each pig is fed 30 ounces of feed. How should the feed be mixed so that each pig receives 3,660 grams of the nutrient?

15. An 18 ft light pole casts a 15 ft shadow, and simultaneously a 40 ft telephone pole casts a 12 ft shadow up a wall. How far is the telephone pole from the wall?

16. A rectangular corral is built beside a barn, using the barn as one side of the corral. If 61 ft of fencing are used for the three sides of the corral, what are the dimensions of the corral if its area must be 450 ft^2?

17. A rectangular piece of sheet metal measuring 9 in by 30 in is cut along a line parallel to its shorter side in order to form two similar but unequal rectangular pieces. What is the area of the smaller piece?

18. Attempting to dilute 20 oz of a 70% solution of acid in water, a chemist adds too much water. To obtain a 60% concentration the chemist must add an additional 4 oz of acid. How much water did the chemist add?

Chapter Review & Test

Chapter Review

2.1 Introduction to Equations (pp. 56–62)

An *equation* is a statement that two expressions represent the same number (e.g., $6x + 4 = -12 + 4x$). Values of the variables that make an equation true are the *solutions* or *roots*. (p. 56)   Equations with the same solution set are called *equivalent equations*. (p. 57)

Addition Property for Equivalent Equations: Adding the same real number (or variable expression) to both sides of an equation produces an equivalent equation. (p. 57)

Multiplication Property for Equivalent Equations: Multiplying both sides of an equation by the same nonzero real number (or variable expression) produces an equivalent equation. (p. 57)

A *linear equation* is an equation containing a first degree monomial, but no monomial of higher degree. A linear equation written in the form $ax + b = 0$, where a and b are constants and $a \neq 0$ is said to be in *standard form*. (p. 58)

A *fractional equation* is an equation containing fractional expressions involving variables. You *must* check solutions of these type of equations. (p. 59)   An *absolute value equation* is an equation containing absolute values of terms involving variables. To solve an absolute value equation of the form $|A| = B$, you must solve the two equations in the statement "$A = B$ **or** $A = -B$." You *must* check solutions. (p. 60)

2.2 Quadratic Equations (pp. 63–69)

A *quadratic equation* is any equation containing a second degree monomial, but no monomial of higher degree. A quadratic equation written in the form $ax^2 + bx + c = 0$, where a, b, and c are constants, and $a \neq 0$, is said to be in *standard form*. (p. 63)

The *Property of Zero Products* is a basic tool in solving quadratic equations. (p. 63)   The method for *completing the square* can be used to solve quadratic equations that cannot be factored. (p. 65)

The *quadratic formula* given below is used to solve quadratic equations. (p. 66)

$$x = \frac{-b \pm \sqrt{b^2 - 4ac}}{2a}$$

2.3 Nonlinear Equations (pp. 70–75)

Many higher degree equations and fractional equations can be solved using techniques such as grouping and the Property of Zero Products. (pp. 70–71) An equation is *quadratic in form* if replacing an expression with a single variable makes the equation quadratic.

Radical equations can be solved using the *Power Property for Equations.* You *must* check solutions when using this property. (p. 73)

2.4 Inequalities (pp. 76–85)

Use the Addition Property for Equivalent Inequalities to solve inequalities like $x + 3 > 4$. Use the Positive Multiplication Property for Equivalent Inequalities to solve $2x > -7$. Use the Negative Multiplication Property for Equivalent Inequalities to solve $-6y < 5$; when using this property, remember to reverse the direction of the inequality. (p. 76)

The solution set of an inequality can be described in four ways: (1) interval notation, $(a, b]$; (2) set notation, $\{x | a < x \leq b\}$; (3) with a graph; and (4) in words. (p. 78)

To solve $|A| > c$ we solve the compound inequality $A > c$ **or** $A < -c$. To solve $|A| < c$ we solve the compound inequality $A < c$ **and** $A > -c$, also written as $-c < A < c$. (p. 80). To solve a quadratic inequality, solve the associated equality and then choose test values to determine when the expression is positive and when it is negative. (p. 82)

2.5 Problem Solving (pp. 86–92)

The essential aim of problem solving is to find a useful strategy from the *Three Problem-Solving Strategies.* (p. 86)   The four stages of problem solving are Analyze, Organize, Model, and Answer. (p. 89)

Chapter Test

2.1A Determine whether -5 is a solution of the following equations.

1. $u^2 + 3u = 40$ **2.** $\dfrac{u}{u + 4} = 5$

Solve each of the following equations for x.

3. $6(x + 1) = x - 1$ **4.** $(4x - 1)x = (2x + 3)(2x - 1)$

5. $3x + a = b$ **6.** $ax - 1 = b(x - 4)$

2 / Equations and Inequalities in One Variable

21.B Solve each of the following equations.

7. $\dfrac{t}{t+2} = 2 - \dfrac{1}{t+2}$

8. $\dfrac{3v}{v-3} - \dfrac{10v+33}{2v^2-5v-3} = \dfrac{6v}{2v+1}$

2.1C Solve each of the following equations.

9. $|4 - 3n| = 8$

10. $|6t - 1| = 8t + 5$

2.2A Solve each of the following equations by factoring.

11. $q^2 - 81 = 0$

12. $9x^2 - 144 = 0$

13. $2u^2 + u - 28 = 0$

14. $3v^2 + 11v - 20 = 0$

2.2B Solve each of the following equations by completing the square.

15. $x^2 - 4x + 21 = 0$

16. $2x^2 - 6x + 1 = 0$

2.2C Solve each of the following equations using the quadratic formula.

17. $5x^2 + 8x + 1 = 0$

18. $x^2 - 3x + 3 = 0$

19. $4x^2 - 20x = -25$

20. $2x^2 = -6x + 3$

2.3A Solve.

21. $x^5 + 3x^4 = x^3$

22. $4p^5 - 9p^3 = 0$

23. $\dfrac{3}{x-1} + \dfrac{4}{x+1} = 1$

24. $\dfrac{2}{y+5} = \dfrac{5}{6y+3} + \dfrac{3}{2y+1}$

2.3B Solve.

25. $6x^6 - 10x^3 - 4 = 0$

26. $p^{-2} - p^{-1} = 0$

27. $4y^4 - 16y^2 = -13$

28. $2c^4 - 7c^2 = -5$

29. $9v^{2/3} - 6v^{1/3} + 1 = 0$

30. $2x^{1/2} - x^{1/4} - 1 = 0$

2.3C Solve.

31. $\sqrt{5x^2 - 8} = 1$

32. $\sqrt{2y - 1} = 2 + \sqrt{y - 4}$

2.4A Describe the given intervals in set notation and sketch the graph.

33. $[-3.5, 2.5)$

34. $(-4, \infty)$

Solve each inequality, and describe the solution set in set notation and sketch the graph.

35. $7 - 4x \leq 39$

36. $3 + 2m > 15 + 3m$

2.4B Graph each of the following sets.

37. $\{x \mid x < -1 \text{ and } x > -6\}$
38. $\{x \mid x \geq 3 \text{ or } x \leq 3\}$
39. $\left\{x \mid -\dfrac{1}{2} \leq x < 5\right\}$

Solve each inequality, and describe the solution set in set notation and sketch the graph.

40. $|4x - 5| < 3$
41. $|3 - 2x| > 8$

2.4C Solve the following inequalities, and describe the solution set in interval notation.

42. $x^2 + 9x \leq -18$
43. $s^2 - 3s - 3 < 7$
44. $\dfrac{11}{x + 3} \leq 2$

2.5A Solve the following word problem

45. A rope 15 ft long is cut into two pieces. The pieces will be used to form two adjacent sides of a rectangle in the corner of a room. The two walls of the room form the other two sides of the rectangle. If the area of the rectangle is to be 44 ft², what are the lengths of the two pieces of rope?

Superset

46. In the figure below, the area of the shaded region is 12 square inches. Find the dimensions of the outer rectangle.

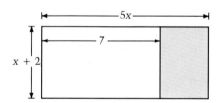

47. Solve the following equations for x. Explain your answer.

(a) $\sqrt{x} - \sqrt{x - 2} = \dfrac{1}{\sqrt{x - 2}}$

(b) $\sqrt{x\sqrt{x - 1}} = \sqrt{-x}$

48. Determine whether the statement is true or false. If false, illustrate with an example.

(a) If $a < 0$ and $b > 0$, then $ab < b$.

(b) For all real numbers a and b, $|a + b| = |a| + |b|$.

(c) For all real numbers x, $\sqrt[3]{x^3} = x$.

Cumulative Test: *Chapters 1 and 2*

In exercises 1–2, translate the following sentence into an equation.

1. Twice the sum of two consecutive integers is eight less than three times the larger integer.

2. The average of two numbers is eleven less than their product.

In exercises 3–4, evaluate the following.

3. $-|-7| + |-4|$

4. $-|8| - |-11|$

In exercises 5–6, simplify the following expressions so that x, y, and z appear at most once, with no powers of powers or negative exponents.

5. $\dfrac{5x^{-2}y^5z^{-3}}{6xy^{-2}z^3}$

6. $\dfrac{13xy^{-1}z^6}{2x^{-2}yz^{-3}}$

In exercises 7–8, perform the indicated operation.

7. $(2x - 3)(x^2 + 7x - 2)$

8. $(4x - 7)(-3x - 1)$

In exercises 9–10, factor completely.

9. $15x^2 - 22x + 8$

10. $7x^2 - 37x + 10$

In exercises 11–12, simplify. Assume that all variables represent positive real numbers.

11. $\sqrt{20x^2y} \cdot \sqrt{45x^3y^5}$

12. $\sqrt{75a^5b} \cdot \sqrt{27ab^6}$

In exercises 13–14, simplify. If your answer involves variables, write it in the form $cx^r y^q$.

13. $(-2x^{1/3}y^{2/5})(4x^{4/3}y)^2$

14. $(x^3y)^{1/2}(2x^{3/4}y^{1/4})^2$

In exercises 15–16, perform the indicated operation.

15. $\dfrac{-2}{x^2 - 4} + \dfrac{2}{x^2 - 2x}$

16. $\dfrac{1}{a} + \dfrac{4a}{1 - 4a}$

In exercises 17–18, simplify the following.

17. $\dfrac{3 + \sqrt{x}}{\sqrt{x} - 1}$

18. $\dfrac{\sqrt{u} - \sqrt{2}}{1 - \sqrt{2u}}$

In exercises 19–24, solve the following equations for x.

19. $5(2x - b) = 3 - (x + b)$

20. $x + 11 = x - 3 - (8 - x)$

21. $\dfrac{x+4}{5x+1} - \dfrac{10-2x^2}{15x^2-7x-2} = \dfrac{x}{3x-2}$ 22. $\dfrac{1}{x} - \dfrac{x}{1+x} = \dfrac{x-3}{1-x}$

23. $|7-3x| = 5$ 24. $|1+4x| = \dfrac{1}{2}$

In exercises 25–30, solve the following equations for x.

25. $7x^2 + 2x - 3 = 0$
26. $9 - x - 10x^2 = 0$
27. $7x^3 + 2x^2 - 28x - 8 = 0$
28. $-2x^3 + 4x^2 - 3x + 6 = 0$
29. $9x^{-2} - 30x^{-1} + 25 = 0$
30. $x^{2/3} - 6x^{1/3} - 16 = 0$

In exercises 31–34, solve the following inequalities, and describe the solution set in set notation and sketch the graph.

31. $5x - 4 > 4 - 15x$
32. $x - 5 \leq 4(1 - 2x)$
33. $|16x + 1| \leq 1$
34. $|-x - 6| \geq 5$

Superset

35. Translate the following problem into a single equation in one variable and then solve the equation.

The sum of two numbers is 14. Five plus three times the smaller number equals one less than three times the larger number. Find the numbers. (Hint: Solve both equations for one of the variables.)

36. Perform the indicated operations.

$\left(\dfrac{2x}{x^2-16} + \dfrac{3x}{x^2-7x+12}\right) \dfrac{x^2-3x}{x+1} \div \dfrac{5x^3+6x^2}{x^2+5x+4}$

37. Determine whether each statement is true or false. If false, illustrate this with an example.

(a) If $x < 0$, then $-(-x) = -x$
(b) For all real numbers a, b, and c, $\ a|b-c| \geq ab - ac$.
(c) For any nonnegative real number a, and for any positive integers m and n, $\sqrt[n]{a^m} - a^{m/n} = 0$.

38. Solve the following equation for x.

$\dfrac{1}{\sqrt{x+1}} + \dfrac{\sqrt{x-1}}{5} = \dfrac{\sqrt{x}}{\sqrt{x+1}}$

3 Graphs and Functions

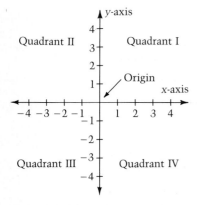

Figure 3.1 The entire system is called the **xy-plane**, and the point where the axes meet is the **origin**.

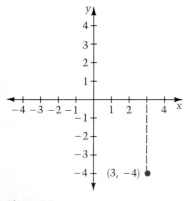

Figure 3.2

3.1 Equations and Graphs

Goal A *to graph equations in a coordinate plane*

In Chapter 2 we studied methods for solving equations containing two variables. An equation can be viewed as a way of associating y-values to x-values. For example, the equation $y = 5x + 1$ determines the y-value 11 for the x-value 2. This pair of numbers is denoted $(2, 11)$, and is called an **ordered pair.** Each value in the pair is called a **coordinate.** Plotting an ordered pair requires two real number lines, one for each coordinate.

The two lines, called **axes,** separate the plane into four **quadrants.** Since we commonly use x and y as variables, we usually call the horizontal axis the **x-axis** and the vertical axis the **y-axis.**

Each ordered pair corresponds to a single point in the plane. Plotting a point requires two moves: one from the origin in the horizontal direction, followed by one in the vertical direction.

Suppose we need to plot the point with coordinates $(3, -4)$. Start at the origin and move 3 units to the right. Then move 4 units downward. Plot the point by drawing a heavy dot.

To graph an equation, we plot ordered pairs determined by the equation. We first consider the straight line graph.

Straight Line

An equation of the form $y = mx + b$, where m and b are real numbers, has a **straight line** as its graph. The graph is completely determined by two points.

In Example 1, we consider an equation having a straight line as its graph.

Example 1 Graph the equation $y = 4x - 5$.

Solution

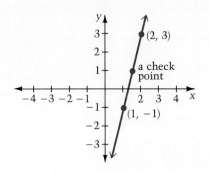

- To begin, find two ordered pairs satisfying the equation.
  $x = 1 \quad y = 4(1) - 5 = -1$
  $x = 2 \quad y = 4(2) - 5 = 3$
- Plot and label the points, and draw the line through them.
- Check: Find another ordered pair of the equation. It should lie on the line.

Figure 3.3

We shall now examine parabolas, and then we shall consider V-shaped graphs.

Parabola

An equation of the form $y = ax^2 + bx + c$, where a, b, and c are real numbers and $a \neq 0$, has a **parabola** as its graph.

In Example 2, we consider the parabola described by the equation $y = x^2$. This parabola will be used to help graph other equations.

Example 2 Graph the following equations. (a) $y = x^2$ (b) $y = -x^2 + 4x + 1$

Solution (a)

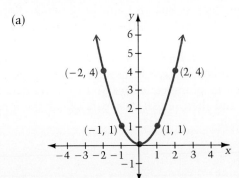

(b)

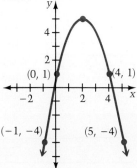

Figure 3.4

V-Shaped Graph

An equation of the form $y = a|x| + k$ where a and k are real numbers and $a \neq 0$, has a **V-shaped graph.**

In Example 3, we consider the V-shaped graph described by the equation $y = |x|$. This V-shaped graph will be used to help graph other equations.

Example 3 Graph the equations. (a) $y = |x|$ (b) $y = -2|x| + 1$

Solution (a) 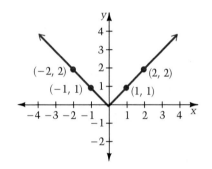 (b)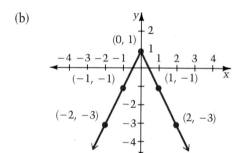

Goal B to reflect a set of points across a line or point

When you look at the reflection of an object in a mirror, the reflection appears to be the same distance behind the mirror as the object is in front of the mirror. The notion of reflection has a useful application to graphing.

Suppose the mirror in Figure 3.5 was turned so that all you could see is the edge l, and suppose the object P was simply a point. Point P', the **reflection of point P across line l** is found by drawing a dotted line through P, perpendicular to l, and then plotting P' the same distance behind l as P is in front of l.

In Example 4, we reflect a point across the x-axis and across the y-axis. In Example 5, we reflect a graph across the x-axis and across the y-axis. In some cases, when graphing equations you can find part of the graph by plotting points, and then reflect that part across the x-axis or y-axis to complete the graph.

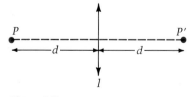

Figure 3.5

Example 4 Plot and label the reflection of the point (2, 3) across the x-axis and across the y-axis.

Solution

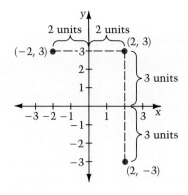

- To reflect (2, 3) across the y-axis, change the sign of the x-coordinate. The reflection is (−2, 3).

- To reflect (2, 3) across the x-axis, change the sign of the y-coordinate. The reflection is (2, −3).

Example 5 Reflect the following graph (a) across the x-axis and (b) across y-axis.

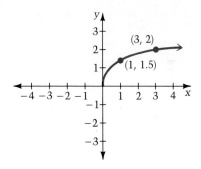

Solution Using (1, 1.5) and (3, 2) as reference points, find their reflections by changing the sign of the proper coordinate. Draw a curve through these points.

(a)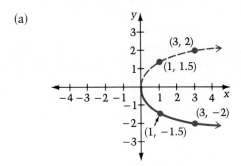

Reflection across x-axis

(b)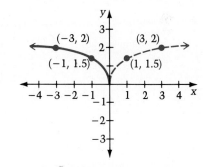

Reflection across y-axis

Figure 3.6

A point can also be reflected through a point. The point P', the **reflection of P through O**, is found by drawing a line through P and O. P' is the same distance from O as P is from O. It is often useful to reflect points through the origin (0, 0).

Example 6 Reflect the graph in Example 5 through the origin.

Solution

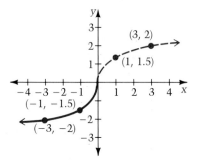

■ Reflect the points (1, 1.5) and (3, 2) through the origin and draw a curve through the points. To reflect a point through the origin, change the sign of both coordinates.

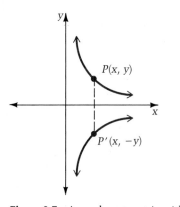

Figure 3.7 A graph symmetric with respect to the *x*-axis.

Goal C *to determine whether a graph is symmetric with respect to the x-axis, y-axis, or the origin, and to use symmetry to graph equations*

We now use the notion of reflection to define an important property of graphs, called **symmetry.** There are three types of symmetry that we shall consider.

Type I. Symmetry with respect to the *x*-axis
In Figure 3.7, the bottom half of the graph is the reflection of the top half across the *x*-axis. A typical point *P* and its reflection *P'* are shown. We say this graph is symmetric with respect to the *x*-axis.

Definition
A graph is **symmetric with respect to the *x*-axis,** if for each point *P* on the graph, the reflection of *P* across the *x*-axis is also on the graph.

Thus, if a graph symmetric with respect to the *x*-axis contains point *P* with coordinates (x, y), then it also contains the point *P'* with coordinates $(x, -y)$. By using this fact, we need only look at the equation of the graph to determine whether the graph is symmetric with respect to the *x*-axis.

Rule
For a graph described by an equation, if replacing *y* with $-y$ produces an equation equivalent to the original equation, then the graph is symmetric with respect to the *x*-axis.

Example 7 Is the graph of the equation $x = y^2 - 1$ symmetric with respect to the *x*-axis?

Solution $\quad x = (-y)^2 - 1\quad$ ■ Replace y with $-y$.
$\qquad\qquad\quad x = y^2 - 1\quad$ ■ Since $(-y)^2 = (-y)(-y) = y^2$, the equation is equivalent to the original equation. Therefore the graph is symmetric with respect to the x-axis.

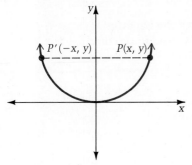

Figure 3.8 A graph symmetric with respect to the y-axis.

Type II. Symmetry with respect to the y-axis

We have seen that symmetry with respect to the x-axis involves vertical reflection. Symmetry with respect to the y-axis involves horizontal reflection. The graph in Figure 3.8 is symmetric with respect to the y-axis, because the left half is a reflection of the right half across the y-axis.

Definition

A graph is **symmetric with respect to the y-axis** if for each point P on the graph, the reflection of P across the y-axis is also on the graph.

Notice that if a point $P(x, y)$ is on this type of graph, then the point $P'(-x, y)$ is also on the graph.

Rule

For a graph described by an equation, if replacing x with $-x$ produces an equation equivalent to the original equation, then the graph is symmetric with respect to the y-axis.

For example, the parabola described by $y = 4x^2 + 9$ is symmetric with respect to the y-axis: when we replace x with $-x$, we obtain $y = 4(-x)^2 + 9$, which is equivalent to $y = 4x^2 + 9$, our original equation.

Type III. Symmetry with respect to the origin

This symmetry involves reflection through the origin. In the graph in Figure 3.9, the bottom half is found by reflecting the top half through the origin.

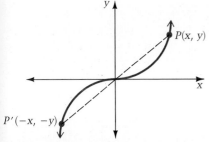

Figure 3.9 A graph symmetric with respect to the origin.

Definition

A graph is **symmetric with respect to the origin,** if for each point P on the graph, the reflection of P through the origin is also on the graph.

Therefore, if a point $P(x, y)$ is on this type of graph, then the point $P'(-x, -y)$ is on the graph.

Rule

For a graph described by an equation, if replacing x with $-x$ *and* y with $-y$ produces an equation equivalent to the original equation, then the graph is symmetric with respect to the origin.

Example 8 Test $y = x^5 - 2x$ for each of the three types of symmetry.

Solution

1. The test for symmetry with respect to the x-axis produces the equation $-y = x^5 - 2x$ or $y = -x^5 + 2x$ which is not equivalent to the original one. The graph is not symmetric with respect to the x-axis.

2. The test for symmetry with respect to the y-axis produces the equation $y = (-x)^5 - 2(-x)$ or $y = -x^5 + 2x$ which is not equivalent to the original one. The graph is not symmetric with respect to the y-axis.

3. Test for symmetry with respect to the origin.

   $-y = (-x)^5 - 2(-x)$ ■ Replace x with $-x$ and y with $-y$.
   $y = x^5 - 2x$ ■ The graph is symmetric with respect to the origin.

When graphing, you should test for symmetry before actually determining ordered pairs. Finding ordered pairs can be very time consuming. If symmetry exists, you can save time by first determining some points on one part of the graph, and then using symmetry to complete the graph.

Example 9 Sketch the graph of the equation $y = x^3 - 4x$.

Solution First we test for symmetry. Replacing x with $-x$ and y with $-y$ produces $-y = (-x)^3 - 4(-x)$, which is equivalent to the given equation. Thus the graph is symmetric with respect to the origin.

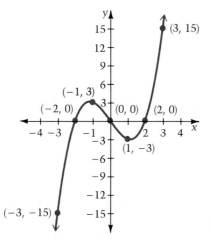

x	y	(x, y)
0	0	(0, 0)
1	−3	(1, −3)
2	0	(2, 0)
3	15	(3, 15)

■ Plot these points, draw a curve through them, and reflect the curve through the origin.

Exercise Set 3.1

Goal A

In exercises 1–18, graph each equation.

1. $y = 2x$
2. $y = \frac{1}{3}x$
3. $y = 3 - \frac{x}{2}$
4. $y = -5x + 1$
5. $y = x^2 - 1$
6. $y = x^2 + 1$
7. $y = -x^2$
8. $y = 1 - x^2$
9. $y = 3x^2$
10. $y = 3x^2 + 4$
11. $2y = x^2 - 2x - 1$
12. $\frac{1}{4}y = x^2 - 8x + 12$
13. $y = |x| + 2$
14. $y = 2|x|$
15. $y = \frac{1}{3}|x|$
16. $y = -|x|$
17. $y = 3 - |x|$
18. $y = 5 - 3|x|$

Goal B

In exercises 19–24, plot and label the reflection of each point (a) across the x-axis, (b) across the y-axis, and (c) through the origin.

19. $(3, -2)$
20. $(-2, 1)$
21. $(4, 3)$
22. $(-4, -8)$
23. $(0, 5)$
24. $(-3, 0)$

In exercises 25–30, reflect the given graph (a) across the x-axis, (b) across the y-axis, and (c) through the origin.

25.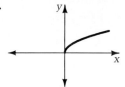
26.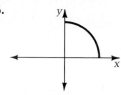
27.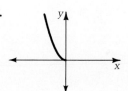
28.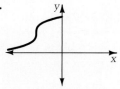
29.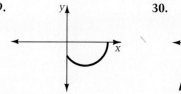
30.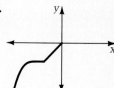

Goal C

In exercises 31–46, test the given relation for each of the three types of symmetry.

31. $y = x^2 - 2$
32. $y^2 = 3 - x$
33. $y^2 = x^2$
34. $y^2 = x^4$
35. $y = (x^2 - 1)^2$
36. $y = x^2 - 4x^4$
37. $x^3 - y^5 = 1$
38. $x^4 - y^2 = 7$
39. $x^2 - y^2 - 4 = 0$
40. $y = 3x - x^2$
41. $xy = 2$
42. $x^2 y^3 = -8$

In exercises 43–50, graph each equation.

43. $y^2 = x - 3$
44. $x = 25 - y^2$
45. $3y = x^3 - x$
46. $2x^3 - 3x + y = 0$
47. $x^2 + y^2 = 25$
48. $3y = x^3$
49. $y^2 = \frac{x + 4}{4}$
50. $\frac{x^2}{16} + \frac{y^2}{4} = 1$

Superset

In exercises 51–56, determine the coordinates of the reflection of the point across the line.

51. $(3, 2)$ across the line $x = 4$.
52. $(2, 7)$ across the line $x = 6$.
53. $(4, -3)$ across the line $x = -1$.
54. $(-2, -5)$ across the line $x = -3$.
55. $(3, 2)$ across the line $y = 5$.
56. $(2, 7)$ across the line $y = -4$.

In exercises 57–60, graph each equation.

57. $x = |y|$
58. $|x| = |y|$
59. $|x| + |y| = 1$
60. $|x| - |y| = 1$

3.2 Introduction to Functions

Goal A *to use a set of ordered pairs to describe a function*

We often can describe a quantity by telling how it is related to, or dependent upon some other quantity. For example, it is said that a child's rate of growth is dependent upon the quality and quantity of food consumed, or that happiness in life is dependent upon the degree of satisfaction you get from your job.

In mathematics we can sometimes express the relationship between things by using the word *function*. Our examples may be restated as

> growth rate is a function of nutrition, and
> happiness is a function of job satisfaction.

The concept of function is one of the most important in mathematics. We shall explore this concept in detail in this chapter.

The graph in Figure 3.10 shows the grades of six students. Suppose the teacher wishes to store the grades on a computer. Then it might be necessary to code the information as a list of pairs, where the first entry of the pair is the student's identification number and the second entry is the grade. This information can be written as the ordered pairs shown at the left. The set of these ordered pairs is an example of what is called a *function*.

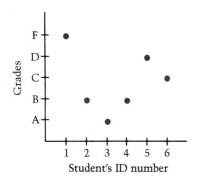

Figure 3.10

(1, F)   (4, B)
(2, B)   (5, D)
(3, A)   (6, C)

Definition
Whenever a process pairs each member of a first set with exactly one member of a second set, the resulting set of ordered pairs is called a **function**. Thus, a function contains no two different ordered pairs having the same first coordinate. The set of all first coordinates is called the **domain** of the function and the set of all second coordinates is called the **range**.

The function described above is the set of ordered pairs

$$\{(1, F), (2, B), (3, A), (4, B), (5, D), (6, C)\}.$$

The domain of the function is the set

$$\{1, 2, 3, 4, 5, 6\}$$

and the range of the function is the set

$$\{A, B, C, D, F\}.$$

Example 1 Use a set of ordered pairs to describe the function represented by the caloric graph shown below. Determine the domain and range of the function.

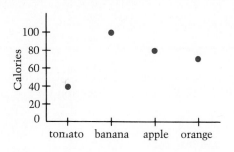

■ As in the graph of students' grades, remember to select the first coordinate from the horizontal scale and the second coordinate from the vertical scale.

Solution The function is the set of ordered pairs: {(tomato, 40), (banana, 100), (apple, 80), (orange, 70)}. The domain is {tomato, banana, apple, orange}, and the range is {40, 70, 80, 100}.

Corn Production (tons per acre)	Nitrogen Fertilizer (lbs per acre)	
12	108	(12, 108)
20	180	(20, 180)
25	220	(25, 220)
30	270	(30, 270)

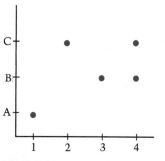

Figure 3.11

Another example of a function is the relationship between crop yield and soil fertility. As corn grows, it leeches nitrogen among other nutrients from the soil. In order to maintain the same level of production from year to year, the nitrogen must be replenished, and fertilizers are used to replace the nitrogen. The table at the left shows the pounds per acre of nitrogen needed to maintain the fertility of the soil for various levels of corn production. This relationship establishes a function given by the following ordered pairs: {(12, 108), (20, 180), (25, 220), (30, 270)}.

Not every set of ordered pairs is a function. The ordered pairs represented in the graph in Figure 3.11 do not form a function, since there are two ordered pairs with the same first coordinate, namely (4, B) and (4, C). In general, we refer to any set of ordered pairs as a **relation;** if no two different ordered pairs have the same first coordinate, we say that the relation is a function.

Goal B *to evaluate functions defined by equations*

At some time you have probably already checked an "ideal weight" chart to determine your "perfect" weight. There are formulas which describe ideal weight as a function of height. For a male over age 25 with a medium frame, the formula might be $w = 4h - 121$, where w represents weight in pounds and h represents height in inches. The formula determines a function since to each value of height there is assigned exactly one ideal weight.

h (in)	w (lb)	(h, w)
62	127	(62, 127)
68	151	(68, 151)
76	183	(76, 183)

Given a value of height h, the ideal weight is found by *evaluating the function*, that is, by replacing h with the given height and then finding the corresponding value of w. The chart at the left lists some ordered pairs for this function: (62, 127), (68, 151), and (76, 183). The second coordinates are called **function values,** because they are found by evaluating the function.

Example 2 The formula for converting the temperature in degrees Fahrenheit (F) to degrees Celsius (C) is given by the function $C = \frac{5}{9}(F - 32)$. Evaluate the function for F = 5, 14, 32, and 50, and display the results in a chart.

Solution For F = 5, we get $C = \frac{5}{9} \cdot (5 - 32) = -15$

F = 14 $C = \frac{5}{9} \cdot (14 - 32) = -10$

F = 32 $C = \frac{5}{9} \cdot (32 - 32) = 0$

F = 50 $C = \frac{5}{9} \cdot (50 - 32) = 10$

F	C	(F, C)
5	−15	(5, −15)
14	−10	(14, −10)
32	0	(32, 0)
50	10	(50, 10)

Example 3 A certain function is described by the formula $y = 7x^2 + 3x - 1$. Evaluate the function for x = −5, −1, 0, and 5, and display the results as ordered pairs.

Solution

$y = 7 \cdot x^2 + 3 \cdot x - 1$

x = −5 $y = 7 \cdot (-5)^2 + 3 \cdot (-5) - 1 = 159$

x = −1 $y = 7 \cdot (-1)^2 + 3 \cdot (-1) - 1 = 3$

x = 0 $y = 7 \cdot (0)^2 + 3 \cdot (0) - 1 = -1$

x = 5 $y = 7 \cdot (5)^2 + 3 \cdot (5) - 1 = 189$

The resulting ordered pairs are (−5, 159), (−1, 3), (0, −1), and (5, 189).

In the two previous examples, we have seen that some functions can be described by equations involving two variables. In Example 2, C is the **dependent variable,** and F is the **independent variable.** C is dependent upon F because we first select a value for F and then use it to find C. We call F the independent variable because we are free to choose any value for it.

We can use letters to name functions. For example,

$$f = \{(-1, 2), (5, 3), (11, 12), (15, 25)\}.$$

In this case, we use $f(x)$ to denote the second coordinate in the ordered pair whose first coordinate is x. Thus, we write

$$f(-1) = 2, \quad f(5) = 3, \quad f(11) = 12, \quad f(15) = 25.$$

$f(-1)$ is read "f of -1," and 2 is the "function value at -1."

The equation $f(x) = 2x^2 + 7x - 4$ determines the same set of ordered pairs (and therefore the same function) as the equation $y = 2x^2 + 7x - 4$. You will see y and $f(x)$ used interchangeably, and you should become comfortable with both these usages.

Example 4 For the function $g(x) = 2x^2 - 4x + 11$ find

(a) $g(3)$ (b) $g(b^3)$ (c) $g(3 + h)$

Solution

(a) $g(3) = 2(3)^2 - 4(3) + 11 = 17$ ■ Replace x with 3.

(b) $g(b^3) = 2(b^3)^2 - 4b^3 + 11$ ■ Replace x with b^3.
$ = 2b^6 - 4b^3 + 11$

(c) $g(3 + h) = 2(3 + h)^2 - 4(3 + h) + 11$ ■ Replace x with $3 + h$.
$ = 2(9 + 6h + h^2) - 12 - 4h + 11$
$ = 18 + 12h + 2h^2 - 12 - 4h + 11$
$ = 2h^2 + 8h + 17$

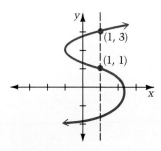

Figure 3.12

Recall that a set of ordered pairs is a function if no two ordered pairs have the same first coordinate. The ordered pairs $(1, 1)$ and $(1, 3)$ have the same first coordinate; thus the curve in Figure 3.12 is not the graph of a function. The fact that such points lie on the same vertical line suggests a quick way of determining when a graph represents a function.

The Vertical Line Test

For any graph in the xy-plane, if some vertical line can be drawn so that it intersects the graph more than once, then the graph does not represent a function of x. If no such line exists, the graph does represent a function of x.

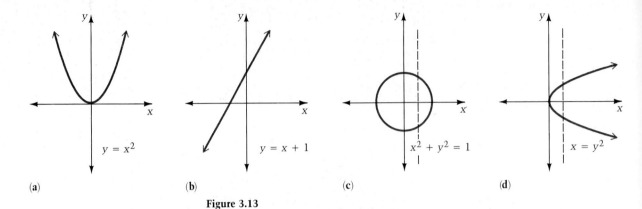

Figure 3.13

In the graphs above, graphs (a) and (b) represent functions of x, while graphs (c) and (d) do not.

Thus far we have discussed functions whose domains are the entire set of real numbers: x can be replaced by any real number. However, it is sometimes necessary to restrict the domain of a function in order to exclude values that produce undefined expressions.

$y = \sqrt{x}$ ■ Since the square root of a negative number is not a real number, the domain of this function is the set of all nonnegative real numbers.

$y = \dfrac{1}{x}$ ■ Since division by zero is undefined, the domain of this function is the set of all real numbers except zero.

Example 8 Graph the function described by the equation $y = \dfrac{1}{x}$.

Solution

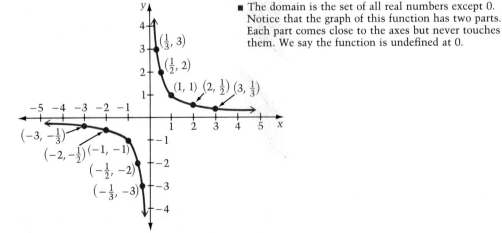

■ The domain is the set of all real numbers except 0. Notice that the graph of this function has two parts. Each part comes close to the axes but never touches them. We say the function is undefined at 0.

Exercise Set 3.2   = Calculator Exercises in the Appendix

Goal A

In exercises 1–4, describe the function with a set of ordered pairs. State the domain and range.

1.
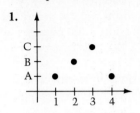

2.
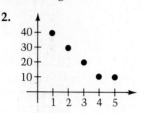

3.
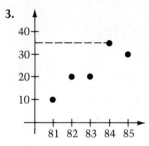

4.
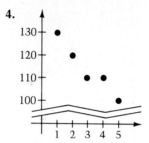

Goal B

In exercises 5–8, evaluate each function for the given values and display the results as a chart.

5. $t = 2h - 1$, $h = -1, 0, 3, 4$
6. $x = -16t^2 + 2t$, $t = -2, -1, 0, 2$
7. $y = x^2 - 3x + 1$, $x = -5, -2, 1, 4$
8. $y = 5 - 4x - x^2$, $x = -10, -5, 5, 10$

In exercises 9–16, find $f(-2)$, $f(0)$, $f(5)$, $f(c^2)$ and $f(2 + h)$ for each function.

9. $f(x) = 8 - 3x$
10. $f(x) = \frac{1}{2}x + 4$
11. $f(x) = 2x^2$
12. $f(x) = 6 - 2x^2$
13. $f(x) = x^2 - 5x + 6$
14. $f(x) = x^2 + 5x - 6$
15. $f(x) = 10$
16. $f(x) = -7$

In exercises 17–26, graph the function described by the equation and state the domain.

17. $f(x) = \sqrt{x + 3}$
18. $g(x) = \sqrt{x - 1}$
19. $y = \sqrt{4 - x}$
20. $y = \sqrt{2 + x}$
21. $y = \frac{1}{3}\sqrt{2x - 9}$
22. $y = \sqrt{\frac{3x - 15}{5}}$
23. $y = \frac{1}{x + 4}$
24. $y = \frac{1}{x - 8}$
25. $f(x) = \frac{1}{x - 7}$
26. $y = \frac{1}{x + 14}$

Superset

27. If $F(x) = x^2 + 1$, find (a) $F(3 + F(1))$, (b) $F(3) + F(1)$, and (c) $F(t^2 - F(t - 1))$.

28. Suppose $f(x) = \sqrt{x}$. Show that
$$\frac{f(3 + h) - f(3)}{h} = \frac{1}{f(3 + h) + f(3)}$$

In exercises 29–30, graph the functions described by the given equations.

29. $g(x) = \begin{cases} x, & \text{if } x < 0 \\ 2, & \text{if } x = 0 \\ x^2, & \text{if } x > 0 \end{cases}$

30. $h(x) = \begin{cases} 3, & \text{if } x < -2 \\ |x|, & \text{if } -2 \le x \le 2 \\ x^2 - 1, & \text{if } x > 2 \end{cases}$

A function f is **even** if for all x in the domain of f, $f(-x) = f(x)$, and **odd** if $f(-x) = -f(x)$. In exercises 31–36, classify each function as even, odd or neither.

31. $f(x) = x$
32. $g(x) = x^2$
33. $h(x) = \sqrt{x}$
34. $F(x) = |x|$
35. $j(x) = (x - 1)(x + 1)$
36. $j(x) = x^5 - x^3 - 6x$

37. What kind of symmetry is exhibited by (a) an odd function? (b) an even function?

38. Suppose f is a function and for all real numbers a and b, $f(a + b) = f(a) + f(b)$.
    (a) By selecting appropriate values for a and b, show that $f(0) = 0$.
    (b) Show that f is an odd function.

3.3 Transformations and Graphing

Goal A *to graph functions by using horizontal or vertical translations*

In this chapter we have been concerned with functions defined by equations. Certain changes in an equation produce simple changes in its graph. These changes in the graph are called **transformations**. There are three general types of transformations: **translations, reflections,** and **changes in shape.**

Type I. Translations. A translation occurs when a graph is moved vertically or horizontally. The shape and size of the graph remain the same; only its position changes.

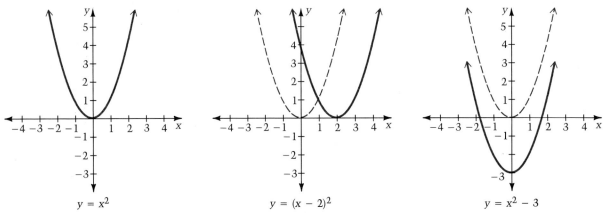

Figure 3.14 The graph of $y = (x - 2)^2$ is a translation of $y = x^2$ two units to the right; $y = x^2 - 3$ is a translation of $y = x^2$ three units down.

Type II. Reflections. The second type of transformation involves the reflection of a graph across the x-axis or y-axis.

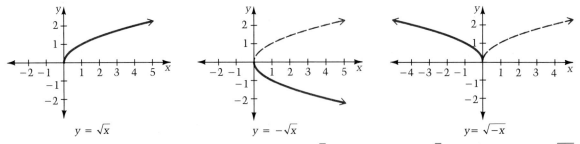

Figure 3.15 The graph of $y = -\sqrt{x}$ is a reflection of $y = \sqrt{x}$ across the x-axis; $y = \sqrt{-x}$ is a reflection of $y = \sqrt{x}$ across the y-axis.

Type III. Changes in Shape. The third type of transformation involves a change in the shape of a graph. A change in shape is also called a **stretching** or **shrinking**.

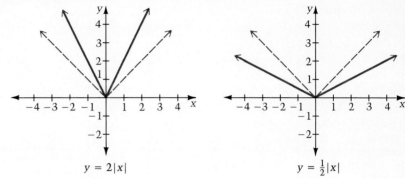

Figure 3.16 The graph of $y = 2|x|$ is a stretching of $y = |x|$ away from the x-axis; $y = \frac{1}{2}|x|$ is a shrinking of $y = |x|$ toward the x-axis.

Now that we have summarized the three general types of transformations, let us consider translations in detail. Compare the charts of ordered pairs for the functions below.

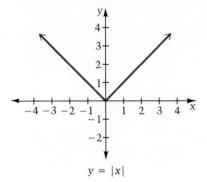

Figure 3.17

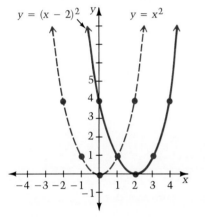

$y = x^2$		
x	y	(x, y)
−2	4	(−2, 4)
−1	1	(−1, 1)
0	0	(0, 0)
1	1	(1, 1)
2	4	(2, 4)

$y = (x - 2)^2$		
x	y	(x, y)
0	4	(0, 4)
1	1	(1, 1)
2	0	(2, 0)
3	1	(3, 1)
4	4	(4, 4)

$y = (x + 1)^2$		
x	y	(x, y)
−3	4	(−3, 4)
−2	1	(−2, 1)
−1	0	(−1, 0)
0	1	(0, 1)
1	4	(1, 4)

Add 2 to x-value to get same y-value

Add −1 to x-value to get same y-value

Notice that by adding 2 to each x-coordinate in the ordered pairs for $y = x^2$, you produce the ordered pairs for $y = (x - 2)^2$. But, adding 2 to each x-coordinate has the effect of moving each point 2 units to the right (positive direction). This suggests that the graph of $y = (x - 2)^2$ will be formed by translating the graph $y = x^2$ two units in the positive x-direction. This is shown in Figure 3.17.

In a similar way, adding -1 to each x-coordinate in the ordered pairs for $y = x^2$ produces ordered pairs for $y = (x + 1)^2$. The graph of $y = (x + 1)^2$ is found by translating the graph $y = x^2$ one unit in the negative x-direction. In this case, think of $(x + 1)^2$ as $(x - (-1))^2$.

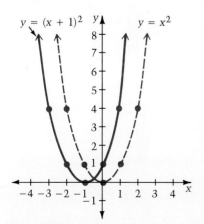

Figure 3.18

Rule for Horizontal Translations

$y = (x - 2)^2$ $y = (x + 3)^2$

▲ $h = 2$ ▲ $h = -3$

h is *subtracted* from x

Suppose h is a real number. Replacing x with $x - h$ in an equation translates a graph horizontally. If $h > 0$, the graph moves $|h|$ units to the right; if $h < 0$, it moves $|h|$ units to the left.

We examine changes in an equation that produce vertical translations.

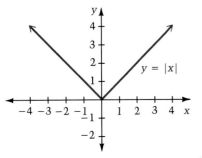
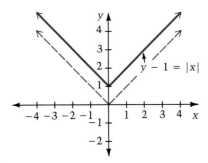
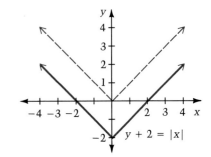

Figure 3.19 Replacing y with $y - 1$ in $y = |x|$ produces a translation 1 unit up. Replacing y with $y + 2$ produces a translation 2 units down.

Rule for Vertical Translations

Suppose k is a real number. Replacing y with $y - k$ in an equation translates the original graph vertically. If $k > 0$, the graph moves $|k|$ units up; if $k < 0$, the graph moves $|k|$ units down.

Example 1   Graph each function by translating $y = \dfrac{1}{x}$: (a) $y = \dfrac{1}{x + 3}$ (b) $y - 2 = \dfrac{1}{x}$

Solution   (a)

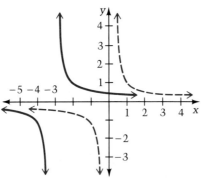

$y = \frac{1}{x}$ is translated 3 units to the left.

(b)

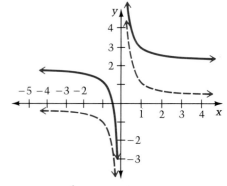

$y = \frac{1}{x}$ is translated 2 units up.

Sometimes we discuss a function f without stating a specific equation. It is customary to use a general equation such as $y = f(x)$ to represent f.

Example 2 Using the given graph of $y = f(x)$, sketch the graphs of the following translations.

(a) $y = f(x - 1)$ (b) $y = f(x) - 1$

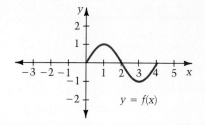

Solution (a) Replacing x with $x - 1$ produces $y = f(x - 1)$. Thus, $y = f(x)$ is translated 1 unit to the right.

(b) Replacing y with $y + 1$ produces $y + 1 = f(x)$ or $y = f(x) - 1$. Thus, $y = f(x)$ is translated 1 unit down.

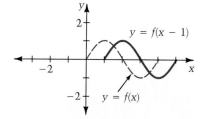 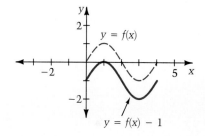

Goal B *to graph functions by using reflections across the x-axis or y-axis*

Compare the charts of ordered pairs for the functions below.

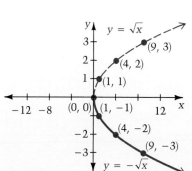

$y = \sqrt{x}$

x	y	(x, y)
0	0	(0, 0)
1	1	(1, 1)
4	2	(4, 2)
9	3	(9, 3)

(a)

$y = -\sqrt{x}$

x	y	(x, y)
0	0	(0, 0)
1	-1	(1, -1)
4	-2	(4, -2)
9	-3	(9, -3)

(b)

$y = \sqrt{-x}$

x	y	(x, y)
0	0	(0, 0)
-1	1	(-1, 1)
-4	2	(-4, 2)
-9	3	(-9, 3)

(a) Each y-coordinate is replaced by its opposite, which causes a reflection of $y = \sqrt{x}$ across the x-axis. (b) Here each x-coordinate is replaced by its opposite, thus reflecting the original function across the y-axis.

Figure 3.20

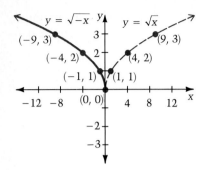

Figure 3.21

Rule for Reflecting Across the y-axis

Replacing x with $-x$ in an equation reflects the graph of the equation across the y-axis.

Rule for Reflecting Across the x-axis

Replacing y with $-y$ in an equation reflects the graph of the equation across the x-axis. If a function is denoted by $y = f(x)$, then the reflection of its graph across the x-axis has the equation $-y = f(x)$ or $y = -f(x)$.

Example 3 Graph $y = |2 - x|$ by reflecting the graph of $y = |x + 2|$.

Solution

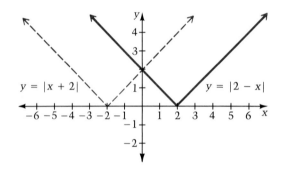

- $y = |x + 2|$ is a horizontal translation of $y = |x|$ two units to the left.

- $y = |2 - x|$ is obtained be replacing x by $-x$ in $y = |x + 2|$. This means we reflect $y = |x + 2|$ across the y-axis.

Goal C *to graph functions by stretching or shrinking*

We now consider changes in the equation of a function that produce changes in the shape of a graph. Compare the charts of ordered pairs for the following functions:

$y = |x|$

x	y	(x, y)
−1	1	(−1, 1)
0	0	(0, 0)
2	2	(2, 2)

(a)

$y = 3|x|$

x	y	(x, y)
−1	3	(−1, 3)
0	0	(0, 0)
2	6	(2, 6)

(b)

$y = \frac{1}{2}|x|$

x	y	(x, y)
−1	$\frac{1}{2}$	$(-1, \frac{1}{2})$
0	0	(0, 0)
2	1	(2, 1)

(a) Each y-coordinate of $y = |x|$ has been multiplied (stretched) by 3. (b) Each y-coordinate of $y = |x|$ has been multiplied (shrunk) by $\frac{1}{2}$.

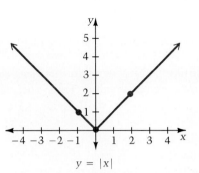 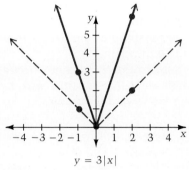 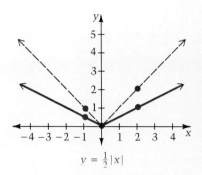

Figure 3.22

The graphs in Figure 3.22 demonstrate the effect of a vertical stretching and shrinking. Notice that in the graph of

$$y = 3|x|,$$

the y-values "increase faster" than the y-values in $y = |x|$. This accounts for the stretch away from the x-axis. The y-values of the graph $y = \frac{1}{2}|x|$ "increase more slowly" than those of $y = |x|$. This accounts for the shrinking toward the x-axis. Such changes in y-values always produce vertical stretching or shrinking. *Points having y-coordinate zero will not be affected by vertical stretching or shrinking.* The transformations above leave the origin fixed.

Rule for Vertical Changes in Shape

Suppose b is a positive real number. If $b > 1$, then the graph of $y = b \cdot f(x)$ is found by stretching the graph of $y = f(x)$ away from the x-axis. If $b < 1$, then $y = b \cdot f(x)$ is found by shrinking $y = f(x)$ down toward the x-axis. In both cases, points having y-coordinate zero are left fixed.

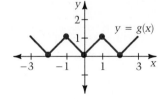

Figure 3.23

A stretching or shrinking also occurs when x is multiplied by a constant. This time the stretching or shrinking is *horizontal*. Consider the function $y = g(x)$ shown at the left. From the graph, a table of ordered pairs can be constructed. You can obtain points for $y = g(\frac{1}{2}x)$ by first finding $\frac{1}{2}x$ and then determining $g(\frac{1}{2}x)$ from the original table.

x	g(x)	(x, g(x))
−2	0	(−2, 0)
−1	1	(−1, 1)
0	0	(0, 0)
1	1	(1, 1)
2	0	(2, 0)

x	$\frac{1}{2}x$	$g(\frac{1}{2}x)$	$(x, g(\frac{1}{2}x))$
−4	−2	0	(−4, 0)
−2	−1	1	(−2, 1)
0	0	0	(0, 0)
2	1	1	(2, 1)
4	2	0	(4, 0)

3 / Graphs and Functions

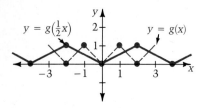

Figure 3.24

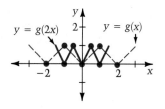

Figure 3.25

Replacing x with $\frac{1}{2}x$ in $y = g(x)$ stretches the graph horizontally. Any point having x-coordinate zero will not be moved. Thus the origin will remain fixed.

Substituting $2x$ for x produces a shrinking of $y = g(x)$. In this case, each original y-value is paired with half the original x-value. Thus, the graph shrinks toward the y-axis.

Rule for Horizontal Changes in Shape

Suppose a is a positive real number. If $a > 1$, then the graph of $y = f(ax)$ is found by shrinking $y = f(x)$ toward the y-axis. If $a < 1$, then the graph of $y = f(ax)$ is found by stretching $y = f(x)$ away from the y-axis. In both cases, points having an x-coordinate of zero are fixed.

Example 4 demonstrates a horizontal stretching and a horizontal shrinking.

Example 4 Using the graph of $y = f(x)$, sketch the following graphs.

(a) $y = 3f(x)$ (b) $y = f(2x)$

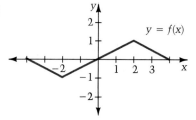

Solution (a) Since $b > 1$, this is a stretching away from the x-axis of $y = f(x)$. Since $b = 3$, each original y-coordinate is tripled.

(b) Since $a > 1$, we have a shrinking toward the y-axis of $y = f(x)$. Since $a = 2$, for each original y-coordinate the original x-coordinate is halved.

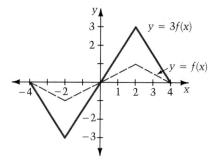

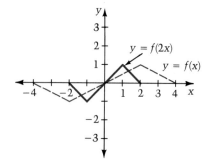

Goal D *to sketch the graph of a function requiring more than one transformation of a known graph*

You have seen that there are three ways of transforming a graph: by translating it, by reflecting it, or by changing its shape (stretching or shrinking). Very often you can produce a desired graph by applying more than one transformation to a known graph, as shown in the examples below.

Example 5 Sketch the following graphs (a) $y = |x - 1| - 2$ (b) $y = -3(x + 1)^2$.

Solution (a)

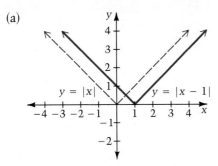

Step 1: Notice that we have a transformation of $y = |x|$. The $x - 1$ suggests translating $y = |x|$ one unit to the right to obtain $y = |x - 1|$.

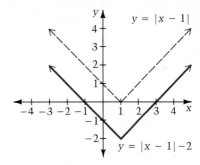

Step 2: Replacing y with $y + 2$ in $y = |x - 1|$ yields $y = |x - 1| - 2$. Thus we have a translation of $y = |x - 1|$ two units down. This is the desired graph.

(b)

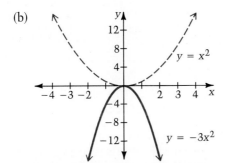

Step 1: This is a transformation of $y = x^2$. The minus sign suggests replacing y with $-y$ to get $y = -x^2$, a reflection across the x-axis. The 3 suggests stretching $y = -x^2$ away from the x-axis to get $y = -3x^2$.

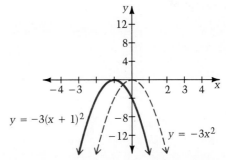

Step 2: The $(x + 1)$ suggests a translation of $y = -3x^2$ one unit to the left to get the graph of $y = -3(x + 1)^2$.

Exercise Set 3.3

Goal A

In exercises 1–14, graph the following functions by translating the graph of $y = x^2$, $y = |x|$ or $y = \frac{1}{x}$.

1. $y = (x - 2)^2$
2. $y = (x - 4)^2$
3. $y = \left(x + \frac{3}{2}\right)^2$
4. $y = \left(x - \frac{5}{2}\right)^2$
5. $y - 2 = x^2$
6. $y = x^2 - 4$
7. $y = |x + 4|$
8. $y = |x - 3|$
9. $y - 3 = |x|$
10. $y = |x| - 3$
11. $y = \dfrac{1}{x - 4}$
12. $y = \dfrac{1}{x - 2}$
13. $y = \dfrac{1}{x} - 4$
14. $y - 3 = \dfrac{1}{x}$

Goal B

In exercises 15–20, graph each function by reflecting the graph of $y = (x - 1)^2$, $y = |x - 1|$, or $y = \sqrt{x - 1}$ across the x- or y-axis.

15. $y = -\sqrt{x - 1}$
16. $y = |-x - 1|$
17. $y = (-x - 1)^2$
18. $y = \sqrt{-x - 1}$
19. $|x - 1| + y = 0$
20. $y = -(x - 1)^2$

Goal C

In exercises 21–26, graph each function by transforming the graph of $y = f(x)$ below.

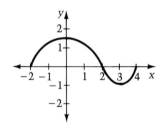

21. $y = f(2x)$
22. $y = f\left(\dfrac{1}{2}x\right)$
23. $y = \dfrac{1}{2}f(x)$
24. $y = f(4x)$
25. $3y = f(x)$
26. $16y = 4f(x)$

Goal D

In exercises 27–38, use transformations to graph the given function.

27. $y = (x - 2)^2 + 1$
28. $y = (x + 3)^2 - 1$
29. $y = 3(x - 2)^2 + 1$
30. $y = -2(x + 1)^2 - 3$
31. $y = -5|x|$
32. $y = -\dfrac{1}{2}|x|$
33. $y = 4|x + 1| - 4$
34. $y = 4|x - 2| + 3$
35. $y = \dfrac{1}{x + 3} + 2$
36. $y = \dfrac{1}{x + 4} + 3$
37. $y = 2 - \dfrac{1}{x - 3}$
38. $y = 3 - \dfrac{1}{x + 4}$

In exercises 39–42, sketch the graph of the given function by transforming the graph used in exercises 21–26.

39. $y = 2f(x - 3)$
40. $y = -2f(x - 1)$
41. $y = -\dfrac{1}{3}f(x + 2)$
42. $y = f\left(\dfrac{1}{3}x + 2\right)$

Superset

The equation $x^2 + y^2 = r^2$ describes a circle with center at the origin and radius r ($r \geq 0$).

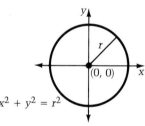

By translating the above graph h units horizontally and k units vertically, we discover that the equation $(x - h)^2 + (y - k)^2 = r^2$ describes a circle with center (h, k) and radius r. In exercises 43–48, graph the circle described by the given equation. Label the center and radius.

43. $x^2 + (y - 3)^2 = 16$
44. $(x - 1)^2 + y^2 = 25$
45. $x^2 + (y + 6)^2 = 10$
46. $(x - 2)^2 + y^2 = 5$
47. $(x + 5)^2 + (y - 2)^2 = 1$
48. $(x - 3)^2 + (y + 2)^2 = 9$

3 / Graphs and Functions

3.4 The Linear Function

Goal A *to determine the slope of a line and to sketch the line*

A function described by an equation of the form

$$y = mx + b$$

is called a **linear function.** In section 3.1 we saw that these functions have straight lines as their graphs. Given any two points on a line, we can determine an important characteristic of the line, its *slope*. A line's slope is a number that measures the slant of the line.

Let us consider the linear function

$$y = \frac{1}{2}x + \frac{3}{2},$$

and a table of x- and y-values.

Δx = Difference between x-values

Δy = Difference between y-values

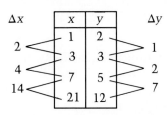

■ No matter which two ordered pairs you choose, the difference between the y-values is always half the difference between the corresponding x-values.

Based on this data, we might hypothesize that

$$\text{difference in } y\text{-values} = \frac{1}{2}(\text{difference in } x\text{-values}),$$

that is,

$$\frac{\text{difference in } y\text{-values}}{\text{difference in } x\text{-values}} = \frac{1}{2}.$$

This ratio is called the slope of the line $y = \frac{1}{2}x + \frac{3}{2}$.

Definition
If $P_1(x_1, y_1)$ and $P_2(x_2, y_2)$ are two different points on a nonvertical line, then the **slope** m of the line is given by the formula

$$m = \frac{y_2 - y_1}{x_2 - x_1}$$

When the equation of a line is in the form $y = mx + b$, the slope of the line is m.

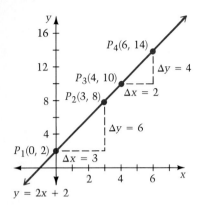

y = 2x + 2

Figure 3.26 y = 2x + 2

The slope of a line can be found geometrically. For the points in Figure 3.26, we calculate the ratio of the change in vertical direction (difference in y-values) to the change in horizontal direction (difference in x-values).

For P_3 and P_4, calculate

$$\frac{\text{vertical change}}{\text{horizontal change}} = \frac{\Delta y}{\Delta x} = \frac{14 - 10}{6 - 4} = \frac{4}{2} = 2.$$

For P_1 and P_2, calculate

$$\frac{\text{vertical change}}{\text{horizontal change}} = \frac{\Delta y}{\Delta x} = \frac{8 - 2}{3 - 0} = \frac{6}{3} = 2.$$

Often the difference in y-values is called the **rise**, and the difference in x-values is called the **run**. Thus, the slope can be thought of as the "rise over the run."

Example 1 Find the slope of the line determined by the following ordered pairs; then sketch the line. (a) (3, 5) and (1, −1)   (b) (−2, 3) and (1, −2)

Solution (a) $m = \dfrac{-1 - 5}{1 - 3} = \dfrac{-6}{-2} = 3$ (b) $m = \dfrac{3 - (-2)}{-2 - 1} = \dfrac{5}{-3} = -\dfrac{5}{3}$

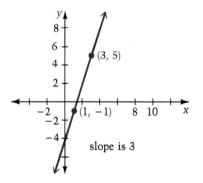

slope is 3

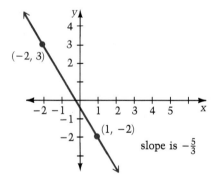

slope is $-\dfrac{5}{3}$

A positive slope means the line slants up from left to right. A negative slope means the line slants down from left to right.

$\dfrac{y_2 - y_1}{x_2 - x_1}$

coordinates   coordinates
of P_2 of P_1

Careful! When setting up the ratio to calculate the slope, make sure that the coordinates of the same ordered pair are aligned. Failure to do so will produce a value for the slope that has the wrong sign.

3 / Graphs and Functions

The graphs below show that the absolute value of the slope is a measure of the steepness of a line: the greater the absolute value, the steeper the line.

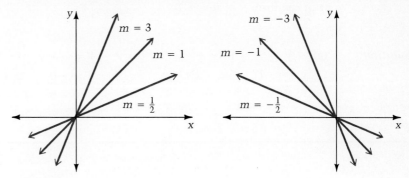

Figure 3.27

Fact

Since every point on a horizontal line has the same y-coordinate, the equation of a horizontal line has the form $y = b$, where b is a real number. The slope of a horizontal line is 0. Note that the equation of the x-axis is $y = 0$.

Fact

Since every point on a vertical line has the same x-coordinate, the equation of a vertical line has the form $x = c$, where c is a real number. The slope of a vertical line is not defined because the difference between the x-values of any two points on a vertical line is always zero, and division by zero is undefined. Note that the equation of the y-axis is $x = 0$.

Goal B *to determine the equation of a line*

Suppose we know that the point (2, 1) lies on a line whose slope is $\frac{3}{4}$. How can we find the equation of this line? Thinking of (2, 1) as (x_1, y_1) and (x, y) as (x_2, y_2), we can use the definition of slope to find the equation of the line.

$$\frac{3}{4} = \frac{y - 1}{x - 2} \qquad \blacksquare \text{ Remember } m = \frac{y_2 - y_1}{x_2 - x_1}.$$

$$y - 1 = \frac{3}{4}(x - 2) \qquad \blacksquare \text{ This is known as the point-slope form.}$$

Straight line: point-slope form

If a line contains the point (x_1, y_1) and has slope m, then the **point-slope form** of the equation of the line is

$$y - y_1 = m(x - x_1).$$

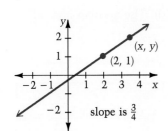

Figure 3.28

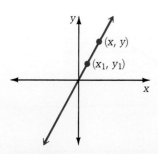

Figure 3.29

Example 2 (a) Write the point-slope form of the equation of the line containing $(3, -1)$ and having slope 5. (b) Find an equation of the line containing $(-2, 4)$ and having slope -1. Write it in the form $y = mx + b$.

Solution (a) $y - y_1 = m(x - x_1)$
$y - (-1) = 5(x - 3)$

(b) $y - 4 = -1(x - (-2))$
$y = -(x + 2) + 4$
$y = -x - 2 + 4$
$y = -x + 2$

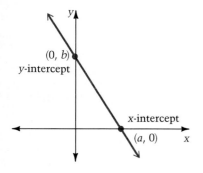

Figure 3.30

An important point on a line is the point where the line intersects the y-axis. The line graphed in Figure 3.30 crosses the y-axis at the point $(0, b)$. The number b is called the **y-intercept.** If we know the line has slope m, we can use the point-slope equation to write $y - b = m(x - 0)$ or simply, $y = mx + b$.

Straight line: slope-intercept form

If a line has a y-intercept b and slope m, then the **slope-intercept form** of the equation of the line is $y = mx + b$.

In Figure 3.30, the line crosses the x-axis at $(a, 0)$. The number a is called the **x-intercept.**

Example 3 Write an equation of the line having y-intercept -1 and slope $\frac{3}{5}$. Sketch the graph.

Solution
$y = mx + b$
$y = \frac{3}{5}x + (-1)$

In order to sketch the line, we need two points. We already know $(0, -1)$ is a point on the line. To find a second point, choose a value for x and find y. Let $x = 5$; then $y = 2$.

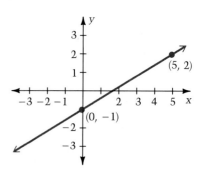

Given two points on a line, you can find the equation of the line by first calculating the slope and then using either the *point-slope form* or the *slope-intercept form*.

Example 4   Determine the equation of the line containing the points $(-1, 3)$ and $(5, 2)$.

Solution

$$m = \frac{3-2}{-1-5} = -\frac{1}{6}$$ ■ First find the slope.

$$y - y_1 = m(x - x_1)$$

$$y - 2 = -\frac{1}{6}(x - 5)$$ ■ We used $m = -\frac{1}{6}$, and (x_1, y_1) is $(5, 2)$.

$$y = -\frac{1}{6}x + \frac{5}{6} + 2$$

$$y = -\frac{1}{6}x + \frac{17}{6}$$

The equation found in Example 4 can be rewritten in the form $\frac{1}{6}x + y - \frac{17}{6} = 0$, known as the general form of the equation of this line.

Straight line: general form

An equation of the form $Ax + By + C = 0$, where A, B and C are real numbers, with A and B not both zero, is called the **general form** of the equation of the line.

Of the three forms of the equation of a line, the point-slope form, the slope-intercept form, and the general form, the slope-intercept form is best suited for comparing lines, because the slope can be read from the equation. In addition, knowing the slope is essential to finding the equation of a line parallel or perpendicular to a given line.

Fact

If two nonvertical lines are parallel, then their slopes are equal.

If two nonvertical lines are perpendicular, then their slopes are negative reciprocals. That is, if m_1 is the slope of a line, then $-\dfrac{1}{m_1}$ is the slope of the line perpendicular to it.

Figure 3.31 $m_2 = m_3$, $m_2 = -\dfrac{1}{m_1}$ and $m_3 = -\dfrac{1}{m_1}$

Example 5 Line $\mathcal{M}$ is described by the equation $2x + 3y - 5 = 0$. Find the equation for the line $\mathcal{P}$ perpendicular to $\mathcal{M}$ which contains the point $(0, -3)$.

Solution

$2x + 3y - 5 = 0$ ■ First, rewrite $\mathcal{M}$ as a slope-intercept equation.

$y = -\frac{2}{3}x + \frac{5}{3}$ ■ We can now conclude $\mathcal{M}$ has slope $-\frac{2}{3}$.

Since $\mathcal{P}$ is perpendicular to $\mathcal{M}$, the slope of $\mathcal{P}$ is the negative reciprocal of $-\frac{2}{3}$, that is, $\frac{3}{2}$.

$y = \frac{3}{2}x + (-3)$ ■ Use the slope-intercept equation $y = mx + b$ because we know the y-intercept is -3.

$\frac{3}{2}x - y - 3 = 0$ ■ Rewrite as a general linear equation.

Exercise Set 3.4 ▣ Calculator Exercises in the Appendix

Goal A

In exercises 1–14, find the slope of the line determined by the following ordered pairs; then sketch the line.

1. $(3, 8)$ and $(4, 13)$
2. $(5, 2)$ and $(7, 10)$
3. $(8, -1)$ and $(-4, 3)$
4. $(-2, 7)$ and $(-7, 4)$
5. $(-5, -6)$ and $(3, -8)$
6. $(8, -5)$ and $(-10, 10)$
7. $(2, 5)$ and $(-7, 5)$
8. $(5, 2)$ and $(5, -7)$
9. $(3, 0)$ and $(1, 0)$
10. $(0, 8)$ and $(0, 11)$
11. $\left(\frac{1}{2}, \frac{1}{8}\right)$ and $\left(\frac{3}{2}, \frac{1}{4}\right)$
12. $\left(\frac{1}{3}, \frac{1}{4}\right)$ and $\left(\frac{2}{3}, \frac{3}{8}\right)$
13. $(0.1, 0.01)$ and $(1, 1)$
14. $(0.2, 1.3)$ and $(1.8, 1.5)$

Goal B

For exercises 15–28, determine an equation of the line passing through the pairs of points given in exercises 1–14. Write your answer in (a) point-slope form, (b) slope-intercept form, and (c) general form.

In exercises 29–46, determine an equation of the line satisfying the stated conditions. Write your answer in (a) point-slope form, (b) slope-intercept form, and (c) general form.

29. slope is -2; passes through $(3, 5)$
30. slope is 3; passes through $(5, -1)$
31. slope is $-\frac{2}{3}$; passes through $(1, 1)$

32. slope is $\frac{1}{2}$; passes through $(-1, 1)$

33. slope is 0; passes through $(2, -2)$

34. slope is not defined; passes through $(2, -2)$

35. y-intercept is 3; slope is $\frac{1}{2}$

36. y-intercept is -2; slope is $\frac{3}{4}$

37. y-intercept is 0; slope is -2

38. y-intercept is -2; slope is 0

39. line is parallel to the graph of $3x - 7y + 8 = 0$ and passes through $(1, -2)$

40. line is parallel to the graph of $2x + y - 10 = 0$ and passes through the origin

41. line is parallel to the x-axis and passes through $(-2, -8)$

42. line is parallel to the y-axis and passes through $(0, 8)$

43. line is perpendicular to the graph of $y = 6x - 10$ and passes through $(5, -2)$

44. line is perpendicular to the graph of $3x - 5y + 7 = 0$ and passes through $(-1, 3)$

45. line is perpendicular to the graph of $x = 2$ and passes through $(5, -10)$

46. line is perpendicular to the graph of $y = -7$ and passes through the origin

Superset

47. Show that the three points $(1, 3)$, $(5, 1)$, and $(6, 3)$ are vertices of a right triangle. (Use slopes to show that two of the sides are perpendicular.)

48. Show that the points $(1, 4)$, $(2, 7)$, and $(-3, -8)$ lie on the same line. (Use slopes.)

49. An ant starts at the point $(0, 4)$ and walks 4 units along the line $y = 3x + 4$ in the first quadrant before stopping. Did the ant pass over either of the points $(1, 5)$ or $(1, 7)$? Explain.

50. Another ant starts at the origin and walks 3 units up a course parallel to the line $y = 3x + 4$ before stopping.
    (a) Use an equation to describe the line along which this ant is walking.
    (b) Did the ant pass over the point $(1, 3)$? Explain.
    (c) How far above the x-axis was the ant when it stopped?

51. A third ant starts at the point $(1, 2)$ when $t = 0$ and follows a path on the xy-plane such that its coordinates at any time t are given by the equations

$$x = 3t + 1$$
$$y = 2 - t$$

By solving each equation for t, and equating these expressions in x and y, determine a function of x that describes the ant's path. What is the domain of the function?

52. Follow the procedure outlined in exercise 51 to determine an equation in slope-intercept form that describes the graph given by each of the following pairs of equations.

    (a) $x = 2t - 3$
        $y = t + 4$
    (b) $x = t$
        $y = 5 - 6t$
    (c) $x = 3t + 2$
        $y = 2 - 3t$
    (d) $x = t^2 - 3$
        $y = t^2 + 4$

53. A teacher has given a test in which the highest grade was 75, and the lowest grade was 40. Determine a linear function, $y = f(x)$, that can be used to distribute the test scores so that the grade of 40 becomes 60 and 75 becomes 100.

54. We say that y *varies directly* with x provided there exists some nonzero constant k such that $y = kx$. Suppose V varies directly with t, and when $t = \frac{1}{2}$, $V = \pi$. For what value of t is V equal to 1?

3.5 The Quadratic Function

Goal A *to analyze the graphs of functions described by equations of the form $f(x) = a(x - h)^2 + k$*

We have seen that the graph of $f(x) = a(x - h)^2 + k$ is a parabola produced by a series of transformations of $f(x) = x^2$. Let us review this method by graphing $f(x) = -2(x - 1)^2 + 3$.

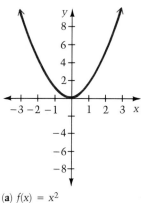

(a) $f(x) = x^2$

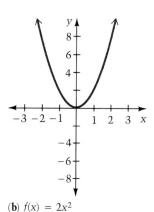

(b) $f(x) = 2x^2$
A vertical stretch of (a)

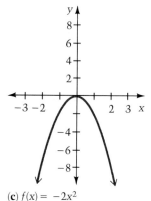

(c) $f(x) = -2x^2$
A reflection of (b) across the x-axis

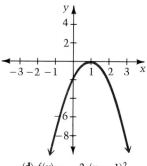

(d) $f(x) = -2(x - 1)^2$
a horizontal translation of (c)

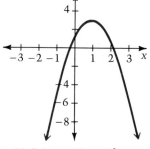

(e) $f(x) = -2(x - 1)^2 + 3$
a vertical translation of (d)

Figure 3.32

Let us examine the graph of

$$f(x) = -2(x - 1)^2 + 3$$

in detail. It is a parabola that opens downward because the coefficient of $(x - 1)^2$ is negative. The graph has its highest point at $(1, 3)$; thus the greatest function value is 3. The graph is symmetric with respect to the line $x = 1$.

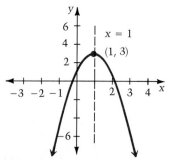

Figure 3.33 $y = -2(x - 1)^2 + 3$.

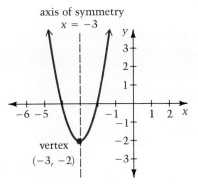

Figure 3.34 $g(x) = 2(x + 3)^2 - 2$.

Now let us consider the graph of the function $g(x) = 2(x + 3)^2 - 2$. Again, we have a parabola, but this graph opens upward because the coefficient of $(x + 3)^2$ is positive. The parabola has its lowest point at $(-3, -2)$; therefore, the least function value is -2. The graph is symmetric with respect to the line $x = -3$.

In general, parabolas have either a high or low point, called the **vertex**. Passing through the vertex is a line of symmetry called the **axis of symmetry.** This information is easily obtained from the **vertex form of a parabola,** which is an equation of the form

$$f(x) = a(x - h)^2 + k$$

■ A negative sign must precede h and a positive sign must precede k.

From this equation we can deduce some features of the parabola:

The graph opens $\begin{cases} \text{upward if } a > 0, \\ \text{downward if } a < 0. \end{cases}$

The graph is $\begin{cases} \text{wider than } f(x) = x^2 \text{ if } |a| < 1, \\ \text{narrower than } f(x) = x^2 \text{ if } |a| > 1. \end{cases}$

The vertex (h, k) is $\begin{cases} \text{a high point if the parabola opens downward,} \\ \text{a low point if the parabola opens upward.} \end{cases}$

The axis of symmetry is the line $x = h$.

Example 1 Write each of the following functions in vertex form. Find the vertex, the axis of symmetry and the direction in which the parabola opens.

(a) $y = -3(x - 2)^2 - 1$ (b) $y = \frac{1}{2}(x + 2)^2 - 3$

Solution (a) The vertex form is $y = -3(x - 2)^2 + (-1)$.
The vertex is $(2, -1)$.   ■ Here, $h = 2$, $k = -1$.
The axis of symmetry is $x = 2$.   ■ The axis is $x = h$.
The parabola opens downward.   ■ Here, $a = -3$; that is, $a < 0$.

(b) The vertex form is $y = \frac{1}{2}(x - (-2))^2 + (-3)$.
The vertex is $(-2, -3)$.   ■ Here, $h = -2$, $k = -3$.
The axis of symmetry is $x = -2$.
The parabola opens upward.   ■ Here, $a = \frac{1}{2}$; that is, $a > 0$.

Goal B *to graph functions of the form* $f(x) = ax^2 + bx + c$

The vertex form $y = a(x - h)^2 + k$ is one way to write an equation of a parabola. Expanding $(x - h)^2$ and simplifying yields an equivalent equation known as the *general form of the quadratic function:*

$$f(x) = a(x - h)^2 + k$$
$$= a(x^2 - 2hx + h^2) + k$$
$$= ax^2 - 2ahx + ah^2 + k \qquad \blacksquare \text{ Replace } -2ah \text{ with } b \text{ and } (ah^2 + k) \text{ with } c$$
$$f(x) = ax^2 + bx + c$$

Definition

A function of the form $f(x) = ax^2 + bx + c$, with a, b, and c real numbers and $a \neq 0$, is called a **quadratic function.** The graph of a quadratic function is a parabola. We call $f(x) = ax^2 + bx + c$ the **general form of the quadratic function.**

When graphing a function, it is useful to determine the **x-intercepts,** that is, the values of x where $f(x) = 0$. Since these are the values of x where $y = 0$, the x-intercepts tell us where the graph intersects the x-axis. For example, the graph of the function $y = \frac{1}{2}(x + 2)^2 - 3$ has two x-intercepts. To find the x-intercepts of a parabola, solve $ax^2 + bx + c = 0$.

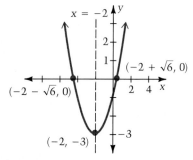

Figure 3.35 $y = \frac{1}{2}(x + 2)^2 - 3$. The x-intercepts are $-2 \pm \sqrt{6}$.

Example 2 Find the x-intercepts of the graph of the following quadratic functions.

(a) $f(x) = 2x^2 - x - 6$ (b) $g(x) = x^2 - x - 10$

Solution (a) Factor and apply the Principle of Zero Products.

$$2x^2 - x - 6 = 0$$
$$(2x + 3)(x - 2) = 0$$
$$(2x + 3) = 0 \quad | \quad (x - 2) = 0$$
$$x = -\frac{3}{2} \quad | \quad x = 2$$

The x-intercepts are $x = -\frac{3}{2}$ and 2.

(b) Use the quadratic formula.

$$x^2 - x - 10 = 0$$
$$x = \frac{-(-1) \pm \sqrt{(-1)^2 - 4(1)(-10)}}{2(1)}$$
$$x = \frac{1 \pm \sqrt{41}}{2}$$

The x-intercepts are

$$x = \frac{1 + \sqrt{41}}{2} \text{ and } \frac{1 - \sqrt{41}}{2}$$

Suppose we need to graph the function $f(x) = 2x^2 - 4x - 6$. We know the graph is a parabola. It would be easier to sketch the graph if the equation were in vertex form. A method called **completing the square,** which we used in Chapter 2 to factor quadratic expressions, can be used to convert a quadratic equation from general form to vertex form.

$$f(x) = 2x^2 - 4x - 6$$
$$= 2(x^2 - 2x + \square) - 6$$

- First, factor the coefficient of x^2 out of the x^2- and x-terms. A space is left for the number that makes the expression in parentheses a perfect square.

$$= 2(x^2 - 2x + 1) - 6 - 2 \cdot 1$$

the square of $\left(\frac{1}{2} \cdot (-2)\right)$

- In the space write the square of half the coefficient of the x-term. Since we have added $2 \cdot 1$, we must subtract $2 \cdot 1$.

$$= 2(x - 1)^2 - 8$$

- This is the vertex form. The vertex is $(1, -8)$.

Example 3 Graph the function $y = 2x^2 - 4x - 6$. Label the vertex, the axis of symmetry, x-intercepts, and two other points.

Solution We have just found the vertex form of this parabola is $f(x) = 2(x - 1)^2 - 8$. Thus, the vertex is $(1, -8)$ and the axis of symmetry is $x = 1$. Since $a = 2$, that is, $a > 0$, the parabola opens upward. To find the x-intercepts, set $2x^2 - 4x - 6$ equal to zero. Solving $2x^2 - 4x - 6 = 0$, we get $x = -1$ and $x = 3$. The x-intercepts are $x = -1$ and 3.

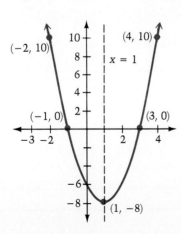

- Finally, we find two more points: Select points where $x = -2$ and 4:

$$x = -2 \quad y = 2(-2)^2 - 4(-2) - 6 = 10$$
$$x = 4 \quad y = 2(4)^2 - 4(4) - 6 = 10$$

Example 4 Graph the function $y = 4x^2 - 16x + 7$. Show the vertex, the axis of symmetry, x-intercepts and two other points.

Solution (1) First, find the x-intercepts.

$$x = \frac{-(-16) \pm \sqrt{(-16)^2 - 4(4)(7)}}{2(4)}$$ ■ Use the quadratic formula.

$$x = \frac{7}{2} \quad \text{and} \quad x = \frac{1}{2}$$ ■ The x-intercepts are $\frac{7}{2}$ and $\frac{1}{2}$.

(2) Find the vertex form.

$$y = 4(x^2 - 4x + \square) + 7$$ ■ Complete the square.
$$y = 4(x^2 - 4x + 2^2) + 7 - 4(2^2)$$
$$y = 4(x - 2)^2 - 9$$ ■ Thus, $h = 2$ and $k = -9$.

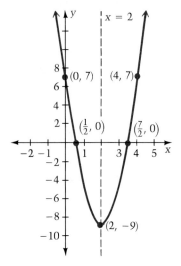

■ The vertex is $(2, -9)$ and the axis of symmetry is $x = 2$.
■ Since $a = 4$, that is, $a > 0$, the parabola opens upward.
■ Finally, we find two more points: Select points where $x = 0$ and 4.
$x = 0 \quad y = 4(0)^2 - 16(0) + 7 = 7$
$x = 4 \quad y = 4(4)^2 - 16(4) + 7 = 7$
■ The two other points are $(0, 7)$ and $(4, 7)$.

Exercise Set 3.5   ▭ Calculator Exercises in the Appendix

Goal A

In exercises 1–18, write the given equation in vertex form. Find the vertex, the axis of symmetry, and the direction in which the parabola opens.

1. $y = 3(x - 1)^2 + 4$
2. $y = -2(x + 3)^2 - 4$
3. $y = -2(x + 6)^2$
4. $y = 5(x - 7)^2$
5. $y = 2(x - 3)^2$
6. $y = -3(x + 7)^2$
7. $y = -(x + 5)^2 - 1$
8. $y = -(x - 2)^2 + 3$
9. $y - 8 = (x - 3)^2 + 2$
10. $y + 7 = 3(x + 4)^2 + 5$
11. $3y = -(x + 1)^2 - 5$

12. $-2y = -(x-4)^2 - 7$

13. $3(y+1) = (x-9)^2$

14. $4(y-6) = (x+2)^2$

15. $7y = -2(x+6)^2$

16. $10y = 3(x+1)^2$

17. $y = x^2 + 4$

18. $2y = 3x^2 - 8$

Goal B

For exercises 19–36, find the *x*-intercepts of the parabolas in exercises 1–18.

In exercises 37–50, graph the given function. Label the vertex, the *x*-intercepts, the axis of symmetry, and two other points.

37. $f(x) = 5x^2 - 3$

38. $g(x) = 4 - 2x^2$

39. $y = x^2 - 4x - 5$

40. $y = x^2 + 6x + 6$

41. $y = 4x - x^2 - 4$

42. $y = 2 - x^2 - x$

43. $y = x^2 - 6x + 9$

44. $y = x^2 + 4x + 4$

45. $f(x) = 4x^2 - 20x + 25$

46. $f(x) = 9x^2 + 48x + 64$

47. $g(x) = 4x^2 - 16x + 15$

48. $f(x) = 4x^2 + 10x + 21$

49. $y = -2x^2 + 20x - 54$

50. $y = 3x^2 - 12x + 20$

Superset

If a parabola opens upward, the *y*-coordinate of the vertex is called the **minimum function value.** If a parabola opens downward, the *y*-coordinate of the vertex is called the **maximum function value.**

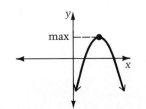
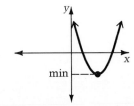

In exercises 51–54, determine the maximum or minimum function value. State whether this value is a maximum or minimum.

51. $y = -3(x+1)^2$

52. $y = 5(x-2)^2$

53. $y = 5 + 4x - x^2$

54. $y = 4x - x^2$

55. Determine b such that the quadratic function $f(x) = x^2 - bx + 6$ has exactly one *x*-intercept.

56. The quadratic function $h(x) = 2x^2 - 3x + C$ has 5 as one of its *x*-intercepts. Determine the value of C and the other zero.

57. We say that *y* varies directly with x^2 provided there exists some nonzero constant *k* such that $y = kx^2$. Suppose *d* varies directly with t^2 and that *d* is 5 when $t = 3$. Find *d* when $t = 5$.

58. The height $h(t)$ of a projectile shot upwards with an initial velocity of 352 ft/sec can be described as a function of time t:

$$h(t) = 352t - 16t^2.$$

(a) By finding the *y*-coordinate of the vertex, determine the maximum height of the projectile.

(b) When does the projectile achieve its maximum height?

(c) When does the projectile return to the ground?

59. The sum of two numbers is 12.

(a) Describe the two numbers using the variable *x*.

(b) Describe the product of the two numbers as a function of *x*.

(c) Determine the two values that make the product as large as possible.

60. The perimeter of a rectangle is 60 ft.

(a) Describe the area of the rectangle as a function of one variable.

(b) Determine the dimensions of the rectangle that produce the largest area, and state the value of this maximum area.

3.6 Composite Functions

Goal A *to evaluate composite functions*

A function may be viewed as a number processor that takes a domain value as input and produces a range value as output. For example, we can treat the functions $f(x) = x^2 + 4$ and $g(x) = 3x - 1$ as processors.

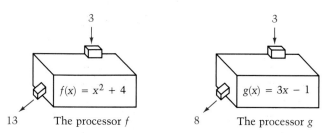

Figure 3.36

The function f takes an x-value from the domain as input, squares it and adds four to produce the function value $y = f(x)$. The function g takes an x-value, multiplies it by three and subtracts one to produce the function value $y = g(x)$.

The functions g and f can be combined to generate the *composite function* $f \circ g$, read "f circle g." This is illustrated below.

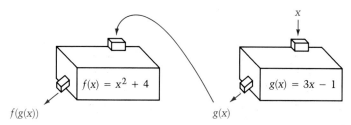

Figure 3.37

The function $f \circ g$ is the composition of g followed by f. First, x is processed by g and then the output $g(x)$ becomes the input for f.

Definition

Let f and g be functions with the range of g contained in the domain of f. For each x in the domain of g, the **composite function** $f \circ g$, is defined as

$$(f \circ g)(x) = f(g(x)).$$

How does the $f \circ g$ processor work? The domain value x is first processed by g to produce $g(x)$. Then $g(x)$ is processed by f to obtain $f(g(x))$. For example, for the functions f and g shown at the left, we can find $(f \circ g)(5)$ as follows.

$$(f \circ g)(5) = f(g(5)) \qquad \blacksquare \text{ First, we evaluate } g(5).$$
$$= f(14) \qquad \blacksquare \; g(5) = 3(5) - 1 = 14.$$
$$= 200$$

Now consider $(g \circ f)(5)$.

$$(g \circ f)(5) = g(f(5)) \qquad \blacksquare \text{ First, evaluate } f(5): f(5) = 5^2 + 4 = 29.$$
$$= g(29)$$
$$= 86$$

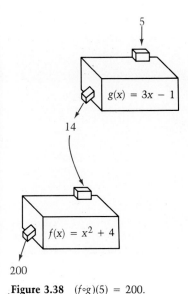

Figure 3.38 $(f \circ g)(5) = 200.$

Notice that $(f \circ g)(5) = 200$, but $(g \circ f)(5) = 86$. In general, $f \circ g$ and $g \circ f$ are not the same function. The order of the functions in a composition is important.

In the definition of $f \circ g$, it is necessary that the range of g be contained in the domain of f so that every output of g is acceptable input for f. For example, suppose the $f(x) = \sqrt{x}$ and $g(x) = x - 10$. To determine $(f \circ g)(3)$, we write $f(g(3)) = f(-7) = \sqrt{-7}$. But the f processor, $\sqrt{}$, cannot accept -7 as input. We say that $(f \circ g)(3)$ is not defined.

Example 1 Suppose $f(x) = 2x^2 - 3$, $g(x) = |x| - 5$, and $h(x) = \sqrt{x}$. Evaluate

(a) $(f \circ g \circ h)(4)$ (b) $(h \circ f)(1)$

Solution

(a) $(f \circ g \circ h)(4) = f(g(h(4)))$ $\blacksquare$ First, we evaluate $h(4)$.
$\qquad\qquad\qquad\quad = f(g(2)) \qquad \blacksquare \; h(4) = \sqrt{4} = 2.$
$\qquad\qquad\qquad\quad = f(-3) \qquad \blacksquare \; g(2) = |2| - 5 = -3.$
$\qquad\qquad\qquad\quad = 15 \qquad\quad\; \blacksquare \; f(-3) = 2(-3)^2 - 3 = 15.$

(b) $(h \circ f)(1) = h(f(1))$ $\blacksquare$ First evaluate $f(1)$.
$\qquad\qquad\quad = h(-1) \qquad \blacksquare \; f(1) = 2(1)^2 - 3 = -1.$
$\qquad\qquad\quad = \sqrt{-1}$

$(h \circ f)(1)$ is not defined because $f(1) = -1$ is not in the domain of h.

Goal B *to find an equation for the composite function $f \circ g$ given equations for f and g*

Suppose we have two functions f and g that are described by the equations $f(x) = 3x^2 + 4x$ and $g(x) = 2x - 1$. To describe $f \circ g$, we determine $(f \circ g)(x)$. (Treat x as you would any number.)

$$\begin{aligned}(f \circ g)(x) &= f(g(x)) \\ &= f(2x - 1) \\ &= 3(2x - 1)^2 + 4(2x - 1) \quad \blacksquare\ f(\Box) = 3(\Box)^2 + 4(\Box) \\ &= 3(4x^2 - 4x + 1) + 4(2x - 1) \\ &= 12x^2 - 4x - 1\end{aligned}$$

Example 2 If $f(x) = 2x + 3$, $g(x) = x^2 - 1$ and $h(x) = \dfrac{1}{x}$, determine

(a) $(h \circ g)(x)$ (b) $(f \circ g)(x)$

Solution (a) $(h \circ g)(x) = h(g(x)) = h(x^2 - 1) = \dfrac{1}{x^2 - 1}$ $\blacksquare\ h(\Box) = \dfrac{1}{(\Box)}$.

(b) $(f \circ g)(x) = f(g(x)) = f(x^2 - 1) = 2(x^2 - 1) + 3 = 2x^2 + 1$ $\blacksquare\ f(\Box) = 2(\Box) + 3$.

Example 3 Suppose $f(x) = 2x + 5$ and $g(x) = \sqrt{x}$. Determine the domain of $(g \circ f)(x)$.

Solution Since $(g \circ f)(x) = g(f(x))$, we need to find those values of x for which $f(x)$ is in the domain of g. We know the domain of g is the set of nonnegative real numbers; thus we need to determine values of x such that $f(x)$ is nonnegative.

$$\begin{aligned} 2x + 5 &\geq 0 \quad\ \blacksquare\ \text{Recall that } f(x) = 2x + 5. \\ 2x &\geq -5 \\ x &\geq -\dfrac{5}{2}\end{aligned}$$

Thus, for all $x \geq -\tfrac{5}{2}$, $f(x)$ is in the domain of g, and $(g \circ f)(x)$ is defined. The domain of $g \circ f$ is $\left[-\dfrac{5}{2}, \infty\right)$.

3 / Graphs and Functions

Goal C *to describe a given function as a composition of other functions*

Given the equations for two functions f and g, we can determine the equation for the composite function $f \circ g$. Let us now consider the reverse process. Suppose we know that $F(x) = (x - 7)^2$. How can we describe $F(x)$ as a composition?

To answer this question, notice that there are two processes operating when we evaluate $F(6)$.

$$F(6) = (6 - 7)^2 \quad \blacksquare \text{ First process: subtract 7.}$$
$$= (-1)^2 \quad \blacksquare \text{ Second process: square.}$$
$$= 1$$

Each process can be represented as a function.

$$f(x) = x - 7 \quad \blacksquare \text{ First function: the function that subtracts 7 from a number}$$
$$g(x) = x^2 \quad \blacksquare \text{ Second function: the function that squares a number}$$

We can therefore write $F(x)$ as the composition of the two functions f and g.

$$F(x) = (g \circ f)(x)$$

first function
second function

When we build the composite function $F(x)$, the first function is placed closest to x because it will process x first. That is, a composition is built from right to left.

To find simple functions used to build a composite function, you must discover a sequence of simple processes that produce the composite function.

Example 4   Express $G(x) = x^3 + 15$ as a composite function.

Solution   Since G cubes and then adds 15, we may write

$$G(x) = (g \circ f)(x)$$

where $f(x) = x^3$ and $g(x) = x + 15$.

Example 5 Consider the following simple functions.

$$f(x) = x + 2 \qquad g(x) = |x| \qquad h(x) = x^2 \qquad k(x) = \frac{1}{x}$$

Write each of the following functions as a composition of two or more of these simple functions.

(a) $F(x) = |x| + 2$ (b) $G(x) = |x + 2|$ (c) $H(x) = \dfrac{1}{x^2 + 2}$

Solution (a) To determine $F(x)$, first take the absolute value of x: $\quad g(x) = |x|$
Then add 2: $\quad f(g(x)) = |x| + 2$
Thus, $F(x) = (f \circ g)(x)$.

(b) To determine $G(x)$, first add 2: $\quad f(x) = x + 2$
Then take the absolute value: $\quad g(f(x)) = |x + 2|$
Thus, $G(x) = (g \circ f)(x)$.

(c) To determine $H(x)$, first form the square: $\quad h(x) = x^2$
Then add 2: $\quad f(h(x)) = x^2 + 2$

Finally, take the reciprocal: $\quad k(f(h(x))) = \dfrac{1}{x^2 + 2}$

Thus, $H(x) = (k \circ f \circ h)(x)$.

Exercise Set 3.6

Goal A

In exercises 1–6, evaluate the given expression for $f(x) = 2x - 1$ and $g(x) = |x|$.

1. $(f \circ g)(-2)$
2. $(g \circ f)(-2)$
3. $(g \circ f)(0)$
4. $(f \circ g)(0)$
5. $(f \circ g)\left(\dfrac{1}{3}\right)$
6. $(g \circ f)\left(\dfrac{1}{3}\right)$

In exercises 7–12, evaluate the given expression for $f(x) = 3$ and $g(x) = 10x - 7$.

7. $(g \circ f)(5)$
8. $(f \circ g)(5)$
9. $(g \circ f)(-5)$
10. $(f \circ g)(-5)$
11. $(f \circ g)(0)$
12. $(g \circ f)(0)$

In exercises 13–26, evaluate the given expression for $f(x) = \dfrac{1}{x - 4}$, $g(x) = 4x^2 - 8$, and $h(x) = \sqrt{x}$.

13. $(g \circ f)(-6)$
14. $(g \circ f)(-2)$
15. $(g \circ f)(0)$
16. $(f \circ g)(0)$
17. $(g \circ f)(-4)$
18. $(f \circ g)\left(\dfrac{1}{2}\right)$
19. $(f \circ f)(0)$
20. $(f \circ g)(-1)$
21. $(f \circ f)(2)$
22. $(g \circ g)(-1)$
23. $(g \circ h)(4)$
24. $(h \circ g)(1)$
25. $(g \circ h)(0)$
26. $(h \circ g)(0)$

Goal B

In exercises 27–36, find an equation that describes $(f \circ g)(x)$ and state the domain of the composite function.

27. $f(x) = x^2 + 1$; $g(x) = x - 3$
28. $f(x) = x + 4$; $g(x) = 2x^2 - 1$
29. $f(x) = |x|$; $g(x) = \dfrac{1}{x-3}$
30. $f(x) = \dfrac{1}{x+3}$; $g(x) = |x+1|$
31. $f(x) = 1 - \sqrt{x}$; $g(x) = x - 5$
32. $f(x) = \sqrt{x-1}$; $g(x) = 5 - x$
33. $f(x) = \dfrac{1}{x+4}$; $g(x) = x^2 + 5$
34. $f(x) = \dfrac{1}{x-1}$; $g(x) = x^2 + 5$
35. $f(x) = \dfrac{1}{x}$; $g(x) = \sqrt{x}$
36. $f(x) = \dfrac{1}{x+1}$; $g(x) = \sqrt{x-3}$

Goal C

In exercises 37–44, write the given function $y = F(x)$ as a composition of two or more of the following:

$f(x) = x + 2$, $j(x) = |x|$, $g(x) = x^2$,
$k(x) = \dfrac{3}{x}$, $h(x) = \sqrt{x}$

37. $F(x) = |x| + 2$
38. $F(x) = \dfrac{3}{|x|}$
39. $F(x) = \sqrt{\dfrac{3}{x}}$
40. $F(x) = \sqrt{x+2}$
41. $F(x) = \dfrac{3}{|x| + 2}$
42. $F(x) = \sqrt{\dfrac{3}{x+2}}$
43. $F(x) = \dfrac{3}{x+4}$
44. $F(x) = x^4 + 2$

Superset

45. Suppose $y = 3x - 2$, $x = 1 - z$, and $z = 5t + 2$. Write y as a function of t.

46. Suppose $x = 3t + 2$, $z = 6 - t$, and $z = 2y + 1$. Write y as a function of x.

47. A physiologist monitored a person moving a heavy object. The subject exerted a force for a period of 50 seconds which caused an increase in pulse rate. Functional relationships observed in this test are shown below. Use the graphs to determine how much time elapsed if the pulse rate reached 120.

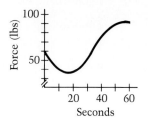
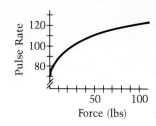

In exercises 48–52, the functions f and g are given together with restricted domains. In each case, find an equation for $(f \circ g)(x)$, and state the domain of the composite function.

48. $f(x) = 3x - 1$; domain: $[-1, 1]$
    $g(x) = x - 5$; domain: $[0, 5]$
49. $f(x) = 2x + 1$; domain: $[0, 2]$
    $g(x) = x + 1$; domain: $[-2, 2]$
50. $f(x) = \sqrt{x}$; domain: $[0, +\infty)$
    $g(x) = x - 4$; domain: $[0, 10]$
51. $f(x) = \sqrt{x}$; domain: $[0, +\infty)$
    $g(x) = x + 3$; domain: $[-5, 0]$
52. $f(x) = x^2$; domain: $[-10, 10]$
    $g(x) = \sqrt{-x}$; domain: $(-\infty, 0]$

Recall that a function f is even if $f(-x) = f(x)$ and odd if $f(-x) = -f(x)$.

53. Show that the composition of two odd functions is an odd function.

54. Show that the composition of an odd and even function is an even function.

Chapter Review & Test

Chapter Review

3.1 Equations and Graphs (pp. 100–107)

We graph ordered pairs of real numbers on two real number lines, called *axes*. The axes separate the plane into four *quadrants*. The horizontal axis is named for the independent variable (i.e., *x-axis*), and the vertical axis is named for the dependent variable (i.e., *y-axis*). The axes meet at the *origin*, and the entire system is the *xy-plane*. (p. 100)

Equation of a *straight line:* $y = mx + b$, where m and b are real numbers.

Equation of a *parabola:* $y = ax^2 + bx + c$, where a, b, and c are real numbers and $a \neq 0$.

Equation of a *V-shaped graph:* $y = a|x| + k$, where a and k are real numbers and $a \neq 0$. (pp. 101–102)

A graph is
- *symmetric with respect to the x-axis,* if for each point P on the graph, the reflection of P across the x-axis is also on the graph. If the graph is described by an equation, then replacing y with $-y$ produces an equation equivalent to the original. (p. 104)
- *symmetric with respect to the y-axis,* if for each point P on the graph, the reflection of P across the y-axis is also on the graph. If the graph is described by an equation, then replacing x with $-x$ produces an equation equivalent to the original. (p. 105)
- *symmetric with respect to the origin,* if for each point P on the graph, the reflection of P through the origin is also on the graph. If the graph is described by an equation, then replacing x with $-x$ and y with $-y$ produces an equation equivalent to the original. (p. 105)

3.2 Introduction to Functions (pp. 108–113)

A *function* is a set of ordered pairs, with no two of them having the same first coordinate and different second coordinate. The set of all first coordinates is the *domain*, and the set of all second coordinates is the *range*. In general, any set of ordered pairs is a *relation*. (pp. 108–109)

For the equation $y = 2x + 5$, we call x the *independent variable* and y the *dependent variable* because we are free to choose any value for x, but the value of y is dependent upon the value of x we choose. (p. 110)

The *Vertical Line Test* is a quick way to determine whether a graph is a function. (p. 111)

3.3 Transformations and Graphing (pp. 114–122)

There are three general types of transformations: *translations, reflections,* and *changes in shape.* (pp. 114–115)

Transformation	Change in Equation	Effect on Graph
Horizontal translation	Replace x with $x - h$	If $h > 0$, graph moves $|h|$ units to the right. If $h < 0$, graph moves $|h|$ units to the left.
Vertical translation	Replace y with $y - k$	If $k > 0$, graph moves $|k|$ units up. If $k < 0$, graph moves $|k|$ units down.
Reflection across y-axis	Replace x with $-x$	Graph is reflected across y-axis.
Reflection across x-axis	Replace y with $-y$	Graph is reflected across x-axis.
Vertical change in shape	$y = f(x)$ becomes $y = b \cdot f(x)$	If $b > 1$, graph is stretched away from x-axis. If $0 < b < 1$, graph is shrunk toward x-axis.
Horizontal change in shape	$y = f(x)$ becomes $y = f(ax)$	If $a > 1$, graph is shrunk toward y-axis. If $0 < a < 1$, graph is stretched away from y-axis.

3.4 The Linear Function (pp. 123–129)

An equation of the form

$$y = mx + b$$

describes a *linear function.* Its graph is a straight line. The *slope m* of any nonvertical line is given by the formula

$$m = \frac{y_2 - y_1}{x_2 - x_1},$$

where (x_1, y_1) and (x_2, y_2) are any two different points on the line. (p. 123)

There are three different types of straight line equations (pp. 125–127):

Point-slope form *Slope-intercept form* *General form*
$y - y_1 = m(x - x_1)$ $y = mx + b$ $Ax + By + C = 0$

If two nonvertical lines are parallel, then their slopes are equal. If two nonvertical lines are perpendicular, then their slopes are negative reciprocals of each other. (p. 127)

3.5 The Quadratic Function (pp. 130–135)

In general, parabolas have either a high or low point, called the *vertex*. The *axis of symmetry* is a line of symmetry that passes through the vertex. An equation of the form $f(x) = a(x - h)^2 + k$ is called the *vertex form of a parabola*. (p. 131)

The equation $f(x) = ax^2 + bx + c$, where a, b, and c are real numbers and $a \neq 0$, is called the *general form of the quadratic function*. Its graph is a parabola. (p. 132)   The method called *completing the square* can be used to convert a quadratic equation from general form into vertex form. (p. 133)

3.6 Composite Functions (p. 136–141)

Let f and g be functions, with the range of g contained in the domain of f. For each x in the domain of g, the *composite function* $f \circ g$ is defined as $(f \circ g)(x) = f(g(x))$. (p. 136)

The order of the functions in a composition is important. In general, $f \circ g$ and $g \circ f$ are not the same function. (p. 137)   When building a composite function, place the first function closest to x because it will process x first. A composite function is built from right to left. (p. 139)

Chapter Test

3.1A Sketch the graph of the function described by each of the following equations.

    1. $y = 10x + 3$ **2.** $f(x) = x^2 - 3x - 8$ **3.** $g(x) = 11 - 4|x|$

3.1B Reflect each of the following graphs (a) across the x-axis, (b) across the y-axis and (c) through the origin.

4.
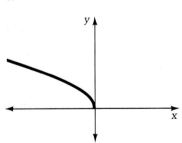

5.
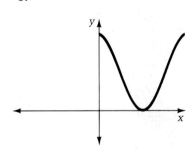

6.
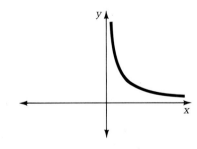

3.1C Test the following relations for each of the three types of symmetry.

    7. $x^2 - y^2 + 8 = 0$ **8.** $y^4 + x^3 = 6$ **9.** $x^3 y = -5$

3.2A Describe each of the following functions by a set of ordered pairs. State the domain and range of each function.

10.

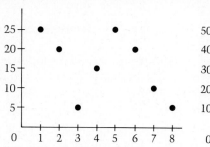

11.

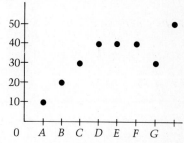

3.2B Determine $f(-2)$, $f(0)$, $f(h^2)$, and $f(b-1)$ for the following functions.

12. $f(x) = -x^2 - 5x + 9$ 13. $f(x) = 4 - x^3$ 14. $f(x) = \frac{1}{3}x - 5$

3.3A Sketch the graph of each of the following functions by translating the graph of $y = x^2$, $y = |x|$, or $y = \frac{1}{x}$.

15. $y = (x - 7)^2$ 16. $y + 2 = |x|$ 17. $y = \frac{1}{x} + 5$

3.3B Sketch the graph of the second function by reflecting the graph of the first function.

18. $y = |x - 3|$; $y = |-3 - x|$ 19. $y = (x + 1)^2$; $y = -(x + 1)^2$

3.3C Sketch the graph of each of the following functions by transforming the graph of $y = f(x)$, shown at the right.

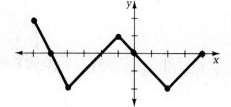

20. $y = f(3x)$ 21. $y = 4f(x)$ 22. $y = \frac{1}{3}f(x)$

3.3D Use transformations to sketch the graph of each of the following functions.

23. $y = -(x + 2)^2 + 1$ 24. $2y = |x - 1| - 2$

3.4A Find the slope of the line determined by each of the following ordered pairs. Then sketch the line.

25. $\left(\frac{1}{6}, \frac{1}{2}\right)$ and $\left(\frac{2}{5}, -\frac{2}{3}\right)$ 26. $(3.4, 0.3)$ and $(-1.2, -2)$

3.4B For each of the following, determine an equation of the line satisfying the stated conditions. Write your answer in (a) point-slope form, (b) slope-intercept form and (c) general form.

27. The slope of the line is -1; it passes through $(-2, 4)$.

28. The line is perpendicular to $y = -5$; it passes through $(6, -1)$.

3.5A Write each of the following equations in vertex form. Find the vertex, the axis of symmetry, and the direction in which the parabola opens.

29. $y = 6(x + 4)^2 + 9$ 30. $-2(y + 3) = (x - 5)^2$

3.5B Sketch the graph of each of the following functions. Label the vertex, the x-intercepts, the axis of symmetry, and two additional points.

31. $y = x^2 - 4x - 21$ 32. $f(x) = 9 - 4x^2$

3.6A Evaluate each of the following expressions for
$f(x) = 3x + 3$, $g(x) = |x| - 2$ and $h(x) = \dfrac{1}{x - 1}$.

33. $(f \circ g)(-2)$ 34. $(g \circ f)(0)$ 35. $(f \circ h)(10)$

3.6B For each of the following, find an equation that describes $(f \circ g)(x)$ and state the domain of the composite function.

36. $f(x) = x^2 - 11$; $g(x) = x + 4$

37. $f(x) = \dfrac{1}{2 - x}$; $g(x) = \sqrt{x + 6}$

3.6C Write each function as a composition of two or more of the following:
$f(x) = x - 3$, $g(x) = |x + 4|$, $h(x) = \sqrt{x}$, $j(x) = x^2$.

38. $F(x) = |x + 4|^2$ 39. $F(x) = (\sqrt{x} - 3)^2$

40. $F(x) = |\sqrt{x - 3} + 4|$

Superset

41. For the function $f(x) = 3x^2 + 2x$, find the following: $\dfrac{f(2 + h) - f(2)}{h}$.

42. Sketch the graph of the following function: $f(x) = \begin{cases} x, & \text{if } x < -3, \\ x^2 + 1, & \text{if } x > 0. \end{cases}$

43. The quadratic function $f(x) = 3x^2 - 5x + C$ has -2 as one of its x-intercepts. Determine the value of C and the other zero.

44. Find two numbers whose difference is 20 and whose product is as small as possible.

4 Polynomial Functions, Rational Functions, and Conic Sections

4.1 Important Features of Polynomial Functions

Goal A *to study nth degree polynomial functions*

In Chapter 3, we saw that the graph of a linear function, such as $f(x) = 2x - 7$, is a straight line, and the graph of a quadratic function, such as $g(x) = 3x^2 - 5x + 4$, is a parabola. Since $2x - 7$ and $3x^2 - 5x + 4$ are polynomials, the functions $f(x) = 2x - 7$ and $g(x) = 3x^2 - 5x + 4$ are called *polynomial functions.* Any function described by a polynomial is a polynomial function. Polynomial functions are classified by the degree of the polynomial.

Polynomial Function	Degree	Common Name
$f(x) = 2$	0	constant function
$g(x) = 2x - 7$	1	linear function
$y = 3x^2 - 5x + 4$	2	quadratic function
$y = -5x^3 + 4x^2 - 2x + 17$	3	cubic function
$p(x) = x^6 - 15x^2 + 10$	6	6th degree polynomial function
$q(x) = 4x^{11} + 2x^6 - 5x^2$	11	11th degree polynomial function

It is customary to write a polynomial so that the degrees of its terms are in descending order. A polynomial function written this way is said to be in **standard form.**

Definition
An ***n*th degree polynomial function in *x*** is a function that can be written in the form

$$f(x) = a_n x^n + a_{n-1} x^{n-1} + \cdots + a_2 x^2 + a_1 x + a_0$$

where $a_n \neq 0$, n is a nonnegative integer, and the coefficients a_n, $a_{n-1}, \ldots a_2, a_1, a_0$ are fixed real numbers.

The term of highest degree of a polynomial function is called the **leading term,** and the coefficient of the leading term is called the **leading coefficient.**

Example 1 Determine whether each of the following is a polynomial function. If so, write it in standard form and identify the leading term.

(a) $f(x) = 2x - 5x^3 + 4$ \qquad (b) $g(x) = x^2 + 3\sqrt{x}$

(c) $F(x) = 3x + \sqrt{2}x^4 - x^2$ \qquad (d) $G(x) = x^4 + 3x + \dfrac{2}{x}$

Solution (a) f is a polynomial function. Its standard form is

$$f(x) = -5x^3 + 2x + 4,$$

and the leading term is $-5x^3$.

(b) g is not a polynomial function. The exponents of the terms of a polynomial function must be nonnegative *integers*. Since $3\sqrt{x} = 3x^{1/2}$, its exponent does not meet this requirement.

(c) F is a polynomial function. Its standard form is

$$F(x) = \sqrt{2}x^4 - x^2 + 3x,$$

and the leading term is $\sqrt{2}x^4$.

(d) G is not a polynomial function. The exponents of the terms of a polynomial function must be *nonnegative* integers. Since $\frac{2}{x} = 2x^{-1}$, its exponent does not meet this requirement.

Goal B *to examine the graph of a polynomial function of the form* $f(x) = ax^n$

Now we shall explore some important features of polynomial functions and their graphs. Our goal is to introduce a few general techniques to help us graph a polynomial function.

Let us begin by considering polynomial functions described by monomials.

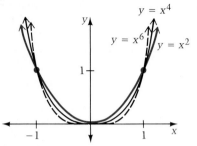

Figure 4.1 Graphs of the form $y = x^{\text{even}}$

$y = x^p$, where p is an even positive integer

Recall that the graph of $y = x^2$ is a parabola. Although the graphs of $y = x^4$ and $y = x^6$ look like parabolas, they are not. They are, however, symmetric with respect to the y-axis and never drop below the x-axis. (Why?) As the exponents increase, the graphs become flatter between -1 and 1 and steeper outside this interval. In all cases, as the x-values get further from zero, the y-values increase more rapidly.

$y = x^q$, where q is an odd positive integer greater than 1

The graphs of $y = x^3$, $y = x^5$, and $y = x^7$ are all symmetric with respect to the origin. (Why?) As the exponents increase, the graphs become flatter between -1 and 1 and steeper outside this interval. In all cases, as the x-values get further to the left of zero, the y-values decrease more rapidly, and as the x-values get further to the right of zero, the y-values increase more rapidly.

From our work with symmetry in Chapter 3, we know that the graph of $y = -x^n$, where n is *any* positive integer, is the reflection of $y = x^n$ across the x-axis.

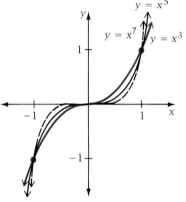

Figure 4.2 Graphs of the form $y = x^{\text{odd}}$

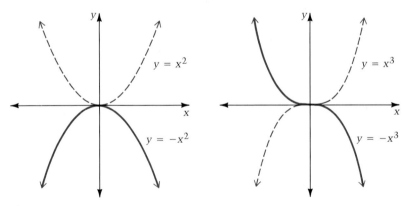

Figure 4.3

Thus, the graph of a polynomial function of the form $f(x) = ax^n$, where a is nonzero, and n is any positive integer greater than 1, will have one of the four shapes shown in Figure 4.4. The basic shape of the graph depends upon the sign of a and whether n is odd or even.

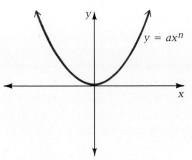
(a) Type I: *a* positive, *n* even

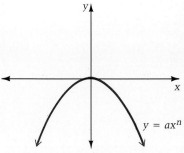
(b) Type II: *a* negative, *n* even

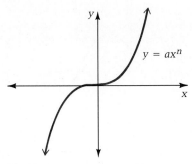
(c) Type III: *a* positive, *n* odd

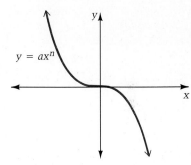
(d) Type IV: *a* negative, *n* odd

Figure 4.4

Example 2 Classify the graph of each polynomial function as one of the types above.
(a) $y = -3x^5$ (b) $f(x) = x^6$ (c) $g(x) = 5x^7$ (d) $y = -3x^4$

Solution (a) Type IV, since $a < 0$ and n is odd.   (b) Type I, since $a > 0$ and n is even.
(c) Type III, since $a > 0$ and n is odd.   (d) Type II, since $a < 0$ and n is even.

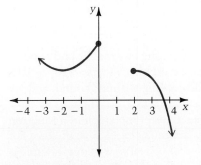

Figure 4.5 Not the graph of a polynomial function.

Goal C *to examine the behavior of a polynomial function for extreme values of x*

We now consider three important features of any polynomial function: its domain, the general shape of its graph, and its behavior for *x*-values to the extreme right or extreme left of 0.

I. Domain

The domain of any polynomial function is the set of all real numbers. The graph continues indefinitely to the right of 0 and to the left of 0 and has no gaps. The graph in Figure 4.5 could not be the graph of a polynomial function since it is not defined for real numbers between 0 and 2.

II. General Shape

We already know how to graph first degree polynomial functions (linear functions) and second degree polynomial functions (quadratic functions). For degrees higher than 2, the graphs of polynomial functions can be characterized crudely by saying that inside the extremes they are composed of smooth "waves," and for extreme values of x, they either rise or fall indefinitely. Furthermore, the graph is always smooth and unbroken: there are no sharp corners, and no jumps or holes of any kind.

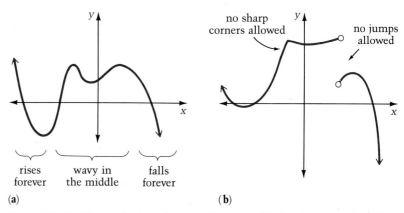

Figure 4.6 (a) The graph of a polynomial function. (b) Not the graph of a polynomial function.

III. Behavior for Extreme Values of x

For any function $y = f(x)$, it is always useful to determine what happens to the y-values as the x-values get extremely large or small. We now consider a way to determine this for polynomial functions. In order to simplify our discussion we introduce the notation $x \to +\infty$ and $x \to -\infty$.

The notation $x \to +\infty$ indicates that x is taking on larger and larger positive values *without bound*. For example, suppose x takes on the values 10,000, 20,000, 30,000, etc. (successive multiples of 10,000). The x-values are getting larger and larger and will eventually be larger than any number you can imagine. We say that x is increasing without bound, and we write $x \to +\infty$.

This notation $x \to -\infty$ indicates that x is taking on smaller and smaller negative values *without bound*. If x takes on the values -10, -100, $-1,000$, $-10,000$, etc., then we say x is decreasing without bound, and we write $x \to -\infty$.

We use this notation to describe the behavior of the graph of a function to the extreme right ($x \to \infty$) or to the extreme left ($x \to -\infty$) of zero. For example, the graph of the function $y = -x^5$ falls without bound as $x \to +\infty$, and it rises without bound as $x \to -\infty$. (See Figure 4.9.)

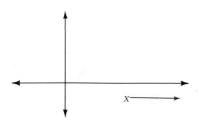

Figure 4.7 $x \to +\infty$

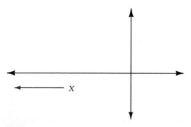

Figure 4.8 $x \to -\infty$

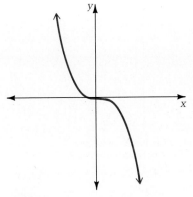

Figure 4.9 $y = -x^5$.

Saying that a graph falls without bound is equivalent to saying that the y-values are decreasing without bound. Thus, for the function $y = -x^5$, we can write

$$\text{as } x \to +\infty, \quad y \to -\infty.$$

Similarly, if a function rises without bound, this means that the y-values are increasing without bound. Thus, for the function $y = -x^5$ we can write

$$\text{as } x \to -\infty, \quad y \to +\infty.$$

To determine the behavior of the graph of a polynomial function to the extreme right or left of the origin, you need only look at the leading term of the polynomial.

First Rule of Polynomial Functions

Consider the polynomial function f, described by the equation

$$f(x) = a_n x^n + a_{n-1} x^{n-1} + \cdots + a_1 x + a_0.$$

As $x \to +\infty$, and as $x \to -\infty$, the graph of f behaves like the graph of $y = a_n x^n$.

For example, knowing what the graph of $y = 2x^3$ looks like towards the extreme left and extreme right tells us what the graph of $y = 2x^3 + 4x^2 - 5x + 13$ looks like towards those extremes.

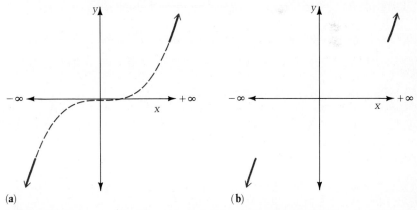

Figure 4.10 (a) $y = 2x^3$. (b) $y = 2x^3 + 4x^2 - 5x + 13$ towards the extremes.

We have seen that there are only four possible ways that a polynomial function of the form $y = ax^n$ ($n > 1$, $a \neq 0$) can behave for extreme values of x. Therefore, by the First Rule of Polynomial Functions, there are only four possible ways (Figure 4.11) that *any* polynomial function of degree greater than one can behave for extreme values of x.

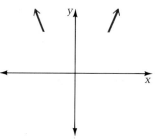
Leading coefficient is positive; degree of the polynomial is even.

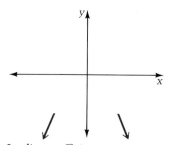
Leading coefficient is negative; degree of the polynomial is even.

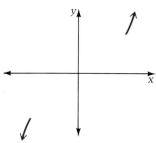
Leading coefficient is positive; degree of the polynomial is odd.

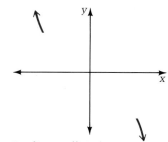
Leading coefficient is negative; degree of the polynomial is odd.

Figure 4.11

Example 3 For each polynomial function below, select the graph that best represents it.

(a) $f(x) = x^9 + 5x^2 + 2$ (b) $f(x) = 2x^8 - 3x^7 + 2$

(c) $f(x) = -x^8 + 5x^2 + 2$ (d) $f(x) = 2x^8 - 3x^9 + 2$

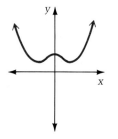
Graph No. 1

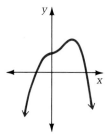
Graph No. 2

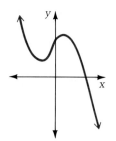
Graph No. 3

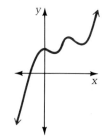
Graph No. 4

Solution
(a) No. 4 since the leading term is x^9. (b) No. 1 since the leading term is $2x^8$.
(c) No. 2 since the leading term is $-x^8$. (d) No. 3 since the leading term is $-3x^9$.

4 / Polynomial Functions and Rational Functions

Exercise Set 4.1

Goal A

In exercises 1–12, determine whether each of the following is a polynomial function. If so, write it in standard form and identify its leading term.

1. $f(x) = 2 + 7x - x^3$
2. $g(x) = 2x^2 - 5x^4 + 9x - 4$
3. $F(x) = 3x^2 - 2x\sqrt{x} + x$
4. $f(x) = 6x - 10x^2 + 3x^3 - x$
5. $H(x) = 3x^3 + 4x^2 - 2x^5 + 5x^2 + 1$
6. $h(x) = x^2 - \dfrac{3}{x} + 5$
7. $g(x) = 8x^2 - 4x + 2 - x^{-1}$
8. $F(x) = 1 - x + x^2 - x^4$
9. $f(x) = (3x + 2)^2 + (x + 2) + 5$
10. $g(x) = 6 - 2x - (5x + 2)^2$
11. $k(x) = 7x + 3x^3 - \sqrt{5}x^4 - 3$
12. $G(x) = x^2 - x + x^{1/3}$

Goal B

In exercises 13–28, classify the graph of each polynomial function as one of the types listed on page 151.

13. $y = 2x^3$
14. $y = -3x^2$
15. $y = -4x^6$
16. $y = -3x^5$
17. $y = 5x^4$
18. $y = 10x^7$
19. $y = -x^2$
20. $y = \dfrac{7}{2}x^{10}$
21. $5y = x^3$
22. $-2y = x^9$
23. $-2y = 3x^7$
24. $\dfrac{1}{3}y = 4x^8$
25. $y + 2x^3 = 0$
26. $3x^4 + y = 0$
27. $x^2 - 2y = 0$
28. $3y - x^4 = 0$

Goal C

In exercises 29–32, for each polynomial function, select the graph that best represents it.

29. $y = x^3 + 2x^2 + 1$
30. $y = 1 - 2x^2 - x^3$
31. $y = 1 - 2x^4$
32. $y = 2x^3 + x^4 + 1$

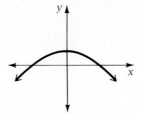

Graph No. 1

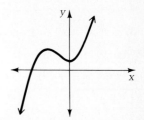

Graph No. 2

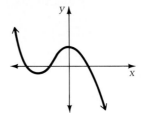

Graph No. 3

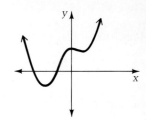

Graph No. 4

In exercises 33–36, for each polynomial function, select the graph that best represents it.

33. $y = 2x^5 + 3x^2 - 2$
34. $y = 2x^4 - x^2 - 2$
35. $y = x^3 - x^4 - 2$
36. $y = 2x^2 - 3x^7 - 2$

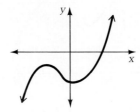

Graph No. 1

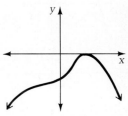

Graph No. 2

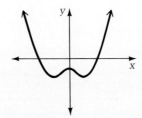

Graph No. 3

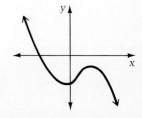

Graph No. 4

Superset

In exercises 37–46, use your knowledge of symmetry and transformations (Sections 3.1 and 3.3) to graph the given polynomial function.

37. $f(x) = -x^3 + 7$
38. $g(x) = -x^4 - 1$
39. $h(x) = 2 + (x - 1)^3$
40. $k(x) = 3 + 2(x + 1)^3$
41. $A(x) = 1 - (x + 2)^4$
42. $B(x) = 2(x + 3)^4 - 20$
43. $F(x) = \frac{1}{4}x^4 + 1$
44. $G(x) = 3x^3 + 2$
45. $f(x) = \frac{(x + 3)^3}{3}$
46. $g(x) = \frac{x^4 - 7}{2} - 2$

In exercises 47–52, determine from their graphs which of the functions could not be a polynomial function. Explain.

47.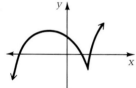
48.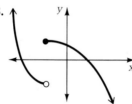
49.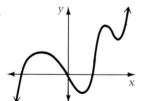
50.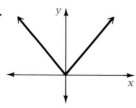
51. 52.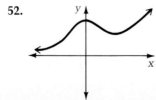

In exercises 53–56, select the graph that best represents the given polynomial function. (You should use the First Rule for Polynomial Functions. Also, determining the y-intercept will help.)

53. $f(x) = -x^3 + x^2 + x - 1$

Graph No. 1   Graph No. 2   Graph No. 3

54. $f(x) = x^3 - 6x^2 + 9x + 5$

Graph No. 1   Graph No. 2   Graph No. 3

55. $y = x^4 - 3x^3 + 3x^2 - x$

Graph No. 1   Graph No. 2   Graph No. 3

56. $f(x) = x^4 - x^3 - 4x^2 + 4x$

Graph No. 1   Graph No. 2   Graph No. 3

4.2 Graphing Polynomial Functions

Goal A *to use x-intercepts to graph a polynomial function*

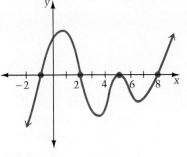

Figure 4.12

The function graphed in Figure 4.12 has x-intercepts at -1, 2, 5, and 8. These four x-values are called **zeros of the function,** because they produce y-values of 0. At a zero of the function, the graph can behave in one of two ways: either the graph crosses the x-axis (as it does at -1, 2, and 8) or the graph touches the x-axis, but does not cross it (as it does at 5).

Recall that a quadratic function $f(x) = ax^2 + bx + c$ can have 0, 1, or 2 real number zeros (which we shall refer to as real zeros) depending upon whether the equation $0 = ax^2 + bx + c$ has 0, 1, or 2 real number solutions.

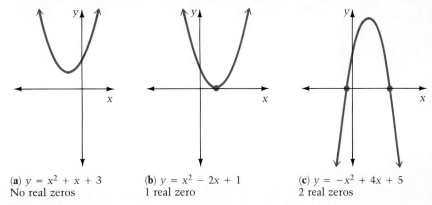

(a) $y = x^2 + x + 3$
No real zeros

(b) $y = x^2 - 2x + 1$
1 real zero

(c) $y = -x^2 + 4x + 5$
2 real zeros

Figure 4.13

A quadratic function has at most two zeros; therefore it cannot intersect the x-axis more than twice.

In general, an nth degree polynomial equation of the form

$$0 = a_n x^n + a_{n-1} x^{n-1} + \cdots + a_1 x + a_0$$

can have *at most n* real number solutions. Since this is the equation used to determine the real zeros of the function

$$f(x) = a_n x^n + a_{n-1} x^{n-1} + \cdots + a_1 x + a_0,$$

we can conclude that an nth degree polynomial function can have at most n real zeros (x-intercepts).

> **Second Rule of Polynomial Functions**
> The graph of an nth degree polynomial function cannot intersect (cross or touch) the x-axis more than n times.

Example 1 Select the graph from the four below that best represents the graph of $f(x) = 2x^4 - 9x^3 + 11x^2 - 4$.

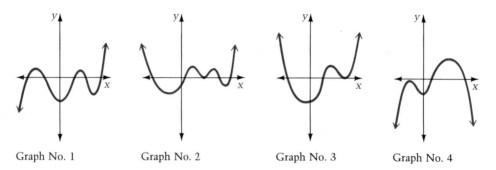

Graph No. 1   Graph No. 2   Graph No. 3   Graph No. 4

Solution By the First Rule of Polynomial Functions, we know that the graph of the given function must look like the graph of $y = 2x^4$ for extreme values of x. This narrows the possibilities to Graphs 2 and 3. By the Second Rule, there can be at most 4 x-intercepts, and so Graph 2 is disqualified. Thus, Graph 3 is the correct choice.

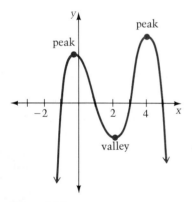

Figure 4.14

At the left is the graph of a fourth degree polynomial function having 4 zeros. To the left of the smallest zero, -1, and to the right of the largest zero, 5, the graph must rise or fall indefinitely (In between -1 and 5 the graph contains waves.) The four zeros divide the x-axis into five intervals, and on three of these intervals the graph must have a peak or a valley. We shall use the term **relative maximum** instead of peak, and the term **relative minimum** instead of valley (plural: maxima and minima).

Figure 4.14 suggests a very useful rule for graphing polynomial functions.

Third Rule of Polynomial Functions

The graph of an nth degree polynomial function can have at most $n - 1$ relative maxima and minima.

Example 2 Select the graph from the four below that best represents the graph of $f(x) = -2x^5 + x^4 - 3x^2 + x - 2$.

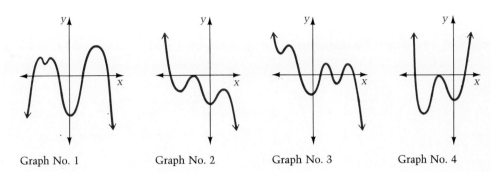

| Graph No. 1 | Graph No. 2 | Graph No. 3 | Graph No. 4 |

Solution By the Third Rule of Polynomial Functions, we know that the graph of the given function has at most 4 relative maxima and minima. This narrows down the possibilities to Graphs 2 and 4. By the First Rule we know that the graph must look like the graph of $y = -2x^5$ towards the extremes; therefore Graph 4 is disqualified. Thus, Graph 2 is the correct choice.

Goal B *to determine the zeros of a given polynomial function*

In order to determine the zeros of the polynomial function $y = f(x)$, it is necessary to solve the equation $f(x) = 0$. For example, for the quadratic function $y = x^2 - 7x - 8$, we follow a three step procedure to find the zeros.

$$x^2 - 7x - 8 = 0$$ ■ The polynomial is set equal to 0.
$$(x - 8)(x + 1) = 0$$ ■ The left side is factored.
$$x - 8 = 0 \quad | \quad x + 1 = 0$$
$$x = 8 \quad | \quad x = -1$$ ■ The property of Zero Products is applied. The zeros are 8 and -1.

We can use the same method to find the zeros of any polynomial function.

Example 3 Determine the zeros of the function $g(x) = x^4 + 5x^3 + 4x^2$.

Solution
$$x^4 + 5x^3 + 4x^2 = 0$$
$$x^2(x^2 + 5x + 4) = 0$$
$$x^2(x + 4)(x + 1) = 0$$

■ Factor the polynomial as completely as possible, and set each factor equal to zero.

The zeros are 0, -4, and -1. (Note, 0 is zero of multiplicity 2, because x^2 is a factor.)

Example 4 Determine the real zeros of the function $h(x) = (3x + 1)(x - 7)(x^2 + 2x + 2)$.

Solution

$$(3x + 1)(x - 7)(x^2 + 2x + 2) = 0$$

$3x + 1 = 0$	$x - 7 = 0$	$x^2 + 2x + 2 = 0$
$x = -\dfrac{1}{3}$	$x = 7$	

■ The discriminant $b^2 - 4ac < 0$; thus there are no real solutions of the equation $x^2 + 2x + 2 = 0$.

The zeros are $-\frac{1}{3}$ and 7.

When applying the Property of Zero Products, you may have observed that if $x - c$ is a factor of a polynomial $f(x)$, then the number c is a zero of the polynomial function $y = f(x)$. The converse of this statement is also true. These two results are summarized in the following theorem.

The Factor Theorem

The number c is a zero of a polynomial function $y = f(x)$ if and only if $x - c$ is one of the factors of $f(x)$.

The Factor Theorem says that if we know one of the zeros of a polynomial function, then we know one of the factors of the polynomial. Knowing one of the factors of an expression can often be the key to factoring the entire expression.

For example, suppose we need to factor 2057 and are told that 17 is one of the factors. The task of finding the other factors is much easier if we first divide 2057 by 17 so that we can write

$$2057 = 17 \cdot 121.$$

Now we factor the number 121, and obtain

$$2057 = 17 \cdot (11)^2.$$

Now consider a related problem. Suppose we need to find all the real zeros of some polynomial function $y = f(x)$, and we know that some real number c is one of them. The following example demonstrates that we can first divide $f(x)$ by $x - c$, and then concentrate on factoring the quotient which may be easier to factor since it is of a lesser degree.

Example 5 Find all the real zeros of the function $f(x) = 2x^3 + 9x^2 - 32x + 21$ given that one of the zeros is -7.

Solution The expression $2x^3 + 9x^2 - 32x + 21$ cannot be factored by any of the standard techniques. Since -7 is a zero of f, we know that $x - (-7) = x + 7$ is a factor of $f(x)$.

$$\begin{array}{r} 2x^2 - 5x + 3 \\ x + 7 \overline{\smash{)}\ 2x^3 + 9x^2 - 32x + 21} \\ \underline{-(2x^3 + 14x^2)} \\ -5x^2 - 32x \\ \underline{-(-5x^2 - 35x)} \\ 3x + 21 \\ \underline{-(3x + 21)} \\ 0 \end{array}$$

■ Divide $f(x)$ by $x + 7$. We want to rewrite $f(x)$ as $(x + 7) \cdot q(x)$, where the polynomial $q(x)$ is the quotient.

We conclude that $f(x) = (x + 7)(2x^2 - 5x + 3)$. To find the zeros, set the polynomial equal to zero, factor the polynomial, and solve.

$$(x + 7)(2x^2 - 5x + 3) = 0$$
$$(x + 7)(2x - 3)(x - 1) = 0$$

$x + 7 = 0$	$2x - 3 = 0$	$x - 1 = 0$
$x = -7$	$x = \dfrac{3}{2}$	$x = 1$

The real zeros are $-7, \frac{3}{2}$, and 1.

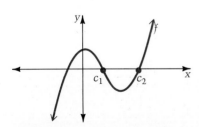

Figure 4.15 c_1 and c_2 are adjacent zeros of f.

Goal C *to sketch the graph of a polynomial function*

We now use the zeros of a polynomial function to obtain additional information about the graph of the function. Let us agree that if c_1 and c_2 are real zeros of a polynomial function f, such that there are no other zeros between c_1 and c_2, then c_1 and c_2 will be called **adjacent zeros** of f.

Fourth Rule of Polynomial Functions

If c_1 and c_2 are two adjacent zeros of a polynomial function f, then the graph of f lies either completely above or completely below the x-axis for all values between c_1 and c_2.

162 4 / Polynomial Functions and Rational Functions

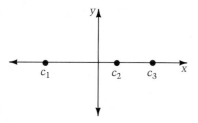

Figure 4.16 Zeros c_1, c_2, and c_3 break the x-axis into 4 intervals: $(-\infty, c_1)$, (c_1, c_2), (c_2, c_3), and (c_3, ∞).

The Fourth Rule is of great use when we sketch the graph of a polynomial function. First, find all the real zeros of the function, and order them from smallest to largest. Next, use the zeros to divide the x-axis into open intervals. Evaluate the function at some convenient test value in each interval. If the function is positive at the test value, then the graph lies completely above the x-axis for all x-values in that interval. If the function is negative at the test value, then the graph lies completely below the x-axis for all x-values in that interval.

We are now in a position to outline a general procedure for sketching the graph of a polynomial function.

Step 1. Find the zeros of the function.
Step 2. Apply the Fourth Rule (select test values).
Step 3. Use the results of Step 2 to begin the sketch.
Step 4. Determine the leading term and apply the First, Second, and Third Rules.

Example 6 Sketch the graph of the function $f(x) = (x - 1)(x + 2)^2(x - 2)$.

Solution *Step 1.* To find the zeros, solve the equation

$$(x - 1)(x + 2)^2(x - 2) = 0.$$

The zeros are -2, 1, and 2 in order from smallest to largest. Plot the zeros.

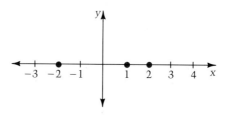

Step 2. To apply the Fourth Rule, we must first determine the intervals having adjacent zeros as endpoints, and then evaluate the function at a convenient test value in each interval.

Interval	Test Value t	$f(t)$	
$(-\infty, -2)$	-3	20	$f(t) > 0$; graph lies above x-axis
$(-2, 1)$	0	8	$f(t) > 0$; graph lies above x-axis
$(1, 2)$	$\frac{3}{2}$	$-\frac{49}{16}$	$f(t) < 0$; graph lies below x-axis
$(2, +\infty)$	$\frac{5}{2}$	$\frac{243}{32}$	$f(t) > 0$; graph lies above x-axis

Step 3. Using the results from Step 2, sketch the graph a little to the right and to the left of each zero.

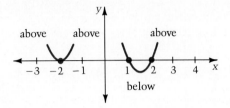

Since the graph lies above the *x*-axis to the left and to the right of −2, it must touch but not cross the *x*-axis at −2. The graph lies above the *x*-axis to the left of 1 and below the *x*-axis to the right of 1, so it must cross at 1. Similar reasoning leads us to conclude that the graph is rising as it crosses at 2.

Step 4. Multiplying the leading term of each of the three factors of the polynomial gives us the leading term (and the degree) of the polynomial: $x \cdot x^2 \cdot x = x^4$. Since the degree is 4, there are at most 3 relative maxima and minima (Third Rule). Since we know one is located at $x = -2$, one must occur between 1 and −2, and another between 1 and 2.

Although we cannot determine exactly where the relative maxima and minima occur without the aid of calculus, we can complete the graph by plotting a few additional points, such as those chosen in Step 2. Notice that towards the extremes the graph behaves like $y = x^4$.

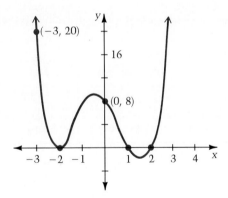

As the previous example illustrates, it is sometimes easier to graph a polynomial function if we first factor the polynomial. Recall that one factoring technique is *u*-substitution, a method that is applied to polynomials that are quadratic in form. In the next example, we shall use *u*-substitution to factor the polynomial.

Example 7 Sketch the graph of $f(x) = x^5 - 3x^4 - 5x^3 + 15x^2 + 4x - 12$ given that 3 is a zero.

Solution *Step 1.* To solve the equation $x^5 - 3x^4 - 5x^3 + 15x^2 + 4x - 12 = 0$, we factor the polynomial. Since 3 is a zero, we know that $x - 3$ is a factor of this polynomial.

$$
\begin{array}{r}
x^4 - 5x^2 + 4 \\
x - 3 \overline{\smash{\big)}\, x^5 - 3x^4 - 5x^3 + 15x^2 + 4x - 12} \\
\underline{-(x^5 - 3x^4)} \\
0 - 5x^3 + 15x^2 \\
\underline{-(-5x^3 + 15x^2)} \\
0 + 4x - 12 \\
\underline{-(4x - 12)} \\
0
\end{array}
$$

■ We divide the polynomial by its factor.

Notice that the quotient is quadratic in x^2. We can use u-substitution to factor it.

$$(x - 3)(x^4 - 5x^2 + 4) = 0$$

$x - 3 = 0$	$x^4 - 5x^2 + 4 = 0$
	$u^2 - 5u + 4 = 0$ ■ Let $u = x^2$.
	$(u - 4)(u - 1) = 0$ ■ Thus, $u = 1$ or 4, i.e. $x^2 = 1$ or 4.

The zeros of this function are $x = -2, -1, 1, 2,$ and 3 from smallest to largest.

Step 2. Now we select a test value in each interval and apply the Fourth Rule.

Interval	Test Value t	$f(t)$	
$(-\infty, -2)$	-3	-240	$f(t) < 0$; graph lies below x-axis
$(-2, -1)$	$-\frac{3}{2}$	$\frac{315}{32}$	$f(t) > 0$; graph lies above x-axis
$(-1, 1)$	0	-12	$f(t) < 0$; graph lies below x-axis
$(1, 2)$	$\frac{3}{2}$	$\frac{105}{32}$	$f(t) > 0$; graph lies above x-axis
$(2, 3)$	$\frac{5}{2}$	$-\frac{189}{32}$	$f(t) < 0$; graph lies below x-axis
$(3, \infty)$	4	180	$f(t) > 0$; graph lies above x-axis

Step 3.

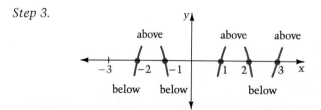

■ Sketch the graph a little to the right and to the left of each zero.

Step 4. Since the degree of the polynomial is 5, there can be at most 4 relative maxima and minima (Third Rule). The relative maxima occur in the intervals $(-2, -1)$, and $(1, 2)$. The relative minima are located in the intervals $(-1, 1)$ and $(2, 3)$.

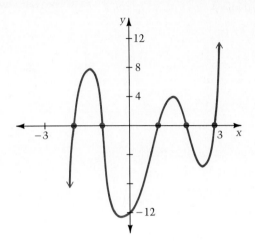

Exercise Set 4.2   =   Calculator Exercises in the Appendix

Goal A

In exercises 1–4, use the First, Second, and Third Rules of Polynomial Functions to select the graph that best represents the function.

In exercises 5–8, use the First, Second, and Third Rules of Polynomial Functions to select the graph that best represents the function.

Graph 1   Graph 2   Graph 3

Graph 1   Graph 2
                                Graph 3

Graph 4   Graph 5   Graph 6

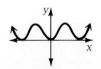
Graph 4   Graph 5   Graph 6

1. $y = 12x^2 - x^3 - x$ 2. $y = x^6 - 7x^4 + 12x^2$ 5. $y = x^3 - 2x$ 6. $y = x^4 - x^3 - x^6 + x$
3. $y = x^5 - 5x^3 + 4x$ 7. $f(x) = 2x^4 - 5x^3 + 4x^2 - x$
4. $y = -x^5 + 10x^3 - 25x + 20$ 8. $f(x) = 2x^4 + x^3 - 1$

Goal B

In exercises 9–14, determine the zeros of the given function.

9. $f(x) = (x + 3)(x - 5)(x + 7)^3$
10. $f(x) = -5x(x + 2)^2(3x - 1)$
11. $f(x) = x(3x - 2)^2(x^2 + 3x + 2)$
12. $f(x) = x^2(2 - x)^3(x^2 + 4x + 3)$
13. $f(x) = (x - 5)^2(2x + 5)^3(x^2 + 2x + 2)$
14. $f(x) = x(7 - 2x)^2(x^2 + x + 1)$

In exercises 15–20, use the given zero to find all the zeros of the function.

15. $y = x^3 - 3x^2 + x + 1$, one zero is 1
16. $y = x^3 + x^2 - 5x - 2$, one zero is 2
17. $y = x^4 + 8x^3 + 19x^2 + 12x$, one zero is -3
18. $y = x^4 + 5x^3 + 2x^2 - 8x$, one zero is 1
19. $y = x^3 - 5x^2 + 4x - 20$, one zero is 5
20. $y = x^3 + x^2 + 2x + 2$, one zero is -1

Goal C

For exercises 21–26, sketch the graphs of the functions in exercises 15–20.

In exercises 27–38, find the zeros of the given function; then sketch the graph.

27. $y = (x - 1)(x - 3)^2$
28. $y = (x + 2)(x - 1)^2$
29. $y = x(x - 1)(x - 2)^2$
30. $y = (2x + 3)(x^2 - 7x + 10)$
31. $y = x^3 - 4x$
32. $y = -2x^3 + 18x$
33. $y = 2x^4 - 7x^2 + 3$
34. $y = x^4 - 9x^2 + 8$
35. $y = x^3 + 2x^2 - x - 2$
36. $y = x^3 + 7x^2 - 4x - 28$
37. $y = x^4 - x^3 - 4x^2 + 4x$; one zero is 1.
38. $y = x^4 + 4x^3 + x^2 - 8x - 6$; one zero is -3.

Superset

The graphs of polynomial functions are unbroken curves. If $f(c_1)$ and $f(c_2)$ have different signs, then f has a zero somewhere between c_1 and c_2.

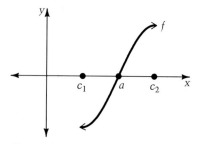

f has a zero at a: $f(a) = 0$

In exercises 39–42, evaluate each function for the integers between -4 and 4 inclusive, and state the interval(s) in which you are sure there is a zero.

39. $f(x) = 8x^3 - 12x^2 - 66x + 35$
40. $f(x) = 400x^3 - 1700x^2 + 1336x - 117$
41. $f(x) = x^5 + 2x^4 - 2x^3 - 4x^2 + x + 2$
42. $f(x) = x^4 - 4x^3 - 17x^2 + 60x$

According to the Rational Root Theorem, if $f(x) = a_n x^n + a_{n-1} x^{n-1} + \cdots + a_1 x + a_0$ has integer coefficients, and $\frac{p}{q}$ is a rational zero of f in lowest terms, then p must be a factor of a_0, and q must be a factor of a_n. For $f(x) = 2x^3 - x^2 - 2x + 3$, we have

$$\text{factors of } a_0: +1, -1, +3, -3$$
$$\text{factors of } a_n: +1, -1, +2, -2$$

Dividing each factor of a_0 by the factors of a_n, we have possible rational roots $\pm 1, \pm\frac{1}{2}, \pm 3, \pm\frac{3}{2}$.

In exercises 43–48, list the possible rational roots and test whether they are zeros.

43. $f(x) = x^3 + 2x^2 - 5x - 6$
44. $f(x) = 2x^3 - 9x^2 - x + 10$
45. $f(x) = 2x^3 - 5x^2 + 2x - 5$
46. $f(x) = 9x^3 - 6x^2 - 5x + 2$
47. $f(x) = x^3 - x^2 - 2x - 12$
48. $f(x) = x^3 + x^2 - 3$

4.3 Rational Functions

$f(x) = \dfrac{x^2 - 3x + 7}{x^2 + 3x - 10}$

$g(x) = \dfrac{1}{x^2 - 1}$

Examples of Rational Functions

Goal A *to determine the horizontal and vertical asymptotes of the graph of a rational function*

Recall that a rational expression is a fraction that has a polynomial in both numerator and denominator. Functions defined by rational expressions are called **rational functions.** Although rational functions are not as well-behaved as polynomial functions, there are a few general techniques which will help us graph rational functions.

The simplest rational function is given by the equation

$$y = \dfrac{1}{x}.$$

Notice that in the figure at the left, the graph gets closer and closer to the x-axis but never touches it. (The graph approaches the y-axis in the same fashion.) When a graph approaches a line in this manner, we say that the line is an **asymptote** of the graph.

> If the line is vertical, it is called a **vertical asymptote;** if it is horizontal, it is called a **horizontal asymptote.**

Consider the following transformations of the graph of $y = \dfrac{1}{x}$. As is customary, asymptotes are indicated with dashed lines.

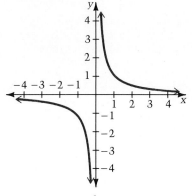

Figure 4.17 $y = \dfrac{1}{x}$.

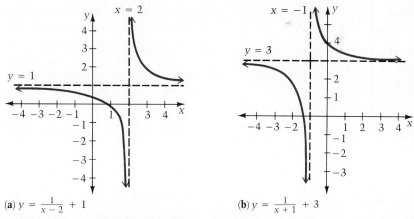

(a) $y = \dfrac{1}{x-2} + 1$ (b) $y = \dfrac{1}{x+1} + 3$

Figure 4.18 (a) Vertical asymptote: $x = 2$, horizontal asymptote: $y = 1$. (b) Vertical asymptote: $x = -1$, horizontal asymptote: $y = 3$.

Not all rational functions are transformations of $y = \dfrac{1}{x}$. In the following example, we consider some rational functions whose graphs do not resemble the graph of $y = \dfrac{1}{x}$.

Example 1 Write the equations of any vertical or horizontal asymptotes of each graph.

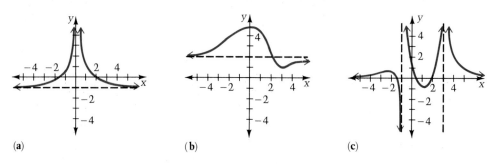

(a) (b) (c)

Solution (a) Vertical asymptote: $x = 0$; horizontal asymptote: $y = -1$.
(b) Vertical asymptote: none; horizontal asymptote: $y = 2$.
(c) Vertical asymptotes: $x = -1$ and $x = 3$; horizontal asymptote: $y = 0$.

Notice that the graph in part (b) crosses the asymptote at $x = 2$, and eventually approaches it from below.

Remember that an asymptote is described by an equation of a line either in the form $x = c$ or $y = d$ (for real numbers c and d).

Asymptotes are characteristic features of the graphs of rational functions and are essential to sketching such graphs. We now consider methods for finding vertical and horizontal asymptotes, given only the equation of the function. On the previous page, we saw that the line $x = -1$ is a vertical asymptote of the graph of the function

$$y = \frac{1}{x+1} + 3 = \frac{3x+4}{x+1}.$$

Evaluating this function at $x = -1$, we find that (i) the numerator is nonzero, and (ii) the denominator is zero. This is true in general for all vertical asymptotes.

First Rule of Rational Functions

If f is a rational function defined by

$$f(x) = \frac{g(x)}{h(x)},$$

then the line $x = c$ is a vertical asymptote of the graph of f provided $g(c) \neq 0$ and $h(c) = 0$.

Example 2 Find all vertical asymptotes of the graph of each function.

(a) $y = \dfrac{3x^2}{x^2 - 1}$ (b) $y = \dfrac{x^3 - 2x^2}{x^2 - 3x + 2}$

Solution (a) $x^2 - 1 = 0$ ■ Determine the values that make the denominator 0.

$(x - 1)(x + 1) = 0$

$x - 1 = 0 \quad | \quad x + 1 = 0$
$x = 1 \quad\;\; | \quad\;\; x = -1$

For $x = 1$, the numerator $3(1)^2 \neq 0$.
For $x = -1$, the numerator $3(-1)^2 \neq 0$.
Thus, the lines $x = 1$ and $x = -1$ are vertical asymptotes.

(b) $x^2 - 3x + 2 = 0$ ■ Determine the values that make the denominator 0.

$(x - 1)(x - 2) = 0$

$x - 1 = 0 \quad | \quad x - 2 = 0$
$x = 1 \quad\;\; | \quad\;\; x = 2$

For $x = 1$, the numerator is $1^3 - 2(1)^2 \neq 0$.
For $x = 2$, the numerator is $2^3 - 2(2)^2 = 0$.
Therefore, the line $x = 1$ is the only vertical asymptote.

The graph of a rational function can have at most one horizontal asymptote. To determine this asymptote, we need only look at the leading terms of the numerator and the denominator.

Second Rule of Rational Functions
If

$$f(x) = \frac{a_n x^n + a_{n-1} x^{n-1} + \cdots + a_2 x^2 + a_1 x + a_0}{b_m x^m + b_{m-1} x^{m-1} + \cdots + b_2 x^2 + b_1 x + b_0}$$

is a rational function such that the numerator has degree n and the denominator has degree m, then the graph of f has

(i) the x-axis ($y = 0$) as a horizontal asymptote if $n < m$;

(ii) the line $y = \dfrac{a_n}{b_m}$ as a horizontal asymptote if $n = m$;

(iii) no horizontal asymptote if $n > m$.

Example 3 For the graph of each function, determine the horizontal asymptote, if it exists.

(a) $y = \dfrac{3x^2}{7x^2 - 1}$ (b) $f(x) = \dfrac{x^3 - 2x^2}{x^2 - 3x + 2}$ (c) $g(x) = \dfrac{3x + 4}{7x^2 - 5}$

Solution (a) The degree of the numerator and denominator are the same. By (ii) of the Second Rule, the horizontal asymptote is $y = \frac{3}{7}$, the ratio of the leading coefficients.
(b) The degree of the numerator is greater than the degree of the denominator. Therefore, by (iii) of the Second Rule, there is no horizontal asymptote.
(c) The degree of the numerator is less than the degree of the denominator. By (i) of the Second Rule, the horizontal asymptote for the graph of g is the x-axis ($y = 0$).

Goal B *to sketch the graph of a rational function*

To graph rational functions, we shall follow a procedure similar to that used to graph polynomial functions. In Section 4.2 we found that the key to graphing polynomial functions was to find the intervals where the graph lies below the x-axis and the intervals where the graph lies above the x-axis. This information, together with the vertical and horizontal asymptotes, will help us graph rational functions.

Suppose we wish to graph the rational function

$$f(x) = \dfrac{x + 4}{(x + 1)(x - 5)}.$$

Using the First and Second Rules of Rational Functions, we find that the graph has vertical asymptotes $x = -1$ and $x = 5$, and has a horizontal asymptote $y = 0$.

Once the asymptotes are found, we can proceed as we did in Section 4.2. We must now find the zeros of the function. (Recall that these values are the x-intercepts of the graph.) To find the zeros, we must solve the equation

$$\dfrac{x + 4}{(x + 1)(x - 5)} = 0.$$

Recall that a fraction is 0 only when its numerator is 0 and its denominator is not zero. Therefore, the solution of the equation is -4; that is, -4 is the only zero of the function. The function is undefined when the denominator is zero, that is, when

$$x = 5 \text{ and } -1.$$

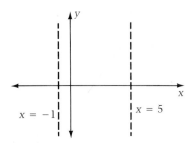

Figure 4.19 Asymptotes of the rational function

$$f(x) = \dfrac{x + 4}{(x + 1)(x - 5)}$$

4 / Polynomial Functions and Rational Functions   171

Recall that when graphing polynomial functions, we use the zeros to divide the x-axis into open intervals in order to determine where the function is positive and where it is negative. For rational functions, the situation is a little different.

Third Rule of Rational Functions

Between any two successive x-values where a rational function f is either 0 *or undefined,* the graph of f lies either completely above or completely below the x-axis.

Since f has a zero at -4, and is undefined at -1 and 5, by this rule we can conclude that on each of the intervals $(-\infty, -4)$, $(-4, -1)$, $(-1, 5)$ and $(5, +\infty)$, the graph lies either completely above or completely below the x-axis. As with polynomial functions, we use test values to determine this.

The graph of the function f is shown in Figure 4.20. Since the graph is above the x-axis to the right of 5 and below the x-axis to the left of 5, we say informally that the graph jumps across the x-axis at 5. Notice that the graph also jumps across the x-axis at -1.

We now summarize the general procedure for graphing rational functions of the form

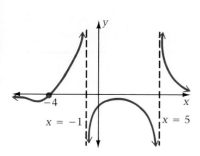

Figure 4.20 $f(x) = \dfrac{x + 4}{(x + 1)(x - 5)}$

$$f(x) = \frac{g(x)}{h(x)}.$$

Step 1. Find the horizontal and vertical asymptotes.
Step 2. Find the zeros of the function and the values where the function is undefined.
Step 3. Apply the Third Rule of Rational Functions (for each interval, select test values to determine whether the graph lies above the x-axis or below the x-axis.)
Step 4. Plot a few additional points and draw a smooth curve. Recall that a graph can cross a horizontal asymptote.

Example 4   Sketch the graph of the rational function $f(x) = \dfrac{2x^2 - 10}{x^2 - 6x + 5}$.

Solution   *Step 1.* To find the horizontal asymptotes, we use the Second Rule. Since the degrees of the numerator and denominator are equal, the line $y = \frac{2}{1}$ (i.e., $y = 2$) is the horizontal asymptote. To determine the vertical asymptotes, rewrite the polynomials in factored form:

$$f(x) = \frac{2(x^2 - 5)}{(x - 5)(x - 1)}.$$

The denominator is 0 when $x = 5$ or $x = 1$, and the numerator is not 0 for either of these values. Thus, the lines $x = 5$ and $x = 1$ are vertical asymptotes.

Step 2. To find the zeros of the function, we determine the values where the numerator is 0 but the denominator is not 0, namely $\sqrt{5}$ and $-\sqrt{5}$. In addition, the function is undefined when $x = 5$ or $x = 1$. Plot the zeros and graph the asymptotes.

Step 3. To apply the Third Rule, we use the zeros and the x-values where the function is undefined to divide the x-axis into open intervals. Select a test value in each interval.

Interval	Test Value t	$f(t)$	
$(-\infty, -\sqrt{5})$	-3	$\frac{1}{4}$	$f(t) > 0$, curve lies above x-axis
$(-\sqrt{5}, 1)$	0	-2	$f(t) < 0$, curve lies below x-axis
$(1, \sqrt{5})$	2	$\frac{2}{3}$	$f(t) > 0$, curve lies above x-axis
$(\sqrt{5}, 5)$	3	-2	$f(t) < 0$, curve lies below x-axis
$(5, +\infty)$	6	$12\frac{2}{5}$	$f(t) > 0$, curve lies above x-axis

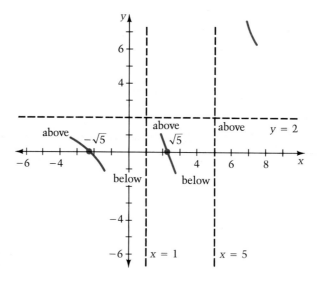

Step 4. Plot a few additional points (use the points chosen in Step 3) and draw a smooth curve.

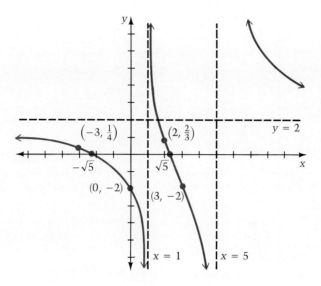

We have said that the zeros of a rational function occur where the numerator $g(x)$ is 0 and the denominator $h(x)$ is not 0. The vertical asymptotes occur where $g(x)$ is not 0 and $h(x)$ is 0. We have not considered what might happen if $g(x) = 0$ and $h(x) = 0$. The following graphs illustrate what could occur.

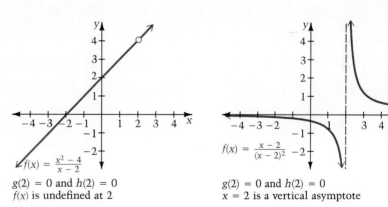

$f(x) = \dfrac{x^2 - 4}{x - 2}$

$g(2) = 0$ and $h(2) = 0$
$f(x)$ is undefined at 2

Figure 4.21

$f(x) = \dfrac{x - 2}{(x - 2)^2}$

$g(2) = 0$ and $h(2) = 0$
$x = 2$ is a vertical asymptote

Exercise Set 4.3

Goal A

In exercises 1–4, write the equations of any vertical or horizontal asymptotes.

1. 2.

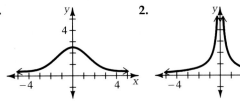

3. 4.

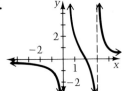 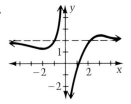

In exercises 5–14, write the equations of any horizontal or vertical asymptotes.

5. $y = \dfrac{1}{x + 9}$
6. $y = \dfrac{1}{x - 3}$
7. $y = \dfrac{2x + 9}{x^2 + 3x + 2}$
8. $y = \dfrac{x - 6}{x^2 - 9}$
9. $y = \dfrac{x^2 - x - 2}{x^2 - 9}$
10. $y = \dfrac{x^2 - 1}{2x^2 - 8}$
11. $y = \dfrac{10x^2 + 2x + 1}{5x^2 + 6x + 1}$
12. $y = \dfrac{x^3 - 8}{x^2 - 2x - 3}$
13. $y = \dfrac{2x^4 - 7x^2 + 4}{(2x - 1)(x + 1)^2}$
14. $y = \dfrac{x^4 + 3x^2 + 4}{x^2 - 4x - 5}$

Goal B

In exercises 15–26, graph each rational function.

15. $y = \dfrac{1}{x + 1}$
16. $y = \dfrac{1}{x - 2}$
17. $y = \dfrac{x - 1}{x - 2}$
18. $y = \dfrac{2x - 1}{4 - x}$
19. $y = \dfrac{4}{x^2 - 2x - 3}$
20. $y = \dfrac{2}{x^2 + x - 6}$
21. $y = \dfrac{1}{x^2 + 4x + 4}$
22. $y = \dfrac{3}{x^2 + 6x + 9}$
23. $y = \dfrac{x^2}{x^2 - 4}$
24. $y = \dfrac{2x}{x^2 - 16}$
25. $y = \dfrac{x^2 - 9}{x^2}$
26. $y = \dfrac{x}{x^2 + 1}$

Superset

In exercises 27–32, sketch the graph of each function and use that graph to sketch the graph of $y = \dfrac{1}{f(x)}$.

27. $f(x) = x$
28. $f(x) = x - 3$
29. $f(x) = x^2 + 1$
30. $f(x) = 1 - x^2$
31. $f(x) = -\sqrt{x^2 + 1}$
32. $f(x) = |x^2 - 1|$

If a rational function f has numerator $g(x)$ whose degree is one more than the degree of the denominator $h(x)$, then the graph of f has an **oblique asymptote,** a line that is neither vertical nor horizontal. Dividing $g(x)$ by $h(x)$ yields a quotient of the form $mx + b$, which describes the oblique asymptote. For example, for the function f below, dividing $x^2 + x - 1$ by x produces the quotient $x + 1$ (the remainder $\frac{1}{x}$ is not important because as $x \to +\infty$, $\frac{1}{x} \to 0$); thus, the equation of the oblique asymptote is $y = x + 1$. Notice that $x = 0$ is a vertical asymptote.

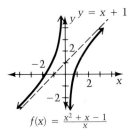

$f(x) = \dfrac{x^2 + x - 1}{x}$

In exercises 33–38, sketch the graph of the given function and show the oblique asymptote.

33. $f(x) = \dfrac{x^2 - 1}{x}$
34. $y = \dfrac{x^2 - 3x + 1}{x}$
35. $f(x) = \dfrac{x^2 + 2x + 1}{x}$
36. $f(x) = \dfrac{2x^2 - 3}{x}$
37. $f(x) = \dfrac{3x^2 - 5x - 1}{x - 2}$
38. $y = \dfrac{2x^2 - 7x + 4}{3 - x}$

4.4 Conic Sections

Goal A *to sketch the graph of a parabola*

In Chapter 3 we found that the graph of a quadratic function is always a parabola. Invariably these kinds of parabolas either open upward or downward: the axis of symmetry is always parallel to the y-axis.

What about parabolas that do not open upward or downward, such as those at the left? Clearly these are not the graphs of functions because they fail the vertical line test. Recall that we refer to such sets of points as *relations*.

In this section, we shall study a special group of relations called **conic sections.** This group consists of the parabola, the circle, the ellipse, and the hyperbola. They are called conic sections because each can be formed by cutting a double cone with a plane as shown in Figure 4.23.

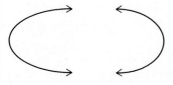

Figure 4.22

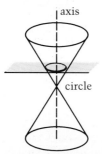

The plane is perpendicular to the axis of the cone.

The plane is tilted.

The plane is parallel to one side of the cone.

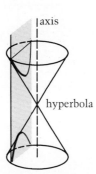

The plane is parallel to the axis of the cone.

Figure 4.23

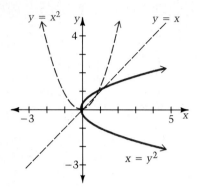

Figure 4.24

We shall begin by considering parabolas. The parabola described by the equation $y = x^2$ is shown in Figure 4.24. Interchanging x and y in this equation produces the equation $x = y^2$, which also describes a parabola. When x and y are interchanged in an equation, the resulting graph is the reflection of the original graph across the line $y = x$. Although $x = y^2$ does not describe a function, the rules of symmetry and transformations still apply. We summarize these rules in the following table.

Change in Equation	Change in the Graph
x is replaced with $x - h$	If $h > 0$, graph moves $|h|$ units to the right. If $h < 0$, graph moves $|h|$ units to the left.
y is replaced with $y - k$	If $k > 0$, graph moves $|k|$ units up. If $k < 0$, graph moves $|k|$ units down.
x is replaced with $-x$	Graph is reflected across the y-axis.
y is replaced with $-y$	Graph is reflected across the x-axis.

For example, two transformations of the graph of

$$x = y^2$$

are shown below. Replacing x with $-\frac{1}{3}x$ reflects the graph of $x = y^2$ across the y-axis and stretches the graph horizontally (the new function is $-\frac{1}{3}x = y^2$ or $x = -3y^2$). This is displayed in Figure 4.25(a). Replacing y with $y - (-3)$ and x with $x - 2$ translates the original graph 3 units down and 2 units to the right (the new function is $x = (y + 3)^2 + 2$). This is shown in Figure 4.25(b).

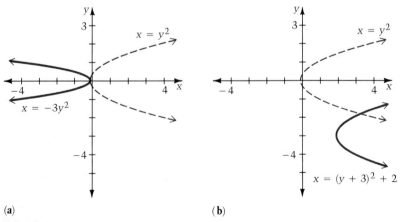

(a)   (b)

Figure 4.25

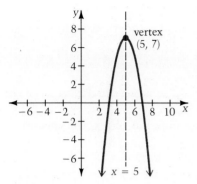

Figure 4.26

Recall that it is easier to graph a quadratic function if it is written in vertex form: $f(x) = a(x - h)^2 + k$. For example, the graph of $f(x) = -2(x - 5)^2 + 7$ has its vertex at (5, 7), the axis of symmetry is the line $x = 5$, and the parabola opens downward since a (in this case -2) is negative. There is an analogous vertex form of the equation of a parabola opening to the right or the left.

Rule

The graph of an equation of the form $x = a(y - k)^2 + h$ is a parabola whose vertex is (h, k) and whose axis of symmetry is the line $y = k$. If $a > 0$, the parabola opens to the right; if $a < 0$ the parabola opens to the left.

Example 1 Sketch the graph of the relation $x = 2(y + 3)^2 - 4$. Determine the vertex, the axis of symmetry, and the x- and y-intercepts.

Solution $x = 2[y - (-3)]^2 + (-4)$ ■ To use the rule, the form of the equation must be $x = a(y - k)^2 + h$.

Since $a > 0$, the parabola opens to the right. The vertex is $(-4, -3)$, and the axis of symmetry is the line $y = -3$. The y-intercepts occur when $x = 0$.

$$0 = 2(y + 3)^2 - 4$$
$$4 = 2(y + 3)^2$$
$$2 = (y + 3)^2$$

Thus, the y-intercepts are $-3 \pm \sqrt{2}$ or approximately -1.6 and -4.4. The x-intercept occurs when $y = 0$, that is, $x = 2(0 + 3)^2 - 4 = 14$.

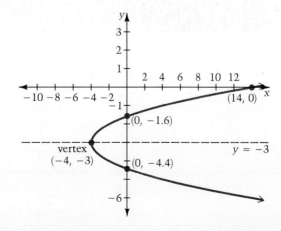

Example 2 Sketch the graph of the relation $x = -3y^2 + 6y + 5$. Determine the vertex, the axis of symmetry, and the x- and y-intercepts.

Solution

$x = -3(y^2 - 2y \quad\quad) + 5$ ■ Begin completing the square by factoring the coefficient of y^2 out of the variable terms.

$x = -3(y^2 - 2y + 1) + 5 + 3$ ■ Take half of the coefficient of the y-term and square it. Add this inside the parentheses. In effect we added -3; thus, we must add 3.

$x = -3(y - 1)^2 + 8$

The vertex is $(8, 1)$. Since $a < 0$ the parabola opens to the left. To find the y-intercepts, use the quadratic formula to solve the equation $0 = -3y^2 + 6y + 5$. The y-intercepts are approximately 2.6 and -0.6. The x-intercept occurs when $y = 0$. Since $x = -3(0 - 1)^2 + 8 = 5$, the x-intercept is 5.

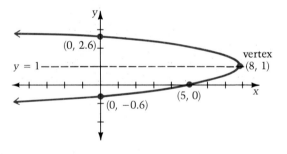

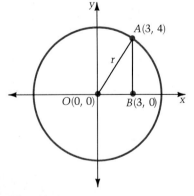

Figure 4.27

Goal B *to sketch the graph of a circle*

Suppose we are given that a circle has its center at $O(0, 0)$ and contains a point $A(3, 4)$. How can we determine r, the radius of the circle, that is, the distance between $O(0, 0)$ and $A(3, 4)$?

Recall that the distance between any two points P and Q on the number line is the absolute value of the difference of their coordinates. To find the distance between $O(0, 0)$ and $A(3, 4)$, which we denote $d(O, A)$, first consider the distance between $O(0, 0)$ and $B(3, 0)$. Since it is a distance in the x-direction only, $d(O, B)$ is just the absolute value of the difference of the x-coordinates:

$$d(O, B) = |3 - 0| = 3.$$

4 / Polynomial Functions and Rational Functions

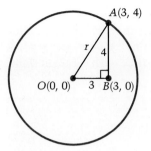

Figure 4.28 $r = \sqrt{3^2 + 4^2}$

Now consider the distance between $B(3, 0)$ and $A(3, 4)$. Since it is a distance in the y-direction only, $d(B, A)$ is the difference of the y-coordinates: $d(B, A) = |4 - 0| = 4$. The two distances, 3 and 4, are the lengths of the two legs of the right triangle $\triangle OBA$. Using the Pythagorean Theorem, the radius is

$$r = d(O, A) = \sqrt{(x\text{-dist.})^2 + (y\text{-dist.})^2} = \sqrt{3^2 + 4^2} = 5.$$

We generalize this method in the following formula.

The Distance Formula

The distance between any two points $A(x_1, y_1)$ and $B(x_2, y_2)$ is given by the formula

$$d(A, B) = \sqrt{(x_2 - x_1)^2 + (y_2 - y_1)^2}.$$

If a circle has its center at $O(0, 0)$, we can use the distance formula to describe the distance between any point $A(x, y)$ on the circle and the center $O(0, 0)$: $d(O, A) = \sqrt{(x - 0)^2 + (y - 0)^2} = \sqrt{x^2 + y^2}$.

Since this distance is the radius, we can write

$$\sqrt{x^2 + y^2} = r \quad \text{or} \quad x^2 + y^2 = r^2.$$

Rule

An equation of the form $x^2 + y^2 = r^2$, for $r > 0$, describes a circle with center at the origin and radius r.

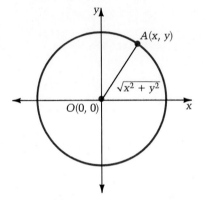

Figure 4.29

Example 3   Sketch the relation described by $x^2 + y^2 = 9$. Determine the center and radius.

Solution

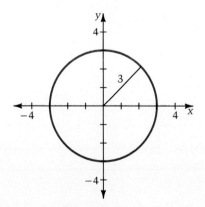

■ The center is at $(0, 0)$. Since $r^2 = 9$, the radius is 3.

Replacing x with $x - h$ and y with $y - k$ in the equation $x^2 + y^2 = r^2$ translates the original circle so that its new center is at (h, k). The radius remains unchanged. The resulting equation, $(x - h)^2 + (y - k)^2 = r^2$, is called the **center/radius form** of the equation of a circle with center at (h, k) and radius r.

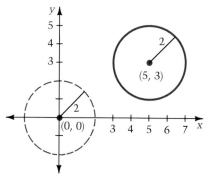

■ The graph of the relation $(x - 5)^2 + (y - 3)^2 = 4$ is a circle with center at $(5, 3)$ and radius 2.

Figure 4.30

Suppose the equation $Ax^2 + Bx + Ay^2 + Cy + D = 0$, describes a circle. To graph this, rewrite the equation in center/radius form.

Example 4 Sketch the graph of the relation $4x^2 - 8x + 4y^2 + 16y + 11 = 0$. Determine the center and radius.

Solution

$(4x^2 - 8x) + (4y^2 + 16y) = -11$

$4(x^2 - 2x) + 4(y^2 + 4y) = -11$

$4(x^2 - 2x + 1) + 4(y^2 + 4y + 4) = -11 + 4 \cdot 1 + 4 \cdot 4$

$(x - 1)^2 + (y + 2)^2 = \dfrac{9}{4}$

■ Prepare to complete the square.

■ Complete the square for x-terms and the y-terms.

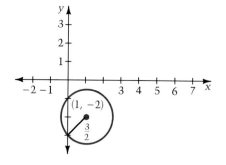

■ The graph is a circle of radius $\frac{3}{2}$ with center at $(1, -2)$.

4 / Polynomial Functions and Rational Functions   181

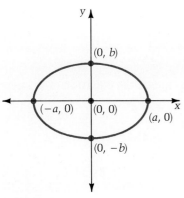

Figure 4.31

Goal C *to sketch the graph of an ellipse*

Suppose we stretched a circle horizontally. This distorted circle, called an **ellipse,** would look like Figure 4.31. The ellipse has its **center** at the origin. The points $(a, 0)$, $(-a, 0)$, $(0, b)$, and $(0, -b)$ are the **vertices.**

Rule

An equation of the form

$$\frac{x^2}{a^2} + \frac{y^2}{b^2} = 1$$

describes an ellipse with center at the origin and x-intercepts $\pm a$ and y-intercepts $\pm b$. The vertices are $(a, 0)$, $(-a, 0)$, $(0, b)$, and $(0, -b)$. Notice that the right side of the equation is $+1$.

Example 5 Sketch the graph of each relation. Determine the center, x- and y-intercepts, and vertices.

(a) $\dfrac{x^2}{16} + \dfrac{y^2}{9} = 1$ (b) $100x^2 + 25y^2 = 100$

Solution

(a) The center is at the origin. The x-intercepts are ± 4, and the y-intercepts are ± 3. The vertices are $(4, 0)$, $(-4, 0)$, $(0, 3)$, and $(0, -3)$.

(b) Divide both sides by 100: $\dfrac{x^2}{1} + \dfrac{y^2}{4} = 1$. The center is at the origin, and the x-intercepts are ± 1; the y-intercepts are ± 2. The vertices are $(1, 0)$, $(-1, 0)$, $(0, 2)$, and $(0, -2)$.

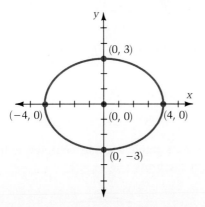
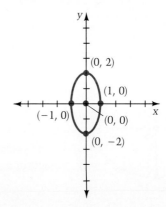

Suppose we translate the graph of

$$\frac{x^2}{a^2} + \frac{y^2}{b^2} = 1,$$

so that its center is at (h, k). As you might expect, the equation of the new ellipse is

$$\frac{(x - h)^2}{a^2} + \frac{(y - k)^2}{b^2} = 1.$$

The point (h, k) corresponds to the old center $(0, 0)$. By adding a or $-a$ to the x-coordinate of the new center, and adding b or $-b$ to the y-coordinate of the new center, we can find the new vertices.

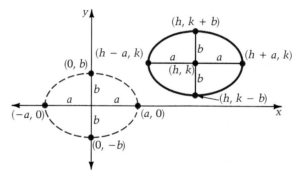

Figure 4.32

For example, the equation

$$\frac{(x - 5)^2}{16} + \frac{(y - 2)^2}{9} = 1$$

describes an ellipse with center at $(5, 2)$. Since $a = 4$ and $b = 3$, the vertices are $(5 + 4, 2)$, $(5 - 4, 2)$, $(5, 2 + 3)$, and $(5, 2 - 3)$ (i.e., $(9, 2)$, $(1, 2)$, $(5, 5)$, and $(5, -1)$).

If you are given an equation of an ellipse in the form

$$Ax^2 + Bx + Cy^2 + Dy + E = 0,$$

and you wish to sketch its graph, begin by completing the square (twice) to obtain an equation of the form

$$\frac{(x - h)^2}{a^2} + \frac{(y - k)^2}{b^2} = 1.$$

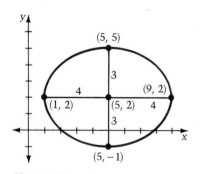

Figure 4.33

Example 6 Sketch the graph of the relation $9x^2 + 18x + 4y^2 - 40y + 73 = 0$. Determine the center and vertices.

Solution

$$(9x^2 + 18x \quad) + (4y^2 - 40y \quad) = -73$$
$$9(x^2 + 2x \quad) + 4(y^2 - 10y \quad) = -73$$
$$9(x^2 + 2x + 1) + 4(y^2 - 10y + 25) = -73 + 9 + 100$$
$$9(x + 1)^2 + 4(y - 5)^2 = 36$$
$$\frac{(x + 1)^2}{4} + \frac{(y - 5)^2}{9} = 1$$

■ Complete the square for x and for y.

■ Now divide by 36 to get the proper form.

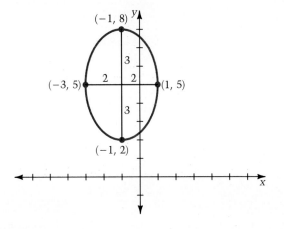

■ The ellipse has its center at $(-1, 5)$, and $a = 2$, and $b = 3$. The vertices are $(-3, 5)$, $(1, 5)$, $(-1, 8)$, and $(-1, 2)$.

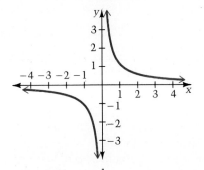

Figure 4.34 $y = \frac{1}{x}$.

Goal D *to sketch the graph of a hyperbola*

Consider the equations

$$\frac{x^2}{a^2} - \frac{y^2}{b^2} = 1 \quad \text{and} \quad \frac{y^2}{b^2} - \frac{x^2}{a^2} = 1.$$

Although these equations differ only slightly from the equation of an ellipse (notice the minus sign), their graphs differ greatly. The graphs of the above equations are **hyperbolas.** The graph of $y = \frac{1}{x}$, which we considered in the last section, is a hyperbola having the x-axis as a horizontal asymptote and the y-axis as a vertical asymptote.

In this section we shall study hyperbolas like those in Figure 4.35 that open in the *x*-direction or the *y*-direction.

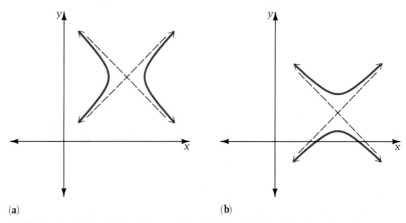

Figure 4.35 (a) Hyperbola opening in *x*-direction. (b) Hyperbola opening in *y*-direction.

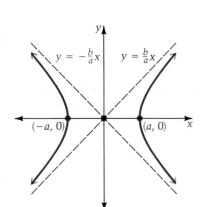

Figure 4.36

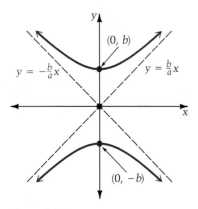

Figure 4.37

We begin by considering hyperbolas whose asymptotes intersect at the origin. We say that such hyperbolas have their centers at the origin.

Rule

The graph of an equation of the form

$$\frac{x^2}{a^2} - \frac{y^2}{b^2} = 1$$

is a **hyperbola** that opens on the *x*-axis. The vertices are $(a, 0)$ and $(-a, 0)$, and the asymptotes are the lines $y = \frac{b}{a}x$ and $y = -\frac{b}{a}x$. There are no *y*-intercepts. The center of the hyperbola is at $(0, 0)$.

Rule

The graph of an equation of the form

$$\frac{y^2}{b^2} - \frac{x^2}{a^2} = 1$$

is a **hyperbola** that opens on the *y*-axis. The vertices are $(0, b)$ and $(0, -b)$, and the asymptotes are the lines $y = \frac{b}{a}x$ and $y = -\frac{b}{a}x$. There are no *x*-intercepts. The center of the hyperbola is at $(0, 0)$.

Notice that the right side of the equation in either of the above forms is $+1$. If the x^2-term is positive, the hyperbola opens on the *x*-axis, and if the y^2-term is positive, the hyperbola opens on the *y*-axis. The two cases are illustrated in Figure 4.38.

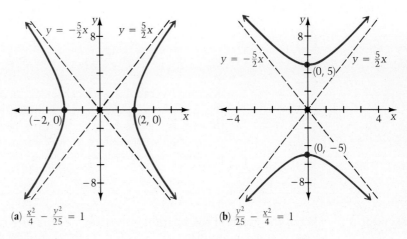

(a) $\frac{x^2}{4} - \frac{y^2}{25} = 1$ (b) $\frac{y^2}{25} - \frac{x^2}{4} = 1$

Figure 4.38

To determine the asymptotes, plot the points (a, b), $(-a, b)$ $(a, -b)$, and $(-a, -b)$. These four points determine the corners of a rectangle. The diagonals of the rectangle, when extended, form the two asymptotes.

Example 7   Sketch the graph of the relation $\frac{x^2}{16} - \frac{y^2}{9} = 1$. Show the center, vertices, and asymptotes.

Solution   Since the x^2-term is positive, this hyperbola opens on the x-axis with vertices $(4, 0)$ and $(-4, 0)$. Since $a = 4$ and $b = 3$, we can find the asymptotes by drawing the rectangle with corners $(4, 3)$, $(-4, 3)$, $(4, -3)$, and $(-4, -3)$, and then using the diagonals to graph the asymptotes.

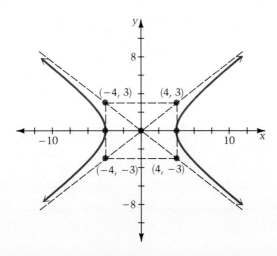

The two equations

$$\frac{(x-h)^2}{a^2} - \frac{(y-k)^2}{b^2} = 1 \quad \text{and} \quad \frac{(y-k)^2}{b^2} - \frac{(x-h)^2}{a^2} = 1$$

describe hyperbolas with centers at (h, k). If the $(x - h)^2$-term is positive, the hyperbola opens in the x-direction with vertices $(h + a, k)$ and $(h - a, k)$. If the $(y - k)^2$-term is positive, the hyperbola opens in the y-direction with vertices $(h, k + b)$ and $(h, k - b)$. The four corners of the rectangle that determine the asymptotes are $(h \pm a, k \pm b)$.

Example 8 Sketch the graph of the relation $9x^2 - 16y^2 + 18x - 64y + 521 = 0$. Show the center, vertices and asymptotes.

Solution

$(9x^2 + 18x \quad) + (-16y^2 - 64y \quad) = -521$ ■ Complete the square for x and y.

$9(x^2 + 2x \quad) - 16(y^2 + 4y \quad) = -521$

$9(x^2 + 2x + 1) - 16(y^2 + 4y + 4) = -521 + 9 - 64$

$9(x + 1)^2 - 16(y + 2)^2 = -576$ ■ Divide by -576 to get the proper form.

$$\frac{(y+2)^2}{36} - \frac{(x+1)^2}{64} = 1$$

Since the $(y + 2)^2$-term is positive, the hyperbola opens in the y-direction. The center of the hyperbola is at $(-1, -2)$. Since $a = 8$ and $b = 6$, the corners of the rectangle used to find the asymptotes are $(-1 \pm 8, -2 \pm 6)$, that is, $(7, -8), (7, 4), (-9, -8)$, and $(-9, 4)$. The vertices are $(-1, -2 + 6)$ and $(-1, -2 - 6)$ or simply $(-1, 4)$ and $(-1, -8)$.

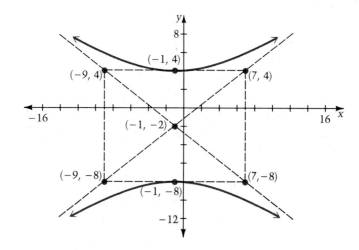

Exercise Set 4.4 ☐ Calculator Exercises in the Appendix

Goal A

In exercises 1–12, sketch the graph of the given relation. Determine the vertex, axis of symmetry, and the x- and y-intercepts.

1. $y^2 = 4x$
2. $2y^2 = -10x$
3. $9y^2 = 3x$
4. $4y^2 = x$
5. $x = -2(y - 3)^2 + 5$
6. $x = 3(y + 1)^2 - 2$
7. $x = 2(y + 1)^2 + \frac{3}{2}$
8. $x = -\frac{1}{2}(y - 2)^2 + 4$
9. $x = -y^2 + 6y - 13$
10. $x = 3 - 4y - y^2$
11. $y^2 + 2y + 3x - 8 = 0$
12. $y^2 - 6y - 2x - 9 = 0$

Goal B

In exercises 13–24, sketch the graph of each of the following relations. Determine the center and radius.

13. $x^2 + y^2 = 16$
14. $x^2 + y^2 = 25$
15. $(x + 1)^2 + (y - 3)^2 = 9$
16. $(x - 3)^2 + (y + 4)^2 = 1$
17. $x^2 + y^2 - 4x = 0$
18. $x^2 + y^2 - 6y + 5 = 0$
19. $x^2 + 4x + y^2 - 2y - 11 = 0$
20. $x^2 - 3x + y^2 - 5y - \frac{1}{2} = 0$
21. $4x^2 + 4y^2 - 4x + 4y - 10 = 0$
22. $9x^2 + 9y^2 - 6x + 18y - 8 = 0$
23. $25x^2 + 25y^2 - 10x - 150y + 1 = 0$
24. $9x^2 + 9y^2 - 3x - 6y - 1 = 0$

Goal C

In exercises 25–38, sketch the graph of each of the given relations. Determine the center and vertices.

25. $\frac{x^2}{9} + \frac{y^2}{16} = 1$
26. $\frac{x^2}{25} + \frac{y^2}{4} = 1$
27. $\frac{(x - 2)^2}{4} + \frac{(y + 3)^2}{100} = 1$
28. $\frac{(x + 1)^2}{36} + \frac{(y - 1)^2}{16} = 1$
29. $\frac{(x - 5)^2}{10} + \frac{y^2}{16} = 1$
30. $\frac{x^2}{3} + (y - 2)^2 = 1$
31. $\frac{25x^2}{121} + \frac{y^2}{16} = 1$
32. $x^2 + \frac{16y^2}{81} = 1$
33. $16(x - 1)^2 + 9(y - 2)^2 = 144$
34. $25(x + 3)^2 + (y + 6)^2 = 100$
35. $8x^2 + y^2 + 48x + 56 = 0$
36. $36x^2 + y^2 - 36x - 27 = 0$
37. $x^2 + 9y^2 + 2x - 18y + 1 = 0$
38. $9x^2 + 4y^2 - 36x + 8y + 4 = 0$

Goal D

In exercises 39–52, sketch the graph of each of the given relations. Show the center, vertices, and asymptotes.

39. $\frac{x^2}{16} - \frac{y^2}{9} = 1$
40. $\frac{y^2}{9} - \frac{x^2}{16} = 1$
41. $\frac{x^2}{2} - \frac{y^2}{10} = 1$
42. $\frac{y^2}{20} - \frac{x^2}{12} = 1$
43. $\frac{(y - 2)^2}{25} - \frac{(x - 1)^2}{4} = 1$

44. $\dfrac{(x-3)^2}{36} - \dfrac{y^2}{16} = 1$

45. $\dfrac{16(x-2)^2}{25} - \dfrac{64(y-3)^2}{49} = 1$

46. $36\left(x - \dfrac{1}{2}\right)^2 - \dfrac{9y^2}{4} = 1$

47. $4x^2 - 25y^2 - 100 = 0$

48. $y^2 - 10x^2 - 10 = 0$

49. $16y^2 - x^2 - 48y - 28 = 0$

50. $16x^2 - y^2 - 56x - 207 = 0$

51. $9y^2 - 4x^2 + 18y - 16x - 43 = 0$

52. $x^2 - 8y^2 - 4x - 32y - 36 = 0$

Superset

In exercises 53–58, identify the graph of each equation as a parabola, circle, ellipse, or hyperbola; then sketch the graph.

53. $x^2 + y^2 - 4x + 2y - 4 = 0$

54. $x^2 + 10x + 2y^2 + 23 = 0$

55. $x = 3y^2 - 12y + 11$

56. $2y^2 - 3x^2 - 24y - 30x - 9 = 0$

57. $x^2 + y^2 + 8x + 2y + 13 = 0$

58. $4x^2 + 3y^2 + 24x - 6y + 27 = 0$

In exercises 59–60, use the fact that if (x_1, y_1) and (x_2, y_2) are the endpoints of a line segment, then the midpoint of the segment has coordinates

$$\left(\dfrac{x_1 + x_2}{2}, \dfrac{y_1 + y_2}{2}\right).$$

59. Find the equation of a circle if one of its diameters has endpoints $(-4, 5)$ and $(2, -1)$.

60. Find the equation of a circle if one of its diameters has endpoints $(-1, 1)$ and $(9, -1)$.

Recall that a line is tangent to a circle if the line touches the circle only once and does not cross it.

61. A circle of radius 3 has its center in the fourth quadrant and is tangent to both axes. Write the equation of the circle.

62. The line tangent to a circle is perpendicular to the radius at the point where the radius and tangent meet. This is illustrated below.

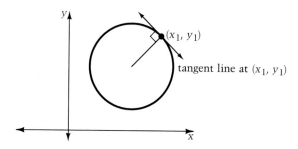

Find the equation of the line tangent to the circle $x^2 + y^2 + 6x - 4y - 12 = 0$ at the point $(1, 5)$.

In exercises 63–70, graph the given relation, and state whether or not it is a function. Begin by squaring both sides. Remember that the graph of the resulting conic section must be restricted in order to obtain the graph described by the given equation.

63. $y = \sqrt{9 - x^2}$

64. $x = -\sqrt{9 - y^2}$

65. $x = \sqrt{4 - y^2}$

66. $y = -\sqrt{4 - x^2}$

67. $y = -\sqrt{x - 2}$

68. $y = \sqrt{2 - x}$

69. $x = \sqrt{y^2 - 1}$

70. $x = -\sqrt{y^2 + 4}$

4 / Polynomial Functions and Rational Functions

Chapter Review & Test

Chapter Review

4.1 Important Features of Polynomial Functions (pp. 148–156)

An *nth degree polynomial function* in x is a function that has the form $f(x) = a_n x^n + a_{n-1} x^{n-1} + \cdots + a_1 x + a_0$, where $a_n \neq 0$, and n is a non-negative integer. (p. 149) Polynomial functions of the form $y = ax^n$ (where n is a positive integer greater than 1) can have one of the following shapes. (p. 151)

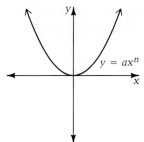
Type I: a positive, n even

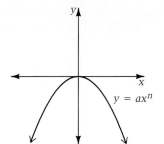
Type II: a negative, n even

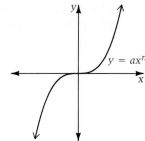
Type III: a positive, n odd

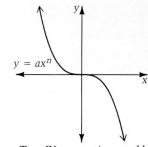
Type IV: a negative, n odd

The First Rule of Polynomial Functions says that as $x \to +\infty$ and as $x \to -\infty$, the graph of a polynomial function described by the equation $f(x) = a_n x^n + a_{n-1} x^{n-1} + \cdots + a_1 x + a_0$ behaves like the graph of $g(x) = a_n x^n$. (p. 153)

4.2 Graphing Polynomial Functions (pp. 157–166)

In general, an nth degree polynomial function of the form $f(x) = a_n x^n + a_{n-1} x^{n-1} + \cdots + a_1 x + a_0$ can have at most n real zeros (x-intercepts). (p. 157)

The Second Rule of Polynomial Functions says that the graph of an nth degree polynomial function cannot intersect (cross or touch) the x-axis more than n times. (p. 157)

The Third Rule of Polynomial Functions says that the graph of an nth degree polynomial function can have at most $n - 1$ relative maxima and minima (peaks and valleys). (p. 158)

The Fourth Rule of Polynomial Functions says that if c_1 and c_2 are adjacent zeros of a polynomial function f, then the graph of f lies either completely above or completely below the x-axis for all values between c_1 and c_2. (p. 161)

4.3 Rational Functions (pp. 167–174)

A *rational function* has the form $f(x) = \dfrac{g(x)}{h(x)}$, where $g(x)$ and $h(x)$ are polynomials. (p. 167)

If a graph gets closer and closer to a certain line but never touches it, then the line is an *asymptote* of the graph (p. 167)   The *First Rule of Rational Functions* says that the line $x = c$ is a *vertical asymptote* of the graph of the rational function f defined above provided $g(c) \neq 0$ and $h(c) = 0$. (p. 168)

The *Second Rule of Rational Functions* says that if f is a rational function whose numerator has degree n and whose denominator has degree m, then (a) if $n < m$, then the x-axis is a *horizontal asymptote*; (b) if $n = m$, then the line $y = c$, where c is the quotient of the leading coefficients, is a horizontal asymptote; or (c) if $n > m$, there is no horizontal asymptote. (p. 169)

The *Third Rule of Rational Functions* says that between any two successive x-values where a rational function f is either 0 or not defined, the graph of f lies either completely above or completely below the x-axis. (p. 171)

4.4 Conic Sections (pp. 175–188)

The equation of a *parabola* that opens to the right or left:
$$x = a(y - k)^2 + h. \qquad \text{(p. 177)}$$

The equation of a *circle* with center at (h, k) and radius r:
$$(x - h)^2 + (y - k)^2 = r^2. \qquad \text{(p. 180)}$$

The equation of an *ellipse* with center at (h, k):
$$\frac{(x - h)^2}{a^2} + \frac{(y - k)^2}{b^2} = 1. \qquad \text{(p. 182)}$$

The equation of a *hyperbola* with center at (h, k) is given by one of the forms
$$\frac{(x - h)^2}{a^2} - \frac{(y - k)^2}{b^2} = 1 \quad \text{and} \quad \frac{(y - k)^2}{b^2} - \frac{(x - h)^2}{a^2} = 1. \qquad \text{(p. 186)}$$

Chapter Test

4.1A Determine whether the given function is a polynomial function. If so, write it in standard form, and state its leading term.

1. $f(x) = 5 - 2x + 3x^2$ **2.** $g(x) = 1 - \dfrac{1}{x} + 3x^3$

3. $h(x) = x^3 + 5x^2 - \sqrt{x} + 1$ **4.** $k(x) = 3x^2 + 2(1 - x^3) + 7$

4.1B Classify the graph of each equation as one of types shown on page 189.

5. $y = 6x^5$
6. $y = -\frac{1}{2}x^6$
7. $y = 3x^4$
8. $2x^3 + y = 0$

4.1C Select the graph from the four below that best represents the given function.

9. $f(x) = x^3 - x^2 - x + 1$
10. $f(x) = 1 - x^3 + x^2 + x$
11. $f(x) = 1 - x^4 + 4x^2$
12. $f(x) = x^4 - 4x^2 + 4$

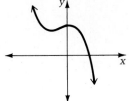 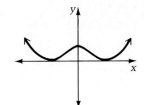 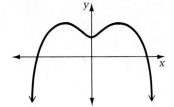 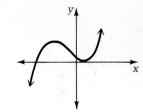

Graph No. 1　　　Graph No. 2　　　Graph No. 3　　　Graph No. 4

4.2A Use the First, Second, and the Third Rules of Polynomial Functions to select the graph from the four below that best represents the function.

13. $f(x) = x^3 - 5x^2$
14. $f(x) = x^3 + 4x^2 - 11x + 6$

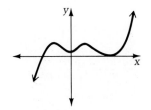 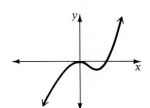 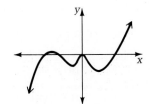 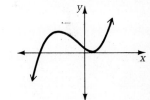

Graph No. 1　　　Graph No. 2　　　Graph No. 3　　　Graph No. 4

4.2B Determine the zeros of each function.

15. $y = x(x - 5)^2(3 - 2x)^3$
16. $f(x) = 4x^2(x + 2)(7 - x)^2$
17. One of the zeros of $f(x) = x^3 + 2x^2 - 5x - 6$ is 2. Find all the other zeros.

4.2C 18. Find the zeros of $f(x) = x(x - 2)(x + 4)^4$, then sketch the graph.

4.3A Write the equations of any horizontal or vertical asymptotes of the graph of the given function.

19. $y = \dfrac{3x^2 - 12}{6x^2 - 5x + 1}$
20. $f(x) = \dfrac{2x}{5x^2 - 45}$

4.3B Sketch the graph of each rational function.

21. $y = \dfrac{4}{x - 4}$
22. $y = \dfrac{1}{x^2 - 4x - 5}$

4.4A Sketch each relation. Determine the vertex, the axis of symmetry and the x- and y-intercepts.

23. $x = 2(y - 1)^2$
24. $y^2 - 6y - x + 5 = 0$

4.4B Sketch each relation. Determine the center and radius.

25. $(x - 2)^2 + (y + 3)^2 = 25$
26. $x^2 + 2x + y^2 - 6y + 7 = 0$

4.4C Sketch each relation. Determine the center and vertices.

27. $\dfrac{(x - 5)^2}{25} + \dfrac{y^2}{4} = 1$
28. $9x^2 - 18x + 5y^2 + 20y = 16$

4.4D Sketch each relation. Determine the center, vertices, and asymptotes.

29. $\dfrac{(y + 2)^2}{24} - \dfrac{(x - 1)^2}{4} = 1$
30. $4x^2 + 24x - 10y^2 + 40y = 44$

Superset

31. Sketch the graphs of the following functions.

(a) $f(x) = 2 + (x - 1)^2$
(b) $g(x) = \dfrac{1}{2 + (x - 1)^2}$
(c) $h(x) = (x + 2)^2 - 3$
(d) $k(x) = |(x + 2)^2 - 3|$

32. A circle has its center in the first quadrant, has radius 5, and passes through the origin. If the y-coordinate of the center is twice the x-coordinate, find the equation of the circle.

33. An ant is following a path in the xy-plane. It starts at the point $(1, 2)$ and its coordinates at any time t are given by $x = t + 1$ and $y = 2 - t^2$.

(a) Sketch the ant's path.
(b) Identify the conic section, part of which the ant is tracing.
(c) At which point on the ant's path is the y-coordinate greatest?

5 Systems of Equations and Inequalities

5.1 Systems of Equations in Two Variables

Goal A *to determine whether a given ordered pair is a solution of a system of equations*

In Chapter 3 we saw that the graph of an equation of the form

$$Ax + By = C$$

(where A and B are not both zero) is a straight line. Now, let us consider a pair of such equations.

$$\begin{cases} 4x + 3y = 11 \\ -3x + 2y = -4. \end{cases}$$

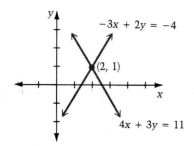

Figure 5.1

These two equations are graphed in Figure 5.1. Each point on these lines is a solution of one equation or the other. Only the point (2, 1) lies on both lines, and thus, only the ordered pair (2, 1) is a solution of both equations.

A collection of two or more equations is called a **system of equations.** The example above is a system of two equations in two variables. Those ordered pairs that are solutions of *every* equation in the system are called **solutions of the system.** We refer to the process of finding solutions of the system as **solving the system** (or solving the system simultaneously).

Example 1 Determine whether the following ordered pairs are solutions of the system

$$\begin{cases} x + 2y + 1 = 0 \\ 3x + 4y - 3 = 0. \end{cases}$$

(a) $(3, -2)$ (b) $(5, -3)$

Solution (a) $x + 2y + 1 = 0$
$3 + 2(-2) + 1 = 0$ ■ $(3, -2)$ is a solution of the first equation.

$3x + 4y - 3 = 0$
$3(3) + 4(-2) - 3 \neq 0$ ■ $(3, -2)$ is not a solution of the second equation.

Thus, $(3, -2)$ is not a solution of the system.

(b) $x + 2y + 1 = 0$
$5 + 2(-3) + 1 = 0$ ■ $(5, -3)$ is a solution of the first equation.

$3x + 4y - 3 = 0$
$3(5) + 4(-3) - 3 = 0$ ■ $(5, -3)$ is a solution of the second equation.

Thus, $(5, -3)$ is a solution of the system.

Goal B *to solve a system of two linear equations in two variables by the addition method*

Recall the strategy for solving a linear equation in one variable. Given an equation such as

$$3(2 - x) + 5x = 18 - 4x,$$

we rewrite the equation as a succession of equivalent equations,

$$6 - 3x + 5x = 18 - 4x$$
$$2x + 6 = 18 - 4x$$
$$6x = 12$$
$$x = 2,$$

until we obtain an equivalent equation (such as $x = 2$) whose solution is obvious.

Our approach to solving a system of equations is similar. We will develop techniques whereby a system such as

$$\begin{cases} 4x + 3y = 11 \\ -3x + 2y = -4 \end{cases}$$

can be written as a simpler, equivalent system, such as

$$\begin{cases} x = 2 \\ y = 1. \end{cases}$$

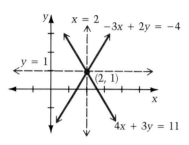

Figure 5.2

Notice that the lines $x = 2$ and $y = 1$ intersect at the point $(2, 1)$. Thus, the lines in the original system also intersect at this point, and the ordered pair $(2, 1)$ is a solution of the system.

The fact below can be used to justify the procedures for writing equivalent systems. (In the statement of the fact, we use the phrase "multiple of the equation" to mean the new equation found by multiplying both sides of the original equation by the same number.)

Fact

Suppose lines $\mathscr{L}$ and $\mathscr{M}$ intersect at point P. Adding any nonzero multiple of the equation of line $\mathscr{L}$ to any nonzero multiple of the equation of line $\mathscr{M}$ produces an equation of a line. This line also passes through point P.

For example, using the system at the top of the page, we have the following:

$$\begin{array}{r} 8x + 6y = 22 \\ + -15x + 10y = -20 \\ \hline -7x + 16y = 2 \end{array}$$

- 2 "times" $4x + 3y = 11$
- 5 "times" $-3x + 2y = -4$
- The "sum" of the two equations.

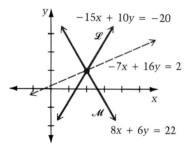

Figure 5.3

The solution $(2, 1)$ also lies on the line $-7x + 16y = 2$.

The first technique we shall use to solve a system of equations is called the **addition method**. This method involves two steps.

Step 1. Form multiples of the equations so that their sum is an equation in one variable and solve this resulting equation.

Step 2. Substitute the value found in Step 1 back into either of the original equations to find the value of the other variable.

Example 2 Use the addition method to solve the system $\begin{cases} 4x + 3y = 11 \\ -3x + 2y = -4. \end{cases}$

5 / Systems of Equations and Inequalities

Solution
$$\begin{cases} 4x + 3y = 11 \\ -3x + 2y = -4 \end{cases}$$

$3(4x + 3y) = 3(11)$
$4(-3x + 2y) = 4(-4)$
- Step 1. Multiply the first equation by 3 and the second equation by 4. These multipliers were chosen in order to eliminate the variable x.

$12x + 9y = 33$
$\underline{+\ -12x + 8y = -16}$
$17y = 17$
- Add the equations.
- Solve the equation for y.

$y = 1$
- Replace one of the equations in the original system with $y = 1$.

$$\begin{cases} 4x + 3y = 11 \\ y = 1 \end{cases}$$

$4x + 3(1) = 11$
$4x = 8$
$x = 2$
- Step 2. Substitute the value of y and solve for x.
- Use the equation $x = 2$ to state a simpler, equivalent system.

$$\begin{cases} x = 2 \\ y = 1 \end{cases}$$

The solution is $(2, 1)$.

We chose the multipliers 3 and 4 in Example 2 so that the sum of the resulting equations would have no x-term. Instead, we could have eliminated the y-term by using the multipliers 2 and -3:

$2(4x + 3y) = 2(11) \longrightarrow 8x + 6y = 22$
$-3(-3x + 2y) = -3(-4) \longrightarrow \underline{+\ 9x - 6y = 12}$
$17x = 34$

There is always more than one way to select the multipliers.

There are three kinds of answers you can get when you solve a system of two linear equations in two variables. These three types are illustrated in Figure 5.4.

In the Figure 5.4(a), lines $\mathscr{L}$ and $\mathscr{M}$ are nonparallel. The corresponding system of equations has exactly one solution. In the Figure 5.4(b), lines $\mathscr{L}$ and $\mathscr{M}$ are different parallel lines. In this case, the system has no solutions. In the Figure 5.4(c), lines $\mathscr{L}$ and $\mathscr{M}$ are the same. Such a system has infinitely many solutions. We looked at the first type of system in Example 2. We will consider the second and third types in Examples 3 and 5.

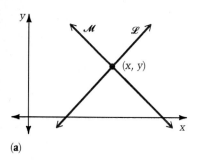 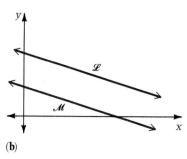 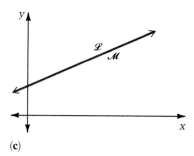

(a) (b) (c)

Figure 5.4

Example 3 Use the addition method to solve the system $\begin{cases} 3x + 4y = 7 \\ 6x + 8y = 5. \end{cases}$

Solution $\begin{cases} 3x + 4y = 7 \\ 6x + 8y = 5 \end{cases}$

$2(3x + 4y) = 2(7)$
$-1(6x + 8y) = -1(5)$

- Step 1. Multiply the first equation by 2 and the second equation by -1. The multipliers were chosen to eliminate the variable x.

$6x + 8y = 14$
$+\ -6x - 8y = -5$
$0 = 9$

- Add the equations.

- Note that both variables have been eliminated. Replace one of the equations in the original system with $0 = 9$.

$\begin{cases} 3x + 4y = 7 \\ 0 = 9 \end{cases}$

The statement $0 = 9$ is never true. A system that includes such an equation can have no solutions. Since the original system is equivalent to this system, it also has no solutions.

To verify our work in Example 3, we rewrite the equations in slope-intercept form.

$$y = -\frac{3}{4}x + \frac{7}{4} \quad \text{and} \quad y = -\frac{3}{4}x + \frac{5}{8}.$$

These are different parallel lines. (Slopes are the same, but y-intercepts are different.) Thus, the lines do not intersect.

5 / Systems of Equations and Inequalities

Goal C *to solve a system of two linear equations in two variables by the substitution method*

The second technique we shall use to solve a system of equations is called the **substitution method**. This method involves three steps, as shown in Example 4.

Example 4 Use the substitution method to solve the system $\begin{cases} 3y + 4x = 11 \\ 2y - 3x = -4. \end{cases}$

Solution $\begin{cases} 3y + 4x = 11 \\ 2y - 3x = -4 \end{cases}$

$$3y + 4x = 11$$
$$4x = -3y + 11$$
$$x = -\frac{3}{4}y + \frac{11}{4}$$

■ Step 1. Solve one of the equations for x. We have selected the first equation.

$\begin{cases} x = -\frac{3}{4}y + \frac{11}{4} \\ 2y - 3x = -4 \end{cases}$

$$2y - 3\left(-\frac{3}{4}y + \frac{11}{4}\right) = -4$$
$$2y + \frac{9}{4}y - \frac{33}{4} = -4$$
$$\frac{17}{4}y = -4 + \frac{33}{4}$$
$$17y = 17$$
$$y = 1$$

■ Step 2. Substitute the expression for x from Step 1 into the second equation, and then solve for y.

$\begin{cases} x = -\frac{3}{4}y + \frac{11}{4} \\ y = 1 \end{cases}$

$$x = -\frac{3}{4}(1) + \frac{11}{4}$$
$$x = 2$$

■ Step 3. Substitute the y-value found in Step 2 into the other equation to obtain a value for x.
■ You should check that this is a solution of the system.

$\begin{cases} x = 2 \\ y = 1 \end{cases}$

The solution of the system is (2, 1).

Careful! In the previous examples, notice that we have restated the system as an equivalent system at various stages. You should follow this practice in order to keep track of the pair of equations you must work on next.

Let us summarize the three steps of the substitution method.

Step 1. Solve one of the equations for one of the variables.
Step 2. Substitute the result from Step 1 into the other equation of the system, and solve for the second variable.
Step 3. Substitute the value found in Step 2 into either one of the original equations and solve for the value of the first variable.

The next example illustrates a system in which the two equations describe the same line. Notice that we never reach the third step of the substitution method.

Example 5 Use the substitution method to solve the system $\begin{cases} 3x + 4y = 7 \\ 6x + 8y = 14. \end{cases}$

Solution $\begin{cases} 3x + 4y = 7 \\ 6x + 8y = 14 \end{cases}$

$$6x + 8y = 14$$
$$6x = -8y + 14$$
$$x = -\frac{4}{3}y + \frac{7}{3}$$

■ Step 1. Solve for x in the second equation.

$\begin{cases} 3x + 4y = 7 \\ x = -\frac{4}{3}y + \frac{7}{3} \end{cases}$

$$3\left(-\frac{4}{3}y + \frac{7}{3}\right) + 4y = 7$$
$$-4y + 7 + 4y = 7$$
$$7 = 7$$

■ Step 2. Substitute the expression for x and (try to) solve for y.

$\begin{cases} 7 = 7 \\ x = -\frac{4}{3}y + \frac{7}{3} \end{cases}$

The equation $7 = 7$ is always true. The solutions of the equation $x = -\frac{4}{3}y + \frac{7}{3}$ are all the points on that line. Therefore, the solution set of the system is the set of all points on the line $x = -\frac{4}{3}y + \frac{7}{3}$.

Goal D *to solve a system of nonlinear equations in two variables*

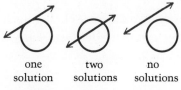

one solution   two solutions   no solutions

Figure 5.5

When solving a system that contains a nonlinear equation we follow the same general strategy we used for linear systems: rewrite the system as a succession of equivalent systems until a solution becomes obvious.

The equations in the next example represent a line and a circle. There are only three types of solutions that may be obtained; these are shown in Figure 5.5. You should never find three or more points of intersection.

Example 6 Solve the system $\begin{cases} y = x - 2 \\ (x-4)^2 + y^2 = 4. \end{cases}$

Solution $\begin{cases} y = x - 2 \\ (x-4)^2 + y^2 = 4 \end{cases}$

$(x-4)^2 + y^2 = 4$
$(x-4)^2 + (x-2)^2 = 4$ ■ Substitute the expression for y from the first equation into the second equation to obtain an equivalent system.

$\begin{cases} y = x - 2 \\ (x-4)^2 + (x-2)^2 = 4 \end{cases}$

$(x-4)^2 + (x-2)^2 = 4$ ■ Solve the second equation for x.
$x^2 - 8x + 16 + x^2 - 4x + 4 = 4$
$2x^2 - 12x + 16 = 0$
$x^2 - 6x + 8 = 0$
$(x-2)(x-4) = 0$
$x = 2$ or $x = 4$ ■ Each of these two equations produces one equivalent system.

$\begin{cases} y = x - 2 \\ x = 2 \end{cases}$ $\begin{cases} y = x - 2 \\ x = 4 \end{cases}$

$y = (2) - 2$ $y = (4) - 2$ ■ Substitute the x-values into the first equation.
$y = 0$ $y = 2$

$$\begin{cases} y = 0 \\ x = 2 \end{cases} \quad \begin{cases} y = 2 \\ x = 4 \end{cases}$$

The given system has two solutions: (2, 0) and (4, 2). You should check that these are, in fact, solutions of each equation of the original system.

In the next example, the equations represent two different circles. Such a system may have no solutions, one solution, or two solutions, as illustrated below.

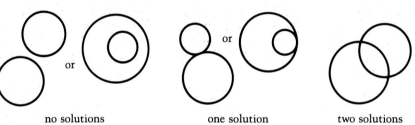

no solutions   one solution   two solutions

Figure 5.6

Example 7   Solve the system $\begin{cases} (x - 2)^2 + (y - 2)^2 = 4 \\ (x - 4)^2 + y^2 = 4. \end{cases}$

Solution $\begin{cases} (x - 2)^2 + (y - 2)^2 = 4 \\ (x - 4)^2 + y^2 = 4 \end{cases}$

$$(x^2 - 4x + 4) + (y^2 - 4y + 4) = 4 \quad \blacksquare \text{ Each expression is expanded.}$$
$$(x^2 - 8x + 16) + y^2 = 4$$

$$\begin{array}{r} x^2 + y^2 - 4x - 4y + 4 = 0 \\ -\underline{ - x^2 + y^2 - 8x + 12 = 0} \\ 4x - 4y - 8 = 0 \\ 4y = 4x - 8 \\ y = x - 2 \end{array}$$

■ Each equation has an x^2- and y^2-term. Therefore, we subtract the second equation from the first.
■ We simplify and then state an equivalent system.

$$\begin{cases} y = x - 2 \\ (x - 4)^2 + y^2 = 4 \end{cases}$$

This is precisely the system we solved in Example 6. Our original system is equivalent to a system for which we already know the solutions: (2, 0) and (4, 2).

Exercise Set 5.1   = Calculator Exercises in the Appendix

Goal A

In exercises 1–4, determine whether each of the ordered pairs is a solution of the system.

1. (a) (3, 2)
   (b) (3, −2)
   $\begin{cases} 5x + 7y = 1 \\ x - 2y = 1 \end{cases}$

2. (a) (4, 1)
   (b) (1, 4)
   $\begin{cases} 3x - y = -1 \\ x + 2y = 9 \end{cases}$

3. (a) (5, 1)
   (b) (−4, −2)
   $\begin{cases} 2x - 6y = 4 \\ 3x - 9y = 6 \end{cases}$

4. (a) (5, 1)
   (b) (3, 0)
   $\begin{cases} 2x - 4y = 6 \\ -3x + 6y = 9 \end{cases}$

Goal B

In exercises 5–12, use the addition method to solve each system of equations.

5. $\begin{cases} 3x + 5y = 1 \\ 4x + 3y = 5 \end{cases}$

6. $\begin{cases} -2x + 7y = 8 \\ 3x - y = 7 \end{cases}$

7. $\begin{cases} 5x - 2y = 2 \\ 7x + 6y = 5 \end{cases}$

8. $\begin{cases} -9x + 6y = 4 \\ 7x - 2y = 2 \end{cases}$

9. $\begin{cases} 12x + 9y = 6 \\ 8x + 6y = 4 \end{cases}$

10. $\begin{cases} 2x + 4y = 5 \\ 3x + 6y = 1 \end{cases}$

11. $\begin{cases} \sqrt{14}x - \sqrt{10}y = 2 \\ \sqrt{7}x - \sqrt{5}y = -3 \end{cases}$

12. $\begin{cases} \sqrt{8}x + 2y = 8 \\ \sqrt{2}x + y = 4 \end{cases}$

Goal C

In exercises 13–20, use the substitution method to solve each system of equations.

13. $\begin{cases} 3x + 6y = 4 \\ x - 3y = -1 \end{cases}$

14. $\begin{cases} 2x - 6y = 5 \\ 6x + 7y = 0 \end{cases}$

15. $\begin{cases} 6x + 2y = 8 \\ 3x + y = 4 \end{cases}$

16. $\begin{cases} 6x - 10y = -4 \\ 3x - 5y = 2 \end{cases}$

17. $\begin{cases} 10x + 5y = 7 \\ 6x + 3y = 4 \end{cases}$

18. $\begin{cases} -15x + 9y = -3 \\ 5x - 3y = 1 \end{cases}$

19. $\begin{cases} \sqrt{3}x - y = 5 \\ x + \sqrt{3}y = -\sqrt{3} \end{cases}$

20. $\begin{cases} 2x + \sqrt{5}y = 7 \\ \sqrt{5}x - 3y = -2\sqrt{5} \end{cases}$

Goal D

In exercises 21–28, solve each system.

21. $\begin{cases} x^2 - 2x + y^2 = 3 \\ x - y = 3 \end{cases}$

22. $\begin{cases} x^2 - 10y + y^2 = 0 \\ 4y - 3x = 20 \end{cases}$

23. $\begin{cases} x^2 - 6x + y^2 = -4 \\ 2x - y = 1 \end{cases}$

24. $\begin{cases} x^2 + 2y + y^2 = 12 \\ 3x - 2y = 7 \end{cases}$

25. $\begin{cases} x^2 - 4x + y^2 + 8y = -19 \\ 4y + 5x = 20 \end{cases}$

26. $\begin{cases} x^2 - 2x + y^2 - 6y = -6 \\ 4y - x = -8 \end{cases}$

27. $\begin{cases} x^2 + 8x + y^2 - 18y = 72 \\ (x - 2)^2 + (y - 9)^2 = 1 \end{cases}$

28. $\begin{cases} x^2 - 10x + y^2 + 6y = -30 \\ (x - 5)^2 + (y + 3)^2 = 4 \end{cases}$

Superset

In exercises 29–32, solve each system.

29. $\begin{cases} (y - 1)^2 = x + 1 \\ x = 4y - 9 \end{cases}$

30. $\begin{cases} y + 3 = (x - 2)^2 \\ y = x - 3 \end{cases}$

31. $\begin{cases} 8(x - 1) = (y + 2)^2 \\ y + 2 = (x - 1)^2 \end{cases}$

32. $\begin{cases} x^2 + y^2 = 4 \\ x = y^2 + 2 \end{cases}$

33. A cash register contains dimes and quarters. There are 9 more quarters than dimes, and the total value of the coins is $5.05. How many quarters are there? How many dimes?

34. A 450 mi plane trip takes 3 hrs with the wind and 5 hrs against the wind. Find the rate of the plane in still air and the rate of the wind.

35. You have invested $1500 in stocks and bonds, offering a return of 15% and 9%, respectively. If you receive $195 in interest, how much is invested in bonds?

36. Find all two-digit numbers such that the sum of the digits is 13, and the number is increased by 45 if the order of the digits is reversed.

5.2 Systems of Inequalities in Two Variables

Goal A *to graph the solution set of a single inequality in two variables*

To solve an inequality in two variables, we begin by looking at the associated equality. For example, in order to solve

$$y - 2x - 3 > 0$$

we first look at

$$y - 2x - 3 = 0.$$

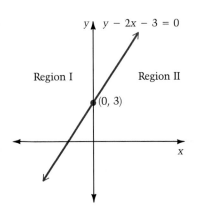

Figure 5.7

The graph of this equation is the line shown at the left. This line divides the plane into two regions; call them Region I and Region II. Since any point on the line satisfies the equation

$$y - 2x - 3 = 0,$$

it seems reasonable that any point not on the line satisfies one of the inequalities

$$y - 2x - 3 < 0$$

or

$$y - 2x - 3 > 0.$$

It turns out that all the points in the same region satisfy the same inequality.

Select a point in Region I, say $(-1, 5)$. Substituting yields

$$5 - 2(-1) - 3 = 4.$$

Thus, $(-1, 5)$ satisfies $y - 2x - 3 > 0$. We conclude that all other points in Region I satisfy this inequality as well.

Now consider the point $(4, 6)$, which lies in Region II. Substituting yields

$$6 - 2(4) - 3 = -5.$$

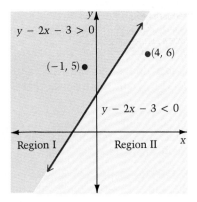

Figure 5.8

Thus, $(4, 6)$ satisfies $y - 2x - 3 < 0$. We conclude that all the other points in Region II satisfy this inequality. The reasoning behind these claims is based on the following fact.

Fact

The graph of an equation divides the *xy*-plane into regions. All the points in any one region must satisfy the same inequality formed from the associated equality. Thus, to determine which inequality holds for a region, you need only test one point from that region.

Example 1 Sketch the graph of each inequality.

(a) $x^2 + y^2 < 9$ (b) $y \geq 2$.

Solution (a)

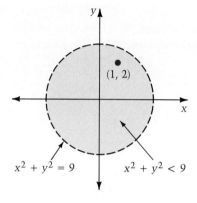

- Begin by graphing the equation $x^2 + y^2 = 9$ (a circle) with dashes to indicate that the points on the circle are not in the solution set of the inequality.
- Select a point outside the circle, say (3, 4). Since $3^2 + 4^2 > 9$, all points outside the circle satisfy the inequality $x^2 + y^2 > 9$.
- Select a point inside the circle, say (1, 2). Since $1^2 + 2^2 < 9$, all points inside the circle satisfy the inequality $x^2 + y^2 < 9$.
- We shade the interior of the circle to indicate the solution set of the inequality.

(b)

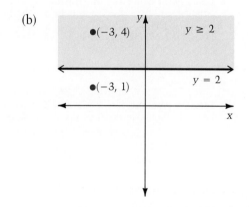

- Begin by graphing the equation $y = 2$ as a solid line to indicate that the points on the line are included in the solution set of the inequality.
- Select a point above the line, say $(-3, 4)$. Since its y-coordinate, 4, is greater than 2, all points above the line satisfy the inequality $y > 2$.
- Select a point below the line, say $(-3, 1)$. Since its y-coordinate, 1, is less than 2, all points below the line satisfy the inequality $y < 2$.
- We graph the solution set of $y \geq 2$ by shading the region above the line.

Careful! Remember that if the inequality symbol is either $\geq$ or $\leq$, we use a solid curve to indicate that the points on the boundary of the region (i.e., the solutions of the associated equation) are included in the solution set of the inequality. If the symbol is $>$ or $<$, we indicate that these points are excluded by using a dashed curve.

Goal B *to graph the solution set of a system of inequalities in two variables*

The following is a system of inequalities in two variables:

$$\begin{cases} y < 3x - 1 \\ y > 3 - x. \end{cases}$$

The graph of such a system lies in the xy-plane and consists of all points satisfying both inequalities. To graph this system, we graph the region of each inequality and then determine the points common to both regions.

Example 2 Graph the solution set of the following system: $\begin{cases} y < 3x - 1 \\ y > 3 - x. \end{cases}$

Solution

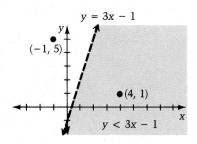

- $y < 3x - 1$

  Begin by graphing $y = 3x - 1$ as a dashed line.
- Test a point above the line, say $(-1, 5)$. Since $(-1, 5)$ satisfies $y > 3x - 1$, all points above the line satisfy $y > 3x - 1$.
- Test a point below the line, say $(4, 1)$. Since $(4, 1)$ satisfies $y < 3x - 1$, all points below the line satisfy $y < 3x - 1$.
- Thus, we shade the region below the line.

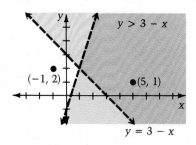

- $y > 3 - x$

- Begin by graphing $y = 3 - x$ as a dashed line. Test a point above the line, say $(5, 1)$. Since $(5, 1)$ satisfies $y > 3 - x$, all points above the line satisfy $y > 3 - x$.
- Test a point below the line, say $(-1, 2)$. Since $(-1, 2)$ satisfies $y < 3 - x$, all points below the line satisfy $y < 3 - x$.
- Thus, we shade the region above the line.

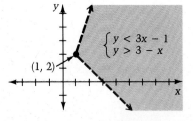

- Finally, the solution set of the system consists of those points common to both regions. These points are represented by the overlap of the two regions. A good point of reference is the "corner point" $(1, 2)$, which is found by solving the system

$$\begin{cases} y = 3x - 1 \\ y = 3 - x. \end{cases}$$

Example 3 Graph the solution set of the following system: $\begin{cases} y \geq 0 \\ x \leq 5 \\ y \leq x - 1. \end{cases}$

Solution

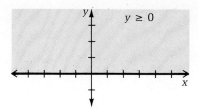

- Begin by graphing the solution set of each of the inequalities on the same pair of coordinate axes. We first graph the solution set of $y \geq 0$.

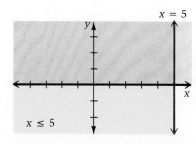

- Next, we graph the solution set of $x \leq 5$.

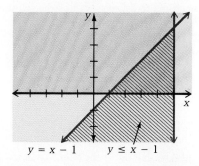

- Now, sketch the solution set of $y \leq x - 1$. The point (3, 1) lies below the line. Since $1 < 3 - 1 = 2$, every point below the line satisfies the inequality $y < x - 1$.
- The point (2, 5) lies above the line. Since $5 > 2 - 1 = 1$, every point above the line satisfies the inequality $y > x - 1$.
- Thus, we shade the region below the line.

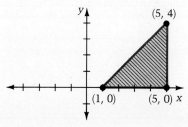

- The region where all three graphs overlap is the graph of the solution set of the system. We find the coordinates of the "corners" by solving the appropriate two-equation systems. For example, we obtain the corner point (5, 4) by solving the system
$$\begin{cases} x = 5 \\ y = x - 1. \end{cases}$$

Example 4 Graph the solution set of the following system: $\begin{cases} y \geq 2 \\ x \geq 3 \\ x > y \\ x + y < 10. \end{cases}$

Solution

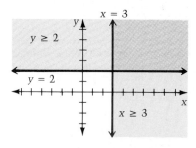

- Begin by graphing the solution set of each of the inequalities on the same pair of coordinate axes. We first graph the solution set of $y \geq 2$, and the solution set of $x \geq 3$.

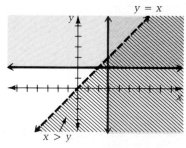

- Now, we graph the solution set of $x > y$.

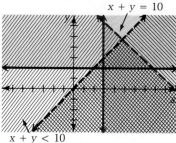

- Now, using test points, we find the solution set of $x + y < 10$.

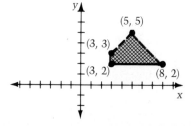

- Finally, we determine the points common to the previous four solution sets. The region where all four graphs overlap is the graph of the solution set of the system.

Exercise Set 5.2

Goal A

In exercises 1–24, sketch the graph of each inequality.

1. $x \leq 0$
2. $y > 0$
3. $x + y > 1$
4. $y < 2x + 3$
5. $y \leq 3x - 1$
6. $y \geq 5 - x$
7. $x^2 + y^2 \leq 9$
8. $x^2 + y^2 \geq 4$
9. $(x - 1)^2 + y^2 \geq 1$
10. $x^2 + (y + 2)^2 < 9$
11. $(x + 4)^2 + (y - 7)^2 < 4$
12. $(x + 3)^2 + (y + 5)^2 > 25$
13. $y \geq x^2$
14. $x < y^2$
15. $x - 2 \leq (y + 3)^2$
16. $y - 1 < (x + 2)^2$
17. $y > |x|$
18. $x \geq -|y|$
19. $x < |6 - y|$
20. $y \leq |x + 3|$
21. $xy \geq 1$
22. $xy < -9$
23. $y(x - 1) < -3$
24. $y(x + 2) \leq 4$

Goal B

In exercises 25–40, graph the solution set of each system.

25. $\begin{cases} x \geq -2 \\ y < 1 \end{cases}$
26. $\begin{cases} x < 3 \\ y \geq -4 \end{cases}$
27. $\begin{cases} x + y > 1 \\ y < 1 \end{cases}$
28. $\begin{cases} x + y < 3 \\ y > 2x - 5 \end{cases}$
29. $\begin{cases} 5y + 2x \geq 10 \\ 3y - x > 3 \end{cases}$
30. $\begin{cases} 7y + 8x \leq 56 \\ 2y - x < -2 \end{cases}$
31. $\begin{cases} y \geq x \\ y \leq 2x \\ y \leq 6 \end{cases}$
32. $\begin{cases} y \leq 3x \\ y \geq -x \\ x \leq 2 \end{cases}$
33. $\begin{cases} y > x - 2 \\ y < 6 - x \\ y < 2x - 3 \end{cases}$
34. $\begin{cases} y < x + 4 \\ y > -x \\ y < 4 - 2x \end{cases}$
35. $\begin{cases} 3y - x - 10 < 0 \\ y + 3x - 10 \leq 0 \\ x + 2y \geq 0 \end{cases}$
36. $\begin{cases} 3y - x - 10 \leq 0 \\ y - 4x + 15 < 0 \\ 2y + 3x - 3 \geq 0 \end{cases}$
37. $\begin{cases} x \geq 1 \\ y \geq -1 \\ y \leq 2x - 1 \\ y \leq 8 - x \end{cases}$
38. $\begin{cases} x \geq -2 \\ y \geq x \\ y \leq x + 4 \\ y \leq 8 - x \end{cases}$
39. $\begin{cases} 2y < 6 - x \\ 2y > x - 6 \\ y < x - 3 \\ y > -x - 3 \end{cases}$
40. $\begin{cases} 4y < 24 - x \\ 4y < 24 + x \\ y > -2x - 3 \\ y > 2x - 3 \end{cases}$

Superset

In exercises 41–48, graph the solution set of each system.

41. $\begin{cases} y \geq x^2 - 4x + 1 \\ y \leq x - 3 \end{cases}$
42. $\begin{cases} x \leq y^2 - 2y \\ x \leq 2 - y \end{cases}$
43. $\begin{cases} (x - 2)^2 + y^2 \leq 9 \\ (x + 2)^2 + y^2 \leq 9 \end{cases}$
44. $\begin{cases} x^2 + (y - 1)^2 \geq 4 \\ x^2 + (y + 1)^2 \geq 4 \end{cases}$
45. $\begin{cases} |x + y| \leq 2 \\ |x| \leq 3 \end{cases}$
46. $\begin{cases} |x + y| \leq 3 \\ |y| \leq 2 \end{cases}$
47. $\begin{cases} y \geq |x| \\ y \leq |2x| - 4 \end{cases}$
48. $\begin{cases} |y - 2x| \leq 4 \\ |y + x| \leq 2 \end{cases}$

In exercises 49 and 50, describe the graphed region by a system of inequalities.

49.

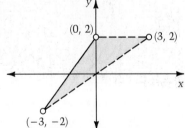

50.

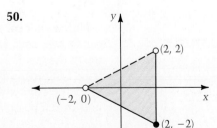

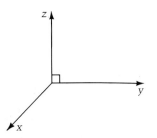

Figure 5.9 The *xyz*-coordinate system

5.3 Systems of Linear Equations in Three Variables

Goal A *to plot ordered triples and graph planes*

Up to this point all graphing has been done in the *xy*-plane. By adding a third coordinate axis, perpendicular to the *xy*-plane, we produce a coordinate system for describing the position of a point in 3-dimensional space. We usually call this third axis the **z-axis** and the entire system, the **xyz-coordinate system.**

In the *xyz*-coordinate system, we graph **ordered triples** that have the form

$$(x, y, z).$$

Plotting a point requires three moves: a move in the *x*-direction, followed by one in the *y*-direction, followed by another in the *z*-direction.

For example, suppose we must plot the point with coordinates (1, 4, 2). Starting at the origin (0, 0, 0), we move 1 unit in the positive *x*-direction, then 4 units in the positive *y*-direction, and finally 2 units in the positive *z*-direction.

To help us understand point-plotting in the *xyz*-coordinate system, let us examine the figures below. Figure 5.11(a) represents a box 3 units long, 2 units wide and 1 unit high. In Figure 5.11(b), we have placed the box in the three-dimensional *xyz*-coordinate system. The corners of the box are labeled with ordered triples of the form (x, y, z).

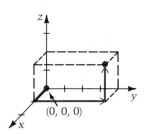

Figure 5.10

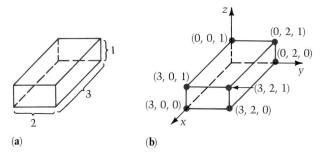

Figure 5.11

Notice that each corner of the bottom of the box lies in the *xy*-plane and can be described as an ordered triple of the form (x, y, 0). Each corner of the top of the box is 1 unit above the *xy*-plane and has coordinates of the form (x, y, 1). Starting at the origin, we find the corner (3, 2, 1) by moving 3 units in the positive *x*-direction, 2 units in the positive *y*-direction, and then 1 unit in the positive *z*-direction.

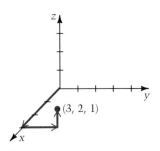

Figure 5.12

Example 1 Plot the following points in the *xyz*-coordinate system.

(a) $A(1, 0, 0)$ (b) $B(0, 0, -1)$ (c) $C(0, -2, 0)$ (d) $D(-2, 1, 3)$

Solution

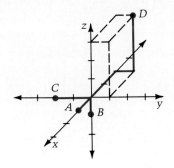

- To plot point A, start at $(0, 0, 0)$ and move 1 unit in the positive x-direction.
- To plot point B, start at $(0, 0, 0)$ and move 1 unit in the negative z-direction.
- To plot point C, start at $(0, 0, 0)$ and move 2 units in the negative y-direction.
- To plot point D, start at $(0, 0, 0)$ and move 2 units in the negative x-direction, 1 unit in the positive y-direction, and 3 units in the positive z-direction.

Our study of ordered triples will now help us understand linear equations in three variables. The graph of a linear equation in three variables is a plane in the *xyz*-coordinate system.

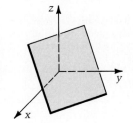

Figure 5.13

Plane: General Form

The **general form** of the equation of a plane in the *xyz*-coordinate system is

$$Ax + By + Cz + D = 0,$$

where A, B, C, and D are real numbers with A, B, and C not all zero.

Some very special planes and their equations deserve mention. As we suggested earlier, every point in the *xy*-plane has coordinates of the form $(x, y, 0)$. Since every point in the *xy*-plane has z-coordinate 0,

$$z = 0 \text{ is the equation of the } xy\text{-plane.}$$

Likewise,

$$x = 0 \text{ is the equation of the } yz\text{-plane, and}$$
$$y = 0 \text{ is the equation of the } xz\text{-plane.}$$

More generally, an equation of the form

$$z = c, \text{ with } c \text{ a constant,}$$

is the equation of a plane that is parallel to the xy-plane and c units above (for $c > 0$) or below (for $c < 0$) the xy-plane. Similarly, the plane $x = c$ is parallel to the yz-plane, and the plane $y = c$ is parallel to the xz-plane.

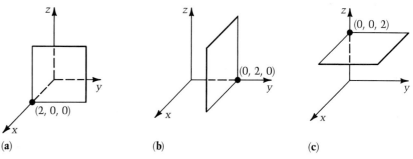

Figure 5.14 (a) The plane $x = 2$. (b) The plane $y = 2$. (c) The plane $z = 2$.

Example 2 Graph the following planes on the same set of coordinate axes and then determine the point where the three planes intersect: $x = 2$, $y = 3$ and $z = 4$.

Solution

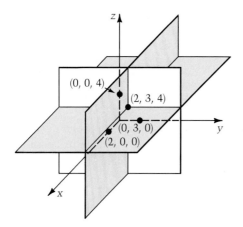

- The plane $x = 2$ is parallel to and 2 units in front of the yz-plane. All points on this plane are of the form $(2, y, z)$, since x must be 2, but y and z can represent any numbers.
- The plane $y = 3$ is parallel to and 3 units to the right of the xz-plane. All points on this plane are of the form $(x, 3, z)$, since y must be 3, but x and z can represent any numbers.
- The plane $z = 4$ is parallel to and 4 units above the xy-plane. All points on this plane are of the form $(x, y, 4)$, since z must be 4, but x and y can represent any numbers.

The three planes intersect at the point $(2, 3, 4)$.

5 / Systems of Equations and Inequalities

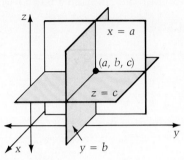

Figure 5.15

Goal B *to solve a system of three linear equations in three variables*

When we are solving a system of three linear equations in three variables, our objective is to produce a succession of equivalent systems until we obtain a system of the form

$$\begin{cases} x = a \\ y = b \\ z = c \end{cases}$$

where a, b, and c are constants. We then conclude that the ordered triple (a, b, c) is the solution of the original system.

You should picture a system of three linear equations in three variables as three planes in space. Then, the solutions of the system are the ordered triples describing the points of intersection of the three planes (Figure 5.16). This is summarized in the table below.

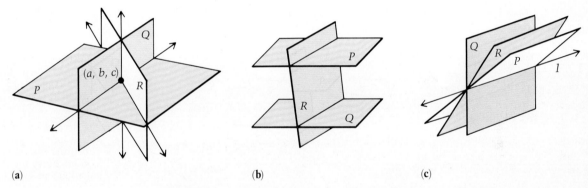

Figure 5.16 (a) One solution: (a, b, c). (b) No solutions. (c) Infinitely many solutions on line l.

Facts About Solutions of Linear Systems	
Form of an Equivalent System	**Number of Solutions**
x = a number y = a number z = a number	one
Includes an equation like $7 = 7$ or $0 = 0$.	infinitely many
Includes an equation like $0 = 9$ or $2 = 3$.	none

Example 3 Use the substitution method to solve the system $\begin{cases} 5x + y - z = 3 \\ x + 2y + z = 0 \\ 3x - 4y - 5z = -2. \end{cases}$

Solution $\begin{cases} 5x + y - z = 3 \\ x + 2y + z = 0 \\ 3x - 4y - 5z = -2 \end{cases}$

$$y = 3 + z - 5x \quad \blacksquare \text{ The first equation is solved for } y.$$

$\blacksquare$ Substitute the expression for y into the other two equations and simplify.

$x + 2y + z = 0$	$3x - 4y - 5z = -2$
$x + 2(3 + z - 5x) + z = 0$	$3x - 4(3 + z - 5x) - 5z = -2$
$x + 6 + 2z - 10x + z = 0$	$3x - 12 - 4z + 20x - 5z = -2$
$3x - z = 2$	$23x - 9z = 10$

$\begin{cases} y = 3 + z - 5x \\ 3x - z = 2 \\ 23x - 9z = 10 \end{cases}$

$\blacksquare$ Using the simplified equations, the original system is restated as an equivalent system.

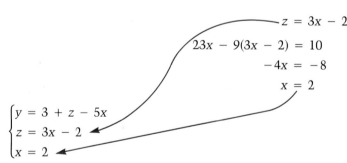

$z = 3x - 2$

$23x - 9(3x - 2) = 10$

$-4x = -8$

$x = 2$

$\blacksquare$ The second equation is solved for z.
$\blacksquare$ Substitute the expression for z into the third equation and solve for x.

$\begin{cases} y = 3 + z - 5x \\ z = 3x - 2 \\ x = 2 \end{cases}$

$z = 3(2) - 2 = 4$

$y = 3 + (4) - 5(2) = -3$

$\blacksquare$ The value of x is substituted into the second equation.
$\blacksquare$ The values of x and z are substituted into the first equation.

$\begin{cases} y = -3 \\ z = 4 \\ x = 2 \end{cases}$

The solution of the system is $(2, -3, 4)$.

Exercise Set 5.3 ▣ Calculator Exercises in the Appendix

Goal A
In exercises 1–6, plot the following points in the xyz-coordinate system.

1. $A(2, 3, 0)$
2. $B(1, -2, 3)$
3. $C(-2, 1, 3)$
4. $D(3, -1, 2)$
5. $E(-1, 3, 2)$
6. $F(3, 0, 2)$

In exercises 7–12, graph each of the following planes.

7. $x = 3$
8. $y = 5$
9. $z = 1$
10. $x = -6$
11. $y = -4$
12. $z = -8$

In exercises 13–18, graph each set of planes on the same set of coordinate axes and then determine the point where the three planes intersect.

13. $x = 4$
    $y = 2$
    $z = 6$
14. $x = -1$
    $y = 5$
    $z = 4$
15. $y = 7$
    $x = -3$
    $z = -1$
16. $y = -3$
    $z = 5$
    $x = -6$
17. $z = 0$
    $x = -2$
    $y = -10$
18. $z = -9$
    $y = 0$
    $x = 12$

Goal B
In exercises 19–36, solve each of the following systems.

19. $\begin{cases} 4x + 2y + 5z = 4 \\ x - y + z = 0 \\ y = 1 \end{cases}$
20. $\begin{cases} 5x + 4y + z = 2 \\ 3x + 2y - z = 0 \\ z = 2 \end{cases}$

21. $\begin{cases} 4x + 2y + z = 8 \\ 2x - y - 3z = 3 \\ x + y - z = 1 \end{cases}$
22. $\begin{cases} -x + 2y - 3z = 5 \\ 4x + 3y + z = 2 \\ x - y - z = 3 \end{cases}$

23. $\begin{cases} 4x - 5y + 2z = -1 \\ x + 2y + z = 4 \\ 2x + 3z = 1 \end{cases}$
24. $\begin{cases} 5x - y + z = -1 \\ x + 3y = 7 \\ y + 2z = -6 \end{cases}$

25. $\begin{cases} 5x - 4y - 3z = -1 \\ x + 2y + 5z = -3 \\ 3x - y - z = 2 \end{cases}$
26. $\begin{cases} 3x + 2y + z = -1 \\ 4x + y + z = 1 \\ x + 4y + z = 2 \end{cases}$

27. $\begin{cases} -5x + 3y + 2z = 2 \\ 3x + y - 4z = 3 \\ x - 2y + z = 0 \end{cases}$
28. $\begin{cases} 3x + 2y + 4z = -6 \\ -x + y + 3z = 5 \\ 2x + y - z = 1 \end{cases}$

29. $\begin{cases} -5x + 3y + 2z = 2 \\ 3x + y - 4z = 8 \\ -x + 2y - z = 5 \end{cases}$
30. $\begin{cases} 3x + y + 2z = 1 \\ 4x - y + 4z = 4 \\ x + y - 2z = -5 \end{cases}$

31. $\begin{cases} 2x + 5y + 4z = 1 \\ 5x + 3y - 3z = -2 \\ 4x + y + 2z = 3 \end{cases}$
32. $\begin{cases} 3x - 6y + 4z = -1 \\ 3x + 2y - 2z = 5 \\ 3x - 2y + z = 2 \end{cases}$

33. $\begin{cases} 3x + 4y + 2z = -8 \\ 3x + 8y + z = -6 \\ 6x - 4y - z = 8 \end{cases}$
34. $\begin{cases} -x + 2y - 7z = 9 \\ 4x + 3y + z = 8 \\ 3x - y - 3z = -7 \end{cases}$

35. $\begin{cases} x + 10y - 11z = 0 \\ 2x - y + 5z = 5 \\ x + 3y - 2z = 1 \end{cases}$
36. $\begin{cases} -3x - 7y + 7z = 0 \\ 3x + 2y - z = 1 \\ 6x - y + 4z = 3 \end{cases}$

Superset

37. Frank is 3 times as old as Ann and 15 years older than Mary. Ann is one year younger than Mary. How old is Frank?

38. A child's bank contains twice as many dimes as nickels and four times as many pennies as nickels. There are 42 coins in the bank. What is the coins' total value?

39. Find all three digit numbers such that the sum of the digits is 7, the number is increased by 99 if the order of the digits is reversed, and the hundreds digit is 3 less than the sum of the other two digits.

40. Find all three digit numbers such that the sum of the digits is 6, the hundreds digit is equal to the sum of the other two digits, and the number is decreased by 297 if the order of the digits is reversed.

41. Bob is ten years older than Ted. Last year Ted was twice as old as Carol was. Three years ago Bob was five times as old as Carol was. How old is Carol today?

42. A child's bank contains twice as many nickels as pennies and two-thirds as many dimes as nickels. The total value of the coins is at least $3.70. Determine the smallest number of coins that could be in the bank.

5.4 Matrices

Goal A *to represent a system of linear equations by means of a matrix and vice versa*

A **matrix** is a rectangular array of numbers. Matrices provide a convenient shorthand for working with systems of equations.

$$\text{row} \begin{array}{c} 1 \\ 2 \\ 3 \end{array} \overset{\overset{\text{column}}{1 \quad 2 \quad 3 \quad 4}}{\begin{bmatrix} 3 & -2 & 1 & 0 \\ 1 & 5 & -3 & 2 \\ 1 & 0 & -4 & 3 \end{bmatrix}} \qquad \begin{bmatrix} 7 & 2 & 1 \\ 4 & 1 & 6 \\ 3 & 1 & -5 \\ 0 & 0 & 0 \end{bmatrix}$$

A matrix with 3 rows and 4 columns   A matrix with 4 rows and 3 columns

Each number in the matrix is called an **entry**. A matrix with m rows and n columns is said to be an $m \times n$ (or m **by** n) **matrix**. If the number of rows equals the number of columns, the matrix is called a **square matrix.**

Recall that when we use the addition method to solve a system of linear equations, we must line up the variables and constant terms to be sure we are combining x-terms with x-terms, y-terms with y-terms, and constants with constants. The linear system below is written in matrix form with the coefficients and the constants as entries. A dotted line in the matrix separates the coefficients from the constants.

$$\begin{cases} 2x + y - 3z = 2 \\ x + 3y = 2 \\ x - y + 4z = 5 \end{cases} \qquad \overset{x \quad y \quad z}{\begin{bmatrix} 2 & 1 & -3 & \vdots & 2 \\ 1 & 3 & 0 & \vdots & 2 \\ 1 & -1 & 4 & \vdots & 5 \end{bmatrix}}$$

A linear system in 3 variables   The 3 × 4 matrix of the system

There is a zero in the second row-third column position in the matrix because the second equation has no z-term. If we remember to associate the first column with x, the second column with y, the third column with z, and the fourth column with the constants, then the matrix gives us the same information the system does.

Example 1 For the matrix below, identify the following:

(a) the first row-second column entry
(b) the second row-first column entry

$$\begin{bmatrix} 3 & 0 & 1 & \vdots & 5 \\ 2 & -3 & 7 & \vdots & -8 \\ 5 & 4 & 6 & \vdots & 10 \end{bmatrix}$$

Solution (a) The first row-second column entry is 0.
(b) The second row-first column entry is 2.

Example 2 Represent the systems with matrices and the matrices with systems.

(a) $\begin{cases} 2x + 3y = 5 \\ x - 4y = 0 \end{cases}$ (b) $\begin{cases} 3x + y = 4 \\ 2x - y + 5z = 3 \\ 0 = 9 \end{cases}$ (c) $\begin{bmatrix} 1 & 0 & | & 10 \\ 0 & 1 & | & -2 \end{bmatrix}$ (d) $\begin{bmatrix} 1 & 0 & 2 & | & -3 \\ 8 & 0 & 10 & | & -5 \\ 12 & 1 & 0 & | & 16 \end{bmatrix}$

Solution (a) $\begin{bmatrix} 2 & 3 & | & 5 \\ 1 & -4 & | & 0 \end{bmatrix}$ (b) $\begin{bmatrix} 3 & 1 & 0 & | & 4 \\ 2 & -1 & 5 & | & 3 \\ 0 & 0 & 0 & | & 9 \end{bmatrix}$ (c) $\begin{cases} x = 10 \\ y = -2 \end{cases}$ (d) $\begin{cases} x + 2z = -3 \\ 8x + 10z = -5 \\ 12x + y = 16 \end{cases}$

$\begin{bmatrix} 2 & 3 & | & 5 \\ 1 & -1 & | & 0 \end{bmatrix}$

Using the matrix at the left as an example, we now define the following matrix terminology.

$\begin{bmatrix} 2 & 3 \\ 1 & -1 \end{bmatrix}$

The entries to the left of the dashed line may be used to form another matrix called the **coefficient matrix**.

$\begin{bmatrix} 5 \\ 0 \end{bmatrix}$

The entries to the right of the dashed line may be used to form another matrix called the **constant matrix**.

$\begin{bmatrix} 2 & 3 & | & 5 \\ 1 & -1 & | & 0 \end{bmatrix}$

The entire matrix, including coefficients and constants, is called the **augmented matrix**.

Example 3 Determine the augmented, coefficient, and constant matrices for the system

$$\begin{cases} x + 3y = 4 \\ 2x + 7y = 6. \end{cases}$$

Solution

$\begin{bmatrix} 1 & 3 & | & 4 \\ 2 & 7 & | & 6 \end{bmatrix}$ $\begin{bmatrix} 1 & 3 \\ 2 & 7 \end{bmatrix}$ $\begin{bmatrix} 4 \\ 6 \end{bmatrix}$

augmented matrix   coefficient matrix   constant matrix

Goal B *to solve linear systems by using matrices*

If we wished to solve the system in Example 3 by the addition method, we would rewrite it as a succession of equivalent systems:

$$\begin{cases} x + 3y = 4 \\ 2x + 7y = 6 \end{cases} \longrightarrow \begin{cases} x + 3y = 4 \\ y = -2 \end{cases} \longrightarrow \begin{cases} x = 10 \\ y = -2. \end{cases}$$

At this point, you should convince yourself that these three equivalent systems may be written in matrix form.

$$\begin{bmatrix} 1 & 3 & | & 4 \\ 2 & 7 & | & 6 \end{bmatrix} \longrightarrow \begin{bmatrix} 1 & 3 & | & 4 \\ 0 & 1 & | & -2 \end{bmatrix} \longrightarrow \begin{bmatrix} 1 & 0 & | & 10 \\ 0 & 1 & | & -2 \end{bmatrix}$$

In the following table, we state the rules for changing the rows of an augmented matrix. These rules correspond to the rules for changing the equations of a system by the addition method. Remember that each row of an augmented matrix corresponds to one of the equations in a system.

Steps of the Addition Method	Corresponding Row Operations for Matrices
Write the equations in a different order.	Interchange the rows of the matrix.
Multiply (or divide) an equation by a nonzero number.	Multiply (or divide) every entry in a row by a nonzero number.
Add (or subtract) a multiple of one equation to (from) another.	Add (or subtract) a multiple of one row to (from) another.

This table states that whatever you do to the equations of a system, you can do to the corresponding rows of its augmented matrix. This process is called **row reduction.** When we use matrices to solve a system of two equations in two variables, our goal is to transform the augmented matrix into the form

$$\begin{bmatrix} 1 & 0 & | & c \\ 0 & 1 & | & d \end{bmatrix}$$

where c and d are real numbers. We can do this only if the system has a unique solution.

Example 4 Solve the following system by the addition method and by row reduction of a matrix.

$$\begin{cases} 2x - 4y = 2 \\ -3x + 7y = -1 \end{cases}$$

Solution   Addition Method   Row Reduction Method

$\begin{cases} 2x - 4y = 2 \\ -3x + 7y = -1 \end{cases}$ $\begin{bmatrix} 2 & -4 & | & 2 \\ -3 & 7 & | & -1 \end{bmatrix}$ ■ Divide the first equation by 2. Divide the first row by 2.

$\begin{cases} x - 2y = 1 \\ -3x + 7y = -1 \end{cases}$ $\begin{bmatrix} 1 & -2 & | & 1 \\ -3 & 7 & | & -1 \end{bmatrix}$ ■ Multiply the first equation or row by 3 and add it to the second equation or row.

$\begin{cases} x - 2y = 1 \\ y = 2 \end{cases}$ $\begin{bmatrix} 1 & -2 & | & 1 \\ 0 & 1 & | & 2 \end{bmatrix}$ ■ Multiply the second equation or row by 2 and add it to the first.

$\begin{cases} x = 5 \\ y = 2 \end{cases}$ $\begin{bmatrix} 1 & 0 & | & 5 \\ 0 & 1 & | & 2 \end{bmatrix}$

The solution of the system (by either method) is (5, 2).

Example 5 Solve the system $\begin{cases} 4x - 4y = 8 \\ 5x - 4y = 2. \end{cases}$

Solution $\begin{bmatrix} 4 & -4 & | & 8 \\ 5 & -4 & | & 2 \end{bmatrix}$ ■ Begin by writing the augmented matrix for the system.

$\begin{bmatrix} 1 & -1 & | & 2 \\ 5 & -4 & | & 2 \end{bmatrix}$ ■ The first row has been divided by 4 so that the first row-first column entry is 1.

$\begin{bmatrix} 1 & -1 & | & 2 \\ 0 & 1 & | & -8 \end{bmatrix}$ ■ -5 times the first row has been added to the second row. The second row-first column entry is now 0.

$\begin{bmatrix} 1 & 0 & | & -6 \\ 0 & 1 & | & -8 \end{bmatrix}$ ■ The second row has been added to the first row to obtain 0 in the first row-second column position.

The solution of the system is $(-6, -8)$.

When solving a system by matrix methods, you should always begin, as in Example 5, by transforming the given augmented matrix so that there is a "1" in the first row-first column position. For example, suppose you wish to solve the system

$$\begin{cases} 3y = 14 \\ 5x - 7y = 8. \end{cases}$$

Since there is a zero in the first row-first column position of the system's augmented matrix, you should begin by interchanging the rows, and then dividing the resulting first row by 5. This produces a "1" in the first row-first column position.

$$\begin{bmatrix} 0 & 3 & | & 14 \\ 5 & -7 & | & 8 \end{bmatrix} \longrightarrow \begin{bmatrix} 5 & -7 & | & 8 \\ 0 & 3 & | & 14 \end{bmatrix} \longrightarrow \begin{bmatrix} 1 & -\frac{7}{5} & | & \frac{8}{5} \\ 0 & 3 & | & 14 \end{bmatrix}$$
$$\text{interchange rows} \qquad \text{divide first row by 5}$$

However, it is not always possible to transform an augmented matrix into the form shown on the left. If the final coefficient matrix has a row of zeros, then the system does not have a unique solution. Two such cases are illustrated in Examples 6 and 7. Example 6 contains a system of equations that has no solution; Example 7 contains a system of equations that has an infinite number of solutions.

$$\begin{bmatrix} 1 & 0 & | & a \\ 0 & 1 & | & b \end{bmatrix}$$

Example 6   Solve the system $\begin{cases} 3x + 4y = 7 \\ 6x + 8y = 5. \end{cases}$

Solution

$\begin{bmatrix} 3 & 4 & | & 7 \\ 6 & 8 & | & 5 \end{bmatrix}$ ■ Begin by writing the augmented matrix for the system.

$\begin{bmatrix} 1 & \frac{4}{3} & | & \frac{7}{3} \\ 6 & 8 & | & 5 \end{bmatrix}$ ■ The first row has been divided by 3 to produce a "1" in the first row-first column position.

$\begin{bmatrix} 1 & \frac{4}{3} & | & \frac{7}{3} \\ 0 & 0 & | & -9 \end{bmatrix}$ ■ -6 times the first row has been added to the second row.

The second row corresponds to the statement $0 = -9$, which is never true. Thus, the system has no solutions.

Example 7 Solve the system $\begin{cases} 3x + 4y = 7 \\ 6x + 8y = 14. \end{cases}$

Solution

$\begin{bmatrix} 3 & 4 & | & 7 \\ 6 & 8 & | & 14 \end{bmatrix}$ ■ Begin by writing an augmented matrix for the system.

$\begin{bmatrix} 1 & \frac{4}{3} & | & \frac{7}{3} \\ 6 & 8 & | & 14 \end{bmatrix}$ ■ The first row has been divided by 3 to produce a "1" in the first row-first column position.

$\begin{bmatrix} 1 & \frac{4}{3} & | & \frac{7}{3} \\ 0 & 0 & | & 0 \end{bmatrix}$ ■ -6 times the first row has been added to the second row.

The second row corresponds to the statement $0 = 0$, which is always true. Thus, the system has infinitely many solutions. The first row corresponds to the statement $x + \frac{4}{3}y = \frac{7}{3}$, and the solutions of this equation are the solutions of the system. Thus, the solutions of the system are all the points on the line $x + \frac{4}{3}y = \frac{7}{3}$.

Another way to indicate the solution set of the system is as follows. Let $x = c$, where c is any real number. Since $y = \frac{7}{4} - \frac{3}{4}x$, we know that $y = \frac{7}{4} - \frac{3}{4}c$. Thus, we can describe the solution set of the system as the set of ordered pairs of the form $(c, \frac{7}{4} - \frac{3}{4}c)$, where c is any real number.

The augmented matrix for the system shown below on the left is given on the right.

$\begin{cases} 2x + 4y + 2z = 6 \\ 2x + y - z = 2 \\ x + y + 3z = -2 \end{cases}$ $\begin{bmatrix} 2 & 4 & 2 & | & 6 \\ 2 & 1 & -1 & | & 2 \\ 1 & 1 & 3 & | & -2 \end{bmatrix}$

When we use matrix methods to solve the system, our goal is to transform, if possible, the augmented matrix shown above into one of the form

$\begin{bmatrix} 1 & 0 & 0 & | & n \\ 0 & 1 & 0 & | & m \\ 0 & 0 & 1 & | & p \end{bmatrix},$

where n, m, and p are real numbers. The following example illustrates the matrix methods for solving a system of three equations in three variables.

Example 8 Solve the following system:

$$2x + 4y + 2z = 6$$
$$2x + y - z = 0$$
$$x + y + 3z = -2.$$

Solution

$\begin{bmatrix} 2 & 4 & 2 & | & 6 \\ 2 & 1 & -1 & | & 0 \\ 1 & 1 & 3 & | & -2 \end{bmatrix}$ ■ Divide the first row by 2 to obtain a 1 in the first row-first column position.

$\begin{bmatrix} 1 & 2 & 1 & | & 3 \\ 2 & 1 & -1 & | & 0 \\ 1 & 1 & 3 & | & -2 \end{bmatrix}$ ■ Add -2 times the first row to the second row.

$\begin{bmatrix} 1 & 2 & 1 & | & 3 \\ 0 & -3 & -3 & | & -6 \\ 1 & 1 & 3 & | & -2 \end{bmatrix}$ ■ Add -1 times the first row to the third row.

$\begin{bmatrix} 1 & 2 & 1 & | & 3 \\ 0 & -3 & -3 & | & -6 \\ 0 & -1 & 2 & | & -5 \end{bmatrix}$ ■ Divide the second row by -3.

$\begin{bmatrix} 1 & 2 & 1 & | & 3 \\ 0 & 1 & 1 & | & 2 \\ 0 & -1 & 2 & | & -5 \end{bmatrix}$ ■ Add -2 times the second row to the first row.

$\begin{bmatrix} 1 & 0 & -1 & | & -1 \\ 0 & 1 & 1 & | & 2 \\ 0 & -1 & 2 & | & -5 \end{bmatrix}$ ■ Add the second row to the third row.

$\begin{bmatrix} 1 & 0 & -1 & | & -1 \\ 0 & 1 & 1 & | & 2 \\ 0 & 0 & 3 & | & -3 \end{bmatrix}$ ■ Divide the third row by 3.

$\begin{bmatrix} 1 & 0 & -1 & | & -1 \\ 0 & 1 & 1 & | & 2 \\ 0 & 0 & 1 & | & -1 \end{bmatrix}$ ■ Add the third row to the first row.

$$\begin{bmatrix} 1 & 0 & 0 & | & -2 \\ 0 & 1 & 1 & | & 2 \\ 0 & 0 & 1 & | & -1 \end{bmatrix}$$ ■ Subtract the third row from the second row.

$$\begin{bmatrix} 1 & 0 & 0 & | & -2 \\ 0 & 1 & 0 & | & 3 \\ 0 & 0 & 1 & | & -1 \end{bmatrix}$$

Thus, the solution of the system is $(-2, 3, -1)$.

A system of three linear equations in three variables may have no solutions, exactly one solution, or infinitely many solutions. We can determine the nature of the solution set from the corresponding augmented matrix by using row reduction.

Fact

Suppose we have a system of n linear equations in n variables. This system will have

- *a unique solution* if row reduction of the augmented matrix does *not* produce a row of zeros in the coefficient matrix.
- *no solutions* if row reduction of the augmented matrix produces a row of zeros in the coefficient matrix, and the corresponding entry in the constant matrix is not zero.
- *infinitely many solutions* if row reduction of the augmented matrix produces one or more rows of zeros in the coefficient matrix, and each corresponding entry in the constant matrix is also zero.

Exercise Set 5.4

Goal A

In exercises 1–4, use the following matrix.

$$\begin{bmatrix} 3 & 8 & -1 & | & -6 \\ 4 & -5 & 2 & | & 7 \\ -2 & 0 & -3 & | & 9 \end{bmatrix}$$

1. Identify the first row-third column entry.
2. Identify the third-row-first column entry.
3. Identify the third row-second column entry.
4. Identify the second row-fourth column entry.

In exercises 5–12, represent the systems with matrices and the matrices with systems.

5. $\begin{cases} 3x + 7y = 8 \\ 2x + 9y = 1 \end{cases}$
6. $\begin{cases} 3x + 4y = 11 \\ 2x - y = 0 \end{cases}$

7. $\begin{cases} 2x - y + z = 2 \\ x + 3y = 5 \\ x - 3z = 1 \end{cases}$
8. $\begin{cases} -x + 2y - 3z = 5 \\ 4x + 3y + z = 2 \\ x - y - z = 3 \end{cases}$

9. $\begin{bmatrix} 12 & 9 & | & 6 \\ 8 & 6 & | & 4 \end{bmatrix}$
10. $\begin{bmatrix} 10 & 5 & | & 7 \\ -6 & -3 & | & -4 \end{bmatrix}$

11. $\begin{bmatrix} -1 & 1 & -2 & | & 1 \\ 1 & 1 & -1 & | & 7 \\ 2 & 0 & 1 & | & 4 \end{bmatrix}$
12. $\begin{bmatrix} -5 & 3 & 2 & | & 2 \\ 3 & 1 & -4 & | & 6 \\ -1 & 2 & -1 & | & 4 \end{bmatrix}$

In exercises 13–26, determine the augmented, coefficient, and constant matrices for each of the following systems.

13. $\begin{cases} -2x + 7y = 8 \\ 3x - y = 7 \end{cases}$
14. $\begin{cases} 2x + 3y = 6 \\ 5x + 7y = 13 \end{cases}$

15. $\begin{cases} 8x + 4y = 5 \\ 6x + 3y = 7 \end{cases}$
16. $\begin{cases} 2x - 3y = 5 \\ x + 2y = 3 \end{cases}$

17. $\begin{cases} 3x + 5y = 1 \\ -5 + 3y = -4x \end{cases}$
18. $\begin{cases} -4x + 10y = -2 \\ 2x - 1 = 5y \end{cases}$

19. $\begin{cases} 5x - y + z = -1 \\ x + 3y = 7 \\ y + 2z = -6 \end{cases}$
20. $\begin{cases} x + 2y + z = 4 \\ 2x + 3z = 1 \\ 5y + 2z = -1 \end{cases}$

21. $\begin{cases} 3x + 2y - z = 6 \\ -x - y + z = 1 \\ x + 2y + 2z = 5 \end{cases}$
22. $\begin{cases} 3x + y + 2z = 1 \\ 4x - y + 4z = 4 \\ x + y - 2z = -5 \end{cases}$

23. $\begin{cases} x - 10y + 14z = 2 \\ -2x - y + 2z = 5 \\ -x + 3y - 4z = 1 \end{cases}$
24. $\begin{cases} 2x - y - 3z = 3 \\ x + y - z = 1 \\ 4x + 2y + z = 8 \end{cases}$

25. $\begin{cases} 3x - 6y + 2z = -1 \\ x + 4y + z = 0 \\ 3x + 2y - 5 = 2z \end{cases}$
26. $\begin{cases} 5x - 3y - 2z = -2 \\ x - 2y + z = 0 \\ 3x - 4z - 3 = -y \end{cases}$

Goal B

For exercises 27–34, solve the systems in exercises 5–12. (Use matrix methods.)

For exercises 35–48, solve the systems in exercises 13–26. (Use matrix methods.)

Superset

In exercises 49–51, refer to the following system.

$$\begin{cases} x + 3y + z = A^2 \\ 2x + 5y + 2Az = 0 \\ x + y + A^2 z = -9 \end{cases}$$

Determine all real numbers A, if any, for which each of the following statements is true.

49. The system has a unique solution.
50. The system has no solutions.
51. The system has infinitely many solutions.

In exercises 52–53, refer to the following system.

$$\begin{cases} x + 2y + z = A \\ 3x + 2y + Az = -A \\ Ax + 2Ay + 3z = 9 \end{cases}$$

Determine all real numbers A, if any, for which each of the following statements is true.

52. The system has a unique solution.
53. The system has no solutions.

5.5 Determinants

Goal A *to use determinants to solve systems of two linear equations in two variables*

Associated with each $n \times n$ square matrix is a number called the **determinant.** We can use this number to determine whether a system of n linear equations in n variables has a unique solution, and to find the unique solution when it exists.

Let us begin with the case $n = 2$. For the 2×2 matrix

$$A = \begin{bmatrix} a & b \\ c & d \end{bmatrix}$$

the value of the associated 2×2 determinant is $ad - bc$. There are several ways to denote this number:

$$\det A, \qquad |A|, \qquad \begin{vmatrix} a & b \\ c & d \end{vmatrix}.$$

Careful! We use square brackets to indicate a matrix and vertical bars to indicate the associated determinant. Thus,

$$\det A = \det \begin{bmatrix} a & b \\ c & d \end{bmatrix} = \begin{vmatrix} a & b \\ c & d \end{vmatrix} = ad - bc.$$

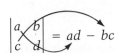

As shown at the left, the determinant of a 2×2 matrix is the difference of the products of the diagonal elements of the matrix.

Example 1 Compute the determinant of each of the following matrices:

(a) $A = \begin{bmatrix} 1 & 2 \\ 3 & 4 \end{bmatrix}$ (b) $B = \begin{bmatrix} 2 & 1 \\ 3 & 4 \end{bmatrix}$ (c) $C = \begin{bmatrix} 2 & 3 \\ 4 & 6 \end{bmatrix}$

Solution

(a) $\det A = \begin{vmatrix} 1 & 2 \\ 3 & 4 \end{vmatrix} = 1 \cdot 4 - 3 \cdot 2 = 4 - 6 = -2$

(b) $\det B = \begin{vmatrix} 2 & 1 \\ 3 & 4 \end{vmatrix} = 2 \cdot 4 - 3 \cdot 1 = 8 - 3 = 5$

(c) $\det C = \begin{vmatrix} 2 & 3 \\ 4 & 6 \end{vmatrix} = 2 \cdot 6 - 4 \cdot 3 = 12 - 12 = 0$

■ Be careful to form the difference of the products of the diagonal entries in proper order. For

$$\begin{vmatrix} a & b \\ c & d \end{vmatrix}$$

we want $ad - bc$, not $bc - ad$.

To see how determinants can help us solve systems of linear equations, we consider the general case of two linear equations in two variables x and y:

$$\begin{cases} ax + by = e \\ cx + dy = f. \end{cases}$$

To simplify the work, let us restrict our attention to the case where none of the coefficients a, b, c, or d is zero. As it turns out, we get the same results in those cases when one or more of the coefficients are zero.

We use the addition method to solve the system. Multiplying the first equation in the system by $-c$ and the second equation by a, we get

$$\begin{cases} -acx - bcy = -ce \\ acx + ady = af. \end{cases}$$

Adding these two equations, we have

$$(ad - bc)y = af - ce.$$

Replacing the second equation in the original system with this last equation yields the equivalent system

$$\begin{cases} ax + by = e \\ (ad - bc)y = af - ce. \end{cases}$$

Solve the second equation in this system for y by multiplying each side by the reciprocal of $ad - bc$ (assuming that $ad - bc \neq 0$). Clearly this system has a unique solution if and only if

$$ad - bc \neq 0.$$

Let us suppose then that $ad - bc \neq 0$ and continue to look for the unique solution. We substitute the value for y,

$$y = \frac{af - ce}{ad - bc},$$

in the other equation

$$ax + by = e$$

or

$$x = \frac{1}{a}(e - by),$$

and solve for x:

$$x = \frac{1}{a}\left[e - b\left(\frac{af - ce}{ad - bc}\right)\right] = \frac{1}{a}\left[\frac{e(ad - bc) - b(af - ce)}{ad - bc}\right]$$

$$= \frac{1}{a}\left[\frac{aed - afb}{ad - bc}\right] = \frac{ed - fb}{ad - bc}.$$

Fact

A system of 2 linear equations in 2 variables

$$\begin{cases} ax + by = e \\ cx + dy = f \end{cases}$$

has a unique solution if and only if $ad - bc \neq 0$. This unique solution, if it exists, is given by

$$x = \frac{ed - fb}{ad - bc}, \quad y = \frac{af - ce}{ad - bc}.$$

Notice that the number $ad - bc$, which plays such an important role in establishing this fact, is the determinant of the coefficient matrix of the system:

$$A = \begin{vmatrix} a & b \\ c & d \end{vmatrix}, \quad \det A = ad - bc.$$

Furthermore, since

$$ed - fb = \begin{vmatrix} e & b \\ f & d \end{vmatrix} \quad \text{and} \quad af - ce = \begin{vmatrix} a & e \\ c & f \end{vmatrix},$$

we notice that when there is a unique solution to the system (i.e., $\det A \neq 0$), then the values of the variables can be written as quotients of determinants:

$$x = \frac{\begin{vmatrix} e & b \\ f & d \end{vmatrix}}{\begin{vmatrix} a & b \\ c & d \end{vmatrix}}$$

- The x-coefficients have been replaced with the constants of the system.
- The denominator is $\det A$.

$$y = \frac{\begin{vmatrix} a & e \\ c & f \end{vmatrix}}{\begin{vmatrix} a & b \\ c & d \end{vmatrix}}$$

- The y-coefficients have been replaced with the constants of the system.
- The denominator is $\det A$.

Example 2 Determine whether the following system has a unique solution, and find the solution if it exists.

$$\begin{cases} 2x - 3y = -1 \\ 4x - 5y = 7 \end{cases}$$

Solution We begin by computing the determinant of the coefficient matrix: $A = \begin{bmatrix} 2 & -3 \\ 4 & -5 \end{bmatrix}$.

$$\det A = \begin{vmatrix} 2 & -3 \\ 4 & -5 \end{vmatrix} = (2)(-5) - (4)(-3) = -10 + 12 = 2.$$

Since $\det A \neq 0$, the system has a unique solution. It is given by

$$x = \frac{\begin{vmatrix} -1 & -3 \\ 7 & -5 \end{vmatrix}}{\det A} = \frac{(-1)(-5) - (7)(-3)}{2} = 13$$

■ The matrix A is modified by replacing the column of x-coefficients with the constants of the system.

$$y = \frac{\begin{vmatrix} 2 & -1 \\ 4 & 7 \end{vmatrix}}{\det A} = \frac{(2)(7) - (4)(-1)}{2} = 9$$

■ The matrix A is modified by replacing the column of y-coefficients with the constants of the system.

The unique solution of the system is (13, 9).

Example 3 Determine those real numbers b, if any, for which the following system does *not* have a unique solution.

$$\begin{cases} 2x + (1 - b)y = 5 \\ bx - 3y = 4 \end{cases}$$

Solution

$$\begin{vmatrix} 2 & 1-b \\ b & -3 \end{vmatrix} = (2)(-3) - (b)(1-b) = b^2 - b - 6$$

■ The determinant of the coefficient matrix is computed.

$$b^2 - b - 6 = 0$$
$$(b - 3)(b + 2) = 0$$

■ The system does not have a unique solution when $\det A = 0$.

$b - 3 = 0 \quad | \quad b + 2 = 0$
$b = 3 \quad \quad | \quad b = -2$

When $b = 3$ or $b = -2$, the system does not have a unique solution.

Goal B *to evaluate the determinant of an n × n matrix by using signed minors*

In order to compute the determinant of a 3 × 3 matrix or, in general, any n × n matrix with n > 2, we must first discuss the idea of a *signed minor*.

Associated with each entry in an n × n matrix is a "subdeterminant" called a *minor*. The **minor** is the determinant of the matrix obtained by deleting the row and column containing the entry. For the matrix

$$\begin{bmatrix} 3 & 2 & 1 \\ 4 & 5 & 6 \\ 9 & 7 & 8 \end{bmatrix}$$

the minor for the entry in the first row-second column is found by deleting the first row and the second column of the matrix,

$$\begin{bmatrix} 3 & 2 & 1 \\ 4 & 5 & 6 \\ 9 & 7 & 8 \end{bmatrix},$$

and forming the determinant of the 2 × 2 matrix of remaining entries:

$$\begin{vmatrix} 4 & 6 \\ 9 & 8 \end{vmatrix}.$$

Similarly, the minor for the second row-first column entry 4 in the original 3 × 3 matrix is

$$\begin{vmatrix} 2 & 1 \\ 7 & 8 \end{vmatrix}.$$

Clearly, in any 3 × 3 matrix, since there are 9 entries, there are 9 minors. For a 4 × 4 matrix there are 16 minors, one for each entry.

We next consider the notion of a **signed minor.** To compute a signed minor we multiply a minor by +1 or −1 depending upon the position of the entry in the matrix from which the minor was formed. The choice of +1 or −1 is determined by a checkerboard pattern that begins with a + for the entry in the first row-first column, as shown in Figure 5.17.

$$\begin{bmatrix} + & - & + & - & \cdots \\ - & + & - & + & \cdots \\ + & - & + & - & \cdots \\ - & + & - & + & \cdots \\ \vdots & \vdots & \vdots & \vdots & \vdots \end{bmatrix}$$

Figure 5.17

$$\begin{bmatrix} 1 & 2 & 3 \\ 4 & 5 & 6 \\ 7 & 8 & 9 \end{bmatrix}$$

For example, for the 3 × 3 matrix at the left, the signed minors for the entries in the first row of the matrix are

$$+\begin{vmatrix} 5 & 6 \\ 8 & 9 \end{vmatrix}, \quad -\begin{vmatrix} 4 & 6 \\ 7 & 9 \end{vmatrix}, \quad +\begin{vmatrix} 4 & 5 \\ 7 & 8 \end{vmatrix}.$$

We now can state a rule that will enable us to evaluate the determinant for any $n \times n$ matrix.

Rule

To compute the determinant of any $n \times n$ matrix, select any row (or column) of the matrix and form the sum of the products of the entries from that row (or column) with their corresponding signed minors.

When we use this fact to evaluate a determinant, we say that we are **expanding along a row (or column)**.

Example 4   Compute det A for the matrix

$$A = \begin{bmatrix} 2 & 3 & 4 \\ 1 & 0 & -2 \\ 5 & -1 & -3 \end{bmatrix}$$

by expanding along (a) the first row, (b) the second row, (c) the third column.

Solution   (a) $\det A = 2\left(+\begin{vmatrix} 0 & -2 \\ -1 & -3 \end{vmatrix}\right) + 3\left(-\begin{vmatrix} 1 & -2 \\ 5 & -3 \end{vmatrix}\right) + 4\left(+\begin{vmatrix} 1 & 0 \\ 5 & -1 \end{vmatrix}\right)$

$= 2(-2) + 3(-7) + 4(-1)$

$= -29$

(b) $\det A = 1\left(-\begin{vmatrix} 3 & 4 \\ -1 & -3 \end{vmatrix}\right) + 0\left(+\begin{vmatrix} 2 & 4 \\ 5 & -3 \end{vmatrix}\right) + (-2)\left(-\begin{vmatrix} 2 & 3 \\ 5 & -1 \end{vmatrix}\right)$

$= 1(5) + 0 - 2(17)$

$= -29$

(c) $\det A = 4\left(+\begin{vmatrix} 1 & 0 \\ 5 & -1 \end{vmatrix}\right) + (-2)\left(-\begin{vmatrix} 2 & 3 \\ 5 & -1 \end{vmatrix}\right) + (-3)\left(+\begin{vmatrix} 2 & 3 \\ 1 & 0 \end{vmatrix}\right)$

$= 4(-1) - 2(17) - 3(-3)$

$= -29$

It is not necessary to use signed minors to compute a 3×3 determinant. There is a "diagonal method" similar to the scheme used for 2×2 determinants. Begin by writing the determinant with the first two columns written over again to the right of the third column. Then, the determinant is the sum of products of the entries on each "falling" diagonal minus the products of the entries on each "rising" diagonal:

$$\begin{vmatrix} a & b & c \\ d & e & f \\ r & s & t \end{vmatrix} \begin{matrix} a & b \\ d & e \\ r & s \end{matrix} = aet + bfr + cds - rec - sfa - tdb$$

Careful! This diagonal method does *not* generalize to 4×4 determinants or larger $n \times n$ determinants.

Example 5 Use the diagonal method to compute the determinant of the 3×3 matrix

$$\begin{vmatrix} 2 & 3 & -1 \\ 0 & -4 & 2 \\ 1 & 5 & 1 \end{vmatrix}.$$

Solution

$$\begin{aligned} &= (2)(-4)(1) + (3)(2)(1) + (-1)(0)(5) \\ &\quad - (1)(-4)(-1) - (5)(2)(2) - (1)(0)(3) \\ &= -8 + 6 + 0 - 4 - 20 - 0 = -26 \end{aligned}$$

Goal C to use Cramer's Rule to solve a system of n linear equations in n variables

The results we found earlier for systems of two equations in two variables can be generalized to systems of n equations in n variables.

Cramer's Rule

A system of n linear equations in n variables has a unique solution if and only if the determinant of the coefficient matrix A is not zero. If a unique solution exists, the values of the variables can be written as quotients of determinants; the denominators are det A and the numerators are the determinant of A modified by replacing the column whose entries are the coefficients of the variable with the constants of the system.

In Example 6, we use Cramer's rule to determine whether a system of 3 linear equations in 3 variables has a unique solution.

Example 6 Determine whether the following system has a unique solution, and find the solution if it exists.

$$\begin{cases} 2x + y = 3 \\ x + y + 2z = -1 \\ x + 3z = 1 \end{cases}$$

Solution We begin by computing the determinant of the coefficient matrix A.

$$\det A = \begin{vmatrix} 2 & 1 & 0 \\ 1 & 1 & 2 \\ 1 & 0 & 3 \end{vmatrix} = \begin{matrix} 2 \cdot 1 \cdot 3 + 1 \cdot 2 \cdot 1 + 0 \cdot 1 \cdot 0 \\ -1 \cdot 1 \cdot 0 - 0 \cdot 2 \cdot 2 - 3 \cdot 1 \cdot 1 = 5 \end{matrix}$$

Since $\det A \neq 0$, the system has a unique solution. It is given by

$$x = \frac{\begin{vmatrix} 3 & 1 & 0 \\ -1 & 1 & 2 \\ 1 & 0 & 3 \end{vmatrix}}{\det A} = \frac{(9 + 2 + 0) - (0 + 0 - 3)}{5} = \frac{14}{5}$$

■ Replace first column of A with the column of constants.

$$y = \frac{\begin{vmatrix} 2 & 3 & 0 \\ 1 & -1 & 2 \\ 1 & 1 & 3 \end{vmatrix}}{\det A} = \frac{(-6 + 6 + 0) - (0 + 4 + 9)}{5} = -\frac{13}{5}$$

■ Replace second column of A with the column of constants.

$$z = \frac{\begin{vmatrix} 2 & 1 & 3 \\ 1 & 1 & -1 \\ 1 & 0 & 1 \end{vmatrix}}{\det A} = \frac{(2 - 1 + 0) - (3 + 0 + 1)}{5} = -\frac{3}{5}$$

■ Replace third column of A with the column of constants.

The unique solution of the system is $\left(\frac{14}{5}, -\frac{13}{5}, -\frac{3}{5}\right)$.

Notice that in Example 6, we used the diagonal method to calculate the determinants. In Example 7, we use signed minors.

Example 7 Determine those real numbers b, if any, for which the following system does *not* have a unique solution.

$$\begin{cases} x + 2y + bz = 3 \\ bx - y = 4 \\ 3x + y + 2z = 0 \end{cases}$$

Solution

$$\begin{vmatrix} 1 & 2 & b \\ b & -1 & 0 \\ 3 & 1 & 2 \end{vmatrix} = b\left(-\begin{vmatrix} 2 & b \\ 1 & 2 \end{vmatrix}\right) + (-1)\left(+\begin{vmatrix} 1 & b \\ 3 & 2 \end{vmatrix}\right)$$

■ The determinant of the coefficient matrix A is computed by expanding along the second row.

$$= b[-(4 - b)] - (2 - 3b)$$
$$= b^2 - b - 2$$
$$= (b - 2)(b + 1)$$

■ Recall that the system does not have a unique solution when $\det A = 0$.

When $b = 2$ or $b = -1$, the determinant is 0 and thus the system does not have a unique solution.

Exercise Set 5.5

Goal A

In exercises 1–8, compute the determinant of each of the following matrices.

1. $\begin{bmatrix} 2 & 5 \\ 1 & 4 \end{bmatrix}$
2. $\begin{bmatrix} 3 & 1 \\ 5 & 3 \end{bmatrix}$
3. $\begin{bmatrix} 2 & 4 \\ 1 & 5 \end{bmatrix}$
4. $\begin{bmatrix} 3 & 3 \\ 5 & 1 \end{bmatrix}$
5. $\begin{bmatrix} 2 & -4 \\ 3 & -6 \end{bmatrix}$
6. $\begin{bmatrix} 2 & -1 \\ 8 & -4 \end{bmatrix}$
7. $\begin{bmatrix} 2 & -4 \\ -3 & 7 \end{bmatrix}$
8. $\begin{bmatrix} 5 & -4 \\ -2 & 3 \end{bmatrix}$

In exercises 9–20, determine whether the system has a unique solution, and find the solution if it exists.

9. $\begin{cases} 2x + 5y = 3 \\ x + 4y = 2 \end{cases}$
10. $\begin{cases} 3x + y = 0 \\ 5x + 3y = -1 \end{cases}$
11. $\begin{cases} 2x + 6y = 4 \\ x + 3y = 2 \end{cases}$
12. $\begin{cases} 2x + 3y = 5 \\ 3x + 4y = 8 \end{cases}$
13. $\begin{cases} 2x + y = 1 \\ x - 4y = 0 \end{cases}$
14. $\begin{cases} 3x - y = 0 \\ -6x + 2y = 0 \end{cases}$
15. $\begin{cases} 2x - 3y = 4 \\ 4x - 6y = -2 \end{cases}$
16. $\begin{cases} 3x - y = 2 \\ -9x + 3y = 6 \end{cases}$
17. $\begin{cases} -2x + y = 3 \\ 3x - 4y = -5 \end{cases}$
18. $\begin{cases} 3x - y = 1 \\ -9x + 4y = -1 \end{cases}$
19. $\begin{cases} 4x + y = 2 \\ 2x - y = 4 \end{cases}$
20. $\begin{cases} 2x + y = 1 \\ x + 4y = 2 \end{cases}$

In exercises 21–26, determine those values of b, if any, for which the system does not have a unique solution.

21. $\begin{cases} 2x + 3y = 1 \\ 3x + by = 5 \end{cases}$
22. $\begin{cases} 3x - by = 3 \\ 4x + y = -2 \end{cases}$

23. $\begin{cases} x - by = 3 \\ bx + y = 2 \end{cases}$

24. $\begin{cases} x + b^2 y = 2 \\ x + by = 3 \end{cases}$

25. $\begin{cases} bx + 5y = 2 \\ 3x + (b-2)y = 1 \end{cases}$

26. $\begin{cases} bx + 3y = 2 \\ x + by = 5 \end{cases}$

Goal B

In exercises 27–32, compute the determinant by expanding along (a) the first row, (b) the second column.

27. $\begin{vmatrix} 2 & 1 & 3 \\ 1 & 0 & 2 \\ 1 & 1 & 1 \end{vmatrix}$

28. $\begin{vmatrix} 1 & 2 & 1 \\ 0 & 1 & 3 \\ 2 & 3 & 4 \end{vmatrix}$

29. $\begin{vmatrix} 2 & 1 & 0 \\ 1 & 3 & 4 \\ 2 & 0 & 1 \end{vmatrix}$

30. $\begin{vmatrix} 1 & 2 & -1 \\ 3 & 1 & 1 \\ -1 & 3 & -3 \end{vmatrix}$

31. $\begin{vmatrix} 1 & -1 & -2 \\ 3 & -2 & 0 \\ -1 & 4 & 2 \end{vmatrix}$

32. $\begin{vmatrix} -1 & 2 & 1 \\ 0 & -1 & 2 \\ 1 & -3 & -2 \end{vmatrix}$

For exercises 33–38, use the diagonal method to compute the determinants in exercises 27–32.

Goal C

For exercises 39–56, using Cramer's Rule, determine whether each system of linear equations in exercises 19–36 in Exercise Set 5.3 has a unique solution. Find the solution if it exists.

Superset

In exercises 57–60, compute the determinant using signed minors.

57. $\begin{vmatrix} 1 & 2 & 0 & 3 \\ -1 & 1 & 2 & 0 \\ 1 & 0 & -3 & 1 \\ 4 & 1 & -3 & 5 \end{vmatrix}$

58. $\begin{vmatrix} 1 & 2 & 0 & 1 \\ 0 & 2 & 1 & -3 \\ 0 & 0 & 3 & 4 \\ 0 & 0 & 0 & -1 \end{vmatrix}$

59. $\begin{vmatrix} 0 & 0 & 0 & 2 \\ 0 & 0 & -1 & 4 \\ 0 & 1 & 1 & 0 \\ 2 & 0 & -3 & 1 \end{vmatrix}$

60. $\begin{vmatrix} 2 & 0 & 1 & 3 \\ 1 & 2 & 0 & -1 \\ 3 & -1 & 0 & 2 \\ 2 & 3 & 2 & 3 \end{vmatrix}$

The following statements are true for every $n \times n$ matrix A.

I. If A contains a row of zeros, then $\det A = 0$.
II. If two rows of A are identical, then $\det A = 0$.
III. If a matrix B is obtained from A by interchanging two rows, then $\det B = -\det A$.
IV. If a matrix B is obtained from A by multiplying each entry in some row of A by a constant k, then $\det B = k \det A$.
V. If a matrix B is obtained from A by adding a multiple of one row of A to another row of A, then $\det B = \det A$.

Statements I–V are also true if "row" is replaced with "column."

61. Using a 3×3 matrix of your choice, verify Statement I.

62. Using a 3×3 matrix of your choice, verify Statement II.

In exercises 63–66, verify the specified statement using the 3×3 matrix

$$A = \begin{bmatrix} -1 & 0 & 4 \\ 3 & 1 & -2 \\ 0 & 1 & 1 \end{bmatrix}$$

63. Statement III, using the first and second rows.

64. Statement IV using 2 times the third row.

65. Statement V, using 2 times the first row added to the third row.

66. Statement V, using -3 times the second row added to the first row.

For exercises 67–72, redo exercises 61–66 with "row" replaced with "column."

73. Verify that the following equation describes a straight line through the points (x_0, y_0) and (x_1, y_1).

$$\begin{vmatrix} 1 & x & y \\ 1 & x_0 & y_0 \\ 1 & x_1 & y_1 \end{vmatrix} = 0$$

74. Use exercise 73 to find an equation that describes a line through the points $(2, 5)$ and $(4, 1)$.

5.6 Linear Programming

Goal A *to solve linear programming problems in two variables*

When trying to solve a real-world problem, one must bear in mind that the solution is very much influenced by certain physical limitations. Resources such as money, labor, raw materials, time, or fuel may be needed and are always limited. A good problem solver tries to allocate resources in such a way that the problem is solved in the most efficient or least wasteful manner.

Often limitations can be translated into a system of linear inequalities, such as

$$\begin{cases} y \geq 0 \\ x \leq 5 \\ y \leq x - 1, \end{cases}$$

and the function to be maximized (or minimized) can be written as a linear equation like

$$P = 17x + 12y.$$

A problem that can be described in this way is called a **linear programming problem.**

In linear programming problems, the inequalities are referred to as **constraints,** the solution set of the system of inequalities is called the **feasible region,** and the ordered pairs in this region are called **feasible solutions.** The **corner points** of the feasible region are especially important. (For the feasible region in Figure 5.18, the corner points are

$$(1, 0), \quad (5, 0), \quad \text{and } (5, 4).$$

Finally, the function to be maximized (or minimized) is called the **objective function.**

The solution of a linear programming problem is that ordered pair in the feasible region which produces the greatest (or least) value of the objective function. In the table at the left, some ordered pairs have been selected from the feasible region sketched above. These pairs have been used to determine corresponding values of the objective function. Of those ordered pairs chosen, the corner point (5, 4) produces the greatest value of P, namely 133. You will soon discover that the solution of a linear programming problem always occurs at one of the corner points.

Before we can state the fundamental fact about objective functions, we must first explore the different types of regions which can exist.

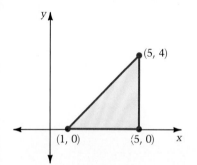

Figure 5.18 The solution set, called the **feasible region,** of the system

$$\begin{cases} y \geq 0 \\ x \leq 5 \\ y \leq x - 1. \end{cases}$$

Sample Feasible Solutions (x, y)	Objective Function $P = 17x + 12y$
(1, 0)	$P = 17$
(2, 1)	$P = 46$
(3, 2)	$P = 75$
(5, 0)	$P = 85$
(5, 4)	$P = 133$

A region is **convex** if, for any two points in the region, the line segment joining the two points is also part of the region.

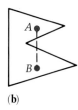

(a)   (b)

Figure 5.19 (a) A convex region. (b) *Not* a convex region.

When we refer to feasible regions, we assume them to be convex.

A feasible region can be **bounded** or **unbounded,** as shown in Figure 5.20. We now state a fact that is very useful in solving linear programming problems.

Fact
If the objective function in a linear programming problem attains a maximum or minimum value, then it does so at one of the corner points of the feasible region.

Figure 5.20 (a) A bounded region. (b) An unbounded region.

Example 1   A small furniture company manufactures two styles of waterbeds: modern and traditional. (All beds are queen-size.) Two machines are used to produce the beds: machine A cuts and forms the wood, and machine B finishes the wood. A modern bed requires two hours of machine A's time and $\frac{1}{2}$ hour of machine B's time. A traditional bed requires one hour on machine A and one hour on machine B. Limitations on the costly finishing materials for the modern bed restrict production to a maximum of 7 modern beds per day. Machines A and B can be operated a maximum of 24 and 18 hours per day, respectively. If the profit is $90 per modern bed and $60 per traditional bed, how many of each type should be produced each day to achieve the maximum possible profit?

Solution   Begin by organizing the data in a table.

	Modern	Traditional	Hours Available
Machine A	2 hrs	1 hr	24 hrs
Machine B	$\frac{1}{2}$ hr	1 hr	18 hrs

Profit per bed	$90	$60

■ Data for the objective function.

Assign Variables

Let x = the number of modern beds produced each day.
Let y = the number of traditional beds produced each day.
Let P = the daily profit on the sale of the two kinds of beds.

Constraints

$$\begin{cases} 2x + y \leq 24 \\ \frac{1}{2}x + y \leq 18 \\ x \leq 7 \\ x \geq 0 \\ y \geq 0 \end{cases}$$

- Line 1 of the table of values.
- Line 2 of the table of values.
- "a maximum of 7 modern beds per day"
- Nonnegativity constraints: the number of beds produced cannot be negative.

Objective Function $P = 90x + 60y$ ■ The function to be maximized.

Feasible Region

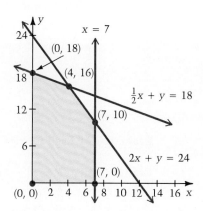

- Determine the feasible region by graphing the system of constraints. Corner points are found by solving systems of two of the five equations. For example, (4, 16) is the solution to the system

$$\begin{cases} 2x + y = 24 \\ \frac{1}{2}x + y = 18. \end{cases}$$

Evaluate P at Each Corner Point

Corner Point	$P = 90x + 60y$	
(0, 0)	$P = 90(0) + 60(0) = 0$	
(0, 18)	$P = 90(0) + 60(18) = 1080$	
(4, 16)	$P = 90(4) + 60(16) = 1320$	■ Maximum value of P.
(7, 10)	$P = 90(7) + 60(10) = 1230$	
(7, 0)	$P = 90(7) + 60(0) = 630$	

The maximum value of the objective function is $1320 and occurs when $x = 4$ and $y = 16$. Thus, for maximum profit ($1320 per day), the company should manufacture 4 modern and 16 traditional beds daily.

Example 2 A health club promises: "our high energy lunch contains at least 36 grams of protein and at least 42 grams of carbohydrates." Suppose one ounce of the club's vegetable rice soup provides 2 grams of protein, 6 grams of carbohydrates, and 1 gram of fat, and one ounce of the club's tuna salad provides 9 grams of protein, 6 grams of carbohydrates, and 4 grams of fat. How many ounces each of vegetable rice soup and tuna salad constitute a lunch that fulfills the club's promise, but does so with minimum fat content?

Solution

	Vegetable rice soup	Tuna salad	Minimum content
Protein	2 gm (per oz)	9 gm (per oz)	36 gm
Carbohydrates	6 gm (per oz)	6 gm (per oz)	42 gm
Fat	1 gm (per oz)	4 gm (per oz)	■ Data for the objective function.

Assign Variables Let x = the number of ounces of soup.
Let y = the number of ounces of tuna salad.
Let F = the number of grams of fat in the entire lunch.

Constraints
$$\begin{cases} 2x + 9y \geq 36 \\ 6x + 6y \geq 42 \\ x \geq 0 \\ y \geq 0 \end{cases}$$

■ Lunch must contain at least 36 gm of protein and 42 gm of carbohydrates.

■ Amount of protein and carbohydrates cannot be negative.

Objective Function $F = x + 4y$ ■ The function to be minimized.

Feasible Region

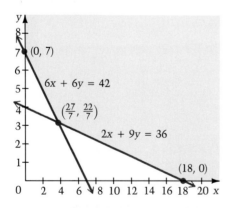

Evaluate F at Each Corner Point

Corner Point	$F = x + 4y$
$(0, 7)$	$F = (0) + 4(7) = 28$
$\left(\frac{27}{7}, \frac{22}{7}\right)$	$F = \left(\frac{27}{7}\right) + 4\left(\frac{22}{7}\right) = 16\frac{3}{7}$
$(18, 0)$	$F = (18) + 4(0) = 18$

The promise is fulfilled with minimum fat content when the lunch is composed of $\frac{27}{7}$ oz of soup and $\frac{22}{7}$ oz of tuna salad (roughly 4 oz of soup and 3 oz of tuna salad).

We now summarize the steps for solving a linear programming problem.

Step 1. Assign variables; then write the constraints and objective function.
Step 2. Determine the feasible region and corner points.
Step 3. Evaluate the objective function at each corner point, and determine which corner point produces the maximum or minimum.
Step 4. Answer the question stated in the problem.

Example 3 Suppose you have $10,000 and wish to invest all or part of this money in stocks and bonds. You would like to invest at least $3000 in bonds, offering a return of 8%, and at least $2000 in stocks, offering a return of 12%. As a precaution, you decide that the investment in bonds should be as much as or more than your investment in stocks. How should the money be invested to maximize your earnings?

Solution **Assign Variables** Let s = the amount invested in stocks (in dollars).
Let b = the amount invested in bonds (in dollars).
Let E = total earnings on the investments.

Constraints
$$\begin{cases} s + b \leq 10{,}000 \\ b \geq 3{,}000 \\ s \geq 2{,}000 \\ b \geq s \end{cases}$$

- You can invest up to $10,000.
- You will invest at least $3,000 in bonds.
- You will invest at least $2,000 in stocks.
- Bond investments are as much as or more than stock investments.

Objective Function $E = 0.08b + 0.12s$

Feasible Region

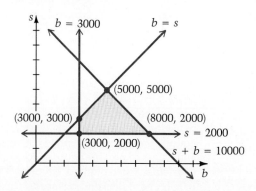

Evaluate E at Each Corner Point

$E = 0.08b + 0.12s$
$E = 0.08(3000) + 0.12(2000) = 480$
$E = 0.08(3000) + 0.12(3000) = 600$
$E = 0.08(5000) + 0.12(5000) = 1000$
$E = 0.08(8000) + 0.12(2000) = 880$

Since the maximum earnings occur when $b = 5{,}000$ and $s = 5{,}000$, you should invest $5,000 in bonds and $5,000 in stocks.

Exercise Set 5.6

Goal A

1. A baker decides to operate a small outlet at the airport and sell jelly donuts and brownies. Jelly donuts will be sold at a profit of 15 cents and brownies at a profit of 12 cents. The shelves can hold up to 300 jelly donuts and 500 brownies. However, the delivery van can bring, at most, 600 items (jelly donuts and brownies) each morning. How should the baker stock the outlet to maximize the profit? What is the maximum profit?

2. A manufacturer of hubcaps makes a profit of $17 on each deluxe hubcap and $11 on each standard hubcap. The manufacturer can produce 700 hubcaps each day, but not more than 400 deluxe or more than 600 standard. How many of each type should the manufacturer produce to maximize profit? What is the maximum profit?

3. The local record store makes a profit of $.37 on each 45 rpm and $1.12 on each LP record it sells. Consumer demand suggests that the store should maintain an inventory of 5000–9000 45's and 2000–3000 LP's. Storage space in the store limits the inventory to, at most, 11,000 records. How many of each type of record should the store have to maximize the profit potential of its inventory? What is the profit?

4. The pro shop at the local country club makes $9 profit for every golfer who uses a caddy and $5 profit for every golfer who rents an electric cart. A golfer who wants to use a caddy will rent a cart if no caddy is available; and a golfer who wants to use a cart will use a caddy if no cart is available. The manager of the shop must decide how many carts and caddies to have available to maximize the profit. The manager can purchase 25–40 carts and contract to have 20–30 caddies on hand each day. The manager also knows that, on any given day, 65 golfers, at most, will want a cart or a caddy. How many carts and caddies should the manager have on hand? What is the profit?

5. A history exam consists of 20 multiple choice questions worth 3 points each, and 25 true/false questions worth 2 points each. You are required to answer at least 5 questions of each type. If you can answer only 30 questions in the time period, how many of each type should you answer to maximize your test score?

6. In Example 2 of Section 5.6, what would the solution have been if the fat content and the protein content of the soup were each 3 gm per ounce?

7. A baseball team in the local league is classified as semipro because some of the players are amateurs and some are professionals. The facilities used for spring training will accommodate between 35 and 60 players. League rules require that at the start of camp, each team must have at least 10 amateurs and at least 20 professionals on its roster. The daily camp costs are $35 per amateur and $55 per professional. How many amateurs and professionals should the team have to minimize its daily costs? How many should it have to maximize its daily costs?

8. A local trucking firm transports the products of two manufacturers, A and B. Each carton of products from manufacturer A weighs 36 pounds and is 8 cubic feet in volume. Each carton of products from manufacturer B weighs 24 pounds and is 10 cubic feet in volume. The trucking firm charges Company A $1.70 and Company B $2.15 for each carton shipped. Each of the trucks has a maximum capacity of 4800 cubic feet and cannot carry more than 20,000 pounds. To keep labor costs down for loading and unloading the trucks, the firm limits each truck to a maximum of 560 cartons. How many cartons from each manufacturer should be loaded on a truck to maximize the charges? What is this total maximum charge?

9. A nursery uses two brands of fertilizer for rose bushes. Fertilizer A costs $4 per pound and provides 250 units of nutrients per pound. Fertilizer B costs $5 per pound and provides 300 units of nutrients per pound. The nursery spends $100 or less for fertilizer and wants to provide between 5200 and 6100 units of nutrients. Since brand A contains a special nutrient that brand B doesn't, the nursery uses at least 5 lbs of brand A. How many pounds of each brand should the nursery use to minimize cost?

10. A tailor has 100 square yards of cotton material and 120 square yards of woolen material. It takes 2 square yards of each type of material to make a sport coat. It takes 1 square yard of cotton and 3 square yards of wool to make a pleated skirt. The tailor already has orders for at least 15 sports coats and 5 skirts. If the tailor sells each coat for $40 and each skirt for $45, how many of each garment should be made in order to fill the orders and maximize income?

11. A large restaurant specializes in chicken dishes and advertises chicken that is "fresh daily." Each day the restaurant uses at least 350 lb of chicken breasts, at least 290 lb of chicken legs, and at least 390 lb of chicken wings. A local poultry supplier offers two chicken-part packages. Package I contains 25 lb of breasts, 15 lb of legs, and 15 lb of wings for $32. Package II contains 20 lb of breasts, 20 lb of legs and 45 lb of wings for $35. How many of each package should the restaurant order to meet its chicken requirement economically?

12. In exercise 11, what would the solution be if Package I cost $30 and Package II cost $40?

13. A farmer has 128 acres available for planting beans and alfalfa. Seed and fertilizer costs per acre are $50 for alfalfa and $70 for beans. Total labor costs per acre are $60 for alfalfa and $120 for beans. The farmer has $7000 to spend for seed and fertilizer and $11,280 to spend for labor. The farmer can realize an income (net profit) of $200 for each acre of alfalfa harvested and $260 for each acre of beans harvested. How many acres of each crop should the farmer plant to maximize profit? What is the profit?

14. The students in a gym class are mastering various gymnastic exercises. The students' grades will be based upon those exercises they are able to demonstrate successfully during a 48 minute gym period. The various exercises have been divided into two groups and assigned point values of 3 and 5 points. Each student must stand in line two minutes before demonstrating mastery of a 3-point exercise and three minutes before demonstrating mastery of a 5-point exercise. Each student is required to demonstrate at least three 3-point exercises. No student may demonstrate more than 20 exercises. How many exercises of each type should be demonstrated to maximize a student's point total? (Assume that the time needed to demonstrate each exercise can be ignored.)

15. A local dairy farmer has two farms, F_1 and F_2, from which milk is shipped to two retail outlet stores, S_1 and S_2. Farm F_1 produces 40 gallons of milk each day; farm F_2 produces 30 gallons each day. Store S_1 orders 25 gallons each day; store S_2 orders 20 gallons each day. The milk that the farmer does not ship to either store is sold to a consortium. The cost of shipping a gallon of milk from a farm to a store is given in the table below.

Farm	Store	Shipping Cost per Gallon
F_1	S_1	7¢
F_1	S_2	9¢
F_2	S_1	10¢
F_2	S_2	13¢

How should the farmer fulfill the stores' orders each day to minimize the shipping costs?

16. A candy manufacturer produces three kinds of fudge. The number of ounces of chocolate and nuts in a pound of each type is shown in the table below.

Type	Ounces of Chocolate	Ounces of Nuts
Smooth	13	2
Tasty	8	5
Crunchy	4	9

A pound of chocolate costs $1.20 and a pound of nuts costs $1.92. The manufacturer intends to market 20 ounce packages containing at least 2 ounces of each type of fudge, and advertises "contains at least as much chocolate as nuts." What is the manufacturer's minimum costs per package of chocolate and nuts?

Superset

17. Refer to Example 3 in Section 5.6. On the given feasible region, sketch the graph of the line
$$800 = 0.08b + 0.12s.$$
What can you say about all the points in the feasible region that lie on this line?

Next, sketch the graph of the line
$$E = 0.08b + 0.12s,$$
with E first as 900 and then as 1000. What do you notice? What happens if $E > 1000$? What happens if $E < 480$?

18. Refer to Example 3 in Section 5.6. Is there an objective function for which the maximum (minimum) value could occur at points other than corner points?

19. Explain the reason that the maximum and minimum values of $E = x^2 + y^2$ for a bounded convex region in the first quadrant must occur on the boundary of the region.

20. For a bounded convex region in the first quadrant, where would you look for the maximum and minimum values of $E = x^3$? Where would you look for $E = y^3$?

21. For a bounded convex region in the first quadrant, where would you look for the minimum value of $E = |x - y|$?

22. For a bounded convex region in the first quadrant, where would you look for the minimum value of $E = |2x - 3y|$?

Exercises 23–24 are problems that require integer solutions. We can use linear programming methods to help us determine these solutions even though they do not occur at corner points.

23. Suppose the only resources a furniture manufacturer needs to produce tables and chairs are wood and labor. A table requires 7 square feet of wood and $3\frac{1}{2}$ hrs of work, and a chair requires 3 square feet of wood and 4 hrs of work. The manufacturer has 40 square feet of wood and 28 hrs of work. If there is a $5 profit on each table produced and $3 profit on each chair, how many chairs and tables should the manufacturer produce to maximize its profit?

24. In exercise 23, what is the solution if the profit is $6 per table and $2 per chair?

5 / Systems of Equations and Inequalities

Chapter Review & Test

Chapter Review

5.1 Systems of Equations in Two Variables (pp. 194–203)

A collection of two or more equations is called a *system of equations*. The *solutions* of a system of two equations in two variables are those ordered pairs that are solutions of *every* equation of the system. (p. 194)   To solve such a system, we rewrite it as a succession of equivalent systems until we obtain an equivalent system whose solution is obvious. To do this, we use either the *addition method* (p. 196) or the *substitution method* (p. 200).

A system of two linear equations in two variables can have no solutions (parallel lines), one solution (nonparallel lines), or infinitely many solutions (the same line). (p. 201)

5.2 Systems of Inequalities in Two Variables (pp. 204–209)

To solve an inequality in two variables, we begin by looking at the associated equality. The graph of this equation divides the *xy*-plane into regions. All the points in any one region must satisfy the same inequality. To determine which inequality holds for a region, you need only test one point from that region. (p. 204)

To graph the solution set of a system of two or more inequalities in two variables, we graph the solution set of each inequality and then identify the points common to all regions. (p. 206)

5.3 Systems of Linear Equations in Three Variables (pp. 210–215)

We use the *xyz-coordinate system* to describe the position of a point in three-dimensional space. In this coordinate system, we graph *ordered triples* of the form (x, y, z). (p. 210)

We can use either the addition method or substitution method to solve a system of three linear equations in three variables. Such a system can have no solutions, one solution, or infinitely many solutions. (p. 213)

5.4 Matrices (pp. 216–224)

A *matrix* is a rectangular array of numbers used to represent a linear system. A matrix with m rows and n columns is called an $m \times n$ *matrix*. (p. 216)   *Row reduction* is a method used to solve linear systems (p. 218)

$$\begin{bmatrix} 1 & 2 & | & 3 \\ 4 & 5 & | & 6 \end{bmatrix} \qquad \begin{bmatrix} 1 & 2 \\ 4 & 5 \end{bmatrix} \qquad \begin{bmatrix} 3 \\ 6 \end{bmatrix}$$

an *augmented matrix*   its *coefficient matrix*   its *constant matrix*

5.5 Determinants (pp. 225–234)

The **determinant** is a number associated with an $n \times n$ square matrix. The determinant of a 2×2 matrix is the product of the diagonals. (p. 225) Signed minors are used to compute determinants of $n \times n$ matrices for $n > 2$. (p. 229)

Cramer's Rule is used to determine whether a system has a unique solution, and to find that solution if it exists. A system has a unique solution if and only if the determinant of the coefficient matrix is *not* zero. (p. 231)

5.6 Linear Programming (pp. 235–242)

To solve a linear programming problem, we

- assign variables; then write the constraints and objective function;
- determine the feasible region and the corner points;
- evaluate the objective function at each corner point, and determine which corner point produces the desired extreme value;
- answer the question stated in the problem.

Chapter Test

5.1A Determine whether the following points are solutions of the system

$$\begin{cases} y + 5x - 3 = 0 \\ 3y - 2x + 6 = 0. \end{cases}$$

1. (3, 12) **2.** (–2, 1) **3.** (1, 2)

5.1B Use the addition method to solve each of the following systems.

4. $\begin{cases} 6x + 2y = 12 \\ 3x - 4y = 6 \end{cases}$ **5.** $\begin{cases} -25x + 10y = 5 \\ 5x - 2y = -1 \end{cases}$

5.1C Use the substitution method to solve each of the following systems.

6. $\begin{cases} 4x + 2y = -5 \\ 6x + 3y = -7 \end{cases}$ **7.** $\begin{cases} 2x + 5y = -7 \\ 4x - 3y = 10 \end{cases}$ **8.** $\begin{cases} -2x + 10y = 6 \\ -x + 5y = 3 \end{cases}$

5.1D Solve each of the following systems.

9. $\begin{cases} x^2 + 4x + y^2 - 2y = -1 \\ y - 5x = -5 \end{cases}$ **10.** $\begin{cases} x^2 - 8x + y^2 = 0 \\ x^2 + y^2 = 9 \end{cases}$

5.2A Sketch the graph of each inequality.

11. $x^2 + y^2 > 4$ **12.** $y \leq 2x + 3$

5.2B Graph the solution set of each of the following systems.

13. $\begin{cases} y \geq 5 - 3x \\ y < x - 2 \end{cases}$
14. $\begin{cases} y \geq 0 \\ x \leq 8 \\ y < 2x - 1 \end{cases}$
15. $\begin{cases} y < -2 \\ x > 1 \\ -y \geq x \\ y \geq x - 10 \end{cases}$

5.3A Graph each of the following planes.

16. $x = -7$
17. $y = 6$
18. $x = -4$

5.3B Solve each of the following systems.

19. $\begin{cases} 10x - 13y - 2z = 1 \\ 2x - 7y - 4z = -1 \\ 3x + 4y + 5z = 2 \end{cases}$
20. $\begin{cases} x + y + z = 5 \\ 2x - z = 0 \\ 5y + z = -1 \end{cases}$

5.4A Determine the augmented, coefficient, and constant matrices for each system.

21. $\begin{cases} 4x - 3y = 7 \\ x + 2y = -3 \end{cases}$
22. $\begin{cases} 2x + y - 3z = -7 \\ x - 2y = 5 \\ 4y + 2z = 0 \end{cases}$

5.4B Solve each of the following systems. (Use matrix methods.)

23. $\begin{cases} x + 2y = 3 \\ 3x - y = 1 \end{cases}$
24. $\begin{cases} 2x - y + 3z = -8 \\ 5x - 2y - z = -1 \\ x + y + z = 3 \end{cases}$

5.5A Compute the determinants of each of the following matrices.

25. $\begin{bmatrix} -1 & 1 \\ 3 & 3 \end{bmatrix}$
26. $\begin{bmatrix} 4 & 3 \\ -2 & -2 \end{bmatrix}$

5.5B Compute the following determinants by expanding along the (a) first row, (b) the second column.

27. $\begin{vmatrix} 2 & 0 & -1 \\ 1 & -1 & 4 \\ 1 & 3 & 1 \end{vmatrix}$
28. $\begin{vmatrix} -1 & -1 & 0 \\ 2 & 3 & 1 \\ 3 & 1 & 1 \end{vmatrix}$

5.5C For exercises 29–30, using Cramer's Rule, determine whether each system of linear equations in exercises 19–20 has a unique solution. Find the solution if it exists.

5.6A Solve the following linear programming problems.

31. A vending firm at the football stadium makes a $0.27 profit on each hamburger and a $0.22 profit on each hotdog sold. The firm sells between 4000 and 7000 hamburgers and between 6000 and 11,000 hotdogs per game. Refrigerator space limits the vendor's inventory to 16,000 items (hamburgers and hotdogs). How many of each type should the vendor have to maximize the potential profit of the inventory?

32. A firm produces two items, each of which requires three manufacturing departments: assembly, inspection, and packing. Item A requires 0.30 hrs of assembly time, 0.20 hrs of inspection time and 0.10 hrs of packing time. Item B requires 0.20 hrs of assembly time, 0.05 hrs of inspection time, and 0.05 hrs of packing time. Daily capacities are 40 hrs of assembly time, 25 hrs of inspection time, and 15 hrs of packing time. The profit per item sold is $5 for A and $4 for B. Any items produced are sold. How many of each item should the manufacturer produce to maximize profit?

Superset

33. Determine the real number values of A, if any, such that the system below has (a) exactly one solution, (b) no solutions, and (c) infinitely many solutions.

$$\begin{cases} x - 5y = A \\ Ax + 10y = -4 \end{cases}$$

34. Graph the solution set of each of the following systems.

(a) $\begin{cases} y \geq x^2 - 3x - 10 \\ y < x \end{cases}$ (b) $\begin{cases} x^2 + (y - 2)^2 \leq 9 \\ x^2 + (y + 2)^2 < 9 \end{cases}$ (c) $\begin{cases} |x + y| \leq 7 \\ |y| \leq 5 \end{cases}$

35. Describe each of the following graphed regions by a system of inequalities.

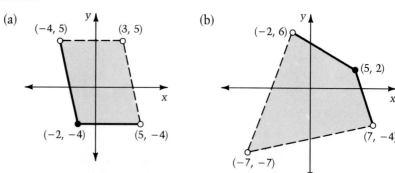

Cumulative Test: *Chapters 3–5*

In exercises 1–2, test the relations for each of the three types of symmetry.

1. $x^2 + y^2 - 3x^2y = 0$ **2.** $x^3 - y^3 = -1$

In exercises 3–4, determine $f(3), f(-2), f(h^2), f(h-4)$ for each function.

3. $f(x) = x^2 - 2x + 3$ **4.** $f(x) = 9 - x^2$

In exercises 5–6, sketch the graph of the second equation by reflecting the graph of the first equation.

5. $y = |x - 1|;\ \ y = -|x - 1|$ **6.** $y = \sqrt{x + 2};\ \ y = \sqrt{2 - x}$

In exercises 7–8, use transformations to sketch the following.

7. $y = -(x - 2)^2 + 2$ **8.** $y = -2|x + 1|$

In exercises 9–10, determine an equation of the line satisfying the stated conditions. Write your answer in (a) point-slope form, (b) slope-intercept form, and (c) general form.

9. The line is perpendicular to the graph of $2x - 3y + 5 = 0$ with y-intercept -2.

10. The line is parallel to the graph of $2y - 6x + 3 = 0$ and passes through $(-3, 0)$.

In exercises 11–12, sketch the graph of each function. Label the vertex, the x-intercepts, the axis of symmetry, and two additional points.

11. $f(x) = x^2 - 4x + 4$ **12.** $f(x) = x^2 - 2x + 5$

In exercises 13–14, find an equation that describes $(f \circ g)(x)$ and state the domain of the composite function.

13. $f(x) = x^2 - 4;\ g(x) = 2x - 5$ **14.** $f(x) = \sqrt{x - 2};\ g(x) = 4x^2 + 1$

In exercises 15–16, find the zeros of the following functions and then sketch the graphs.

15. $f(x) = x(x^2 - x - 2)$ **16.** $x^4 + 3x^2 - 4$

In exercises 17–18, sketch the graph of the following rational functions.

17. $y = \dfrac{x + 1}{x - 2}$ **18.** $y = \dfrac{2x + 1}{x + 1}$

In exercises 19–20, sketch the graph of the relation. Determine the vertex, the axis of symmetry and the x- and y-intercepts.

19. $x = 2(y - 2)^2 - 2$ **20.** $x = -y^2 - 6y - 10$

In exercises 21–22, sketch the graph of the relation. Determine the center, vertices, and asymptotes.

21. $16x^2 - 32x - 9y^2 + 54y = 209$ 22. $x^2 - y^2 - 10x - 2y = -15$

In exercises 23–24, use the addition method to solve each system.

23. $\begin{cases} 2x + 5y = -1 \\ 3x - 2y = -11 \end{cases}$ 24. $\begin{cases} 2x + 3y = 12 \\ 4x + 6y = -3 \end{cases}$

In exercises 25–26, graph the solution of each system.

25. $\begin{cases} 3y \leq 4x \\ 3y < -4x + 24 \\ y \leq 0 \end{cases}$ 26. $\begin{cases} y \geq x \\ 2y + x < 5 \\ y + 2 \geq 0 \end{cases}$

For exercises 27–28, solve the systems in exercises 23–24 using matrix notation.

In exercise 29, solve the following linear programming problem.

29. A newlywed couple has received $7000 from their parents. They want to invest all or part of this money in a savings account and in stocks. The savings account offers a return of 9.5%, and the stocks will earn 11%. They decide to invest at least as much in the savings account as they do in stocks, and they want to put at least $2000 into their savings and at least $1500 into stock. To get the greatest return on their investments, how much money should be placed into each investment?

Superset

30. A circle with center (2, 3) has an x-intercept at 2. Find the equation of the circle and sketch its graph.

31. Sketch the graph of the following functions and show all the asymptotes.

(a) $f(x) = \dfrac{x^2 + 3x + 4}{x}$ (b) $f(x) = \dfrac{2x^2 + x + 1}{x + 1}$

32. Use matrix notation to solve the following: Find all two-digit numbers such that the sum of the digits is 16, and the number is decreased by 18 if the order of the digits is reversed.

33. Graph the solution set of each system.

(a) $\begin{cases} |y| \leq 2 \\ y \geq 1 - |x| \end{cases}$ (b) $\begin{cases} y \geq x^2 - 1 \\ y \leq -x^2 + 1 \\ x \geq 0 \end{cases}$

6

Exponential and Logarithmic Functions

6.1 Exponential Functions

Goal A *to evaluate and simplify expressions containing exponents*

Our work with the functions in this chapter will rely upon an understanding of the rules of exponents. These rules are listed below. Even though we have defined b^x for rational exponents only, the expression b^x makes sense even when x is not a rational number.

Rules of Real Number Exponents

Rule 1 $b^0 = 1$ Rule 2 $b^1 = b$

Rule 3 $b^x b^y = b^{x+y}$ Rule 4 $(b^x)^t = b^{xt}$

Rule 5 $\dfrac{b^x}{b^y} = b^{x-y}$ Rule 6 $b^{-x} = \dfrac{1}{b^x}$

where the base b is positive, and the exponents x, y, and t represent any real numbers.

Example 1 Simplify the following expressions.

(a) $2(2^x)^3$ (b) $\dfrac{3^{x+1}}{3^2}$ (c) $4^{x+5} \cdot 8^{1-x}$

Solution (a) $2(2^x)^3 = 2(2^{3x}) = 2^{3x+1}$ ■ Rules 4 and 3 of Real Number Exponents are used. Note that $(2^x)^3$ is not the same as 2^{x+3}.

(b) $\dfrac{3^{x+1}}{3^2} = 3^{(x+1)-2} = 3^{x-1}$ ■ Rule 5 of Real Number Exponents is used.

(c) $4^{x+5} \cdot 8^{1-x} = (2^2)^{x+5}(2^3)^{1-x}$ ■ Begin by rewriting each factor with the same base.
$= 2^{2x+10} 2^{3-3x}$
$= 2^{-x+13}$

Goal B to sketch the graph of an exponential function

We now consider functions that are described by equations containing exponential expressions. These functions are used to solve problems involving population growth, compound interest, and radioactive decay.

Definition

An **exponential function** is a function of the form

$$f(x) = b^x,$$

where the base b is any positive real number except 1.

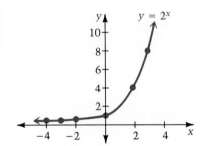

Figure 6.1

To sketch the graph of the exponential function $f(x) = 2^x$, we begin with a table of values.

x	-4	-3	-2	-1	0	1	2	3
$f(x)$	$\frac{1}{16}$	$\frac{1}{8}$	$\frac{1}{4}$	$\frac{1}{2}$	1	2	4	8

We plot these points and draw a smooth curve as shown in Figure 6.1. Notice that as $x \to -\infty$, the corresponding values of $f(x)$ remain positive but get very close to zero. Thus, the x-axis is a horizontal asymptote.

Next we sketch the graph of $f(x) = (\tfrac{1}{2})^x$. By Rule 6 of Exponents we know that

$$\left(\frac{1}{2}\right)^x = \frac{1}{2^x} = 2^{-x}$$

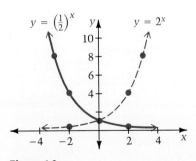

Figure 6.2

How does the graph of $y = 2^{-x}$ compare with the graph of $y = 2^x$? Recall that in general, replacing x with $-x$ in an equation reflects a graph across the y-axis. Thus, $f(x) = (\tfrac{1}{2})^x$ (or $y = 2^{-x}$) is the reflection of $y = 2^x$ across the y-axis. We generalize these graphing results in Figure 6.3 for any positive base b not equal to 1.

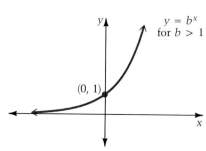 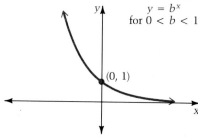

(a) Domain: $(-\infty, +\infty)$, Range: $(0, +\infty)$ (b) Domain: $(-\infty, +\infty)$, Range: $(0, +\infty)$

Figure 6.3

In either case, the point $(0, 1)$ is on the graph of $y = b^x$.

Example 2 Sketch the graphs of each set of equations on a single set of coordinate axes.

(a) $y = 2^x$, $y = 3^x$, and $y = 10^x$ (b) $y = \left(\dfrac{1}{2}\right)^x$ and $y = \left(\dfrac{1}{3}\right)^x$

Solution (a)

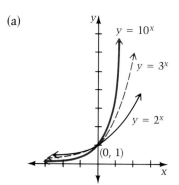 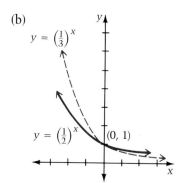

These graphs illustrate four important features of the function $f(x) = b^x$, where $b > 0$ and $b \neq 1$.

- If $x > 0$, then as the base b increases, b^x also increases.
- If $x < 0$, then as the base b increases, b^x decreases.
- For any base b, $b^0 = 1$.
- The x-axis is a horizontal asymptote.

Example 3 Sketch the graph of $y = 2^x - 4$.

Solution

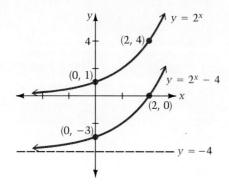

- Recall that the graph of $y = f(x) - 4$ is a translation of $y = f(x)$ four units down. The graph of $y = 2^x$ is lowered four units to obtain $y = 2^x - 4$.
- The line $y = -4$ is a horizontal asymptote.

Goal C *to solve exponential equations*

Before we use the Rules of Real Number Exponents to simplify and solve exponential equations, we must write the exponential expressions with the same base. Once this has been accomplished, we can use the following fact to write a simpler equivalent equation.

Fact
If $b > 0$, and $b \neq 1$, then $b^a = b^c$ is equivalent to $a = c$.

Example 4 Solve the following equation for x: $2^{6-x} = \frac{1}{4}(8^x)$.

Solution

$2^{6-x} = \frac{1}{4}(8^x)$ ■ Begin by writing each term with the same base.

$2^{6-x} = \frac{1}{2^2}(2^3)^x$

$2^{6-x} = 2^{-2} \cdot 2^{3x}$ ■ Rules 6 and 4 of Real Number Exponents are used.

$2^{6-x} = 2^{-2+3x}$ ■ Since the bases are equal, by the fact above, the exponents are equal.

$6 - x = -2 + 3x$

$x = 2$

Exercise Set 6.1   = Calculator Exercises in the Appendix

Goal A

In exercises 1–14, simplify each of the expressions.

1. $(2^x)^{-3}$
2. $(3^x)^{-2}$
3. $5^{x-7} \cdot 5^{x+3}$
4. $7^{4-x} \cdot 7^{x-9}$
5. $2^{x+2} \cdot 4^{-x}$
6. $3^{x-2} \cdot 9^x$
7. $\dfrac{3^{x+2}}{3^{4-x}}$
8. $\dfrac{2^{1-x}}{2^{5+x}}$
9. $4^{x+1} \cdot 8^{2-x}$
10. $9^{2-3x} \cdot 27^{x-4}$
11. $\dfrac{27^{2x-1}}{9^{x+2}}$
12. $\dfrac{8^{x-1}}{4^{3+2x}}$
13. $2^x + 4^x$
14. $9^x - 3^x$

Goal B

In exercises 15–26, sketch the graph of each of the following equations.

15. $y = 4^x$
16. $y = 5^x$
17. $y = 3^{-x}$
18. $y = 5^{-x}$
19. $y = 3^x - 1$
20. $y = 2^x + 5$
21. $y = 3^{x-1}$
22. $y = 2^{x+5}$
23. $y = \left(\dfrac{1}{2}\right)^x - 1$
24. $y = 1 - \left(\dfrac{1}{3}\right)^x$
25. $y = -\left(\dfrac{1}{3}\right)^{x-2}$
26. $y = -\left(\dfrac{1}{2}\right)^{x+1}$

Goal C

In exercises 27–38, solve each of the following equations for x.

27. $2^{x+1} = 16$
28. $3^{7-x} = 81$
29. $8^x = 4$
30. $9^x = 27$
31. $3^x = 9^{x+4}$
32. $2^x = 8^{5-x}$
33. $3^x = 9(3^{5-x})$
34. $2^x = 2^{3x}(4^2)$
35. $2^x = -4$
36. $3^{x-2} = 0$
37. $4^x = \dfrac{1}{2}(8^{x+1})$
38. $9^x = \dfrac{1}{27}(3^{1-x})$

Superset

In exercises 39–44, solve each equation for x.

39. $2^{x^2} = 4$
40. $2^{x^3} = \dfrac{1}{2}$
41. $2^x(2^x - 1) = 0$
42. $3^x(9 - 3^x) = 0$
43. $2^{2x} - 5 \cdot 2^x + 4 = 0$
44. $3^{2x} - 12 \cdot 3^x + 27 = 0$
45. Carefully graph $y = 2^x$ and use your graph to estimate the value of:
    (a) $2^{\sqrt{2}}$ (b) 2^π (c) $2^{-\sqrt{3}}$
46. Carefully graph $y = 10^x$ and use your graph to estimate the value of x when:
    (a) $y = 50$ (b) $y = 75$ (c) $y = 5$
47. Suppose some quantity Q is related to another quantity k by the relationship $Q = 3^k$. What happens to Q if the value of k is doubled? tripled? increased tenfold?
48. Does the equation $3^x = x$ have any solutions? (Hint: graph $y = 3^x$ and $y = x$ on the same set of axes.)

In exercises 49–54, sketch the graph of each of the following equations.

49. $y = 2^{|x|}$
50. $y = |2^x|$
51. $y = 2^{x^2}$
52. $y = 2^{1/x}$
53. $y = -2^{x^2}$
54. $y = 2^{-|x|}$

In exercises 55–58, determine whether each of the following functions is symmetric with respect to the y-axis or the origin.

55. $y = 2^x$
56. $y = 2^{x^2}$
57. $y = 2^x + 2^{-x}$
58. $y = 2^x - 2^{-x}$

6.2 Logarithmic Functions

Goal A *to sketch the graph of an inverse function*

We now consider two questions that may be asked about a function $y = f(x)$. The first is straightforward: given a value of x, what is the value of y? To answer this, we simply evaluate the function. For example,

$$\text{if } y = 2^x \text{ and } x = 3, \text{ then } y = 8.$$

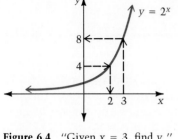

Figure 6.4 "Given $x = 3$, find y."
"Given $y = 4$, find x."

The second question reverses the first: given a value of y, what is the value of x? Sometimes we can answer this question easily. For instance,

$$\text{if } y = 2^x \text{ and } y = 4, \text{ then } x = 2.$$

Sometimes, however, the second question does not have a unique answer. For example,

$$\text{if } y = x^2 \text{ and } y = 16, \text{ then } x = 4 \text{ or } x = -4.$$

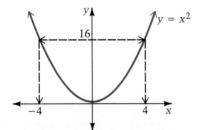

Figure 6.5 "Given $y = 16$, find x."
The answer is not unique.

If, for a given function f, the answer to "given y, find x" is always unique, then f is said to be *one-to-one*. That is, a function is **one-to-one** if each y-value in the range of f corresponds to exactly one x-value in the domain.

If a function f is one-to-one, then there exists an *inverse function*, denoted f^{-1}, associated with f. The inverse function f^{-1} reverses the process f; that is, if f processes 3 and produces 10, then f^{-1} processes 10 and produces 3.

Definition

If a function f is one-to-one, then there exists a unique function f^{-1}, called the **inverse function of f,** such that

$$(f^{-1} \circ f)(x) = x \quad \text{for any } x \text{ in the domain of } f.$$
$$(f \circ f^{-1})(x) = x \quad \text{for any } x \text{ in the domain of } f^{-1}.$$

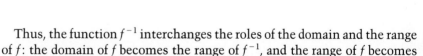

domain of f
range of f^{-1}

range of f
domain of f^{-1}

Figure 6.6

Thus, the function f^{-1} interchanges the roles of the domain and the range of f: the domain of f becomes the range of f^{-1}, and the range of f becomes the domain of f^{-1}.

Example 1 The function $f(x) = 3x - 2$ is one-to-one. Find its inverse function.

Solution To find the equation describing the inverse function, first interchange x and y. Then, since we usually write x as the independent variable and y as the dependent variable, we solve for y in terms of x.

$$y = 3x - 2$$
■ Interchange x and y to obtain the inverse.
$$x = 3y - 2$$
■ Solve for y in terms of x.
$$y = f^{-1}(x) = \frac{x + 2}{3}$$

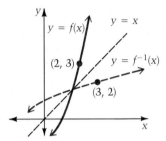

Figure 6.7

The graphs of a function f and its inverse f^{-1} are displayed in Figure 6.7. Note that the graph of $y = f^{-1}(x)$ is the reflection of $y = f(x)$ across the line $y = x$. Thus, to graph f^{-1}, we interchange the x- and y-coordinates of each ordered pair of f and graph the resulting ordered pairs.

Fact

If a function f has an inverse function f^{-1}, then we reflect the graph of $y = f(x)$ across the line $y = x$ to obtain the graph of $y = f^{-1}(x)$.

Example 2 The function $f(x) = x^3$, with restricted domain $-1 \leq x \leq 2$, is one-to-one. Find the equation describing the inverse function and sketch the inverse.

Solution

$$y = x^3$$
$$x = y^3$$
$$y = f(x)^{-1} = x^{1/3}$$

■ To find the equation describing the inverse function, interchange x and y and then solve for y.

■ Sketch the function $y = f(x)$. Then reflect it across the line $y = x$ to obtain the graph of the inverse function.

■ Domain of f: $[-1, 2]$
  Range of f: $[-1, 8]$

■ Domain of f^{-1}: $[-1, 8]$
  Range of f^{-1}: $[-1, 2]$

Goal B *to rewrite an exponential equation as a logarithmic equation and vice versa*

To find the inverse of $y = b^x$, we use the same techniques we used in Goal A. The statement "given y, find x" for the function $y = b^x$ is the same as the statement "given x, find y" for the inverse function $x = b^y$ (interchange x and y). The graphs of the inverse functions $x = b^y$ in Figure 6.8—one for the case where $b > 1$, and one for the case where $0 < b < 1$—are found by reflecting $y = b^x$ across the line $y = x$.

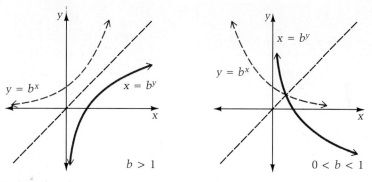

Figure 6.8 $x = b^y$: domain is $(0, +\infty)$, range is $(-\infty, +\infty)$.

In order to describe these inverse functions in the usual form $y = f^{-1}(x)$, it is necessary to write the equation $x = b^y$ with y in terms of x. We can describe y in words:

y is the power to which you raise b in order to obtain x.

We define *logarithm* as an abbreviation for this description of y.

> **Definition**
> For $b > 0$ and $b \neq 1$, the function $f(x) = b^x$ has an inverse function, denoted $f^{-1}(x) = \log_b x$. This notation is an abbreviation for the **logarithm of x to the base b.** The domain of the logarithmic function is $(0, +\infty)$ and the range is $(-\infty, +\infty)$.

Notice that we have taken y in the equation $x = b^y$ and defined it as a logarithm. Therefore, a logarithm is an exponent, and every exponential statement can be rewritten as an equivalent statement about logarithms. This equivalence is summarized below.

> **Log/Exp Principle**
> For $b > 0$ and $b \neq 1$, $b^A = C$ is equivalent to $\log_b C = A$.

Example 3 Rewrite the following as logarithmic equations: (a) $x^3 = 64$ (b) $9^x = 3$.

Solution (a) $x^3 = 64$ ■ Use the Log/Exp Principle.
$\log_x 64 = 3$

(b) $9^x = 3$ ■ Use the Log/Exp Principle.
$\log_9 3 = x$

Example 4 Solve each of the following equations for x.

(a) $\log_3 x = 2$ (b) $\log_8 x = \dfrac{1}{3}$ (c) $\log_x 64 = 3$ (d) $\log_9 3 = x$

Solution
(a) $\log_3 x = 2$
$3^2 = x$
$x = 9$

(b) $\log_8 x = \dfrac{1}{3}$
$8^{1/3} = x$
$x = 2$

(c) $\log_x 64 = 3$
$x^3 = 64$
$x = 64^{1/3}$
$x = 4$

(d) $\log_9 3 = x$
$9^x = 3$
$x = \dfrac{1}{2}$

You should keep in mind that a logarithm is just an exponent. When we say $\log_3 9 = 2$, we are saying that 2 is the exponent to which we raise 3 in order to get 9. That is, $3^2 = 9$.

Goal C *to graph logarithmic functions*

The two general graphs of the function $f(x) = \log_b x$ are shown in Figure 6.9. Looking at these graphs, we recall the following important information about the logarithmic function:

- the domain is $(0, +\infty)$ and the range is $(-\infty, +\infty)$;
- $\log_b 1 = 0$ and $\log_b b = 1$;
- the function is one-to-one.

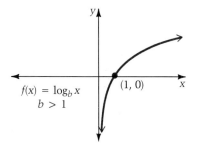
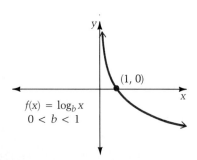

Figure 6.9

Fact
Since the logarithmic function is one-to-one, no two x-values correspond to the same y-value. Thus,

if $\log_b A = \log_b C$, then $A = C$.

This fact will be useful when we solve logarithmic equations.

Example 5 Sketch the graph of (a) $y = \log_2 x$ (b) $y = \log_2(x - 1)$.

Solution (a)

x	$\frac{1}{4}$	$\frac{1}{2}$	1	2	4
y	-2	-1	0	1	2

- To compute a table of values, we use the Log/Exp Principle to write $y = \log_2 x$ as $x = 2^y$.
- Plot the points and sketch the curve. The base for the logarithm in this example is $b = 2$. Note that this sketch is similar to the graph given on the previous page for the general case $b > 1$.

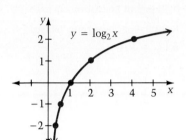

(b)

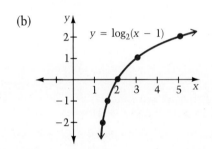

- Recall that the graph of $y = f(x - 1)$ is a translation of $y = f(x)$ one unit to the right. Thus, the graph of $y = \log_2 x$ is translated one unit to the right to obtain $y = \log_2(x - 1)$.

Goal D *to simplify and evaluate logarithmic expressions*

We use the following rules to simplify and evaluate logarithmic expressions. The proofs of these rules will be postponed until you have seen how they work.

Rules of Logarithms

Rule 1 $\log_b 1 = 0$ **Rule 2** $\log_b b = 1$

Rule 3 $\log_b(xy) = \log_b x + \log_b y$ **Rule 4** $\log_b(x^t) = t \log_b x$

Rule 5 $\log_b\left(\dfrac{x}{y}\right) = \log_b x - \log_b y$

Rule 6 $\log_b\left(\dfrac{1}{x}\right) = -\log_b x$

where x and y are positive real numbers, t is any real number, and base b is positive, but not equal to 1.

Example 6 Given that $\log_2 x = 3.1$ and $\log_2 y = 1.6$, evaluate $\log_2 \dfrac{8x}{y^3}$.

Solution

$\log_2 \dfrac{8x}{y^3} = \log_2 8x - \log_2 y^3$ ■ Rule 5 of Logarithms is used.

$\qquad = \log_2 8 + \log_2 x - \log_2 y^3$ ■ Rule 3 of Logarithms is used.

$\qquad = \log_2 8 + \log_2 x - 3 \log_2 y$ ■ Rule 4 of Logarithms is used.

$\qquad = 3 + 3.1 - 3(1.6)$ ■ $\log_2 8 = 3$.

$\qquad = 1.3$

As the next example illustrates, you *must* check solutions of logarithmic equations. Since logarithms are defined for positive numbers only, you must discard any solution that requires taking the logarithm of zero or a negative number.

Example 7 Solve the following equation for x: $2 \log_2 x = \log_2(3x + 4)$.

Solution

$2 \log_2 x = \log_2(3x + 4)$ ■ Use Rule 4 of Logarithms.

$\log_2 x^2 = \log_2(3x + 4)$ ■ Use the fact that if $\log_b A = \log_b C$, then $A = C$.

$x^2 = 3x + 4$

$x^2 - 3x - 4 = 0$ ■ Use the Property of Zero Products.

$(x - 4)(x + 1) = 0$

$x - 4 = 0 \quad | \quad x + 1 = 0$

$x = 4 \quad | \quad x = -1$ ■ Now, we *must* check solutions.

Must Check:

$x = 4$

$2 \log_2 4 = \log_2(3 \cdot 4 + 4)$

$\log_2 4^2 = \log_2 16$ ■ True.

$x = -1$

$2 \log_2(-1) = \log_2(3(-1) + 4)$ ■ $\log_2(-1)$ is undefined.

The only solution is $x = 4$.

The first fact below is a restatement of the definition of logarithm. The second fact is just a special case of Rule 4 of Logarithms. These two facts provide another way of saying that the logarithmic and exponential functions are inverse functions of one another.

Fact

1   $b^{\log_b x} = x$ 2 $\log_b(b^x) = x$

We conclude by proving Rules 1, 3, and 4 of Logarithms. The proofs of Rules 2, 5, and 6 are left as exercises.

Proof of Rule 1: $\log_b 1 = 0$

Rule 1 of Real Number Exponents states that	$b^0 = 1$
Then, by the Log/Exp Principle	$\log_b 1 = 0$ ∎

Proof of Rule 3: $\log_b(xy) = \log_b x + \log_b y$

Begin by letting	$r = \log_b x$ and $s = \log_b y$
By the Log/Exp Principle	$x = b^r$ and $y = b^s$
By Rule 3 of Real Number Exponents	$xy = b^{r+s}$
By the Log/Exp Principle	$\log_b(xy) = r + s$
Replacing r with $\log_b x$ and s with $\log_b y$, we obtain	$\log_b(xy) = \log_b x + \log_b y$ ∎

Proof of Rule 4: $\log_b(x^t) = t \cdot \log_b x$

Begin by letting	$r = \log_b x$
By the Log/Exp Principle	$x = b^r$
By Rule 4 of Real Number Exponents	$x^t = b^{rt}$
By the Log/Exp Principle	$\log_b(x^t) = rt$
Replacing r with $\log_b x$, we obtain	$\log_b(x^t) = t \cdot \log_b x$ ∎

Exercise Set 6.2

Goal A

In exercises 1–8, the given function $y = f(x)$ is one-to-one. Find the equation describing the inverse function and sketch the inverse.

1. $f(x) = 6 - 2x, \quad 1 \leq x \leq 3$
2. $f(x) = 3x - 8, \quad 0 \leq x \leq 6$
3. $f(x) = x^2, \quad 0 < x \leq 3$
4. $f(x) = x^3, \quad -3 \leq x < 3$
5. $f(x) = \sqrt{4 - x}, \quad x \leq 4$
6. $f(x) = 5 - x^2, \quad 0 < x < 2$
7. $f(x) = \frac{1}{3}(x - 1)^3, \quad 0 \leq x < 3$
8. $f(x) = \frac{1}{4}x^4, \quad 0 < x < 3$

For exercises 9–16, identify the domain and range of the inverse function $y = f^{-1}(x)$ for each of the functions in Exercises 1–8.

Goal B

In exercises 17–22, write each of the following as a logarithmic equation.

17. $x^2 = 25$
18. $x^4 = 16$
19. $16^x = 4$
20. $27^x = 3$
21. $5^3 = x$
22. $10^5 = x$

In exercises 23–32, use the Log/Exp Principle to solve each of the following equations for x.

23. $\log_2 x = 3$
24. $\log_4 x = 3$
25. $\log_8 x = \dfrac{1}{2}$
26. $\log_9 x = \dfrac{1}{2}$
27. $\log_x 4 = 2$
28. $\log_x 8 = 3$
29. $\log_8 4 = x$
30. $\log_9 27 = x$
31. $\log_x 3 = -2$
32. $\log_x 16 = -4$

Goal C

In exercises 33–42, sketch the graph of each of the following equations.

33. $y = \log_3 x$
34. $y = \log_4 x$
35. $y = \log_{1/2} x$
36. $y = \log_{1/3} x$
37. $y = \log_2(x + 1)$
38. $y = \log_3(x - 2)$
39. $y = \log_3(-x)$
40. $y = \log_2(-x)$
41. $y = -\log_2(x - 2)$
42. $y = -\log_3(x + 1)$

Goal D

In exercises 43–48, use the values $\log_2 x = 3.1$, $\log_2 y = 1.6$, and $\log_2 z = -2.7$ to evaluate each of the following expressions.

43. $\log_2(xyz)$
44. $\log_2\left(\dfrac{4}{xyz}\right)$
45. $\log_2\left(\dfrac{16x}{yz^3}\right)$
46. $\log_2\left(\dfrac{xy^2}{4z}\right)$
47. $\log_2\left(\dfrac{x}{y}\right) - \dfrac{\log_2 x}{\log_2 y}$
48. $\log_2 xy - (\log_2 x \cdot \log_2 y)$

In exercises 49–58, solve each of the following equations for x.

49. $4^{\log_8 x} = 2$
50. $9^{\log_3 x} = 3$
51. $\log_3(9^x) = -4$
52. $\log_4(2^x) = -3$
53. $\log_2(\log_3 x) = 2$
54. $\log_3(\log_2 x) = 2$
55. $\log_2(x^3) = \log_2(8x)$
56. $\log_2(x^2 - 6) = \log_2(-x)$
57. $\log_2(x^2) = 2 \cdot \log_2 x$
58. $\log_2 5x = \log_2 5 + \log_2 x$

Superset

In exercises 59–64, sketch the graph of each of the following equations.

59. $y = \log_2(x^2)$
60. $y = 1 - \dfrac{1}{3}\log_2(x^3)$
61. $y = \log_2(8^x)$
62. $y = \log_8(2^x)$
63. $y = 4^{\log_2 x}$
64. $y = \log_2 |x - 4|$

In exercises 65–70, solve each of the following equations for x.

65. $\log_4(x + 7) = \log_2(x + 1)$
66. $\log_3(3x) = \log_3 27 \cdot \log_3 x$
67. $(3 - \log_2 x)\log_2 x = 0$
68. $\log_x 1 = 0$
69. $4 \log_2 x = \log_2(5x^2 - 4)$
70. $\log_2[\log_2(\log_2 x)] = 0$
71. Prove Rule 2 of Logarithms.
72. Prove Rule 5 of Logarithms.
73. Prove Rule 6 of Logarithms.

6.3 Common Logarithms

Goal A *to use a table of values to estimate common logarithms*

Frequently 10 is used as a base for logarithms because we use a base 10 system of enumeration. Base 10 logarithms are called **common logarithms.** To simplify the notation we shall write $\log_{10} x$ as $\log x$.

If $y = \log x$, then y is the exponent to which you must raise 10 in order to get x. In this case, x is called the **common antilogarithm** of y; that is, x is the number you get when you raise 10 to the power y. The table at the left illustrates how we compute $\log x$ for certain special values of x. The graph of $y = \log x$ is shown in Figure 6.10.

A table of logarithmic values can be a valuable aid in performing complicated computations. Table 1 in the Appendix lists values for $\log x$. A small segment of the table is reproduced below.

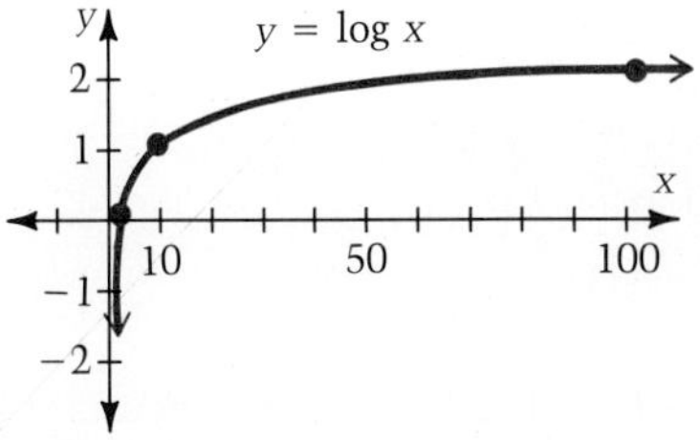

x	$\log x$
$0.001 = 10^{-3}$	-3
$0.01 = 10^{-2}$	-2
$0.1 = 10^{-1}$	-1
$1 = 10^{0}$	0
$10 = 10^{1}$	1

x	0	1	2	3	4	5	6	7	8	9
1.0	.0000	.0043	.0086	.0128	.0170	.0212	.0253	.0294	.0334	.0374
1.1	.0414	.0453	.0492	.0531	.0569	.0607	.0645	.0682	.0719	.0755
1.2	.0792	.0828	.0864	.0899	.0934	.0969	.1004	.1038	.1072	.1106
1.3	.1139	.1173	.1206	.1239	.1271	.1303	.1335	.1367	.1399	.1430
1.4	.1461	.1492	.1523	.1553	.1584	.1614	.1644	.1673	.1703	.1732
1.5	.1761	.1790	.1818	.1847	.1875	.1903	.1931	.1959	.1987	.2014
1.6	.2041	.2068	.2095	.2122	.2148	.2175	.2201	.2227	.2253	.2279

Figure 6.10 The Common Logarithmic Function.

The values of x lie between 1 and 10 and are accurate to two decimal places. The values of $\log x$ lie between 0 and 1 and are accurate to four decimal places. Remember that $\log 1 = 0$ and $\log 10 = 1$. The first two digits of x are given in the leftmost column, and the third digit is one of the column headings. The values of $\log x$ are the entries in the body of the table. To determine $\log 1.54$, we read down the leftmost column to 1.5, and then read across to the column labeled 4. The entry at this position in the table is 0.1875. Since the values in the table are approximations, $\log 1.54$ is approximately 0.1875, and we write $\log 1.54 \approx 0.1875$.

Example 1 Use Table 1 in the Appendix to find (a) $\log 2.95$ (b) $\log 3.13$ (c) $\log 3.04$.

Solution (a) $\log 2.95 \approx 0.4698$ (b) $\log 3.13 \approx 0.4955$ (c) $\log 3.04 \approx 0.4829$

We can use Table 1 to find the common logarithm of a number, even when the number is not between 1 and 10. To do so, we first write x in **scientific notation.** That is, we write

$$x = \begin{bmatrix} \text{a number} \\ \text{between 1 and 10} \end{bmatrix} \times \begin{bmatrix} \text{a power} \\ \text{of ten} \end{bmatrix} = A \times 10^t.$$

We then use the Rules of Logarithms to write

$$\log x = \log(A \times 10^t) = \log A + \log 10^t = \log A + t.$$

For example, if $x = 178.0$, then in scientific notation $x = 1.78 \times 10^2$, and

$$\log x = \log(1.78 \times 10^2) = \log 1.78 + 2 \approx 2.2504.$$

Example 2 Use the fact that $\log 3.14 \approx 0.4969$ to compute: (a) $\log 31.4$ (b) $\log 0.00314$.

Solution
(a) $\log 31.4 = \log(3.14 \times 10^1)$
$ = \log 3.14 + \log 10$
$ \approx 0.4969 + 1$
$ = 1.4969$

(b) $\log 0.00314 = \log(3.14 \times 10^{-3})$
$ = \log 3.14 + \log 10^{-3}$
$ \approx 0.4969 - 3$
$ = -2.5031$

Goal B *to use a table to estimate common antilogarithms*

To find a common antilogarithm, we follow our procedure for finding common logarithms in reverse. Suppose we are given that $\log x = 0.4843$. To find x, we must find the entry 0.4843 in Table 1. This entry appears in the row labeled 3.0 and the column labeled 5. Thus, if $\log x = 0.4843$, then $x \approx 3.05$.

Example 3 Use Table 1 to solve the following for x: (a) $\log x = 0.9175$ (b) $\log x = 0.0792$.

Solution (a) if $\log x = 0.9175$, then $x \approx 8.27$ (b) if $\log x = 0.0792$, then $x \approx 1.20$

To estimate the antilogarithm of a number that is not between 0 and 1, we first write the number as a sum of the form:

$$y = \text{(an integer)} + \text{(a number between 0 and 1)}.$$

For any number there is only one way to form this sum. For example,

$$2.4969 = 2 + 0.4969 \qquad -1.4969 = -2 + 0.5031$$
$$5.4969 = 5 + 0.4969 \qquad -3.4969 = -4 + 0.5031$$

We can then use Table 1 to find the antilogarithm of the decimal part of the sum. (The integer becomes the power of 10 in the answer.) For example, suppose we must solve $\log x = 2.4969$ for x. First we write

$$\log x = 2 + 0.4969.$$

We know that $2 = \log 10^2$ and using Table 1, we determine that $0.4969 = \log 3.14$. Therefore,

$$\log x \approx \log 10^2 + \log 3.14.$$

Then by Rule 3 of Logarithms, we have

$$\log x \approx \log(10^2 \times 3.14).$$

Thus, $x \approx 3.14 \times 10^2 = 314$.

Example 4 Find the antilogarithm of 1.1430. That is, solve $\log x = 1.1430$ for x.

Solution
$\log x = 1.1430$
$ = 1 + 0.1430$
$ \approx \log 10^1 + \log 1.39$ ■ From Table 1 we found that $\log 1.39 = 0.1430$.
$ = \log(10^1 \times 1.39)$ ■ Rule 3 of Logarithms is used.
$x \approx 1.39 \times 10^1 = 13.9$ ■ $\log A = \log C$ implies $A = C$.

Example 5 Find the antilogarithm of -2.7033. That is, solve $\log x = -2.7033$ for x.

Solution $\log x = -2.7033 = -3 + 0.2967 \approx \log 10^{-3} + \log 1.98 = \log(10^{-3} \times 1.98)$

Thus, $x \approx 1.98 \times 10^{-3}$ or 0.00198.

Exercise Set 6.3

Goal A

In exercises 1–12, use Table 1 in the Appendix to find the following.

1. $\log 4.83$
2. $\log 2.57$
3. $\log 8.81$
4. $\log 9.00$
5. $\log 1.87$
6. $\log 5.99$
7. $\log 3.26$
8. $\log 6.71$
9. $\log 2.33$
10. $\log 4.00$
11. $\log 1.60$
12. $\log 7.04$

In exercises 13–24, use scientific notation to rewrite each of the following numbers.

13. 47.24
14. 0.1008
15. 0.3502
16. 32.15
17. 0.015
18. 348.9
19. 0.0908
20. 0.0029
21. 3294.1
22. 6804.5
23. 0.0003
24. 0.00001

In exercises 25–36, use Table 1 to compute each of the following.

25. $\log 32.9$
26. $\log 0.142$
27. $\log 0.68$
28. $\log 75.1$
29. $\log 0.057$
30. $\log 920$
31. $\log 349$
32. $\log 0.010$
33. $\log 5160$
34. $\log 0.0082$
35. $\log 14{,}600$
36. $\log 0.0045$

Goal B

In exercises 37–54, use Table 1 to solve each of the following equations for x.

37. $\log x = 0.5490$
38. $\log x = 0.6702$
39. $\log x = 0.9304$
40. $\log x = 0.4771$
41. $\log x = 0.7396$
42. $\log x = 0.2122$
43. $\log x = 2.7520$
44. $\log x = 7.9800$
45. $\log x = 2.4800$
46. $\log x = 1.3054$
47. $\log x = 4.6637$
48. $\log x = 3.0414$
49. $\log x = -0.7520$
50. $\log x = -1.3054$
51. $\log x = -2.4609$
52. $\log x = -2.5391$
53. $\log x = -0.9957$
54. $\log x = -3.0044$

Superset

In exercises 55–64, determine whether each statement is true or false. If false, illustrate this with an example.

55. $\log x + \log y = \log(x + y)$
56. $\log x - \log y = \log(x - y)$
57. $(\log x)(\log y) = \log(xy)$
58. $(\log x)^2 = 2 \log x$
59. $\log x^2 = 2 \log x$
60. $x + \log x^{-1} = x - \log x$
61. $\log 100x = 100 \log x$
62. $\log 10^x = x$
63. $\log 10^x = \log x^{10}$
64. $\log(10 + x) = 1 + \log x$

In exercises 65–68, use the fact that

$$\log_b x = \frac{\log x}{\log b}$$

to estimate the given expression.

65. $\log_3 10$
66. $\log_5 15$
67. $\log_2 5$
68. $\log_2 10$

69. Suppose some quantity Q is related to another quantity x by the relationship $Q = \log_2 x$. What happens to the value of Q if x is doubled? tripled? quadrupled?

70. Suppose some quantity Q is related to another quantity x by the relationship $Q = \log_b x$. What happens to the value of Q if x is doubled? tripled? quadrupled?

6.4 Linear Interpolation

Goal A *to use linear interpolation to approximate function values*

x	f(x)
−1	1.9
1	3.4
3	4.8

Frequently a table of values (such as Table 1 in the Appendix) does not include the specific value we need. For example, suppose we are given the table on the left and we want to determine the value of $f(2)$. Although the table provides no function value for $x = 2$, we can use the information we have to estimate $f(2)$. Having no other information about the function f, we can reasonably assume that, since 2 is midway between 1 and 3, the value $f(2)$ is midway between $f(1)$ and $f(3)$.

As shown in Figure 6.11, the geometric meaning of our assumption is that the point $(2, f(2))$ is near the midpoint of the line segment that joins the points $(1, f(1))$ and $(3, f(3))$. As you can see, the midpoint of the line segment and the point $(2, f(2))$ do not coincide, but the y-coordinate of the midpoint is a good approximation of $f(2)$.

The use of a line to approximate function values is called **linear interpolation.** For the example above, the equation of the line is found by using the points $(1, 3.4)$ and $(3, 4.8)$. We find the approximation by substituting $x = 2$ into the equation of this line.

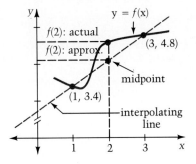

Figure 6.11

Example 1 Use linear interpolation and the function values given in the table at the right to approximate $f(6.2)$ to the nearest tenth.

x	1	3	5	7
f(x)	3.4	8.7	15.2	23.5

Solution Since 6.2 is an x-value lying between 5 and 7 in the table, we use the points $(5, 15.2)$ and $(7, 23.5)$ to determine the equation of the interpolating line.

$$\text{slope} = \frac{23.5 - 15.2}{7 - 5} = \frac{8.3}{2} = 4.15$$

■ The slope of the interpolating line is computed. Now use the point-slope form to write an equation of the interpolating line.

$$y - 15.2 = 4.15(x - 5)$$
$$y - 15.2 = 4.15(6.2 - 5)$$

■ $x = 6.2$ is substituted into the equation.

$$y = 4.15(1.2) + 15.2$$
$$y = 20.18$$

The approximate value of $f(6.2)$ to the nearest tenth is 20.2.

This method also works when we are given a value of y and are looking for x. We now consider such an example.

Example 2 Given that $f(x) = 4.3$, use linear interpolation and the function values given in the table in Example 1 to approximate x to the nearest tenth.

Solution Since 4.3 is a y-value lying between 3.4 and 8.7 in the table, we use the points $(1, 3.4)$ and $(3, 8.7)$ to determine the equation of the interpolating line.

$$\text{slope} = \frac{8.7 - 3.4}{3 - 1} = \frac{5.3}{2} = 2.65$$

$y - 3.4 = 2.65(x - 1)$ ■ The point-slope form is used.
$4.3 - 3.4 = 2.65(x - 1)$ ■ $y = 4.3$ is substituted into the equation.
$x \approx 1.34$

The approximate value of x to the nearest tenth is 1.3.

Goal B *to use linear interpolation to estimate common logarithms and antilogarithms*

When we apply linear interpolation to logarithms, we shall use the values in Table 1 in the Appendix. For a given problem, it is helpful to make a small table of those values that bracket the value of interest.

Example 3 Find log 85.37.

Solution $\log 85.37 = \log(8.537 \times 10^1) = \log 8.537 + \log 10 = \log 8.537 + 1$

Interpolation

x	8.53	8.537	8.54
$\log x$	0.9309	?	0.9315

■ Consult Table 1, and record the bracketing values in a small table.

$$\text{slope} = \frac{0.9315 - 0.9309}{8.54 - 8.53} = 0.06$$

$y - 0.9309 = 0.06(x - 8.53)$
$\quad y = 0.06(0.007) + 0.9309$
$\quad y = 0.93132$

■ The point-slope form is used.

$\log 85.37 \approx 0.9313 + 1 = 1.9313$

■ Thus, 0.9313 is the approximation for log 8.537.

In the following example, the interpolation is done off to the side of the problem, much as auxiliary computations are often done.

Example 4 Solve $\log A = -1.4132$ for A.

Solution Remember that we must first write -1.4132 as the sum of an integer and a number between 0 and 1.

$$-1.4132 = -2 + 0.5868$$

x	3.86	?	3.87
$\log x$	0.5866	0.5868	0.5877

Interpolation

$$\text{slope} = \frac{0.5877 - 0.5866}{3.87 - 3.86} = 0.11$$

$$y - 0.5866 = 0.11(x - 3.86)$$
$$0.5868 - 0.5866 = 0.11(x - 3.86)$$
$$0.0002 = 0.11(x - 3.86)$$
$$x \approx 3.862$$

$$\log A \approx \log 10^{-2} + \log 3.862 = \log(10^{-2} \times 3.862)$$

Thus, $A \approx 3.862 \times 10^{-2}$ or 0.03862.

Exercise Set 6.4 $\boxed{=}$ Calculator Exercises in the Appendix

Goal A

In exercises 1–8, use linear interpolation and the function values given in the table to approximate A to the nearest tenth.

1.

x	1	3	7	13
$f(x)$	2.8	14.3	38.5	64.2

(a) $A = f(5.1)$ (b) $f(A) = 14.3$
(c) $A = f(2.8)$ (d) $f(A) = 53.1$

2.

x	2	5	7	10
$f(x)$	6.1	18.3	27.5	31.2

(a) $A = f(2.7)$ (b) $f(A) = 19$
(c) $A = f(9.9)$ (d) $f(A) = 28.3$

3.

x	2.1	3.8	5.6	8.1
$f(x)$	-7.1	-0.4	6.9	15.2

(a) $A = f(3)$ (b) $f(A) = 8$
(c) $A = f(4.1)$ (d) $f(A) = -0.1$

4.

x	-7.3	-2.4	5.9	12.1
$f(x)$	-17.0	3.8	9.6	16.3

(a) $A = f(-5.1)$ (b) $f(A) = 0$
(c) $A = f(1.7)$ (d) $f(A) = 14.6$

5.

x	-2	1	5	10
$f(x)$	8.6	1.4	-4.3	-7.8

(a) $A = f(8)$ (b) $f(A) = 0$
(c) $A = f(0.2)$ (d) $f(A) = 5$

6.

x	-15.3	4.7	11.5	21.6
$f(x)$	5.2	1.8	-3.6	-10.4

(a) $A = f(-0.6)$ (b) $f(A) = -2.7$
(c) $A = f(20)$ (d) $f(A) = 4.4$

7.

x	-4.1	-1.9	-0.1	1.3
$f(x)$	7.3	0.8	-3.4	-4.9

(a) $A = f(-2.7)$ (b) $f(A) = 6.5$
(c) $A = f(0)$ (d) $f(A) = -0.6$

8.

x	-9.8	-3.2	-0.2	4.5
$f(x)$	11.7	2.1	-5.2	-7.6

(a) $A = f(-1.5)$ (b) $f(A) = 3.9$
(c) $A = f(1.3)$ (d) $f(A) = -1.9$

Goal B

In exercises 9–38, use Table 1 in the Appendix and linear interpolation to solve each of the following equations for A.

9. $\log 7.513 = A$
10. $\log 1.395 = A$
11. $\log 4.739 = A$
12. $\log 6.952 = A$
13. $\log 0.3045 = A$
14. $\log 52.36 = A$
15. $\log 27.98 = A$
16. $\log 0.4861 = A$
17. $\log 478.2 = A$
18. $\log 0.02948 = A$
19. $\log 0.05173 = A$
20. $\log 332.2 = A$
21. $\log 0.051042 = A$
22. $\log 0.60158 = A$
23. $\log 0.91375 = A$
24. $\log 0.01112 = A$
25. $\log A = 0.9501$
26. $\log A = 0.8778$
27. $\log A = 0.3292$
28. $\log A = 0.6131$
29. $\log A = 3.9844$
30. $\log A = 4.3299$
31. $\log A = -0.6573$
32. $\log A = -0.7313$
33. $\log A = -0.2397$
34. $\log A = -0.4658$
35. $\log A = -1.5797$
36. $\log A = -9.2412$
37. $\log A = -3.2846$
38. $\log A = -5.2108$

Superset

In exercises 39–52, use logarithms to solve each of the following equations for x to the nearest hundredth.

39. $x^2 = 2$
40. $x^5 = 2$
41. $8^x = 2$
42. $3^x = 8$
43. $12^x = 5$
44. $15^x = 9$
45. $x^3 = 18$
46. $x^2 = 5$
47. $2^x = \dfrac{1}{3}$
48. $x = (2.7)^{3.1}$
49. $x = \left(1 + \dfrac{1}{10}\right)^{10}$
50. $3^x = \dfrac{1}{2}$
51. $x = \dfrac{(32.8)(17.1)}{53.2}$
52. $x = \dfrac{(14.3)(0.28)}{(31.5)(0.036)}$

53. The product of the first n positive integers is denoted by $n!$ (read "n factorial"):

$$n! = 1 \cdot 2 \cdot 3 \cdot 4 \cdots (n-2)(n-1)n.$$

For example, $7! = 1 \cdot 2 \cdot 3 \cdot 4 \cdot 5 \cdot 6 \cdot 7 = 5040$. Use the fact that

$$\log n! = \log 1 + \log 2 + \log 3 + \cdots + \log n$$

and the logarithm table to estimate the number of digits in the number:

(a) 5!   (b) 10!   (c) 20!

54. Complete the following table, where

$$f(x) = x - \dfrac{x^2}{2} + \dfrac{x^3}{3}$$

x	0.05	0.10	0.15	0.20	0.25
$\log(1 + x)$					
$f(x)$					
$0.4343 \cdot f(x)$					

What do you notice?

6.5 Applications and the Natural Logarithm

Goal A *to use common logarithms to solve applied problems*

t	N(t)
0	1
1	2
2	4
3	8
4	16
5	32

Exponential and logarithmic functions are used to solve a variety of applied problems. Interest on savings and loans, population growth, radioactive decay, and earthquake intensity are some of those subject areas employing exponential or logarithmic models.

Consider the following hypothetical example of population growth. Suppose we are studying the growth of bacteria, and we begin with a single cell that splits after an hour of growth to form two cells. After another hour each of the two cells splits to form two more cells, making a total of four cells. This doubling process continues, and at the end of the third hour there is a total of eight cells. The data for the first 5 hours are shown in the table at the left (t is the number of hours elapsed, and $N(t)$ is the number of cells at time t). A graph for $0 \leq t \leq 10$ is shown in Figure 6.12.

As both the graph and the table suggest, the function that models this situation is $N(t) = 2^t$, and its domain is the set of positive integers. We show a smooth curve to indicate the exponential nature of the process. Note that in this situation, the initial population was 1 cell.

If there are B_0 cells initially, and if the population doubles every hour, then after t hours, the size of the population, $N(t)$, is given by the equation

$$N(t) = B_0 2^t.$$

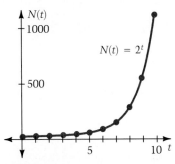

Figure 6.12

Example 1 The bacteria *E. coli* is found in many organisms, including humans. Suppose that, under certain conditions, the number of these bacteria present in an experimental colony is given by the equation $N(t) = B_0 2^t$. How many *E. coli* are present 6 hours after the start of the experiment, if there were 1,500,000 initially?

Solution
$N(t) = B_0 2^t$
$N(6) = 1,500,000 \cdot 2^6$ ■ Values are substituted for B_0 and t.
$N(6) = 96,000,000$

Example 2 How long will it take the colony in Example 1 to grow to 20,000,000 bacteria?

Solution
$N(t) = B_0 2^t$
$20,000,000 = 1,500,000 \cdot 2^t$ ■ Values are substituted for $N(t)$ and B_0.

$$13.3 \approx 2^t$$
$$\log 13.3 \approx \log 2^t$$
$$\log 13.3 \approx t \cdot \log 2$$
$$\frac{\log 13.3}{\log 2} \approx t$$

■ We take the common log of both sides and then use Rule 4 of Logarithms to solve for t.

■ $\log 13.3 \approx 1.1239$ and $\log 2 \approx 0.3010$.

$$t \approx \frac{1.1239}{0.3010} \approx 3.7$$

There will be 20,000,000 bacteria in the colony after approximately 3.7 hours.

Example 2 demonstrates a technique for solving for exponents: first take the logarithm of both sides of the equation; then use Rule 4 of Logarithms to write the exponent as a coefficient in the equation.

Exponential and logarithmic models are used to solve problems concerning investments and loans. If P dollars are invested at an interest rate i (expressed as a decimal), and the interest is compounded n times per year, then the amount, A, of money in the account after t years is

$$A = P\left(1 + \frac{i}{n}\right)^{nt}.$$

Example 3 Suppose $2000 is deposited in an account that advertises a 9.2% interest rate compounded quarterly. What will the value of the account be in 25 years?

Solution
$$A = P\left(1 + \frac{i}{n}\right)^{nt}$$
$$A = 2000\left(1 + \frac{0.092}{4}\right)^{4(25)}$$
$$A = 2000(1.023)^{100}$$
$$\log A = \log[2000(1.023)^{100}]$$
$$\log A = \log 2000 + 100 \log 1.023$$
$$\log A \approx 3.3010 + 100(0.0099)$$
$$\log A \approx 4.291$$
$$A \approx 1.95 \times 10^4$$

■ We shall use logarithms to perform this complicated computation. If you have a calculator with a y^x key, you can find $(1.023)^{100}$ directly.

■ We took the antilogarithm of both sides to find A.

The value of the account will be approximately $19,500 in 25 years.

6 / Exponential and Logarithmic Functions

Goal B *to use natural logarithms to solve applied problems*

Let us perform an experiment to determine the effect of increasing the number of times interest is compounded each year. To make things simple, suppose that $1.00 is invested for one year at an interest rate of 100% (i.e., $P = 1$, $t = 1$, and $i = 1$). In the table below, we show the value of the investment after one year as we vary the number of times the interest is compounded. (It is customary to use a 360-day year when compounding interest.)

n	Interest is compounded:	$A = \left(1 + \dfrac{1}{n}\right)^n$
1	annually	2
2	semiannually	2.25
4	quarterly	≈ 2.441406
12	monthly	≈ 2.613035
52	weekly	≈ 2.692597
360	daily	≈ 2.714516
8,640	hourly	≈ 2.718117
518,400	each minute	≈ 2.718262
31,104,000	each second	≈ 2.718282

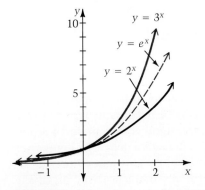

Figure 6.13

Notice that as n increases, the value of the investment approaches some constant value. Like π, this constant value is an irrational number. It occurs so frequently that we give it a name. We call it **e:**

$$e \approx 2.718282.$$

The function $y = e^x$ is called the **natural exponential function,** and the function $y = \log_e x$, usually written $y = \ln x$, is called the **natural logarithmic function.** Values for e^x, e^{-x}, and $\ln x$ are given in Tables 2 and 3 in the Appendix. As you might expect, since e is between 2 and 3, the graph of $y = e^x$ lies between the graphs of $y = 2^x$ and $y = 3^x$.

When interest is advertised as being "compounded continuously", the model used to determine the value of an account after t years is

$$A = Pe^{it,}$$

where P is the amount initially invested, and i is the annual interest rate.

Example 4 An amount of $12,000 is placed in an account advertising an 8% annual interest rate, compounded continuously.

(a) How much is the account worth after 5 years have passed?
(b) How long is it before the account is worth twice the initial investment?

Solution

(a) $A = Pe^{it}$ ■ Substitute values for P, i, and t.
$A = 12,000 \cdot e^{(0.08)(5)}$
$A \approx 12,000 \cdot 1.49182$ ■ $e^{0.4} \approx 1.49182$ (see Table 2).
$A \approx 17,901$

After 5 years, the value of the account is approximately $17,901.

(b) $24,000 = 12,000 \cdot e^{0.08t}$
$2 = e^{0.08t}$
$\ln 2 = \ln e^{0.08t}$
$\ln 2 = 0.08t \ln e$ ■ Rule 4 of Logarithms is used.
$\ln 2 = (0.08t)(1)$ ■ $\ln e = \log_e e = 1$ by Rule 2 of Logarithms.
$\dfrac{0.6931}{0.08} \approx t$ ■ $\ln 2 \approx 0.6931$ (see Table 3).
$t \approx 8.66$

The amount in the account will double its value in roughly $8\frac{2}{3}$ years.

The previous example describes an **exponential growth** model because the value of the account is increasing. Of equal interest is the **exponential decay** model. The most common exponential decay problems involve radioactive substances.

Carbon-14 is a radioactive substance present in all living matter. While an organism is alive, the amount of Carbon-14 it contains remains constant. Once the organism dies, however, the number of Carbon-14 atoms begins to decrease. This process is called **radioactive decay**. By comparing the level of radiation in a fossil with the level present in a similar living sample, we can estimate how long ago the fossil was a living organism. Essential to such an investigation is the formula

$$y = y_0 e^{kt},$$

where y_0 is the amount of radioactive substance present initially, y is the amount present after t years, and k is the *decay constant*.

If T is the half-life in years of the radioactive substance under study (i.e., the time it takes for a given sample to reduce to half its size), then

$$k = \frac{-1}{T}\ln 2 \approx \frac{-0.6931}{T}.$$

Example 5 Carbon-14 has a half-life of 5750 years. Suppose that 100 grams of Carbon-14 were present in an organism when it lived 3000 years ago. How much Carbon-14 would remain now?

Solution

$k \approx \dfrac{-0.6931}{5750}$ ■ Begin by determining the decay constant.

$k \approx -0.00012$

$y \approx y_0 e^{(-0.00012)t}$ ■ The value of k is substituted into the radioactive decay model.

$y \approx 100 e^{(-0.00012)3000}$ ■ Values of y_0 and t are substituted.

$y \approx 100 e^{-0.36}$

$y \approx 100(0.69768)$ ■ $e^{-0.36} \approx 0.69768$ (see Table 2).

$y \approx 69.768$

There would be approximately 69.8 grams of Carbon-14 present today.

Exercise Set 6.5   = Calculator Exercises in the Appendix

Goal A

1. (Refer to Example 1) How many bacteria are present 7 hours after the start of the experiment, if there were 1350 initially?

2. (Refer to Example 1) How many bacteria are present 4 hours after the start of the experiment, if there were 58,000 initially?

3. (Refer to Example 1) How long will it take a colony of 5000 bacteria to triple in number?

4. (Refer to Example 1) How long ago were there less than 1000 bacteria in the colony, if there are 7500 present now?

5. If $1000 is placed in an account that earns 8% compounded quarterly, what is the value of the account 10 years later?

6. If $500 is placed in an account that earns 6% compounded semiannually, what is the value of the account 12 years later?

7. How much money must be placed in an account that earns 7% compounded semiannually, so that there will be $2500 in the account 5 years from now?

8. How much money must be placed in an account that earns 6% compounded quarterly, so that there will be $1000 in the account 8 years from now?

9. If you put 1¢ in your bank account today, 2¢ tomorrow, 4¢ the day after tomorrow, 8¢ the next day, and so on, in how many days will you have to deposit at least $100?

10. (Refer to exercise 9) Suppose instead, you deposit 1¢, 3¢, 9¢, 27¢, and so on. In how many days will you have to deposit at least $100?

Goal B

11. If $1000 is placed in an account that earns 8% compounded continuously, what is the value of the account 10 years later?

12. If $500 is placed in an account that earns 6% compounded continuously, what is the value of the account 12 years later?

13. Suppose $1762 is placed in an account that earns 7% compounded continuously. How long will it be before the account is worth $2500?

14. Suppose $619 is placed in an account that earns 6% compounded continuously. How long will it be before the account is worth $1000?

15. Suppose there were 150 gm of Carbon-14 in an organism when it lived 10,000 years ago. How much Carbon-14 remains today?

16. Suppose there are 20 gm of Carbon-14 remaining in an organism known to have lived 2500 years ago. How many grams of Carbon-14 were in the organism when it lived?

17. Suppose we know that there were 180 gm of Carbon-14 in an organism when it lived and only 25 gm remain today. How long ago did the organism live?

18. Suppose there are 200 gm of Carbon-14 in an organism that dies today. In how many years will there be less than 75 gm left in the organism?

19. If inflation persists at 6% annually compounded continuously, how much will an item that costs $10 today cost in 5 years?

20. The population of the United States tends to grow at a rate proportional to the size of the population. Thus, we may represent the population size by an exponential model:

$$A(t) = 180e^{0.013t}$$

where t is the number of years since 1960 and $A(t)$ is the population in millions. What will the population of the United States be in the year 2000?

Superset

21. How much money must be placed in an account that earns 6% compounded continuously in order that $1000 may be withdrawn at the end of each year forever?

22. (Refer to exercise 21) How much must be placed in the account if the interest is compounded quarterly?

23. The **effective annual rate** of interest in an account that holds $A(t)$ dollars after t years is defined by

$$\frac{[A(t) - A(0)]}{A(0)} \times 100\%.$$

What is the effective annual rate of interest on money invested at 6% compounded quarterly?

24. (Refer to exercise 23) What is the effective annual rate of interest on money invested at 6% compounded continuously?

25. At what rate, compounded continuously, must a sum of money be invested so that the actual earnings are 8% per year?

26. (Refer to exercise 20) In what year will the population of the United States exceed the current world population of 4 billion?

Chapter Review & Test

Chapter Review

6.1 Exponential Functions (pp. 250–254)

Suppose x, y, and t are any real numbers and b is positive. Then the *Rules of Real Number Exponents* are as follows (p. 250):

$$b^0 = 1 \quad b^1 = b \quad b^x b^y = b^{x+y} \quad (b^x)^t = b^{xt} \quad \frac{b^x}{b^y} = b^{x-y} \quad b^{-x} = \frac{1}{b^x}$$

A function of the form $y = b^x$, where b is any positive constant except 1, is called an *exponential function*; b is called the *base*. (p. 251) Four important features of the exponential function $f(x) = b^x$ are (p. 252):

- If $x > 0$, then as b increases, b^x increases.
- If $x < 0$, then as b increases, b^x decreases.
- For any base b, $b^0 = 1$.
- The x-axis is a horixontal asymptote.

6.2 Logarithmic Functions (pp. 255–262)

A function f is *one-to-one* if each y-value in the range of f corresponds to exactly one x-value in the domain. If a function f is one-to-one, then there exists a unique *inverse function* f^{-1} such that $(f^{-1} \circ f)(x) = x$ for any x in the domain of f, and $(f \circ f^{-1})(x) = x$ for any x in the domain of f^{-1}. The domain of f becomes the range of f^{-1}, and the range of f becomes the domain of f^{-1}. (p. 255) The graph of $y = f^{-1}(x)$ is the reflection of $y = f(x)$ across the line $y = x$. (p. 256)

The exponential function $f(x) = b^x$ has an inverse called the *logarithmic function with base b* and denoted $f(x) = \log_b x$. Thus, a logarithm is an exponent. The domain of the logarithmic function is $(0, +\infty)$, and the range is $(-\infty, +\infty)$. (p. 257) Since the logarithmic function is one-to-one, the equation $\log_b A = \log_b C$ implies $A = C$. (p. 258)

Log/Exp Principle: For $b > 0$ and $b \neq 1$, $b^A = C$ is equivalent to $\log_b C = A$. (p. 257)

Suppose x and y are positive real numbers, t is any real number, and base b is positive, but not equal to 1. The *Rules of Logarithms* are as follows (p. 259):

$$\log_b 1 = 0 \qquad \log_b b = 1 \qquad \log_b(xy) = \log_b x + \log_b y$$

$$\log_b(x^t) = t \log_b x \qquad \log_b\left(\frac{x}{y}\right) = \log_b x - \log_b y \qquad \log_b\left(\frac{1}{x}\right) = -\log_b x$$

Also remember that: $b^{\log_b x} = x$ and $\log_b b^x = x$. (p. 261)

6.3 Common Logarithms (pp. 263–266)

Base 10 logarithms are called *common logarithms;* we write $\log_{10} x$ as $\log x$. (p. 263) (Table 1 in the Appendix lists values of $\log x$.)

If $y = \log x$, then x is called the *common antilogarithm* of y; that is, x is the number you get when you raise 10 to the power y: $x = 10^y$. (p. 263)

6.4 Linear Interpolation (pp. 267–270)

The use of a line to approximate function values is called *linear interpolation.* (p. 267) We can use linear interpolation to estimate common logarithms and antilogarithms. (p. 268)

6.5 Applications and the Natural Logarithm (pp. 271–276)

Exponential and logarithmic functions serve as models for problems involving population growth, compound interest, and radioactive decay. (p. 271)

The function $y = e^x$, where $e \approx 2.718282$, is called the *natural exponential function;* the function $y = \log_e x$, usually written $y = \ln x$, is called the *natural logarithmic function.* (p. 273) (Values of e^x, e^{-x}, and $\ln x$ are given in Tables 2 and 3 in the Appendix.)

Chapter Test

6.1A Simplify each of the following expressions.

1. $4^{x-1} \cdot 8^{2-3x}$
2. $\dfrac{9^{x+2}}{27^{2x}}$

6.1B Sketch the graph of each of the following equations.

3. $y = 3^{x-2}$
4. $y = \left(\dfrac{1}{2}\right)^x - 4$

6.1C Solve each of the following equations for x.

5. $3^{x+1} = 0$
6. $4^{-2x} = 8^{5x-2}$

6.2A The given function $y = f(x)$ is one-to-one. Find the equation describing the inverse function, sketch the inverse, and identify its domain and range.

7. $f(x) = 7 - 2x$, $1 \le x \le 4$
8. $f(x) = \sqrt{9 - x}$, $x \le 9$

6.2B Write each of the following as a logarithmic equation.

9. $x^3 = 125$ 10. $4^x = 64$ 11. $7^4 = x$

Use the Log/Exp Principle to solve each of the following equations for x.

12. $\log_3 27 = x$ 13. $\log_x 16 = -2$

6.2C Sketch the graph of each of the following equations.

14. $y = \log_2(x - 3)$ 15. $y = \log_{1/4} x$

6.2D Use the values $\log_2 x = 4.3$, $\log_2 y = 2.8$, and $\log_2 z = -3.1$ to evaluate each expression.

16. $\log_2\left(\dfrac{8x}{yz}\right)$ 17. $\log_2(xy^2z^3)$

Solve each of the following equations for x.

18. $\log_4 2^x = -5$ 19. $\log_3(\log_2 x) = 0$
20. $\log_2 x^3 = \log_2 9x$ 21. $\log_2 8x = \log_2 8 + \log_2 x$

6.3A Use Table 1 in the Appendix to find each of the following.

22. $\log 7.14$ 23. $\log 3.96$

Use scientific notation to rewrite each of the following numbers.

24. 473.89 25. 0.00451

Use Table 1 to compute each of the following.

26. $\log 27.4$ 27. $\log 0.947$

6.3B Use Table 1 to solve each of the following equations for x.

28. $\log x = 0.6096$ 29. $\log x = 1.4609$ 30. $\log x = -1.5391$

6.4A Use linear interpolation and the function values given in the table below to approximate A to the nearest tenth.

x	-3	2	8	15
$f(x)$	4.2	1.7	-2.4	-6.3

31. $f(3) = A$ 32. $f(A) = -3$

6.4B Use Table 1 and linear interpolation to solve each of the following equations for A.

33. $\log 0.2635 = A$

34. $\log 4728 = A$

35. $\log 539.6 = A$

36. $\log A = 1.9452$

37. $\log A = -2.3921$

38. $\log A = -3.3929$

6.5A Solve each of the following.

39. The number of bacteria present in an experimental colony is given by the equation $N(t) = B_0 2^t$, where t is measured in hours. If there are 1,600,000 bacteria cells present initially, how long will it take the colony to grow to 2,400,000 bacteria?

40. If $500 is placed in an account that earns 8% compounded quarterly, what is the value of the account 6 years later?

6.5B Solve each of the following.

41. If $500 is placed in an account that earns 8% compounded continuously, what is the value of the account 6 years later?

42. Suppose there were 80 gm of Carbon-14 in an organism when it lived 300,000 years ago. How much Carbon-14 remains today? (Use the formula $y = y_0 e^{kt}$ where $k = -\frac{1}{7} \ln 2$.)

Superset

43. Solve $3^{x^5} = -1$ for x.

44. Solve $\log_2(x^2 - 5) = \log_2(4x)$ for x.

45. Sketch the graph of $y = 9^{\log_3 x}$.

46. Suppose some quantity Q is related to another quantity x by the relationship $Q = \log_3 x$. What happens to the value of Q if x is doubled? tripled? quadrupled?

47. Use logarithms to solve each of the following equations for x to the nearest hundredth.

(a) $x^2 = 3$ (b) $2^x = 6$

48. What is the effective annual rate of interest on money invested at 10% compounded continuously?

7 Trigonometry: An Introduction

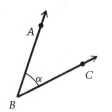

Figure 7.1

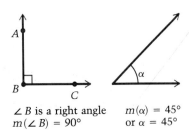

∠B is a right angle $m(\alpha) = 45°$
$m(\angle B) = 90°$ or $\alpha = 45°$

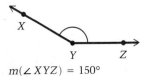

$m(\angle XYZ) = 150°$

Figure 7.2

7.1 Triangles

Goal A *to determine sides and angles of special types of triangles*

Trigonometry is a branch of mathematics that developed as a tool for solving problems involving triangles. In fact, the word "trigonometry" is derived from the Greek words for "measurement of triangles." It is appropriate, therefore, to begin our study of trigonometry by reviewing some basic information about triangles.

An **angle** is a figure consisting of two rays having the same endpoint. This common endpoint is called the **vertex.** For the angle in Figure 7.1, B is the vertex. This angle can be described in three ways: ∠ABC, ∠B, and α. The symbol α is the Greek letter *alpha*. Other Greek letters often used in describing angles are θ (*theta*), φ (*phi*), and β (*beta*).

A common unit for measuring angles is the degree (°). Angles of various measures are shown in Figure 7.2. The symbols $m(\angle B)$ and $m(\alpha)$ are used to indicate the measures of angles B and α, respectively. However, the "m" is often omitted. Thus, you may see $m(\theta) = 45°$ or simply $\theta = 45°$.

An angle with measure 90° is called a **right angle.** The special symbol □, signifies a right angle when placed at a vertex as in Figure 7.2. An angle with measure between 0° and 90° is called **acute,** and an angle with measure between 90° and 180° is called **obtuse.**

7 / Trigonometry: An Introduction   283

Recall that a triangle has 3 angles and 3 sides. The triangle below is referred to as $\triangle ABC$. Points A, B, and C are called **vertices** of the triangle.

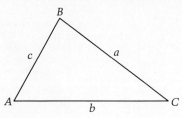

- When we label a triangle, it is common to let a stand for the side opposite vertex A, b for the side opposite vertex B, and c for the side opposite vertex C.

Figure 7.3

Fact

In any triangle, the sum of the measures of the angles is 180°.

Certain types of triangles are of special importance to the development of trigonometry. An **isosceles** triangle is a triangle with two sides of equal length. In an isosceles triangle, the angles opposite the sides of equal length have the same measure.

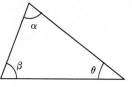

Figure 7.4 $\alpha + \beta + \theta = 180°$ or $m(\alpha) + m(\beta) + m(\theta) = 180°$.

Example 1 In $\triangle ABC$, $a = 5$, $b = 5$, and $m(\angle A) = 65°$. Find $m(\angle C)$.

Solution

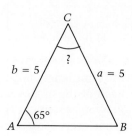

- Begin by drawing a triangle and labeling its known parts.
- Since $a = b$, the triangle is isosceles. Thus $m(\angle B) = m(\angle A) = 65°$.

$$m(\angle A) + m(\angle B) + m(\angle C) = 180°$$
$$65° + 65° + m(\angle C) = 180°$$
$$m(\angle C) = 180° - 65° - 65°$$
$$m(\angle C) = 50°$$

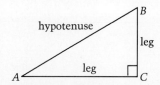

Figure 7.5 A right triangle.

A **right triangle** is a triangle in which one of the angles is a right angle. In a right triangle, it is common to label the right angle C. The longest side of a right triangle is opposite the right angle and is called the **hypotenuse**. The other two sides are called **legs**.

Example 2 $\triangle ABC$ is an isosceles right triangle with $m(\angle C) = 90°$. Find $m(\angle A)$.

Solution

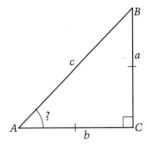

- Begin by drawing a triangle and labeling the known parts.
- Sides (or angles) having equal measure are denoted by the same hash marks. The hash marks in this figure indicate that $a = b$.

$$m(\angle A) + m(\angle B) + m(\angle C) = 180°$$
$$m(\angle A) + m(\angle B) + 90° = 180°$$
$$m(\angle A) + m(\angle A) = 90°$$
$$m(\angle A) = 45°$$

- True for any triangle.
- C is a right angle.
- $m(\angle B) = m(\angle A)$ since $\triangle ABC$ is isosceles.

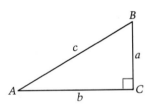

Figure 7.6 $c^2 = a^2 + b^2$

One useful fact about a right triangle is that the square of the hypotenuse is equal to the sum of the squares of the legs. This fact is known as the Pythagorean Theorem.

Pythagorean Theorem

If ABC is a right triangle, with a and b the lengths of the legs and c the length of the hypotenuse, then $c^2 = a^2 + b^2$.

A triangle whose three sides have the same length is called an **equilateral triangle.** In an equilateral triangle, each of the angles measures 60°. You should convince yourself that the following statement is true: every equilateral triangle is also isosceles, but not every isosceles triangle is equilateral.

Example 3 Given an equilateral triangle with side of length 1, find the length of the altitude.

Solution

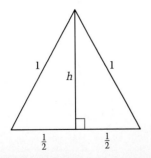

- Begin by drawing an equilateral triangle with altitude h. The **altitude** is a perpendicular from a vertex to the opposite side. In an equilateral triangle, the altitude cuts the opposite side into two equal lengths.

$1^2 = h^2 + \left(\frac{1}{2}\right)^2$ ■ Apply the Pythagorean Theorem to either of the two smaller triangles. They are congruent.

$1 = h^2 + \frac{1}{4}$

$h = \sqrt{\frac{3}{4}} = \frac{\sqrt{3}}{2}$ ■ Length is nonnegative, thus the value $-\frac{\sqrt{3}}{2}$ has been discarded.

The altitude is $\frac{\sqrt{3}}{2}$.

Goal B *to solve problems involving similar triangles*

We call triangles **similar** if they have the same shape, even though they may have different sizes. This means that one of the triangles is a "magnification" or a "reduction" of the other.

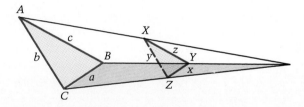

Figure 7.7

In Figure 7.7, triangle ABC and triangle XYZ are similar. This is written $\triangle ABC \sim \triangle XYZ$. When we say that $\triangle ABC$ is similar to $\triangle XYZ$, we imply a special correspondence between the two triangles such that the measures of the corresponding angles are equal and the corresponding sides of the triangles are proportional. Thus,

$$m(\angle A) = m(\angle X), \quad m(\angle B) = m(\angle Y), \quad \text{and} \quad m(\angle C) = m(\angle Z),$$

and

$$\frac{a}{x} = \frac{b}{y} = \frac{c}{z}.$$

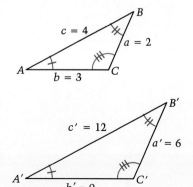

Figure 7.8

In Figure 7.8, $\triangle A'B'C' \sim \triangle ABC$. Notice that

$$\frac{a}{a'} = \frac{2}{6} = \frac{1}{3}, \quad \frac{b}{b'} = \frac{3}{9} = \frac{1}{3}, \quad \frac{c}{c'} = \frac{4}{12} = \frac{1}{3}.$$

Thus, the ratio of the corresponding sides is $\frac{1}{3}$.

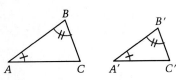

Figure 7.9 △ABC ~ △A'B'C'

You may recall from geometry that to prove two triangles are similar, you must find two pairs of corresponding angles having the same measure.

Angle-Angle (A-A) Similarity Theorem
Two triangles are similar if two angles of one triangle have the same measure as two angles of the other triangle.

Example 4 In the triangles below, $m(\angle A) = m(\angle X)$, $m(\angle B) = m(\angle Y)$, and the lengths of some sides are given. Find c.

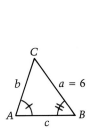
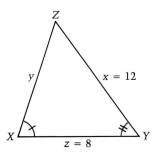

Solution △ABC ~ △XYZ   ■ The triangles are similar by the A-A Similarity Theorem.

$$\frac{6}{12} = \frac{c}{8}$$ ■ By similarity, $\frac{a}{x} = \frac{c}{z}$.

$12c = 6 \cdot 8$ ■ The result of cross-multiplication.

$c = 4$

Example 5 In the right triangles below $m(\angle A) = m(\angle A')$, $a = 3$, $b = 4$, and $a' = 9$. Find c'.

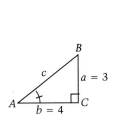
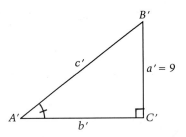

Solution △ABC ~ △A'B'C'   ■ The A-A Similarity Theorem is used.

$$\frac{a}{a'} = \frac{b}{b'} = \frac{c}{c'}$$ ■ This proportion follows from similarity. To find c', we must first find c.

7 / Trigonometry: An Introduction

$$c^2 = a^2 + b^2$$
$$c = \sqrt{a^2 + b^2} = \sqrt{3^2 + 4^2}$$
$$c = 5$$

■ Now that c has been determined, we use the proportion to find c'.

$$\frac{a}{a'} = \frac{c}{c'}$$
$$\frac{3}{9} = \frac{5}{c'}$$
$$c' = 15$$

Example 6   A building casts a shadow 84 ft long. At the same instant a nearby fence casts a shadow 14 ft long. If the fence is 10 ft high, what is the height of the building?

Solution

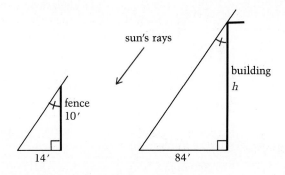

■ Draw a figure. If we assume that the building and fence are perpendicular to the ground, and the sun's rays are parallel, then the two triangles are similar.

$$\frac{h}{10} = \frac{84}{14}$$
$$14h = 10 \cdot 84$$
$$h = 60$$

The height of the building is 60 ft.

■ Set up a proportion using corresponding sides of the two triangles.

Exercise Set 7.1   ≡   Calculator Exercises in the Appendix

Goal A

In exercises 1–6, use the given information about $\triangle ABC$ to solve for the missing part.

1. $m(\angle A) = 40°, m(\angle B) = 73°, m(\angle C) = ?$
2. $m(\angle A) = 58°, m(\angle C) = 16°, m(\angle B) = ?$
3. $a = b, m(\angle A) = 30°, m(\angle B) = ?$
4. $a = b, m(\angle C) = 82°, m(\angle B) = ?$
5. $m(\angle A) = m(\angle B), a = 18, b = ?$
6. $m(\angle C) = 90°, a = b, m(\angle A) = ?$

In exercises 7–18, △ABC is a right triangle, c is the length of the hypotenuse, and a and b are the lengths of the legs. Find the length of the third side given the lengths of two sides.

7. $a = 6, b = 8$
8. $a = 12, c = 15$
9. $b = 15, c = 39$
10. $a = 15, c = 17$
11. $b = 5, c = 13$
12. $a = 12, c = 20$
13. $a = 2, b = 3$
14. $a = 5, b = 5$
15. $b = 1, c = 1\frac{1}{4}$
16. $a = \frac{6}{5}, b = \frac{8}{5}$
17. $a = 1.4, b = 4.8$
18. $a = 7.5, b = 4$

Goal B

In exercises 19–22, △ABC and △A'B'C' are two triangles such that $m(\angle A) = m(\angle A')$ and $m(\angle B) = m(\angle B')$. Given the lengths of some sides, find the length of the indicated side.

19. $a = 10, b = 8, a' = 8, b' = ?$
20. $a = 12, b = 9, a' = ?, b' = 6$
21. $a = 18, b = ?, a' = 9, b' = 5$
22. $a = ?, b = 27, a' = 6, b' = 12$

In exercises 23–28, △ABC is a right triangle and PQ is perpendicular to AC (see the figure below). Find the length of the indicated side.

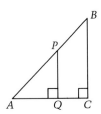

23. $AP = 8, AQ = 6, AC = 18, AB = ?$
24. $AQ = 10, PQ = 8, BC = 12, AC = ?$
25. $AP = 6, PQ = 4, AC = 18, BC = ?$
26. $AQ = 10, PQ = 6, BC = 36, AB = ?$
27. $\frac{PQ}{AQ} = \frac{1}{2}, AB = 18, BC = ?$
28. $\frac{PQ}{AP} = \frac{\sqrt{3}}{3}, BC = 10\sqrt{3}, AC = ?$

29. A 6 ft pole perpendicular to the ground casts a shadow 8 ft long at the same time that a telephone pole casts a shadow 56 ft long. What is the height of the telephone pole?

30. The top of a church spire casts a shadow 200 ft long. At the same time a nearby 8 ft wall casts a shadow 25 ft long. How high is the top of the spire?

Superset

31. A boat travels 8 mi due east and then 15 mi due north. How far is the boat from its starting point?

32. A 40 ft ladder is placed against a wall, with its foot 24 ft from the base of the wall. At what height does the ladder touch the wall?

33. A television set has a square picture screen with a 19 in diagonal. What are the dimensions of the screen? What is its area?

In exercises 34–41, △ABC is an equilateral triangle, BD is the altitude to side AC and the length of AB is 5. Find the following.

34. $m(\angle ADB)$
35. $m(\angle ABD)$
36. $m(\angle DBC)$
37. BD
38. AD
39. DC
40. area of △ABC
41. area of △ABD

42. Find the length of a side of a square having the same area as a rectangle with base of length 12 and diagonal of length 15.

43. Find the area of a right triangle with one leg of length 16 and hypotenuse of length 34.

44. Find the length of a diagonal of a rectangle with sides of lengths 32 and 24.

45. Find the length of a diagonal of a square with side of length 12.

46. Find the length of a side of an equilateral triangle with altitude of length h.

7.2 Trigonometric Ratios of Acute Angles

Goal A *to determine the trigonometric ratios of an acute angle in a right triangle*

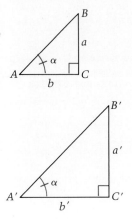

Figure 7.10 shows two right triangles ABC and $A'B'C'$ with

$$m(\angle A) = m(\angle A').$$

Since each triangle has a right angle and each triangle has an acute angle with the same measure, the triangles are similar by the A-A Similarity Theorem. Because of the similarity we can write

$$\frac{a}{a'} = \frac{b}{b'}.$$

Multiplying each side of the above equation by $\frac{a'}{b}$ produces

$$\frac{a}{b} = \frac{a'}{b'}.$$

This last equation tells us that a ratio of sides of $\triangle ABC$ is equal to the ratio of corresponding sides in $\triangle A'B'C'$. These ratios are equal because the equality of $m(\angle A)$ and $m(\angle A')$ implies that the two right triangles are similar. Moreover, it is the value of α that determines the value of this ratio.

For example, suppose that in the right triangle at the left, $\angle A$ has measure φ. Notice that the ratio $\frac{a}{b}$ is $\frac{3}{2}$ in this triangle. Now consider the right triangles in Figure 7.12. Because the corresponding acute angle has measure φ, the corresponding ratio in each of these triangles is also $\frac{3}{2}$.

Figure 7.10

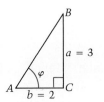

Figure 7.11

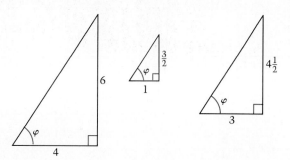

Figure 7.12

As we see in Figure 7.12, a particular acute angle (here φ) always produces the same ratio of opposite to adjacent sides (in this case the ratio is $\frac{3}{2}$). Therefore, we can think of the angle as a domain value for a function having the ratio as the function value.

Let us adopt a way of describing the sides of a right triangle in terms of an acute angle θ.

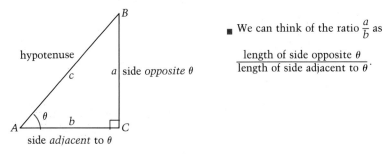

Figure 7.13

■ We can think of the ratio $\frac{a}{b}$ as
$$\frac{\text{length of side opposite } \theta}{\text{length of side adjacent to } \theta}.$$

The opposite/adjacent ratio is called the **tangent** of the angle. For an acute angle θ we write

$$\text{tangent}(\theta) = \frac{\text{length of side opposite } \theta}{\text{length of side adjacent to } \theta}.$$

For any right triangle ABC, there are six possible ratios of sides that can be formed. Each ratio can be defined as a function of an acute angle θ.

Definition

Let θ be an acute angle of a right triangle. The trigonometric functions of θ are defined by the following ratios:

$$\text{sine } \theta = \frac{\text{length of side \textbf{opposite} } \theta}{\text{length of \textbf{hypotenuse}}}, \quad \text{denoted } \sin \theta$$

$$\text{cosine } \theta = \frac{\text{length of side \textbf{adjacent to} } \theta}{\text{length of \textbf{hypotenuse}}}, \quad \text{denoted } \cos \theta$$

$$\text{tangent } \theta = \frac{\text{length of side \textbf{opposite} } \theta}{\text{length of side \textbf{adjacent to} } \theta}, \quad \text{denoted } \tan \theta$$

$$\text{cotangent } \theta = \frac{\text{length of side \textbf{adjacent to} } \theta}{\text{length of side \textbf{opposite} } \theta}, \quad \text{denoted } \cot \theta$$

$$\text{secant } \theta = \frac{\text{length of \textbf{hypotenuse}}}{\text{length of side \textbf{adjacent to} } \theta}, \quad \text{denoted } \sec \theta$$

$$\text{cosecant } \theta = \frac{\text{length of \textbf{hypotenuse}}}{\text{length of side \textbf{opposite} } \theta}, \quad \text{denoted } \csc \theta$$

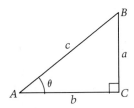

Figure 7.14

$\sin \theta = \dfrac{a}{c} \qquad \csc \theta = \dfrac{c}{a}$

$\cos \theta = \dfrac{b}{c} \qquad \sec \theta = \dfrac{c}{b}$

$\tan \theta = \dfrac{a}{b} \qquad \cot \theta = \dfrac{b}{a}$

Example 1 Find the six trigonometric ratios of α in the right triangle below.

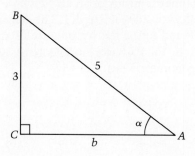

Solution
$5^2 = 3^2 + b^2$
$b^2 = 25 - 9 = 16$
$b = 4$

■ To determine the six trigonometric ratios of α, we need to know the lengths of the three sides of the triangle. The Pythagorean Theorem is used to find b.

$$\sin \alpha = \frac{\text{opposite}}{\text{hypotenuse}} = \frac{3}{5} \qquad \cos \alpha = \frac{\text{adjacent}}{\text{hypotenuse}} = \frac{4}{5} \qquad \tan \alpha = \frac{\text{opposite}}{\text{adjacent}} = \frac{3}{4}$$

$$\csc \alpha = \frac{\text{hypotenuse}}{\text{opposite}} = \frac{5}{3} \qquad \sec \alpha = \frac{\text{hypotenuse}}{\text{adjacent}} = \frac{5}{4} \qquad \cot \alpha = \frac{\text{adjacent}}{\text{opposite}} = \frac{4}{3}$$

Note that in Example 1, the function values of $\sin \alpha$ and $\csc \alpha$ are reciprocals of one another. This is also the case for $\cos \alpha$ and $\sec \alpha$, and for $\tan \alpha$ and $\cot \alpha$. Note also that we do not know the measure of α, even though we know its six trigonometric function values.

Example 2 Angles α and β are the two acute angles in the right triangle shown at the right. Show that

$$\sin \alpha = \cos \beta$$

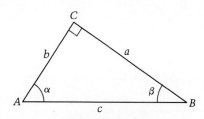

Solution
$$\sin \alpha = \frac{\text{side opposite } \alpha}{\text{hypotenuse}} = \frac{a}{c} \qquad \cos \beta = \frac{\text{side adjacent to } \beta}{\text{hypotenuse}} = \frac{a}{c}$$

Since $\sin \alpha$ and $\cos \beta$ each equal $\frac{a}{c}$, $\sin \alpha = \cos \beta$.

Goal B *to determine the trigonometric ratios of special angles (angles with measure 30°, 45°, or 60°)*

To determine the trigonometric ratios of an angle whose measure is 45°, consider the square *PQRS* with sides of length 1. The diagonal of the square bisects the right angle *P* and forms the hypotenuse of an isosceles right triangle. Its length is $\sqrt{2}$ (by the Pythagorean Theorem). We can use $\triangle PRS$ to determine the six trigonometric ratios of 45°.

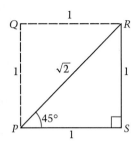

Figure 7.15

$$\sin 45° = \frac{1}{\sqrt{2}} = \frac{\sqrt{2}}{2} \qquad \cos 45° = \frac{1}{\sqrt{2}} = \frac{\sqrt{2}}{2} \qquad \tan 45° = \frac{1}{1} = 1$$

$$\csc 45° = \frac{\sqrt{2}}{1} = \sqrt{2} \qquad \sec 45° = \frac{\sqrt{2}}{1} = \sqrt{2} \qquad \cot 45° = \frac{1}{1} = 1$$

The trigonometric ratios of 30° and 60° can be found by using an equilateral triangle with sides of length 1. Recall that the altitude from a vertex to the opposite side bisects both the angle and the side. Thus, we can use $\triangle ABC$ to determine the six trigonometric ratios of 30° and 60°.

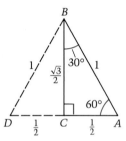

Figure 7.16

$$\sin 30° = \frac{1}{2} \qquad \cos 30° = \frac{\sqrt{3}}{2} \qquad \tan 30° = \frac{1}{\sqrt{3}} = \frac{\sqrt{3}}{3}$$

$$\csc 30° = 2 \qquad \sec 30° = \frac{2}{\sqrt{3}} = \frac{2\sqrt{3}}{3} \qquad \cot 30° = \sqrt{3}$$

$$\sin 60° = \frac{\sqrt{3}}{2} \qquad \cos 60° = \frac{1}{2} \qquad \tan 60° = \sqrt{3}$$

$$\csc 60° = \frac{2}{\sqrt{3}} = \frac{2\sqrt{3}}{3} \qquad \sec 60° = 2 \qquad \cot 60° = \frac{1}{\sqrt{3}} = \frac{\sqrt{3}}{3}$$

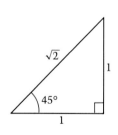

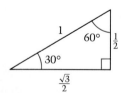

Figure 7.17

$f(\theta)$ θ	30°	45°	60°
$\sin \theta$	$\frac{1}{2}$	$\frac{\sqrt{2}}{2}$	$\frac{\sqrt{3}}{2}$
$\cos \theta$	$\frac{\sqrt{3}}{2}$	$\frac{\sqrt{2}}{2}$	$\frac{1}{2}$
$\tan \theta$	$\frac{\sqrt{3}}{3}$	1	$\sqrt{3}$
$\cot \theta$	$\sqrt{3}$	1	$\frac{\sqrt{3}}{3}$
$\sec \theta$	$\frac{2\sqrt{3}}{3}$	$\sqrt{2}$	2
$\csc \theta$	2	$\sqrt{2}$	$\frac{2\sqrt{3}}{3}$

Example 3 Given a right triangle *FED* with $m(\angle F) = 30°$ and hypotenuse of length 40, find the length of side *f*.

Solution

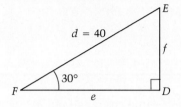

$\sin 30° = \dfrac{f}{d} = \dfrac{1}{2}$

$\dfrac{f}{40} = \dfrac{1}{2}$

$f = 20$

- Begin by drawing a diagram and labeling the parts.
- To find *f*, we choose a trigonometric ratio involving *f* and some known side.
- $\sin \theta = \dfrac{\text{side opposite } \theta}{\text{hypotenuse}}$ and $\sin 30° = \dfrac{1}{2}$.
- The known value of *d* is substituted.

Goal C *to determine trigonometric ratios of an angle given one ratio*

The trigonometric ratios depend upon one another. For an acute angle, if one trigonometric ratio is known, we can find the other five ratios by using the properties of right triangles.

Example 4 Given that $\sin \beta = \dfrac{5}{7}$ and β is an acute angle, find $\cos \beta$.

Solution

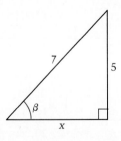

$7^2 = x^2 + 5^2$

$x^2 = 7^2 - 5^2 = 24$

$x = \sqrt{24} = 2\sqrt{6}$

$\cos \beta = \dfrac{2\sqrt{6}}{7}$

- Since $\sin \beta$ is the ratio of the side opposite β to the hypotenuse, we begin by drawing a right triangle in which this ratio is $\frac{5}{7}$. The side opposite β is labeled 5 and the hypotenuse is labeled 7. The side adjacent to β is labeled *x* and will be needed in determining $\cos \beta$.
- The Pythagorean Theorem is used.
- Remember to choose the positive value only.
- $\cos \beta = \dfrac{\text{side adjacent to } \beta}{\text{hypotenuse}}$.

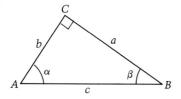

Figure 7.18 α and β are complementary.

Two angles are **complementary** if the sum of their measures is 90°. In any right triangle, the acute angles are complementary. This is illustrated in Figure 7.18 which shows that $\alpha + \beta = 90°$. In Example 2, we discovered that $\sin \alpha = \cos \beta$, that is, sine of α = cosine of the complement of α:

$$\sin \alpha = \cos(90° - \alpha).$$

Example 5 If α and β are complementary, and $\cos \alpha = \dfrac{\sqrt{11}}{6}$, find $\cos \beta$ and $\tan \beta$.

Solution

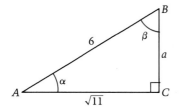

- Since α and β are complementary, we can draw a right triangle having α and β as acute angles.
- $\cos \alpha = \dfrac{\text{side adjacent to } \alpha}{\text{hypotenuse}} = \dfrac{\sqrt{11}}{6}$, so side AC is labeled $\sqrt{11}$ and AB is labeled 6.
- Pythagorean Theorem: $6^2 = (\sqrt{11})^2 + a^2$ implies $a^2 = 36 - 11 = 25$. Thus, $a = 5$.

$$\cos \beta = \frac{\text{side adjacent to } \beta}{\text{hypotenuse}} = \frac{5}{6} \qquad \tan \beta = \frac{\text{side opposite } \beta}{\text{side adjacent to } \beta} = \frac{\sqrt{11}}{5}$$

Exercise Set 7.2   ▭ Calculator Exercises in the Appendix

Goal A

In exercises 1–16, use the given information about the right triangle ABC to find the six trigonometric ratios of α.

1. $a = 6, b = 8$
2. $a = 7, b = 24$
3. $a = 8, c = 17$
4. $b = 10, c = 26$
5. $c = 37, a = 12$
6. $b = 48, c = 50$
7. $a = 21, c = 35$
8. $c = 41, b = 9$
9. $a = 1, c = \sqrt{2}$
10. $a = 2, b = 2$
11. $a = 1, c = 2$
12. $b = \sqrt{3}, a = 1$
13. $b = \dfrac{1}{2}, a = \dfrac{\sqrt{3}}{2}$
14. $c = 2, b = 1$
15. $a = 1, b = \sqrt{3}$
16. $c = 2, a = \sqrt{3}$

In exercises 17–20, α and β are the acute angles in right triangle ABC. Show the following.

17. $\tan \alpha = \cot \beta$
18. $\csc \alpha = \sec \beta$
19. $\sin \beta = \cos \alpha$
20. $\tan \beta = \cot \alpha$

Goal B

In exercises 21–26, given an equilateral triangle $\triangle ABC$ with side of length s and altitude of length h, find the indicated part.

21. If $s = 6$, find h
22. If $s = 12$, find h

23. If $h = 6$, find s
24. If $h = 15$, find s
25. If $h = 5\sqrt{3}$, find the perimeter of $\triangle ABC$
26. If $h = 9$, find the area of $\triangle ABC$

In exercises 27–36, $\triangle ABC$ has right angle at C and sides of length a, b, and c. Given the following information, find the indicated part.

27. $m(\angle A) = 30°$, $b = 21$, find c
28. $m(\angle B) = 30°$, $c = 58$, find a
29. $m(\angle B) = 45°$, $b = 24$, find c
30. $m(\angle A) = 45°$, $c = 50$, find b
31. $m(\angle B) = 60°$, $b = 15\sqrt{3}$, find a
32. $m(\angle B) = 60°$, $a = 24$, find b
33. $a = 12$, $b = 12$, find $m(\angle A)$
34. $b = 30$, $c = 60$, find $m(\angle A)$
35. $b = 24$, $a = 24\sqrt{3}$, find $m(\angle A)$
36. $a = 10\sqrt{3}$, $c = 20$, find $m(\angle A)$

Goal C

In exercises 37–46, α and β are complementary. Given the following information, find the indicated trigonometric ratios.

37. $\cos \alpha = \frac{3}{5}$, find $\sin \alpha$ and $\tan \beta$
38. $\sin \alpha = \frac{5}{13}$, find $\cos \alpha$ and $\cos \beta$
39. $\csc \alpha = \frac{17}{15}$, find $\tan \alpha$ and $\cos \beta$
40. $\sec \alpha = \frac{29}{21}$, find $\tan \beta$ and $\tan \alpha$
41. $\tan \alpha = \frac{3}{4}$, find $\sin \beta$ and $\cos \alpha$
42. $\tan \alpha = \frac{4}{3}$, find $\sin \alpha$ and $\sec \beta$
43. $\sin \alpha = \frac{7}{25}$, find $\cos \alpha$ and $\tan \alpha$
44. $\cos \alpha = \frac{35}{37}$, find $\tan \alpha$ and $\sin \beta$
45. $\cos \alpha = 0.9$, find $\cos \beta$ and $\sin \beta$
46. $\sin \alpha = 0.3$, find $\cos \beta$ and $\tan \alpha$

Superset

In exercises 47–50, use the trigonometric ratios of special angles to solve each problem.

47. Show that in an equilateral triangle with side of length s that the altitude is $\frac{\sqrt{3}}{2}s$.
48. If the diagonal of a square has length $6\sqrt{2}$, what is the area of the square?
49. In an isosceles right triangle with hypotenuse 12, what is the length of the altitude to the hypotenuse?
50. Find the lengths of the sides of $\triangle ABC$ if $m(\angle A) = 30°$, $m(\angle B) = 60°$, and the altitude from C has length 8.

In exercises 51–52, find the area of the triangle.

51.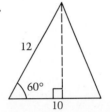
52.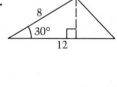

53. The local fire department's longest ladder measures 72 ft. If the angle between the ground and the ladder must be 60°, how high can the ladder reach? How far from a building should the foot of the ladder be?

54. One of the world's tallest flagpoles is 256 ft high. Guy wires are used to support the pole as shown in the diagram below. If the guy wires are anchored 60 ft from the foot of the flagpole, how long is each wire? How high up the pole are the wires fastened?

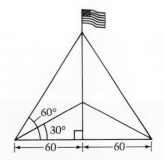

7.3 Angles of Rotation

Goal A *to sketch an arbitrary angle as a rotation*

Thus far we have defined the trigonometric functions for angles between 0° and 90°. In this section we shall extend our definitions to include *all* angles. In order to do this, it is useful to think of an angle as being formed by rotating a ray.

- Between twelve midnight and 4 A.M. the hour hand of a clock rotates to produce the angle shown in the figure. The ray pointing to 12 is called the **initial side** of the angle, and the ray pointing to 4 is called the **terminal side.**

Figure 7.19

An angle is in **standard position** in the xy-plane, if its vertex is at the origin and its initial side lies along the positive x-axis. An angle in standard position is usually named by the quadrant in which the terminal side lies. Each of the angles below is in standard position. Notice that an angle may be the result of one or more complete rotations about the vertex, as shown in Figure 7.20(b).

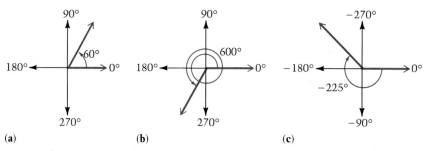

Figure 7.20 (a) A first quadrant angle. (b) A third quadrant angle. (c) A second quadrant angle.

We agree that a counterclockwise rotation produces a positive angle as shown Figure 7.20(a) and Figure 7.20(b). A clockwise rotation produces a negative angle as shown in Figure 7.20(c).

An angle whose terminal side lies on the x- or y-axis is called a **quadrantal angle.** When angles in standard position have the same terminal side, they are called **coterminal.** Angles α, β, and θ in Figure 7.21 are coterminal.

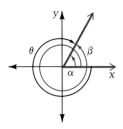

Figure 7.21

Example 1 Sketch each of the following angles in standard position.

(a) 120°   (b) 210°   (c) −60°   (d) 315°

Solution

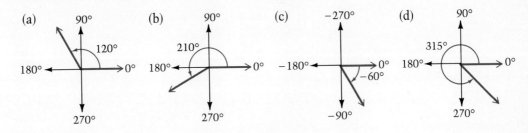

Example 2 Sketch each of the following angles in standard position. Determine the measure of the angle between 0° and 360° that is coterminal with each angle.

(a) −30°   (b) 398°   (c) 810°

Solution

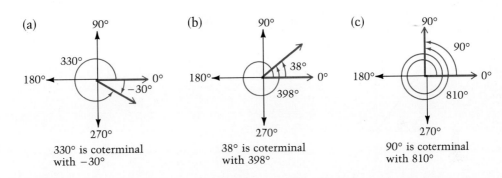

330° is coterminal with −30°   38° is coterminal with 398°   90° is coterminal with 810°

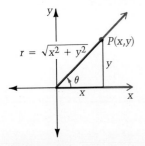

Figure 7.22

Goal B *to determine trigonometric function values of arbitrary angles*

Suppose we have an acute angle θ in standard position, and we choose a point $P(x, y)$ on the terminal side of θ as shown in Figure 7.22. If a perpendicular is drawn from P to the x-axis, we can then define the trigonometric functions of θ in terms of the sides of the resulting right triangle.

$$\sin \theta = \frac{y}{r} \qquad \cos \theta = \frac{x}{r} \qquad \tan \theta = \frac{y}{x}$$

$$\csc \theta = \frac{r}{y} \qquad \sec \theta = \frac{r}{x} \qquad \cot \theta = \frac{x}{y}$$

Note that the trigonometric functions of θ have been defined in terms of the coordinates of a point on the terminal side.

This method suggests a way of defining the values of the trigonometric functions of an angle in any quadrant. We simply need to know the x- and y-coordinates of a point on the terminal side of the angle.

Definition

Let θ be an angle in standard position with $P(x, y)$ a point on its terminal side. Then

$$\sin\theta = \frac{y}{r}, \qquad \cos\theta = \frac{x}{r}, \qquad \tan\theta = \frac{y}{x} \ (x \neq 0),$$

$$\csc\theta = \frac{r}{y} \ (y \neq 0), \qquad \sec\theta = \frac{r}{x} \ (x \neq 0), \qquad \cot\theta = \frac{x}{y} \ (y \neq 0),$$

where $r = \sqrt{x^2 + y^2}$ is the distance from P to the origin.

Example 3 If $(-3, 4)$ is a point on the terminal side of an angle α in standard position, determine the values of the six trigonometric functions of α.

Solution

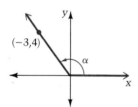

■ Begin by plotting the point $(-3, 4)$ and sketching the angle α.

To determine the trigonometric function values, we must first find the values of x, y, and r.

$$x = -3, \ y = 4, \text{ and } r = \sqrt{x^2 + y^2} = \sqrt{(-3)^2 + (4)^2} = \sqrt{25} = 5.$$

$$\sin\alpha = \frac{y}{r} = \frac{4}{5} \qquad \cos\alpha = \frac{x}{r} = -\frac{3}{5} \qquad \tan\alpha = \frac{y}{x} = -\frac{4}{3}$$

$$\csc\alpha = \frac{r}{y} = \frac{5}{4} \qquad \sec\alpha = \frac{r}{x} = -\frac{5}{3} \qquad \cot\alpha = \frac{x}{y} = -\frac{3}{4}$$

Thus far we have not computed the values of the trigonometric functions of 0°, 90°, or any of the other quadrantal angles. To determine these values, first select a point (x, y) on the terminal side of the angle, and then apply the definition of the trigonometric functions.

7 / Trigonometry: An Introduction

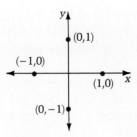

Figure 7.23

- There are only four possible terminal sides for a quadrantal angle: the positive x-axis, the positive y-axis, the negative x-axis, and the negative y-axis. Use the points plotted in Figure 7.23 to determine the values of the trigonometric functions of quadrantal angles.

For example, an angle of 90° has the positive y-axis as its terminal side. The point (0, 1) can be used to determine trigonometric function values of 90°: $x = 0$, $y = 1$, $r = \sqrt{0^2 + 1^2} = 1$.

$$\sin 90° = \frac{1}{1} = 1 \qquad \cos 90° = \frac{0}{1} = 0 \qquad \tan 90° = \frac{1}{0} \quad \blacksquare \text{ undefined}$$

$$\csc 90° = \frac{1}{1} = 1 \qquad \sec 90° = \frac{1}{0} \quad \blacksquare \text{ undefined} \qquad \cot 90° = \frac{0}{1} = 0$$

On the other hand, an angle of 0° has the positive x-axis as its terminal (and initial) side, so the point (1, 0) is used to determine the function values: $x = 1$, $y = 0$, $r = \sqrt{1^2 + 0^2} = 1$.

$$\sin 0° = \frac{0}{1} = 0 \qquad \cos 0° = \frac{1}{1} = 1 \qquad \tan 0° = \frac{0}{1} = 0$$

$$\csc 0° = \frac{1}{0} \quad \blacksquare \text{ undefined} \qquad \sec 0° = \frac{1}{1} = 1 \qquad \cot 0° = \frac{1}{0} \quad \blacksquare \text{ undefined}$$

Example 4   Compute the following: (a) cos 270°   (b) tan (−270°)   (c) sec 540°

Solution   (a) 270° has its terminal side on the negative y-axis, so use (0, −1).

$x = 0$, $y = -1$, $r = \sqrt{0^2 + (-1)^2} = 1$; $\cos 270° = \frac{x}{r} = \frac{0}{1} = 0$

(b) −270° has its terminal side on the positive y-axis, so use (0, 1).

$x = 0$, $y = 1$, $r = \sqrt{0^2 + 1^2} = 1$; $\tan (-270°) = \frac{y}{x} = \frac{1}{0}$ ■ undefined

(c) 540° has its terminal side on the negative x-axis, so use (−1, 0).

$x = -1$, $y = 0$, $r = \sqrt{(-1)^2 + 0^2} = 1$; $\sec 540° = \frac{r}{x} = \frac{1}{-1} = -1$

We add the angles 0° and 90° to our list of special angles, and we expand the table given in the previous section to include these two angles.

Table of Special Angles

$f(\theta)$ \ θ	0°	30°	45°	60°	90°
$\sin \theta$	0	$\frac{1}{2}$	$\frac{\sqrt{2}}{2}$	$\frac{\sqrt{3}}{2}$	1
$\cos \theta$	1	$\frac{\sqrt{3}}{2}$	$\frac{\sqrt{2}}{2}$	$\frac{1}{2}$	0
$\tan \theta$	0	$\frac{\sqrt{3}}{3}$	1	$\sqrt{3}$	*
$\cot \theta$	*	$\sqrt{3}$	1	$\frac{\sqrt{3}}{3}$	0
$\sec \theta$	1	$\frac{2\sqrt{3}}{3}$	$\sqrt{2}$	2	*
$\csc \theta$	*	2	$\sqrt{2}$	$\frac{2\sqrt{3}}{3}$	1

*undefined

Exercise Set 7.3   = Calculator Exercises in the Appendix

Goal A

In exercises 1–12, sketch each of the following angles in standard position.

1. 60°
2. 135°
3. 330°
4. 255°
5. 390°
6. 580°
7. −30°
8. −240°
9. −390°
10. −415°
11. −725°
12. −1142°

In exercises 13–20, determine the measure of the angle between 0° and 360° that is coterminal with each angle.

13. 405°
14. 485°
15. 723°
16. 990°
17. −38°
18. −180°
19. −660°
20. −1689°

In exercises 21–34, determine a positive angle between 0° and 360° inclusive and a negative angle between 0° and −360° inclusive that are coterminal with the given angle.

21. 20°
22. 102°
23. 225°
24. 270°
25. −410°
26. −450°
27. −1351°
28. 1300°
29. 0°
30. 1080°
31. 1575°
32. −940°
33. $\frac{1°}{2}$
34. $-\frac{1°}{2}$

Goal B

In exercises 35–38, determine the values of the six trigonometric functions of α.

35.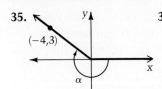

36.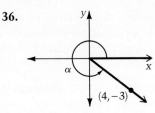

37.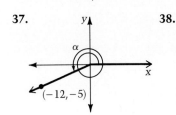

38.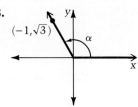

In exercises 39–48, the given point is on the terminal side of an angle α. Determine the values of the six trigonometric functions of α.

39. $(-8, 6)$
40. $(-5, 12)$
41. $(1, 1)$
42. $(2, 6)$
43. $(3, -3)$
44. $(-3, -9)$
45. $(1, -2)$
46. $(-\sqrt{3}, -1)$
47. $(\sqrt{5}, -2)$
48. $(2\sqrt{3}, 4)$

In exercises 49–60, evaluate each of the following.

49. $\sin 90°$
50. $\cos 180°$
51. $\cot 180°$
52. $\tan 270°$
53. $\sec 450°$
54. $\sin 810°$
55. $\cos(-90°)$
56. $\csc(-630°)$
57. $\tan(-540°)$
58. $\sec(-1350°)$
59. $\csc(-720°)$
60. $\cot(-900°)$

Superset

In exercises 61–64, determine the measure of the given angle.

61. φ is half of a complete revolution in a counterclockwise direction.
62. φ is half of a complete revolution in a clockwise direction.
63. φ is three tenths of a complete revolution in a counterclockwise direction.
64. φ is one and one quarter complete revolutions in a clockwise direction.

In exercises 65–72, the value of one trigonometric function of φ is given. Assuming that φ is a second quadrant angle, find the values of the other five trigonometric functions of φ.

65. $\tan \varphi = -5$
66. $\sin \varphi = \dfrac{4}{5}$
67. $\csc \varphi = \dfrac{5}{3}$
68. $\cos \varphi = -\dfrac{3}{4}$
69. $\cot \varphi = -\dfrac{5}{3}$
70. $\tan \varphi = -1$
71. $\cos \varphi = -\dfrac{3}{5}$
72. $\sec \varphi = -\sqrt{3}$

In exercises 73–80, the value of one trigonometric function of φ is given. If the terminal side of φ lies in the given quadrant, find the values of the other five trigonometric functions of φ.

73. third quadrant; $\sin \varphi = -\dfrac{2}{5}$
74. second quadrant; $\sin \varphi = \dfrac{3}{5}$
75. third quadrant; $\cos \varphi = -\dfrac{3}{5}$
76. fourth quadrant; $\tan \varphi = -1$
77. fourth quadrant; $\csc \varphi = -\dfrac{5}{3}$
78. third quadrant; $\tan \varphi = 1$
79. fourth quadrant; $\cos \varphi = 0.7$
80. second quadrant; $\cot \varphi = -1.5$

7.4 Identities and Tables

Goal A *to use basic trigonometric identities*

An **identity** is an equation that is true for all values of the variables for which both sides of the equation are defined. Several trigonometric identities are easily derived from the definitions of the trigonometric functions.

Reciprocal Identities Derivation

(1) $\sin \theta = \dfrac{1}{\csc \theta}$ $\sin \theta = \dfrac{y}{r} = \dfrac{1}{\frac{r}{y}} = \dfrac{1}{\csc \theta}$

(2) $\cos \theta = \dfrac{1}{\sec \theta}$ $\cos \theta = \dfrac{x}{r} = \dfrac{1}{\frac{r}{x}} = \dfrac{1}{\sec \theta}$

(3) $\tan \theta = \dfrac{1}{\cot \theta}$ $\tan \theta = \dfrac{y}{x} = \dfrac{1}{\frac{x}{y}} = \dfrac{1}{\cot \theta}$

(4) $\cot \theta = \dfrac{1}{\tan \theta}$ $\cot \theta = \dfrac{x}{y} = \dfrac{1}{\frac{y}{x}} = \dfrac{1}{\tan \theta}$

(5) $\sec \theta = \dfrac{1}{\cos \theta}$ $\sec \theta = \dfrac{r}{x} = \dfrac{1}{\frac{x}{r}} = \dfrac{1}{\cos \theta}$

(6) $\csc \theta = \dfrac{1}{\sin \theta}$ $\csc \theta = \dfrac{r}{y} = \dfrac{1}{\frac{y}{r}} = \dfrac{1}{\sin \theta}$

Quotient Identities Derivation

(7) $\dfrac{\sin \theta}{\cos \theta} = \tan \theta$ $\dfrac{\sin \theta}{\cos \theta} = \dfrac{\frac{y}{r}}{\frac{x}{r}} = \dfrac{y}{r} \cdot \dfrac{r}{x} = \dfrac{y}{x} = \tan \theta$

(8) $\dfrac{\cos \theta}{\sin \theta} = \cot \theta$ $\dfrac{\cos \theta}{\sin \theta} = \dfrac{\frac{x}{r}}{\frac{y}{r}} = \dfrac{x}{r} \cdot \dfrac{r}{y} = \dfrac{x}{y} = \cot \theta$

Example 1 Given $\sin \beta = 0.324$ and $\cos \beta = 0.946$, find the remaining trigonometric functions of β. Round the answer to three decimal places.

Solution $\tan \beta = \dfrac{\sin \beta}{\cos \beta}$

$= \dfrac{0.324}{0.946} \approx 0.342$ ■ 0.342 is an approximation for tan β, and we write tan $\beta \approx 0.342$.

$\cot \beta = \dfrac{1}{\tan \beta}$

$= \dfrac{1}{0.342} \approx 2.924$ ■ If the identity $\dfrac{\cos \beta}{\sin \beta} = \cot \beta$ is used, the answer is 2.920. The difference is due to rounding error.

$\sec \beta = \dfrac{1}{\cos \beta}$

$= \dfrac{1}{0.946} \approx 1.057$

$\csc \beta = \dfrac{1}{\sin \beta}$

$= \dfrac{1}{0.324} \approx 3.086$

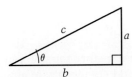

Figure 7.24

The next group of identities that we will consider is referred to as the Pythagorean Identities because their derivation depends on the Pythagorean Theorem. Using the right triangle at the left, we can make the following statements.

$$a^2 + b^2 = c^2$$ ■ This is the Pythagorean Theorem.

$$\dfrac{a^2}{c^2} + \dfrac{b^2}{c^2} = \dfrac{c^2}{c^2}$$ ■ Both sides divided by c^2.

$$\left(\dfrac{a}{c}\right)^2 + \left(\dfrac{b}{c}\right)^2 = 1$$

$$(\sin \theta)^2 + (\cos \theta)^2 = 1$$ ■ $\sin \theta = \dfrac{a}{c}$; $\cos \theta = \dfrac{b}{c}$.

It is common to write $(\sin \theta)^2$ as $\sin^2 \theta$ and $(\cos \theta)^2$ as $\cos^2 \theta$.

Pythagorean Identities   Derivation

(9) $\sin^2 \theta + \cos^2 \theta = 1$ Shown above.

(10) $\tan^2 \theta + 1 = \sec^2 \theta$ Divide both sides of (9) by $\cos^2 \theta$. Then use the Quotient and Reciprocal Identities.

(11) $1 + \cot^2 \theta = \csc^2 \theta$ Divide both sides of (9) by $\sin^2 \theta$. Use the Quotient and Reciprocal Identities.

Example 2 In each of the following, θ is an acute angle.

(a) If $\sin \theta = \dfrac{1}{3}$, find $\cos \theta$.   (b) If $\tan \theta = 1.6$, find $\sec \theta$.

Solution (a) $\sin^2\theta + \cos^2\theta = 1$ ■ Identity (9).

$$\left(\frac{1}{3}\right)^2 + \cos^2\theta = 1$$

$$\cos^2\theta = 1 - \frac{1}{9}$$

$$\cos \theta = +\sqrt{\frac{8}{9}} = \frac{2\sqrt{2}}{3}$$ ■ For an acute angle all trigonometric function values are positive.

(b) $\tan^2\theta + 1 = \sec^2\theta$ ■ Identity (10).

$(1.6)^2 + 1 = \sec^2\theta$

$\sec^2\theta = 3.56$

$\sec \theta = +\sqrt{3.56} \approx 1.89$ ■ Again, select the positive root.

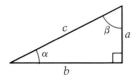

Figure 7.25 Since $\sin \alpha = \frac{a}{c}$ and $\cos \beta = \frac{a}{c}$, $\sin \alpha = \cos \beta$.

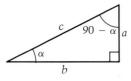

Figure 7.26 Since $\cos \alpha = \frac{b}{c}$ and $\sin(90° - \alpha) = \frac{b}{c}$, $\cos \alpha = \sin(90° - \alpha)$.

Recall that if α and β are complementary angles, then $\alpha + \beta = 90°$. Figure 7.25 shows that in this case $\sin \alpha = \cos \beta$. Since $\beta = 90° - \alpha$, we can write $\sin \alpha = \cos(90° - \alpha)$. This is called a **Cofunction Identity** since the sine and cosine are cofunctions of each other. The tangent and cotangent are also cofunctions of each other, as are the secant and cosecant. We now list the cofunction identities.

Cofunction Identities	Derivation
$\sin \alpha = \cos(90° - \alpha)$	See Figure 2.25.
$\cos \alpha = \sin(90° - \alpha)$	See Figure 2.26.
$\tan \alpha = \cot(90° - \alpha)$	$\tan \alpha = \dfrac{\sin \alpha}{\cos \alpha} = \dfrac{\cos(90° - \alpha)}{\sin(90° - \alpha)}$ $= \cot(90° - \alpha)$
$\cot \alpha = \tan(90° - \alpha)$	$\cot \alpha = \dfrac{1}{\tan \alpha} = \dfrac{1}{\cot(90° - \alpha)}$ $= \tan(90° - \alpha)$
$\sec \alpha = \csc(90° - \alpha)$	$\sec \alpha = \dfrac{1}{\cos \alpha} = \dfrac{1}{\sin(90° - \alpha)}$ $= \csc(90° - \alpha)$
$\csc \alpha = \sec(90° - \alpha)$	$\csc \alpha = \dfrac{1}{\sin \alpha} = \dfrac{1}{\cos(90° - \alpha)}$ $= \sec(90° - \alpha)$

In words, the cofunction identities say that the trigonometric function of an angle is equal to the cofunction of the complement of the angle.

Example 3 Complete the following statements with an acute angle.

(a) $\sin 37° = \cos$ ____ (b) $\cot 52° = \tan$ ____ (c) $\csc 45° = \sec$ ____

Solution (a) $\sin 37° = \cos(90° - 37°) = \cos 53°$
(b) $\cot 52° = \tan(90° - 52°) = \tan 38°$
(c) $\csc 45° = \sec(90° - 45°) = \sec 45°$

Example 4 Given $\sin 70° \approx 0.940$ and $\cos 70° \approx 0.342$, find (a) $\cot 20°$ (b) $\csc 20°$. Round the answer to three decimal places.

Solution (a) $\cot 20° = \tan(90° - 20°) = \tan 70° = \dfrac{\sin 70°}{\cos 70°} \approx \dfrac{0.940}{0.342} \approx 2.749$

(b) $\csc 20° = \dfrac{1}{\sin 20°} = \dfrac{1}{\cos 70°} \approx \dfrac{1}{0.342} \approx 2.924$

Goal B *to use the tables of trigonometric function values*

We know the trigonometric function values of the special angles, but what about $\cos 37°$ or $\sin 72°$? Because such values are not easy to compute, Table 4 in the Appendix lists values of the six trigonometric functions of angles between 0° and 90°. Although the values in this table are approximations, we agree to write $\sin 45° = 0.7071$ (from Table 4) even though the exact value of $\sin 45°$ is $\frac{\sqrt{2}}{2}$, and 0.7071 is only an approximation for $\frac{\sqrt{2}}{2}$.

Fractional parts of a degree are typically measured in minutes, with 60 minutes in a degree (1° = 60'). For example, $24\frac{1}{2}° = 24°30'$ and $42\frac{1}{3}° = 42°20'$. Sometimes it is necessary to record degree measure in decimal form. For example, $58°15' = 58.25°$, since $15' = \frac{1}{4}° = 0.25°$.

Example 5 Use Table 4 to determine (a) $\tan 39°20'$ (b) $\cos 62°40'$

Solution (a) $\tan 39°20' = 0.8195$
- For angles between 0° and 45°, start at the *top left* of Table 4 and read down the extreme left columns until you find 39°20'. Then move across to the column having tan α at the top.

(b) $\cos 62°40' = 0.4592$
- For angles between 45° and 90°, start at the *bottom right* of Table 4 and read up the extreme right columns until you reach 62°20'. Then move across to the column having cos α at the bottom.

Example 6   Use Table 4 to determine acute angle α if csc α = 1.211.

Solution $\alpha = 55°40'$

- Read up the column in Table 4 having csc α at the bottom. Move across to the right-hand column to find 55°40'.

Table 4 only contains angles that are multiples of ten minutes. To compute sin 36°27', we find sin 36°20' and sin 36°30' from the table. The value of sin 36°27' is between these values. To obtain an approximation, we use a method called **linear interpolation.** This method involves setting up a proportion using the 4 differences d_1, d_2, d_3, and d_4 shown below.

	α	sin α	
d_1	36°20'	0.5925	d_3
d_2	36°27'	?	
	36°30'	0.5948	d_4

$$\frac{d_1}{d_2} = \frac{d_3}{d_4}$$

$$\sin 36°27' = 0.5925 + d_3$$

- 36°20' and 36°30' are called **bracketing values** for 36°27'.

- The proportion is solved for d_3, the only unknown value.
- d_3 is added to the upper trigonometric function value.

Example 7   Find   (a) sin 36°27'   (b) cos 73°48'

Solution   (a)

	α	sin α	
7	36°20'	0.5925	x
	36°27'	?	
10	36°30'	0.5948	0.0023

$$\frac{7}{10} = \frac{x}{0.0023}$$

$$x \approx 0.0016$$

$$\sin 36°27' = 0.5925 + 0.0016 = 0.5941$$

- Set up a chart involving the bracketing values. Determine the three differences, and let x represent the unknown difference.

- Set up the proportion and solve. Since the values are given to four decimal places, we approximate the value of x to 4 decimal places.
- x is added to the upper function value.

(b)

α	cos α	
73°40′	0.2812	
73°48′	?	
73°50′	0.2784	

with 8, 10 bracket on left and x, −0.0028 bracket on right

■ Set up the proportion:
$\frac{8}{10} = \frac{x}{-0.0028}$. Solve to get $x = -0.0022$.

$\cos 73°48' = 0.2812 + (-0.0022) = 0.2790$

Example 8 If $\tan \varphi = 0.8915$ and φ is an acute angle, determine φ.

Solution

φ	tan φ	
41°40′	0.8899	0.0016
?	0.8915	
41°50′	0.8952	0.0053

with x, 10 bracket on left

■ The value 0.8915 is between 0.8899 and 0.8952.

$\frac{x}{10'} = \frac{0.0016}{0.0053}$

$x = 3.019'$

$x \approx 3'$

$\varphi = 41°40' + 3' = 41°43'$

■ Set up the proportion and solve.

■ Round to the nearest minute.

■ Add the value of x to the upper angle value.

Exercise Set 7.4 $\boxed{=}$ Calculator Exercises in the Appendix

Goal A

In exercises 1–4, two approximate trigonometric function values of an angle are given. Find the other four trigonometric function values rounded to two decimal places.

1. $\sin 32° = 0.53$, $\cos 32° = 0.85$
2. $\tan 55° = 1.43$, $\sin 55° = 0.82$
3. $\tan 66° = 2.25$, $\sec 66° = 2.46$
4. $\tan 23° = 0.42$, $\cos 23° = 0.92$

In exercises 5–10, complete the following statements with an acute angle.

5. $\sin 54° = \cos$ _____
6. $\cos 12° = \sin$ _____
7. $\tan 38° = \cot$ _____
8. $\cot 81° = \tan$ _____
9. $\sec 45° = \csc$ _____
10. $\csc 71° = \sec$ _____

In exercises 11–18, rewrite each expression as a trigonometric function of an angle between 45° and 90°.

11. $\cos 27°$
12. $\sec 12°$
13. $\cot 38°$
14. $\tan 41°$
15. $\csc 3°$
16. $\cos 43°$
17. $\sin 19°$
18. $\sin 23°$

In exercises 19–24, approximate the function values given that $\sin 32° = 0.53$, $\cos 32° = 0.85$, $\sin 23° = 0.39$, and $\tan 23° = 0.42$. Round your answer to two decimal places.

19. $\sin 58°$
20. $\sec 58°$
21. $\sec 67°$
22. $\tan 58°$
23. $\tan 67°$
24. $\cot 67°$

In exercises 25–32, use the Reciprocal and Pythagorean Identities to determine the values of the other five trigonometric functions of the acute angle α.

25. $\sin \alpha = \dfrac{1}{2}$
26. $\cos \alpha = \dfrac{3}{10}$
27. $\tan \alpha = \dfrac{3}{2}$
28. $\csc \alpha = 3$
29. $\cot \alpha = \dfrac{1}{2}$
30. $\sec \alpha = 10$
31. $\cos \alpha = \dfrac{4}{5}$
32. $\sin \alpha = \dfrac{7}{10}$

Goal B

In exercises 33–44, use Table 4 in the Appendix to determine the following.

33. $\sin 31°10'$
34. $\cos 43°50'$
35. $\tan 21°40'$
36. $\tan 49°20'$
37. $\sec 56°50'$
38. $\sec 10°30'$
39. $\cot 48°10'$
40. $\csc 15°40'$
41. $\cos 52°20'$
42. $\sin 80°10'$
43. $\csc 45°30'$
44. $\cot 78°50'$

In exercises 45–52, use Table 4 to determine the acute angle α.

45. $\sin \alpha = 0.4695$
46. $\cos \alpha = 0.8480$
47. $\cot \alpha = 1.072$
48. $\tan \alpha = 3.078$
49. $\cos \alpha = 0.7698$
50. $\sin \alpha = 0.4669$
51. $\sec \alpha = 1.086$
52. $\csc \alpha = 1.061$

In exercises 53–62, find the value of each of the following.

53. $\sin 39°33'$
54. $\tan 12°15'$
55. $\tan 31°23'$
56. $\cos 19°3'$
57. $\cos 71°27'$
58. $\sin 63°11'$
59. $\cot 82°18'$
60. $\cot 75°52'$
61. $\sec 87°47'$
62. $\csc 45°36'$

In exercises 63–68, determine the acute angle α.

63. $\sin \alpha = 0.0600$
64. $\cos \alpha = 0.9727$
65. $\cos \alpha = 0.3490$
66. $\tan \alpha = 1.7030$
67. $\tan \alpha = 0.5000$
68. $\sin \alpha = 0.8500$

Superset

In exercises 69–76, determine which statements are identities.

69. $\cos \alpha \tan \alpha = \sin \alpha$
70. $\cos \alpha \sec \alpha = \cot \alpha$
71. $\sec \alpha = \tan \alpha \csc \alpha$
72. $\sin \alpha \cot \alpha = \cos \alpha$
73. $\dfrac{\tan \alpha}{\sin \alpha} = \sec \alpha$
74. $\dfrac{\csc \alpha}{\sec \alpha} = \cot \alpha$
75. $\dfrac{1}{\tan \alpha} = \dfrac{\cos \alpha}{\sin \alpha}$
76. $\cot \alpha = \dfrac{\sec \alpha}{\csc \alpha}$

In exercises 77–80, evaluate each pair of expressions by using the values of the special angles.

77. $\sin(90° - 30°)$, $\sin 90° - \sin 30°$
78. $\sin(60° + 30°)$, $\sin 60° + \sin 30°$
79. $\cot 90° - \cot 30°$, $\cot(90° - 30°)$
80. $\cos(45° + 45°)$, $\cos 45° + \cos 45°$

In exercises 81–84, verify that each of the following is true.

81. $\sin(60° + 30°) = \sin 60° \cos 30° + \sin 30° \cos 60°$
82. $\sin 60° \cos 30° = \dfrac{1}{2}(\sin 90° + \sin 30°)$
83. $\cos 90° = (\cos 45°)(\cos 45°) - (\sin 45°)(\sin 45°)$
84. $\tan 60° = \dfrac{2 \tan 30°}{1 - \tan^2 30°}$

7.5 Reference Angles

Goal A *to determine the sign of a trigonometric function value*

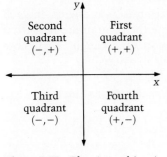

Figure 7.27 The signs of (x, y)

We have defined the six trigonometric functions of an angle θ in terms of x, y, and r, where x and y are coordinates of a point on the terminal side of θ, and r is the distance between that point and the origin. Since r must be positive, the sign of a trigonometric function value, such as $\cos 210°$, is determined by the signs of x and y. For example, since $\cos \theta = \frac{x}{r}$ and r is positive, the sign of $\cos \theta$ is determined by the sign of x, which is positive in the first and fourth quadrants and negative in the second and third quadrants.

The chart below summarizes information needed to determine the sign of a trigonometric function value.

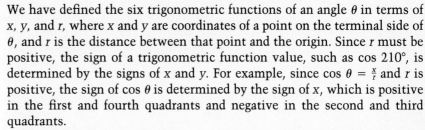

- The memory device ASTC (Always Study Trigonometry Carefully) is often used as a reminder of which trigonometric functions are positive in which quadrants.

positive functions	A	S	T	C
quadrant	1st	2nd	3rd	4th

Figure 7.28 The quadrants in which the trigonometric functions are positive.

Example 1 Determine the sign of the following:

(a) $\sin 190°$ (b) $\tan 135°$ (c) $\cos(-50°)$

Solution

(a) $190°$ is a third quadrant angle. $\sin 190°$ is negative.

Think: $\begin{array}{c|c|c|c} A & S & T & C \\ \hline 1 & 2 & 3 & 4 \end{array}$

■ In the third quadrant, only tan and cot are positive.

(b) $135°$ is a second quadrant angle. $\tan 135°$ is negative.

Think: $\begin{array}{c|c|c|c} A & S & T & C \\ \hline 1 & 2 & 3 & 4 \end{array}$

■ In the second quadrant, only sin and csc are positive.

(c) $-50°$ is a fourth quadrant angle. $\cos(-50°)$ is positive.

Think: $\begin{array}{c|c|c|c} A & S & T & C \\ \hline 1 & 2 & 3 & 4 \end{array}$

■ In the fourth quadrant, only cos and sec are positive.

Goal B *to determine the reference angle for an angle of any measure*

Suppose you wish to evaluate $\cos 210°$. You already know one part of the answer: the *sign* of $\cos 210°$ is negative. Thus,

$$\cos 210° = -\underline{}.$$

To complete the statement $\cos 210° = -\underline{}$, we use a *reference angle*. A **reference angle** for a given angle θ is a first quadrant angle whose trigonometric function values are numerically the same as the trigonometric function values of θ—only the signs may differ.

Figure 7.29 presents second, third, and fourth quadrant angles. In each case, the reference angle is found by performing the appropriate reflection of the terminal side necessary to make it a first quadrant angle. We denote the reference angle of any angle θ by attaching an asterisk (θ^*).

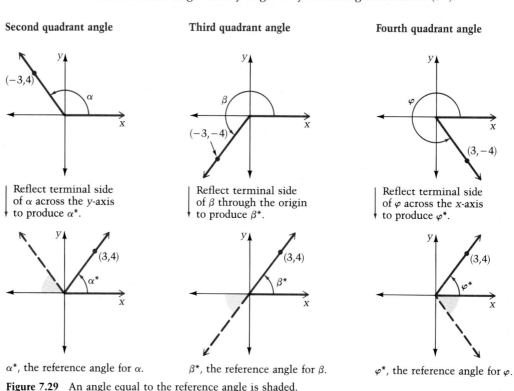

Figure 7.29 An angle equal to the reference angle is shaded.

Figure 7.29 illustrates the following useful rule.

Rule for Finding a Reference Angle

Suppose θ is an angle between 0° and 360° inclusive and θ^* is the reference angle of θ.

If θ is a first quadrant angle, $\theta^* = \theta$.
If θ is a second quadrant angle, $\theta^* = 180° - \theta$.
If θ is a third quadrant angle, $\theta^* = \theta - 180°$.
If θ is a fourth quadrant angle, $\theta^* = 360° - \theta$.

If you are given an angle greater than 360° or less than 0°, add or subtract an appropriate multiple of 360° so that the measure of the resulting angle is between 0° and 360°. Then use the above rule.

Example 2 Determine the reference angle for each of the following angles.
(a) $\theta = 240°$ (b) $\theta = 1000°$ (c) $\theta = -200°$

Solution (a) $240°$ is a third quadrant angle. Therefore $\theta^* = 240° - 180° = 60°$. The reference angle is $60°$.
(b) $1000°$ is greater than $360°$. Subtract $2 \cdot 360°$ from $1000°$ to get $280°$. Now, $280°$ is a fourth quadrant angle. Thus, $\theta^* = 360° - 280° = 80°$. The reference angle is $80°$.
(c) $-200°$ is less than $0°$. Add $1 \cdot 360°$ to get $160°$. Now, $160°$ is a second quadrant angle. Thus, $\theta^* = 180° - 160° = 20°$. The reference angle is $20°$.

Goal C *to use reference angles to determine trigonometric function values*

Because of the way we defined a reference angle, the trigonometric function values of a given angle are the same as the trigonometric function values of the reference angle, except maybe for the sign. We use the ASTC chart in Goal A to determine the correct sign.

To evaluate an expression like $\tan 240°$, we follow a three step procedure:

Step 1. Determine the sign of $\tan \theta$: $\tan 240°$ is positive
Step 2. Find the reference angle θ^*: $240° - 180° = 60°$
Step 3. Evaluate $\tan \theta^*$: $\tan 60° = \sqrt{3}$

Thus, $\tan 240° = +\tan 60° = \sqrt{3}$.

Example 3 Evaluate   (a) $\cos 120°$ (b) $\sin 280°20'$.

Solution (a) $\cos 120° =$ sign $\cos$ reference angle of $120°$

$= -\cos$ reference angle of $120°$ ■ $120°$ is a second quadrant angle; cos is negative there.
$= -\cos 60°$ ■ $\theta^* = 180° - 120° = 60°$.
$= -\dfrac{1}{2}$

(b) $\sin 280°20' =$ sign $\sin$ reference angle of $280°20'$

$= -\sin$ reference angle of $280°20'$ ■ $280°20'$ is a fourth quadrant angle; sin is negative there.
$= -\sin 79°40' = -0.9838$ ■ $\theta^* = 360° - 280°20' = 79°40'$

Exercise Set 7.5

Goal A

In exercises 1–14, determine the sign of the following.

1. cot 200°
2. sec 175°
3. cos 315°
4. tan 285°
5. sin 457°
6. cot 703°
7. csc(1210°)
8. sin 1568°
9. tan(−279°)
10. csc(−112°)
11. sec(−763°)
12. cos(−581°)
13. sin 196°36′
14. tan(−185°42′)

Goal B

In exercises 15–28, determine the reference angle for each of the following angles.

15. 200°
16. 185°
17. 320°
18. 197°
19. 485°
20. 696°
21. −250°
22. −185°
23. −444°
24. −715°
25. −1445°
26. −1081°
27. 265°35′
28. −95°58′

Goal C

In exercises 29–60, evaluate the following.

29. cos 240°
30. csc 240°
31. tan 300°
32. sin 120°
33. cot 225°
34. sec 135°
35. sec 330°
36. tan 180°
37. csc 300°
38. sec 300°
39. tan 750°
40. cos 1290°
41. sin(−210°)
42. sin(−1290°)
43. tan 158°
44. sin 345°
45. cos 408°
46. cot 497°
47. cot 320°
48. tan 212°
49. cos 1258°
50. cos 485°
51. sec(−280°)
52. csc(−272°)
53. sin(−1134°)
54. tan(−666°)
55. csc(−138°)
56. sec(−295°)
57. sin 95°20′
58. cot 245°40′
59. cot(−100°30′)
60. cos 250°10′

Superset

In exercises 61–70, assume that the angle θ terminates in the given quadrant. Find the sign of the given trigonometric function value.

61. second quadrant; cos θ
62. third quadrant; sin θ
63. fourth quadrant; tan θ
64. fourth quadrant; sec θ
65. third quadrant; cos θ
66. fourth quadrant; cot θ
67. second quadrant; sec θ
68. second quadrant; tan θ
69. fourth quadrant; sin θ
70. second quadrant; csc θ

In exercises 71–76, assume that the angle θ has measure between 180° and 270°. Find the sign of the following trigonometric function values.

71. $\sin \frac{\theta}{2}$
72. $\sin 2\theta$
73. $\sin(90° + \theta)$
74. $\sin(90° - \theta)$
75. $\sin(180° - \theta)$
76. $\sin(180° + \theta)$

Chapter Review & Test

Chapter Review

7.1 Triangles (pp. 282–288)

In any triangle, the sum of the measures of the angles is 180°. (p. 283)

An *isosceles triangle* is a triangle with two sides of equal length. In any isosceles triangle the angles opposite the two sides of equal length have the same measure. (p. 283) A *right triangle* is a triangle in which one of the angles is a right angle. (p. 283) An *equilateral triangle* is a triangle with three sides of equal length. Each angle of an equilateral triangle has measure 60°. (p. 284)

In *similar triangles*, corresponding angles have the same measure and corresponding sides are proportional. (p. 285) *A-A Similarity Theorem:* Two triangles are similar if two angles of one triangle have the same measure as two angles of the other triangle. (p. 286)

7.2 Trigonometric Ratios of Acute Angles (pp. 289–295)

Let α be an acute angle in a right triangle ABC. Then the trigonometric functions of α are defined by the following ratios (p. 290):

$$\sin \alpha = \frac{a}{c} \qquad \cos \alpha = \frac{b}{c} \qquad \tan \alpha = \frac{a}{b}$$

$$\csc \alpha = \frac{c}{a} \qquad \sec \alpha = \frac{c}{b} \qquad \cot \alpha = \frac{b}{a}$$

7.3 Angles of Rotation (pp. 296–301)

An angle is in *standard position* in the *xy*-plane if its vertex is at the origin, and its initial side lies on the positive *x*-axis. The angle is *positive* if the rotation is counterclockwise and *negative* if the rotation is clockwise. (p. 296)

Let θ be an angle in standard position with $P(x, y)$ a point on its terminal side. Then

$$\sin \theta = \frac{y}{r}, \qquad \cos \theta = \frac{x}{r}, \qquad \tan \theta = \frac{y}{x} \quad (x \neq 0),$$

$$\csc \theta = \frac{r}{y} \quad (y \neq 0), \qquad \sec \theta = \frac{r}{x} \quad (x \neq 0), \qquad \cot \theta = \frac{x}{y} \quad (y \neq 0),$$

where $r = \sqrt{x^2 + y^2}$ is the distance from P to the origin. (p. 297)

7.4 Identities and Tables (pp. 302–308)

Reciprocal Identities

$$\sin\theta = \frac{1}{\csc\theta} \quad \cos\theta = \frac{1}{\sec\theta} \quad \tan\theta = \frac{1}{\cot\theta}$$

$$\csc\theta = \frac{1}{\sin\theta} \quad \sec\theta = \frac{1}{\cos\theta} \quad \cot\theta = \frac{1}{\tan\theta}$$

Quotient Identities

$$\frac{\sin\theta}{\cos\theta} = \tan\theta$$

$$\frac{\cos\theta}{\sin\theta} = \cot\theta$$

Cofunction Identities

$\sin\theta = \cos(90° - \theta) \quad \cos\theta = \sin(90° - \theta)$
$\tan\theta = \cot(90° - \theta) \quad \cot\theta = \tan(90° - \theta)$
$\sec\theta = \csc(90° - \theta) \quad \csc\theta = \sec(90° - \theta)$

Pythagorean Identities

$\sin^2\theta + \cos^2\theta = 1$
$\tan^2\theta + 1 = \sec^2\theta$
$1 + \cot^2\theta = \csc^2\theta$

7.5 Reference Angles (pp. 309–312)

The sign of a trigonometric function value depends on the quadrant in which the angle terminates. It is found by using the ASTC chart:

positive functions	A	S	T	C
quadrant	1	2	3	4

A reference angle θ^* for a given angle θ is a first quadrant angle. The trigonometric function values of θ and θ^* are numerically the same. Only the signs may differ. (p. 310)

Chapter Test

7.1A Use the given information about $\triangle ABC$ to solve for the missing part.

1. $m(\angle A) = 22°, m(\angle B) = 95°, m(\angle C) = ?$
2. $m(\angle B) = 79°, m(\angle C) = 79°, m(\angle A) = ?$

Assume that $\triangle ABC$ is a right triangle, c is the length of the hypotenuse and a and b are the lengths of the legs. Find the length of the third side given the lengths of two sides.

3. $a = 12, b = 16$
4. $a = 10, c = 26$

7.1B Assume that $\triangle ABC$ and $\triangle A'B'C'$ are two triangles with $m(\angle A) = m(\angle A')$ and $m(\angle B) = m(\angle B')$. Given the lengths of some sides, find the length of the indicated side.

5. $a = 2, b = 5, a' = 12, b' = ?$
6. $a = 7, b = 8, a' = ?, b' = 20$

7.2A Use the given information about the right triangle ABC to find the values of the six trigonometric ratios of α.

7. 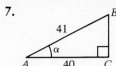   8.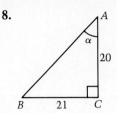

7.2B $\triangle ABC$ has right angle at C and sides of length a, b, and c. Given the following information, find the indicated part.

9. $m(\angle A) = 30°$, $a = 24$, find c 10. $m(\angle A) = 60°$, $c = 18$, find a

7.2C The angles α and β are complementary. Given the following information, find $\tan \alpha$ and $\tan \beta$.

11. $\sin \alpha = \dfrac{3}{7}$ 12. $\cos \alpha = \dfrac{5}{7}$

7.3A Sketch each of the following angles in standard position.

13. $150°$ 14. $-135°$ 15. $-215°$ 16. $325°$

Determine the measure of the angle between $0°$ and $360°$ which is coterminal with the given angle.

17. $435°$ 18. $1255°$ 19. $-435°$ 20. $-1255°$

7.3B The given point is located on the terminal side of angle θ. Find $\sin \theta$ and $\tan \theta$.

21. $(-10, 24)$ 22. $(-12, -9)$

Evaluate each of the following.

23. $\cos(-270°)$ 24. $\sin 180°$

7.4A Complete the following with an acute angle.

25. $\sin 77° = \cos$ _____   26. $\cot 37° = \tan$ _____

If $\sin 72° = 0.95$ and $\cos 72° = 0.31$, find the following function values. Round your answer to two decimal places.

27. $\sec 18°$ 28. $\cot 18°$

Use the Reciprocal and Pythagorean Identities to determine the values of the other five trigonometric functions of acute angle α.

29. $\tan \alpha = \dfrac{3}{2}$ **30.** $\sin \alpha = \dfrac{5}{6}$

7.4B Use Table 4 in the Appendix to determine the following.

31. $\cos 42°23'$ **32.** $\sec 63°48'$

7.5A Determine the sign of the following.

33. $\cos 200°$ **34.** $\cot(-200°)$

35. $\tan(-300°)$ **36.** $\csc 250°$

7.5B Determine the reference angle for each of the following angles.

37. $215°$ **38.** $175°$

7.5C Evaluate each of the following.

39. $\tan 135°$ **40.** $\cot(-240°)$

41. $\cos 348°20'$ **42.** $\sin 178°50'$

Superset

43. Find the side of the square that has the same area as a rectangle with one side of length 16 and diagonal 20.

44. Between midnight and noon, how many times do the hands of a clock make an angle of 90°? an angle of 180°?

In exercises 45–46, the value of one trigonometric function of φ is given. If the terminal side of φ lies in the given quadrant, find the values of the other five trigonometric functions.

45. third quadrant; $\cos \varphi = -\dfrac{12}{13}$ **46.** fourth quadrant; $\cot \varphi = -\dfrac{9}{40}$

In exercises 47–48, verify that each of the following is true.

47. $\sin(2 \cdot 120°) = 2\sin 120° \cos 120°$

48. $\cos(135° - 45°) = \cos 135° \cos 45° + \sin 135° \sin 45°$

In exercises 49–50, assume that the angle θ terminates in the given quadrant. Find the sign of the trigonometric function value.

49. second quadrant; $\tan \theta$ **50.** fourth quadrant; $\csc \theta$

8 Trigonometric Functions

8.1 Radian Measure and the Unit Circle

Goal A *to determine points and standard arcs on the unit circle*

Up to this point, our study of trigonometry has focused on angles and triangles. In this chapter we shall develop a way of defining the trigonometric functions so that their domains consist of real numbers, not just angle measurements. This allows us to apply trigonometric functions to a much wider variety of problems. Many natural phenomena such as the motion of planets, the ocean's tides, and the beat of a human heart, are *periodic*, that is, certain behavior is repeated at regular intervals. When defined as functions of real numbers, trigonometric functions provide an excellent means of describing periodic behavior in nature.

The circle of radius 1 with center at the origin is called the **unit circle.** In Figure 8.1 we have labeled the horizontal axis the u-axis and the vertical axis the v-axis. An equation for the unit circle is easily derived by means of the distance formula. Since any point $T(u, v)$ on the unit circle is at a distance 1 from the origin $(0, 0)$, we have, by the distance formula,

$$\sqrt{(u - 0)^2 + (v - 0)^2} = 1.$$

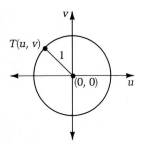

Figure 8.1 The unit circle in the uv-plane.

Upon squaring both sides of this equation, we get

$$u^2 + v^2 = 1,$$

which is the common way of describing the unit circle in the uv-plane.

Example 1 Determine whether each of the following points is on the unit circle.

(a) $\left(\dfrac{1}{2}, \dfrac{\sqrt{3}}{2}\right)$ (b) $\left(-\dfrac{\sqrt{2}}{2}, \dfrac{\sqrt{2}}{2}\right)$ (c) $\left(\dfrac{1}{3}, \dfrac{2}{3}\right)$

Solution (a) $\left(\dfrac{1}{2}\right)^2 + \left(\dfrac{\sqrt{3}}{2}\right)^2 = \dfrac{1}{4} + \dfrac{3}{4} = 1$ ■ To determine whether a point (u, v) is on the unit circle, check to see if its coordinates satisfy the equation $u^2 + v^2 = 1$.

Thus, $\left(\dfrac{1}{2}, \dfrac{\sqrt{3}}{2}\right)$ is on the unit circle.

(b) $\left(\dfrac{-\sqrt{2}}{2}\right)^2 + \left(\dfrac{\sqrt{2}}{2}\right)^2 = \dfrac{2}{4} + \dfrac{2}{4} = 1$

Thus, $\left(-\dfrac{\sqrt{2}}{2}, \dfrac{\sqrt{2}}{2}\right)$ is on the unit circle.

(c) $\left(\dfrac{1}{3}\right)^2 + \left(\dfrac{2}{3}\right)^2 = \dfrac{1}{9} + \dfrac{4}{9} = \dfrac{5}{9} \neq 1$

Thus, $\left(\dfrac{1}{3}, \dfrac{2}{3}\right)$ is not on the unit circle.

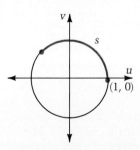

Figure 8.2 The standard arc s.

Let us define a **standard arc** s as an arc on the unit circle which starts at the point $(1, 0)$ and travels s units counterclockwise if s is positive, and s units clockwise if s is negative. Since the circumference C of the unit circle is

$$C = 2\pi r = 2\pi(1) = 2\pi,$$

arcs whose lengths are rational multiples of π (e.g., $\dfrac{\pi}{2}, -\dfrac{3\pi}{4}, 5\pi$) are easy to visualize.

Example 2 Represent each of the following real numbers as a standard arc.

(a) π (b) $\dfrac{\pi}{2}$ (c) $\dfrac{\pi}{3}$ (d) $-\dfrac{3\pi}{2}$ (e) $-\dfrac{2\pi}{3}$ (f) $\dfrac{13\pi}{6}$

Solution

(a)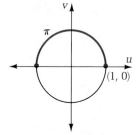

$\pi = \tfrac{1}{2}(2\pi)$, which is $\tfrac{1}{2}$ of the circumference.

(b)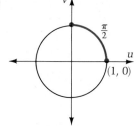

$\dfrac{\pi}{2} = \tfrac{1}{4}(2\pi)$, which is $\tfrac{1}{4}$ of the circumference.

(c)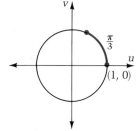

$\dfrac{\pi}{3} = \tfrac{1}{6}(2\pi)$, which is $\tfrac{1}{6}$ of the circumference.

(d)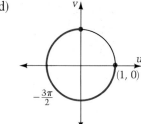

$\dfrac{3\pi}{2}$ is $\tfrac{3}{4}$ of the circumference. Since $-\dfrac{3\pi}{2} < 0$, the arc travels clockwise.

(e)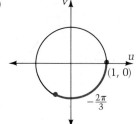

$\dfrac{2\pi}{3}$ is $\tfrac{1}{3}$ of the circumference. Since $-\dfrac{2\pi}{3} < 0$, the arc travels clockwise.

(f)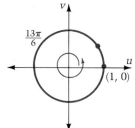

$\dfrac{13\pi}{6}$ is $1\tfrac{1}{12}$ of the circumference.

Goal B *to determine the radian measure of an angle*

We have discussed the trigonometric functions of angles measured in degrees and used these functions to solve some interesting problems. However, to use trigonometric functions to solve other types of problems, we need to define them in such a way that their domains consist of real numbers (with no units attached).

8 / Trigonometric Functions

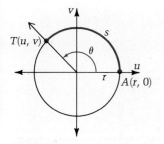

Figure 8.3

Figure 8.3 shows a positive angle θ in standard position, and a circle of radius r with center at the origin. The angle θ is called a **central angle** since its vertex is the center of the circle. We say that angle θ "subtends" or "cuts" an arc of the circle. Let s be the length of the arc.

Notice that $A(r, 0)$ is the point where the initial side of angle θ intersects the circle. Furthermore, $T(u, v)$ is the point where the terminal side of angle θ intersects the circle, and s is the *length* of the arc cut by angle θ. We use s, r, and θ, as just described, to make the following definition.

Definition

The **radian measure** of a positive central angle θ is defined as the ratio of the arclength s to the radius r:

$$\theta = \frac{s}{r}.$$

Since both s and r have the same units of length, their ratio θ will have no units. However, we will say that the angle has a measure of θ "radians" to remind us that it was measured by using the radius. If θ is a negative angle, its radian measure is defined to be negative as Figure 8.4 suggests.

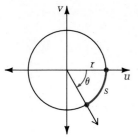

Figure 8.4 $\theta = -\frac{s}{r}$

In a unit circle the radian measure of a positive central angle is equal to the *length* of the arc that it cuts. This is shown in Figure 8.5.

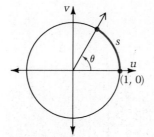

■ Since $r = 1$, $\theta = \frac{s}{r} = \frac{s}{1}$. Thus, $\theta = s$.

Figure 8.5

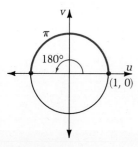

Figure 8.6

Now consider a central angle of 180° in a unit circle as shown in Figure 8.6. Since an angle of 180° cuts an arc of length π (half the circumference), the above statement suggests that an angle of 180° has a radian measure of π. That is,

$$180° = \pi \text{ radians.}$$

We use this relationship in setting up a proportion for converting the degree measure of an angle to radian measure and vice versa.

degrees	radians
0°	0
30°	$\frac{\pi}{6}$
45°	$\frac{\pi}{4}$
90°	$\frac{\pi}{2}$

Fact

If d is the degree measure of an angle and α is its radian measure, then

$$\frac{d}{180°} = \frac{\alpha}{\pi}.$$

Example 3 (a) Convert 165° to radians.
(b) Convert $-\frac{3\pi}{4}$ radians to degrees.
(c) Convert 3 radians to degrees.
(d) Convert 40°36′ to radians.

Solution (a) $\frac{165°}{180°} = \frac{\alpha}{\pi}$ ■ Set up the proportion.

$$\alpha = \frac{165}{180}\pi$$

$$= \frac{11}{12}\pi \text{ radians}$$

(b) $\frac{d}{180°} = \frac{-\frac{3\pi}{4}}{\pi}$

$$d = -\frac{3\pi}{4} \cdot \frac{1}{\pi} \cdot 180° = -135°$$

(c) $\frac{d}{180°} = \frac{3}{\pi}$

$$d = \frac{3 \times 180°}{\pi} \approx 172°$$ ■ $\pi \approx 3.14$.

(d) Since 1° = 60′, we have $36' = \left(\frac{36'}{60'}\right)° = (0.6)°$. Thus, $40°36' = (40 + 0.6)° = 40.6°$.

$$\frac{d}{180°} = \frac{\alpha}{\pi}$$

$$\frac{40.6°}{180°} = \frac{\alpha}{\pi}$$

$$\alpha = \frac{40.6}{180} \cdot \pi \approx 0.71 \text{ radians}$$

8 / Trigonometric Functions

Goal C *to solve problems using radian measure*

Since the definition of radian measure depends on the radius and arc of a circle, a variety of problems involving circular arcs can be solved using radian measure.

Example 4 Determine the length of an arc cut by a central angle of 60° in a circle of radius 2 yd.

Solution $\theta = \dfrac{s}{r}$ ■ To use this formula, 60° must first be converted to radians:
$\dfrac{60°}{180°} = \dfrac{\theta}{\pi}$, thus $\theta = \dfrac{60}{180}\pi = \dfrac{\pi}{3}$.

$\dfrac{\pi}{3} = \dfrac{s}{2}$ ■ $\dfrac{\pi}{3}$ is substituted for θ.

$s = \dfrac{2\pi}{3}$

The arc has length $\dfrac{2\pi}{3}$ yd $\approx$ 2.09 yd (rounded to two decimal places).

Example 5 A wheel of radius 80 cm rolls along the ground without slipping and rotates through an angle of 45°. How far does the wheel move?

Solution

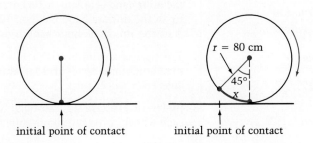

initial point of contact   initial point of contact

The distance that the wheel advances is equal to the length of the arc shown in the right-hand figure above. We use this fact to find s.

$\theta = \dfrac{s}{r}$ ■ To use this formula, θ must be measured in radians.
$\dfrac{45°}{180°} = \dfrac{\theta}{\pi}$, thus $\theta = \dfrac{45}{180}\pi = \dfrac{\pi}{4}$.

$\dfrac{\pi}{4} = \dfrac{s}{80}$

$s = 80 \cdot \dfrac{\pi}{4} = 20\pi$

The wheel has moved 20π cm ($\approx$63 cm) to the right.

Exercise Set 8.1   ▣ Calculator Exercises in the Appendix

Goal A

In exercises 1–4, determine whether the point is on the unit circle.

1. $\left(-\frac{\sqrt{3}}{2}, -\frac{1}{2}\right)$
2. $\left(-\frac{\sqrt{3}}{3}, -\frac{1}{3}\right)$
3. $\left(\frac{\sqrt{2}}{2}, -\frac{1}{2}\right)$
4. $\left(-\frac{\sqrt{3}}{2}, \frac{\sqrt{2}}{2}\right)$

In exercises 5–12, represent the real number as a standard arc.

5. $\frac{\pi}{4}$
6. $\frac{3\pi}{2}$
7. $\frac{2\pi}{3}$
8. $\frac{5\pi}{4}$
9. $-\frac{\pi}{4}$
10. $-\frac{7\pi}{4}$
11. $\frac{25\pi}{6}$
12. $-\frac{11\pi}{3}$

Goal B

In exercises 13–34, convert the measure of the angle from degrees to radians or vice versa.

13. 60°
14. 45°
15. 315°
16. 360°
17. 210°
18. 330°
19. −30°
20. −15°
21. 75°
22. 300°
23. −540°
24. −200°
25. $\frac{3\pi}{2}$
26. $\frac{\pi}{3}$
27. $\frac{5\pi}{6}$
28. $\frac{5\pi}{2}$
29. −3π
30. $-\frac{3\pi}{4}$
31. $-\frac{25\pi}{6}$
32. $-\frac{19\pi}{9}$
33. 1.5
34. 5

Goal C

35. Determine the length of an arc cut by a central angle of 90° in a circle of radius 4 in.

36. Determine the length of an arc cut by a central angle of 30° in a circle of radius 2 cm.

37. A central angle of 1.5 radians cuts an arc of length 12 m. Find the radius of the circle.

38. A central angle of 2.5 radians cuts an arc of 25 ft. Find the radius of the circle.

39. What is the measure in radians of the angle through which the minute hand of a clock turns in 42 min?

40. What is the measure of the angle in degrees through which the minute hand of a clock turns between 1:30 P.M. and 2:20 P.M. of the same day?

Superset

41. Find the number of radians through which each of the hands of a clock move in (a) 12 hours, (b) in one hour, (c) in 30 min, (d) in 5 min.

42. After midnight, at what time are the hands of a clock first perpendicular to each other?

43. The minute hand of a clock is of length 5 cm and the hour hand is of length 3.6 cm. How far does the tip of the minute hand move in 12 hours? How far does the tip of the hour hand move in one hour?

44. The tires of an automobile are 24 in in diameter. If the automobile backs up 20 ft, through what angle does each wheel turn?

45. (Refer to exercise 44) Through what angle does each wheel turn if the automobile travels one mile?

46. Through what angle does a water wheel of radius 32 ft turn if it rotates counterclockwise at the rate of 2 mi per hour for 1 min?

47. Suppose θ is an angle in standard position. Determine the coordinates of the point of intersection of the terminal side of θ with the unit circle if (a) $\theta = \frac{\pi}{6}$, (b) $\theta = \frac{\pi}{4}$, (c) $\theta = \frac{\pi}{3}$.

8.2 Trigonometric Functions of Real Numbers

Goal A *to evaluate trigonometric functions whose domains are real numbers*

One of the central ideas in mathematics is the concept of function. Of special interest to us are those functions having the set of real numbers as domain. Radian measure provides a means for defining the trigonometric functions so that their domains are sets of real numbers.

The transition from angle measurements to real numbers as domain values is made possible by noticing the following:

> every real number x can be associated uniquely with the central angle θ that cuts standard arc x on the unit circle.

This result is illustrated by the following procedure. To determine the angle θ associated with the real number x, begin by representing x as a standard arc. This arc terminates at a unique point $T(u, v)$, and is cut by a unique central angle θ, having $T(u, v)$ on its terminal side. We have thus associated angle θ with the real number x.

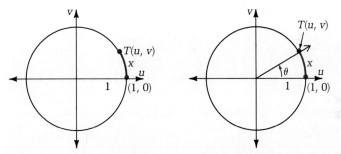

(a) Represent x as a standard arc.

(b) θ is the unique angle that cuts the standard arc x.

Figure 8.7

Recall that we previously defined the trigonometric functions of θ in terms of the coordinates of the point $T(u, v)$. Since $r = 1$, we have

$$\theta = \frac{x}{1} = x.$$

We define the trigonometric functions of the real number x to be the same as the trigonometric functions of angle θ.

Definition

Let x be any real number, let θ be the central angle in standard position in the unit circle having radian measure x, and let $T(u, v)$ be the terminal point on standard arc x. Then

$\sin x = \sin \theta = v$ domain: all real numbers

$\cos x = \cos \theta = u$ domain: all real numbers

$\tan x = \tan \theta = \dfrac{v}{u}$ domain: all real numbers except odd multiples of $\dfrac{\pi}{2}$ (which make denominator $u = 0$).

$\cot x = \cot \theta = \dfrac{u}{v}$ domain: all real numbers except integral multiples of π (which make denominator $v = 0$).

$\sec x = \sec \theta = \dfrac{1}{u}$ domain: all real numbers except odd multiples of $\dfrac{\pi}{2}$.

$\csc x = \csc \theta = \dfrac{1}{v}$ domain: all real numbers except integral multiples of π.

Figure 8.8

Example 1   Evaluate each of the following expressions.

(a) $\sin \dfrac{\pi}{2}$ (b) $\cos(-\pi)$ (c) $\tan \dfrac{3\pi}{2}$ (d) $\sin 0$

Solution   (a)

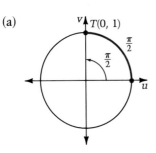

- Draw a unit circle and standard arc $\dfrac{\pi}{2}$. (Note: $\dfrac{\pi}{2}$ is $\dfrac{1}{4}$ of the circumference.)
- $\sin \dfrac{\pi}{2}$ = the second coordinate of $T(0, 1) = 1$.

(b)

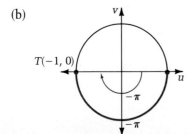

- Draw a unit circle and standard arc $-\pi$.
- $\cos(-\pi)$ = the first coordinate of $T(-1, 0)$, $= -1$.

(c)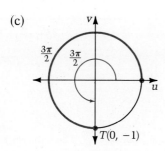

- Draw a unit circle and standard arc $\frac{3\pi}{2}$.
- $\tan \frac{3\pi}{2} = \frac{\text{the second coordinate of } T(0, -1)}{\text{the first coordinate of } T(0, -1)} = \frac{-1}{0}$ which is not defined.

(d) The point (1, 0) is the terminal point of standard arc 0 on the unit circle.

$$\sin 0 = \text{the second coordinate of } T(1, 0) = 0$$

In Example 1 we saw that trigonometric function values of integral multiples of $\frac{\pi}{2}$ are easily determined, since the corresponding arcs terminate at one of the four points (1, 0), (0, 1), (−1, 0), or (0, −1). Three other special arcs and their terminal points are shown on the unit circles below.

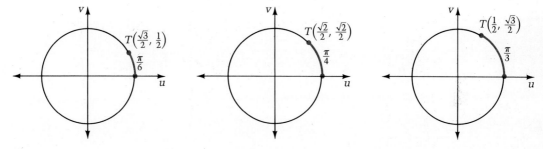

Figure 8.9

Knowing the coordinates of these terminal points allows us to determine trigonometric function values of $\frac{\pi}{6}, \frac{\pi}{4}$, and $\frac{\pi}{3}$. For example,

$$\sin \frac{\pi}{4} = \text{the second coordinate of } \left(\frac{\sqrt{2}}{2}, \frac{\sqrt{2}}{2}\right) = \frac{\sqrt{2}}{2},$$

$$\sec \frac{\pi}{3} = \frac{1}{\text{the first coordinate of } \left(\frac{1}{2}, \frac{\sqrt{3}}{2}\right)} = \frac{1}{\frac{1}{2}} = 2.$$

In addition, we can determine trigonometric function values of some other real numbers by reflecting these three points through the origin or across either axis. This is demonstrated in the next example.

Example 2 Evaluate each of the following expressions.

(a) $\cos \dfrac{5\pi}{6}$ (b) $\sin\left(-\dfrac{\pi}{4}\right)$ (c) $\tan \dfrac{4\pi}{3}$

Solution (a)

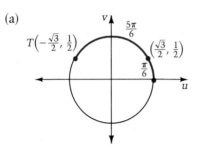

- Draw the standard arc $\dfrac{5\pi}{6}$. The terminal point is found by reflecting the terminal point of $\dfrac{\pi}{6}$ across the vertical axis. Thus,

$$\cos \dfrac{5\pi}{6} = \text{the first coordinate of } \left(-\dfrac{\sqrt{3}}{2}, \dfrac{1}{2}\right) = -\dfrac{\sqrt{3}}{2}.$$

Thus, $\cos \dfrac{5\pi}{6} = -\dfrac{\sqrt{3}}{2}$.

(b)

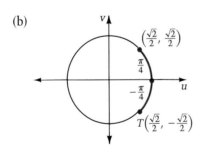

- Draw the standard arc $-\dfrac{\pi}{4}$. The terminal point is found by reflecting the terminal point of $\dfrac{\pi}{4}$ across the horizontal axis. Thus, $\sin\left(-\dfrac{\pi}{4}\right) = $ the second coordinate of $\left(\dfrac{\sqrt{2}}{2}, -\dfrac{\sqrt{2}}{2}\right) = -\dfrac{\sqrt{2}}{2}$.

Thus, $\sin\left(-\dfrac{\pi}{4}\right) = -\dfrac{\sqrt{2}}{2}$.

(c)

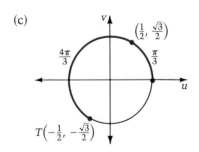

- Draw the standard arc $\dfrac{4\pi}{3}$. The terminal point is found by reflecting the terminal point of $\dfrac{\pi}{3}$ through the origin. Thus,

$$\tan \dfrac{4\pi}{3} = \dfrac{\text{the second coordinate of } \left(-\dfrac{1}{2}, -\dfrac{\sqrt{3}}{2}\right)}{\text{the first coordinate of } \left(-\dfrac{1}{2}, -\dfrac{\sqrt{3}}{2}\right)}$$

$$= \dfrac{-\dfrac{\sqrt{3}}{2}}{-\dfrac{1}{2}} = \sqrt{3}$$

Thus, $\tan \dfrac{4\pi}{3} = \sqrt{3}$.

By virtue of the method used in the previous two examples, we can update the Table of Special Angles from the previous chapter.

Table of Special Values

x $f(x)$	0 (or 0°)	$\frac{\pi}{6}$ (or 30°)	$\frac{\pi}{4}$ (or 45°)	$\frac{\pi}{3}$ (or 60°)	$\frac{\pi}{2}$ (or 90°)
sin x	0	$\frac{1}{2}$	$\frac{\sqrt{2}}{2}$	$\frac{\sqrt{3}}{2}$	1
cos x	1	$\frac{\sqrt{3}}{2}$	$\frac{\sqrt{2}}{2}$	$\frac{1}{2}$	0
tan x	0	$\frac{\sqrt{3}}{3}$	1	$\sqrt{3}$	*
cot x	*	$\sqrt{3}$	1	$\frac{\sqrt{3}}{3}$	0
sec x	1	$\frac{2\sqrt{3}}{3}$	$\sqrt{2}$	2	*
csc x	*	2	$\sqrt{2}$	$\frac{2\sqrt{3}}{3}$	1

* undefined

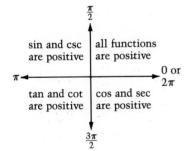

Figure 8.10

With a few simple changes, you can use the "reference angle" technique developed in the previous chapter to determine trigonometric function values of numbers outside the interval $[0, \frac{\pi}{2}]$. To find the trigonometric function value, attach the appropriate sign to the trigonometric function value of the reference value of x. The sign of your answer will depend on the quadrant in which the standard arc x terminates.

For any x between 0 and 2π, the reference value for x is determined as follows. (Recall that $180° = \pi$ radians.)

x	reference value for x
between 0 and $\frac{\pi}{2}$	x
between $\frac{\pi}{2}$ and π	$\pi - x$
between π and $\frac{3\pi}{2}$	$x - \pi$
between $\frac{3\pi}{2}$ and 2π	$2\pi - x$

8 / Trigonometric Functions

To determine the trigonometric function value of a number outside the interval $[0, 2\pi)$, begin by adding or subtracting an appropriate multiple of 2π to produce a value between 0 and 2π.

Example 3 Determine each of the following function values.

(a) $\cos \dfrac{7\pi}{6}$ (b) $\tan\left(-\dfrac{\pi}{4}\right)$ (c) $\csc \dfrac{20\pi}{3}$

Solution

(a) $\cos \dfrac{7\pi}{6} = \boxed{\text{sign}}\cos \boxed{\text{reference value for } \dfrac{7\pi}{6}}$

$= -\cos \boxed{\text{reference value for } \dfrac{7\pi}{6}}$ ■ Standard arc $\dfrac{7\pi}{6}$ terminates in the third quadrant, where the cosine is negative.

$= -\cos \dfrac{\pi}{6} = -\dfrac{\sqrt{3}}{2}$ ■ Reference value: $\dfrac{7\pi}{6} - \pi = \dfrac{\pi}{6}$.

(b) $\tan\left(-\dfrac{\pi}{4}\right) = \tan \dfrac{7\pi}{4}$ ■ Add 2π to $-\dfrac{\pi}{4}$ to produce a value between 0 and 2π.

$= \boxed{\text{sign}}\tan \boxed{\text{reference value for } \dfrac{7\pi}{4}}$ ■ Standard arc $\dfrac{7\pi}{4}$ terminates in the fourth quadrant, where the tangent is negative.

$= -\tan \dfrac{\pi}{4} = -1$ ■ Reference value: $2\pi - \dfrac{7\pi}{4} = \dfrac{\pi}{4}$.

(c) $\csc \dfrac{20\pi}{3} = \csc \dfrac{2\pi}{3}$ ■ Subtract $3 \cdot 2\pi$ from $\dfrac{20\pi}{3}$ to produce a value between 0 and 2π.

$= \boxed{\text{sign}}\csc \boxed{\text{reference value for } \dfrac{2\pi}{3}}$

$= +\csc \boxed{\text{reference value for } \dfrac{2\pi}{3}}$ ■ Standard arc $\dfrac{2\pi}{3}$ terminates in the second quadrant, where the cosecant is positive.

$= +\csc \dfrac{\pi}{3} = \dfrac{2\sqrt{3}}{3}$ ■ Reference value: $\pi - \dfrac{2\pi}{3} = \dfrac{\pi}{3}$.

8 / Trigonometric Functions

To evaluate the trigonometric function values of a real number x that is not associated with a special arc, we can use Table 4. The column headed by the word "radians" contains the values of x.

Example 4 Determine each of the following function values. Round your answer to two decimal places.

(a) $\cos 3.3$ (b) $\sec \dfrac{21\pi}{5}$

Solution

(a) $\cos 3.3 = \boxed{\text{sign}} \cos \boxed{\text{reference value for 3.3}}$
- $\cos 3.3$ means "the cosine of 3.3 radians."

$= -\cos \boxed{\text{reference value for 3.3}}$
- Since 3.3 is between π (≈ 3.14) and $\dfrac{3\pi}{2}$ (≈ 4.71), standard arc 3.3 terminates in the third quadrant, where cosine is negative.

$\approx -\cos 0.16$
- Reference value: $3.3 - \pi \approx 0.16$.

≈ -0.99
- Table 4

(b) $\sec \dfrac{21\pi}{5} = \sec \dfrac{\pi}{5}$
- Subtract $2 \cdot 2\pi$ from $\dfrac{21\pi}{5}$ to produce a value between 0 and 2π.

$= \boxed{\text{sign}} \sec \boxed{\text{reference value for } \dfrac{\pi}{5}}$

$= +\sec \boxed{\text{reference value for } \dfrac{\pi}{5}}$
- Standard arc $\dfrac{\pi}{5}$ terminates in the first quadrant where all trigonometric functions are positive.

$= +\sec \dfrac{\pi}{5}$
- Reference value: $\dfrac{\pi}{5}$

$\approx \sec 0.63$
- $\dfrac{\pi}{5} \approx 0.63$

≈ 1.24
- Table 4

Exercise Set 8.2 = Calculator Exercises in the Appendix

Goal A
In exercises 1–10, determine the function value.

1. $\cot 0$
2. $\csc \dfrac{\pi}{2}$
3. $\tan 2\pi$
4. $\csc 2\pi$
5. $\sin \dfrac{\pi}{3}$
6. $\tan \dfrac{\pi}{4}$
7. $\csc \dfrac{\pi}{6}$
8. $\sec \dfrac{\pi}{6}$
9. $\cot \dfrac{\pi}{4}$
10. $\cos \dfrac{\pi}{3}$

In exercises 11–30, determine the function value.

11. $\cos \dfrac{2\pi}{3}$
12. $\cot \dfrac{5\pi}{3}$
13. $\sec\left(-\dfrac{\pi}{6}\right)$
14. $\tan\left(-\dfrac{5\pi}{6}\right)$
15. $\tan \dfrac{3\pi}{4}$
16. $\csc \dfrac{5\pi}{4}$
17. $\sin\left(-\dfrac{3\pi}{4}\right)$
18. $\sec\left(-\dfrac{5\pi}{4}\right)$
19. $\cot\left(-\dfrac{7\pi}{6}\right)$
20. $\sin\left(-\dfrac{4\pi}{3}\right)$
21. $\csc\left(-\dfrac{7\pi}{4}\right)$
22. $\sec\left(-\dfrac{11\pi}{6}\right)$
23. $\cos \dfrac{5\pi}{2}$
24. $\sin 5\pi$
25. $\cot \dfrac{10\pi}{3}$
26. $\csc \dfrac{9\pi}{4}$
27. $\tan\left(-\dfrac{7\pi}{3}\right)$
28. $\csc\left(-\dfrac{11\pi}{4}\right)$
29. $\sec \dfrac{25\pi}{6}$
30. $\cos\left(-\dfrac{21\pi}{4}\right)$

In exercises 31–38, determine the function value. Round to two decimal places.

31. $\sin 3.7$
32. $\tan 5$
33. $\sec 8.5$
34. $\cos 10.5$
35. $\csc \dfrac{12\pi}{5}$
36. $\sin \dfrac{5\pi}{7}$
37. $\tan -\dfrac{\pi}{9}$
38. $\cot -\dfrac{4\pi}{7}$

Superset
In exercises 39–44, verify that the given statement is true.

39. $\sin \dfrac{13\pi}{6} \sec \dfrac{5\pi}{3} = \tan \dfrac{9\pi}{4}$
40. $\tan \dfrac{4\pi}{3} \tan \dfrac{7\pi}{6} = \tan \dfrac{13\pi}{4}$
41. $\tan \dfrac{\pi}{3} = \dfrac{1 - \cos \dfrac{4\pi}{3}}{\sin \dfrac{\pi}{3}}$
42. $\tan \dfrac{\pi}{6} = \dfrac{1 - \cos \dfrac{\pi}{3}}{\sin \dfrac{\pi}{3}}$
43. $\cos \dfrac{5\pi}{6} = \cos \dfrac{\pi}{2} \cos \dfrac{\pi}{3} - \sin \dfrac{\pi}{2} \sin \dfrac{\pi}{3}$
44. $\sin \dfrac{\pi}{3} = 2 \sin \dfrac{\pi}{6} \cos \dfrac{\pi}{6}$

In exercises 45–50, solve for the smallest positive value of x.

45. $\cot x = -1$
46. $\csc x = -\sqrt{2}$
47. $\sec x = -\dfrac{2\sqrt{3}}{3}$
48. $\tan x = -\sqrt{3}$
49. $\sec(-x) = \sqrt{2}$
50. $\csc(-x) = -2$

8 / Trigonometric Functions

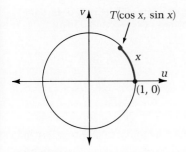

Figure 8.11

8.3 The Trigonometric Functions: Basic Graphs and Properties

Goal A *to graph the sine and cosine functions*

We have seen that every real number x uniquely determines a standard arc on the unit circle. The coordinates of the terminal point T of this arc are used to define the six trigonometric functions of the real number x. The first coordinate of T is $\cos x$ and the second coordinate is $\sin x$. There are many different standard arcs (and thus many different values of x) that are associated with the same terminal point T. Fortunately, these values of x occur in a predictable, or *periodic* way.

Figure 8.12 demonstrates that $\frac{\pi}{3}$, $\frac{7\pi}{3}$, and $\frac{13\pi}{3}$ all determine the same terminal point $(\frac{1}{2}, \frac{\sqrt{3}}{2})$. Adding any positive or negative multiple of 2π to $\frac{\pi}{3}$ adds one or more complete revolutions to the arc. The resulting standard arc still has $(\frac{1}{2}, \frac{\sqrt{3}}{2})$ as its terminal point. This means that $\sin \frac{\pi}{3}$, $\sin \frac{7\pi}{3}$, and $\sin \frac{13\pi}{3}$ are the same: $\frac{\sqrt{3}}{2}$. Thus, the sine function repeats the same value each time the domain value changes by 2π. Functions exhibiting such repetitive behavior are called *periodic*.

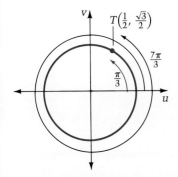

Definition

A function f is called a **periodic function** if there is a positive real number p such that

$$f(x) = f(x + p)$$

for all x in the domain of f; the smallest such positive number p is called the **period** of the function.

Both the sine and cosine functions are periodic with period 2π. For that reason we write

$$\sin x = \sin(x + n \cdot 2\pi), \quad \text{for any integer } n, \text{ and}$$
$$\cos x = \cos(x + n \cdot 2\pi), \quad \text{for any integer } n.$$

To graph the sine function, we shall first study its behavior for x-values from 0 to 2π, and thus produce one complete cycle of the graph. Then, since the function has period 2π, the pattern observed in the graph for values between 0 and 2π will be repeated over and over.

Recall that the sine function is defined by means of the second coordinate of a standard arc's terminal point T. Thus, values on the horizontal axis of the graph of the sine function correspond to standard arcs, and values on the vertical axis correspond to the second coordinates of the arcs' terminal points.

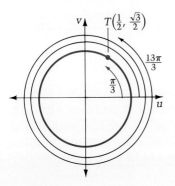

Figure 8.12

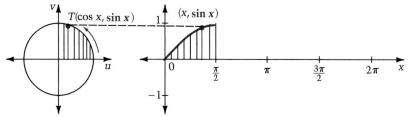
- As x increases from 0 to $\frac{\pi}{2}$, second coordinates ("heights") of terminal points increase from 0 to 1.

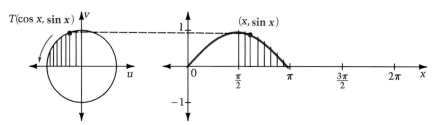
- As x increases from $\frac{\pi}{2}$ to π, second coordinates decrease from 1 to 0.

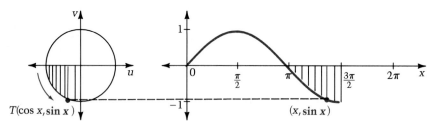
- As x increases from π to $\frac{3\pi}{2}$, second coordinates decrease from 0 to -1.

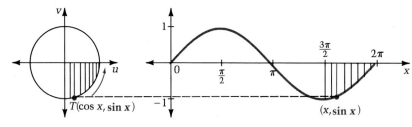
- As x increases from $\frac{3\pi}{2}$ to 2π, second coordinates increase from -1 to 0.

Figure 8.13 Constructing one cycle of the graph of $y = \sin x$.

To get a more precise graph of the sine function, we construct a table of special values of x, plot the corresponding ordered pairs, and draw a smooth curve through the points. Since the sine function is periodic with period 2π, the portion of the graph between 0 and 2π will repeat every 2π units. The complete graph is shown below. You should be able to sketch this graph quickly from memory.

8 / Trigonometric Functions

x	0	$\frac{\pi}{6}$	$\frac{\pi}{4}$	$\frac{\pi}{3}$	$\frac{\pi}{2}$	$\frac{3\pi}{4}$	π	$\frac{5\pi}{4}$	$\frac{3\pi}{2}$	$\frac{7\pi}{4}$	2π
$\sin x$	0	$\frac{1}{2}$	$\frac{\sqrt{2}}{2}$	$\frac{\sqrt{3}}{2}$	1	$\frac{\sqrt{2}}{2}$	0	$-\frac{\sqrt{2}}{2}$	-1	$-\frac{\sqrt{2}}{2}$	0

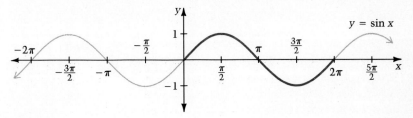

Figure 8.14

Properties of the Sine Function
Domain   the set of all real numbers
Range   the set of real numbers between -1 and 1 inclusive
Period 2π
Symmetry with respect to the origin; thus sine is an "odd" function, that is $\sin(-x) = -\sin x$.

We now turn our attention to the cosine function. Recall that the cosine is defined by means of the first coordinate of an arc's terminal point. Thus, values on the horizontal axis of the graph of the cosine function correspond to standard arcs, and values on the vertical axis correspond to the first coordinates of the arcs' terminal points. For example, to graph the cosine function for values of x between 0 and $\frac{\pi}{2}$, we notice that as standard arc x increases from 0 to $\frac{\pi}{2}$, first coordinates of the terminal points *decrease* from 1 to 0.

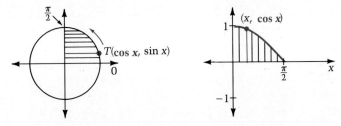

Figure 8.15 Constructing the graph of $y = \cos x$ for $0 \le x \le \frac{\pi}{2}$.

To get a more precise graph of the cosine function, we construct a table of special values of x, plot the corresponding ordered pairs, and draw a smooth curve through the points. We use the fact that the cosine function is periodic with period 2π to complete the graph. Note that the graph of the cosine function is symmetric with respect to the y-axis.

x	0	$\dfrac{\pi}{6}$	$\dfrac{\pi}{4}$	$\dfrac{\pi}{3}$	$\dfrac{\pi}{2}$	$\dfrac{3\pi}{4}$	π	$\dfrac{5\pi}{4}$	$\dfrac{3\pi}{2}$	$\dfrac{7\pi}{4}$	2π
$\cos x$	1	$\dfrac{\sqrt{3}}{2}$	$\dfrac{\sqrt{2}}{2}$	$\dfrac{1}{2}$	0	$-\dfrac{\sqrt{2}}{2}$	-1	$-\dfrac{\sqrt{2}}{2}$	0	$\dfrac{\sqrt{2}}{2}$	1

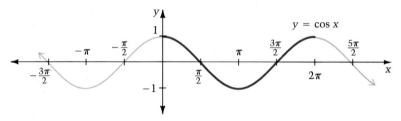

Figure 8.16

Properties of the Cosine Function

Domain   the set of all real numbers
Range   the set of real numbers between -1 and 1 inclusive
Period 2π
Symmetry with respect to the y-axis; thus the cosine is an "even" function, that is, $\cos(-x) = \cos x$.

Example 1   (a) Graph $y = \sin x$ and $y = \sin(-x)$ on the same set of axes.

(b) Graph $y = \cos x$ and $y = \cos\left(x + \dfrac{\pi}{2}\right)$ on the same set of axes.

Solution   (a) To obtain the graph of $y = \sin(-x)$, reflect the graph of $y = \sin x$ across the y-axis.

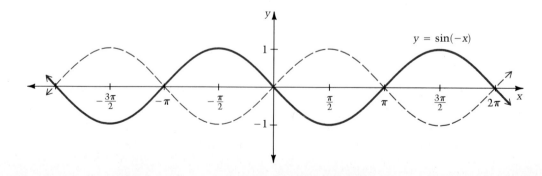

(b) To obtain the graph of $y = \cos\left(x + \frac{\pi}{2}\right)$, translate the graph of $y = \cos x$ $\frac{\pi}{2}$ units to the left.

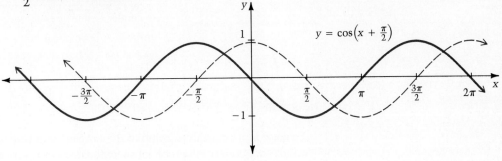

Goal B   *to graph the other trigonometric functions*

The graphs of the tangent and cotangent functions look very different from those of the sine and cosine functions. To get an idea of what the graph of $y = \tan x$ looks like, recall that

$$\tan x = \frac{\sin x}{\cos x},$$

and thus, the tangent function will not be defined when the denominator, $\cos x$, is 0. Since $\cos x$ is 0 for all x in the set

$$\left\{\ldots -\frac{5\pi}{2}, -\frac{3\pi}{2}, -\frac{\pi}{2}, \frac{\pi}{2}, \frac{3\pi}{2}, \frac{5\pi}{2}, \ldots\right\},$$

the tangent function is not defined for any of these values. We begin the graph of the tangent function by drawing the vertical asymptotes at each of these values.

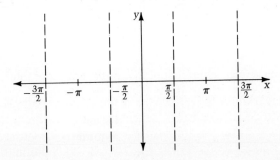

Figure 8.17

Notice that the tangent function is defined for all numbers between $-\frac{\pi}{2}$ and $\frac{\pi}{2}$. Let us consider what the graph will look like on that interval.

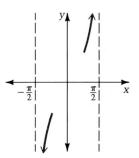

- As x increases to $\frac{\pi}{2}$, sin x is nearly 1, and cos x is nearly 0, so that tan x $\left(\text{which equals } \frac{\sin x}{\cos x}\right)$ becomes a very large positive number.
- A similar argument establishes that as x decreases towards $-\frac{\pi}{2}$, sin x is nearly -1, cos x is nearly 0, and so tan x takes on negative numbers with large absolute values.

Figure 8.18

To get a more complete graph of the tangent function for values between $-\frac{\pi}{2}$ and $\frac{\pi}{2}$, we construct a table of special values of x, plot the corresponding ordered pairs, and draw a smooth curve through the points. This curve is then repeated within each pair of adjacent asymptotes.

x	$-\frac{\pi}{2}$	$-\frac{\pi}{3}$	$-\frac{\pi}{4}$	$-\frac{\pi}{6}$	0	$\frac{\pi}{6}$	$\frac{\pi}{4}$	$\frac{\pi}{3}$	$\frac{\pi}{2}$
tan x	*	$-\sqrt{3}$	-1	$-\frac{\sqrt{3}}{3}$	0	$\frac{\sqrt{3}}{3}$	1	$\sqrt{3}$	*

*undefined

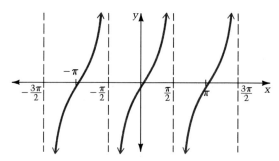

Figure 8.19 $y = \tan x$

Properties of the Tangent Function

Domain all real numbers *except* $x = \frac{\pi}{2} + k\pi$, for any integer k

Range all real numbers

Period π

Symmetry with respect to the origin; tangent is an "odd" function, thus $\tan(-x) = -\tan x$.

Asymptotes vertical asymptotes $x = \frac{\pi}{2} + k\pi$, for any integer k.

8 / Trigonometric Functions

The graph of $y = \tan x$ is periodic, but unlike the sine and cosine functions, the period is π. Thus, the portion of the graph between $-\frac{\pi}{2}$ and $\frac{\pi}{2}$ repeats itself over and over as suggested in Figure 8.19.

The graph of the cotangent function can be found by using a method similar to that used to derive the graph of $y = \tan x$. Recall that

$$\cot x = \frac{1}{\tan x}.$$

This fact suggests that the graph of $y = \cot x$ becomes infinite (has a vertical asymptote) where $\tan x$ is 0, and is zero where the graph of $y = \tan x$ has vertical asymptotes. The complete graph is shown in Figure 8.20. Like the tangent function, the cotangent function has period π. In addition, the graph of the cotangent function, like that of the tangent function, is symmetric with respect to the origin.

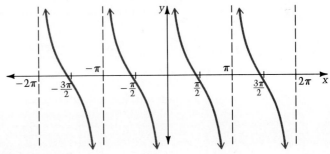

Figure 8.20 $y = \cot x$

Properties of the Cotangent Function
Domain   all real numbers *except* $x = k\pi$, for any integer k
Range   all real numbers
Period π
Symmetry   with respect to the origin; cotangent is an "odd" function, thus $\cot(-x) = -\cot x$.
Asymptotes vertical asymptotes $x = k\pi$, for any integer k.

Example 2   Verify that the graphs of the tangent and cotangent functions are symmetric with respect to the origin.

Solution   Recall that to verify that the graph of a function $y = f(x)$ is symmetric with respect to the origin, we must show that when y is replaced with $-y$ and x is replaced with $-x$, the resulting equation is equivalent to $y = f(x)$.

$$y = \tan x = \frac{\sin x}{\cos x}$$

$$-y = \frac{\sin(-x)}{\cos(-x)}$$ ■ x is replaced with $-x$ and y is replaced with $-y$.

$$-y = \frac{-\sin x}{\cos x}$$ ■ Recall that $\sin(-x) = -\sin x$ and $\cos(-x) = \cos x$.

$$y = \frac{\sin x}{\cos x}$$ ■ Both sides have been multiplied by -1.

$$y = \tan x$$

Thus, the graph of $y = \tan x$ is symmetric with respect to the origin. Symmetry with respect to the origin for the graph of $y = \cot x$ is established similarly.

The graph of the secant function can be easily produced by recalling that

$$\sec x = \frac{1}{\cos x}, \text{ for } \cos x \neq 0.$$

Since the secant function is not defined for values of x where $\cos x$ is 0, its domain does not include any value in the set

$$\left\{ \ldots, -\frac{5\pi}{2}, -\frac{3\pi}{2}, -\frac{\pi}{2}, \frac{\pi}{2}, \frac{3\pi}{2}, \frac{5\pi}{2}, \ldots \right\},$$

and its graph has vertical asymptotes at these values. Since the secant function can be thought of as the reciprocal of the cosine function, it is helpful to graph the cosine function as a dashed curve, and then generate the graph of the secant function by viewing the y-values of the secant function as reciprocals of the y-values of the cosine function.

■ Notice that wherever the cosine function is 0, the secant function is undefined, and wherever the cosine function is 1, so is the secant function. Furthermore, whenever the cosine function is positive (negative), the secant function is positive (negative).

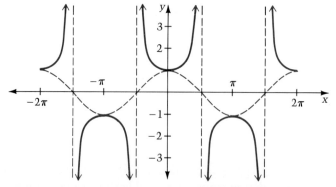

Figure 8.21 $y = \sec x$ ($y = \cos x$ is shown as a dashed curve).

8 / Trigonometric Functions

Properties of the Secant Function

Domain all real numbers *except* $x = \frac{\pi}{2} + k\pi$, for any integer k

Range all real numbers y such that $y \leq -1$ or $y \geq 1$

Period 2π

Symmetry symmetry with respect to the y-axis (an even function, like its reciprocal, cosine)

Asymptotes vertical asymptotes $x = \frac{\pi}{2} + k\pi$ (values where cosine is 0)

Note that the secant function has period 2π, the same as that of its reciprocal, the cosine function. Moreover, like the cosine function, the graph of the secant function is symmetric with respect to the y-axis.

The graph of the cosecant function is given in Figure 8.22. The cosecant function is the reciprocal of the sine function:

$$y = \frac{1}{\sin x}, \text{ for } \sin x \neq 0.$$

(The graph of the sine function is shown as a dashed curve.) Like the sine function, the cosecant function has period 2π, and is symmetric with respect to the origin.

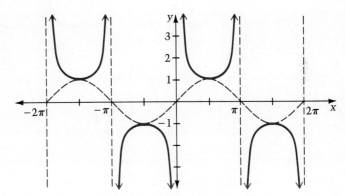

Figure 8.22 $y = \csc x$ ($y = \sin x$ is shown as a dashed curve).

Properties of the Cosecant Function

Domain all real numbers *except* $x = k\pi$, for any integer k

Range all real numbers y such that $y \leq -1$ or $y \geq 1$

Period 2π

Symmetry symmetry with respect to the origin (an odd function, like its reciprocal, sine)

Asymptotes vertical asymptotes $x = k\pi$ (values where sine is 0)

Example 3 Use the graphs of $y = \csc\left(x - \frac{\pi}{2}\right)$ and $y = -\sec x$ to determine whether $\csc\left(x - \frac{\pi}{2}\right) = -\sec x$ is an identity.

Solution

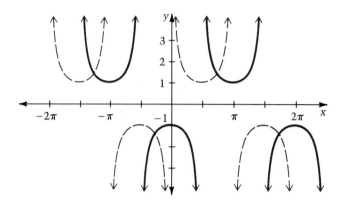

- $y = \csc\left(x - \frac{\pi}{2}\right)$. The graph of $y = \csc x$ (dashed curve) is translated $\frac{\pi}{2}$ units to the right.

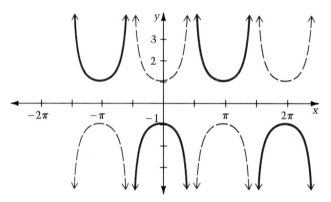

- $y = -\sec x$. The graph of $y = \sec x$ (dashed curve) is reflected across the x-axis.
- Since the graphs of $y = \csc\left(x - \frac{\pi}{2}\right)$ and $y = -\sec x$ are identical, the equation $\csc\left(x - \frac{\pi}{2}\right) = -\sec x$ is an identity.

Example 4 Answer True or False.

(a) $\frac{7\pi}{2}$ is not in the domain of the tangent function.

(b) $\frac{3}{2}$ is in the range of the cosine function.

(c) The sine, cosine, and tangent functions all have the same period.

Solution (a) True ■ $\tan x = \frac{\sin x}{\cos x}$, and $\cos \frac{7\pi}{2} = 0$.
(b) False ■ $-1 \leq \cos x \leq 1$ for all real x.
(c) False ■ Tangent has period π; sine and cosine have period 2π.

Exercise Set 8.3   = Calculator Exercises in the Appendix

Goal A

In exercises 1–12, graph each pair of functions on the same axes.

1. $y = \sin x$, $y = \sin(x + \pi)$
2. $y = \cos x$, $y = \cos(x - \pi)$
3. $y = \sin x$, $y = \sin\left(\frac{\pi}{2} + x\right)$
4. $y = \sin x$, $y = \sin\left(x - \frac{\pi}{2}\right)$
5. $y = \cos x$, $y = \cos(-x)$
6. $y = \sin x$, $y = -\sin x$
7. $y = \sin x$, $y = \sin\left(\frac{\pi}{4} - x\right)$
8. $y = \cos x$, $y = \cos\left(-x - \frac{\pi}{4}\right)$
9. $y = \cos(-x)$, $y = -\cos(-x - \pi)$
10. $y = -\sin x$, $y = -\sin(-x)$
11. $y = \sin x$, $y = \sin(x - 2\pi)$
12. $y = \cos x$, $y = \cos(x + 4\pi)$

Goal B

In exercises 13–22, graph each pair of functions on the same set of axes.

13. $y = \tan x$, $y = \tan\left(x - \frac{\pi}{2}\right)$
14. $y = \sec x$, $y = \sec\left(x + \frac{\pi}{3}\right)$
15. $y = \cot x$, $y = \cot\left(x - \frac{\pi}{6}\right)$
16. $y = \tan x$, $y = \tan(-x)$
17. $y = \csc x$, $y = -\csc x$
18. $y = \cot x$, $y = -\cot x$
19. $y = \sec x$, $y = \sec\left(\frac{\pi}{4} - x\right)$
20. $y = \tan x$, $y = \tan\left(\frac{\pi}{2} - x\right)$
21. $y = \csc(-x)$, $y = -\csc(-x)$
22. $y = \csc(-x)$, $y = \csc\left(-x - \frac{\pi}{2}\right)$

In exercises 23–30, determine whether the statement is True or False.

23. $-\frac{\pi}{2}$ is not in the domain of $y = \cot x$.
24. 7π is in the domain of $y = \csc x$.
25. 2π is in the range of $y = \cot x$.
26. $x = 2\pi$ is an asymptote of $y = \cot x$.
27. Secant and sine have the same period.
28. All trigonometric functions except for tangent and cotangent have period 2π.
29. $\csc(-x) = \sin x$ is an identity.
30. $\cot(-x) = -\tan x$ is an identity.

Superset

In exercises 31–38, sketch the graph of each function.

31. $y = |\sin x|$
32. $y = |\cot x|$
33. $y = \tan|x|$
34. $y = \sec|x|$
35. $y = \csc|x + \pi|$
36. $y = \sin|x - \pi|$
37. $y = \cos\left|\frac{\pi}{4} - x\right|$
38. $y = \csc\left|\frac{\pi}{3} + x\right|$

In exercises 39–42, sketch the graphs of each pair of equations on the same axes.

39. $y = \sin x$, $x = \sin y$
40. $y = \cos x$, $x = \cos y$
41. $y = \tan x$, $x = \tan y$, $-\frac{\pi}{2} < x < \frac{\pi}{2}$
42. $y = \cot x$, $x = \cot y$, $0 < x < \pi$

8.4 Transformations of the Trigonometric Functions

Goal A *to graph functions of the form $y = a \sin bx$ and $y = a \cos bx$*

In this section we will rely heavily on our knowledge of transformations to develop an efficient way of graphing periodic functions. Essentially we will be concerned with translations, stretchings, and shrinkings of the basic trigonometric graphs.

For example, consider the function $y = 3 \sin x$. For this function, each y-coordinate is three times the corresponding y-coordinate of the function $y = \sin x$.

x	0	$\dfrac{\pi}{6}$	$\dfrac{\pi}{4}$	$\dfrac{\pi}{3}$	$\dfrac{\pi}{2}$	$\dfrac{2\pi}{3}$	$\dfrac{3\pi}{4}$	π	$\dfrac{3\pi}{2}$	2π
$\sin x$	0	$\dfrac{1}{2}$	$\dfrac{\sqrt{2}}{2}$	$\dfrac{\sqrt{3}}{2}$	1	$\dfrac{\sqrt{3}}{2}$	$\dfrac{\sqrt{2}}{2}$	0	-1	0
$3 \sin x$	0	$\dfrac{3}{2}$	$\dfrac{3\sqrt{2}}{2}$	$\dfrac{3\sqrt{3}}{2}$	3	$\dfrac{3\sqrt{3}}{2}$	$\dfrac{3\sqrt{2}}{2}$	0	-3	0

In Figure 8.23, the graph of $y = 3 \sin x$ is shown as the result of stretching the graph of $y = \sin x$ vertically (away from the x-axis). Recall that, in general, the graph of $y = k \cdot f(x)$ is found by vertically stretching or shrinking the graph of $y = f(x)$. If $k < 0$, the stretching or shrinking is accompanied by a reflection across the x-axis.

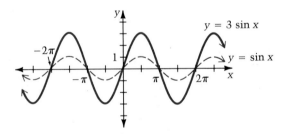

Figure 8.23

Notice that the range of the function $y = \sin x$ is the set of real numbers from -1 to 1 inclusive, whereas the range of $y = 3 \sin x$ is the set of real numbers from -3 to 3 inclusive. We call 3 the *amplitude* of $y = 3 \sin x$.

> **Definition**
>
> For functions of the form $y = a \sin x$ and $y = a \cos x$, the number $|a|$ is called the **amplitude.** The amplitude of a periodic function is one-half the difference between the maximum and minimum values of the function.

Example 1   Sketch the graph of

(a) $y = \frac{1}{2} \cos x$ (b) $y = -2 \sin x$

Solution   (a)

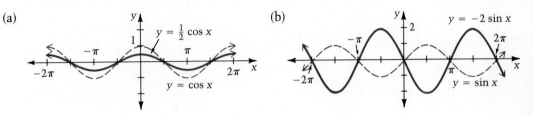

In (a), each y-coordinate for the graph of $y = \frac{1}{2} \cos x$ is $\frac{1}{2}$ the corresponding y-coordinate for the graph of $y = \cos x$. In (b), the graph of $y = -2 \sin x$ involves both a vertical stretching and a reflection across the x-axis.

Next we consider what happens to the graph of $y = \sin x$ when x is multiplied by 3. That is, we wish to determine how the graph of $y = \sin x$ is transformed to produce the graph of $y = \sin 3x$. Consider the following table of values.

x	0	$\frac{\pi}{6}$	$\frac{\pi}{4}$	$\frac{\pi}{3}$	$\frac{\pi}{2}$	$\frac{2\pi}{3}$	$\frac{3\pi}{4}$	π	$\frac{3\pi}{2}$	2π
$3x$	0	$\frac{\pi}{2}$	$\frac{3\pi}{4}$	π	$\frac{3\pi}{2}$	2π	$\frac{9\pi}{4}$	3π	$\frac{9\pi}{2}$	6π
$\sin 3x$	0	1	$\frac{\sqrt{2}}{2}$	0	-1	0	$\frac{\sqrt{2}}{2}$	0	1	0

Notice that as x takes on values from 0 to 2π, $3x$ takes on values from 0 to 6π. As a result, the graph of $y = \sin 3x$ completes *three* full cycles as x goes from 0 to 2π, with one full cycle as x goes from 0 to $\frac{2\pi}{3}$. Recall that the period is the length of the interval over which a periodic function makes one complete cycle. We therefore conclude that the period of $y = \sin 3x$ is $\frac{2\pi}{3}$, or one-third the period of $y = \sin x$. Graphs of

$$y = \sin 3x \quad \text{and} \quad y = \sin x$$

are shown below.

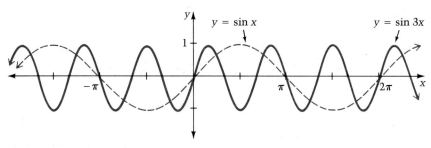

Figure 8.24

The graphs above suggest the following generalization.

Fact

Functions of the form $y = \sin bx$ and $y = \cos bx$ have period equal to $\left|\dfrac{2\pi}{b}\right|$.

Example 2 Sketch the graph of $y = \cos 2x$.

Solution

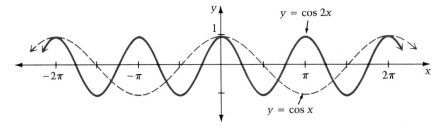

The period of $y = \cos 2x$ is $\left|\dfrac{2\pi}{2}\right| = \pi$, thus the given function completes one full cycle every π units.

Let us summarize our results for functions of the following two forms:

$$y = a \sin bx \quad \text{or} \quad y = a \cos bx.$$

- The amplitude $|a|$ indicates a vertical stretching of the basic sine or cosine curve if $|a| > 1$, and a vertical shrinking if $0 < |a| < 1$.
- The value $|b|$ indicates a horizontal stretching of the basic sine or cosine curve if $0 < |b| < 1$, and a horizontal shrinking if $|b| > 1$.

Example 3 Sketch the graph of (a) $y = -2 \sin \frac{1}{3}x$ (b) $y = 3 \sin(-4x)$.

Solution (a) We begin by sketching the graph of $y = \sin \frac{1}{3}x$ as a dashed curve. It is then stretched vertically and reflected across the x-axis to produce the graph of $y = -2 \sin \frac{1}{3}x$.

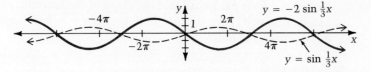

■ The period is $\left|\dfrac{2\pi}{\frac{1}{3}}\right| = 6\pi$.

(b) Recall that the graph of $y = f(-x)$ is the result of reflecting $y = f(x)$ across the y-axis. We begin by sketching the graph of $y = 3 \sin(4x)$ as a dashed curve, then reflect it across the y-axis to produce $y = 3 \sin(-4x)$.

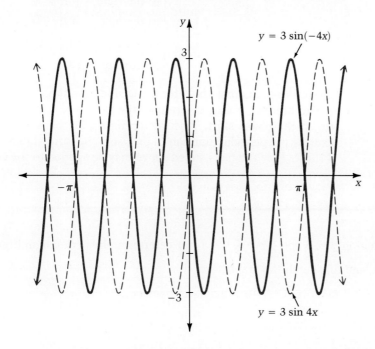

■ The period is $\left|\dfrac{2\pi}{-4}\right| = \dfrac{\pi}{2}$.

Goal B to graph functions of the form $y = k + a \sin b(x - h)$ and $y = k + a \cos b(x - h)$

Consider the function $y = 1 + 5 \sin 6(x - \frac{\pi}{4})$. If this were simply the function $y = 5 \sin 6x$, then we could use results already discussed in this section to conclude that the amplitude is equal to 5 and the period is equal to $\left|\frac{2\pi}{6}\right| = \frac{\pi}{3}$. As it turns out, the function $y = 1 + 5 \sin 6(x - \frac{\pi}{4})$ has the same period and amplitude as $y = 5 \sin 6x$.

Fact

The function $y = k + a \sin b(x - h)$ has the same amplitude $|a|$ and period $\left|\dfrac{2\pi}{b}\right|$ as the function $y = a \sin bx$.

The question then is how the graph of $y = 1 + 5 \sin 6(x - \frac{\pi}{4})$ differs from that of $y = 5 \sin 6x$. Recall that the graph of the function $y = k + f(x - h)$ can be produced by translating the graph of $y = f(x)$ horizontally h units and vertically k units.

Fact

The graph of $y = k + a \sin b(x - h)$ is produced by translating the graph of $y = a \sin bx$:

$$|h| \text{ units} \begin{cases} \text{to the right,} & \text{if } h > 0, \\ \text{to the left,} & \text{if } h < 0, \end{cases}$$

$$|k| \text{ units} \begin{cases} \text{upward,} & \text{if } k > 0, \\ \text{downward,} & \text{if } k < 0. \end{cases}$$

(Similar statements hold for the cosine function.)

The number h that determines the extent of horizontal translation is called the **phase shift.** When determining the phase shift, be careful to express the periodic function in precisely the form stated above. Example 4 shows how the facts stated above may be applied to the cosine function.

Example 4 For the function $y = 3 \cos\left(\dfrac{1}{2}x - \dfrac{\pi}{8}\right) - 1$, (a) determine the period, amplitude, and phase shift, and (b) sketch the graph.

Solution (a) $y = -1 + 3 \cos \dfrac{1}{2}\left(x - \dfrac{\pi}{4}\right)$ ■ Begin by restating as $y = k + a \cos b(x - h)$.

The period is $\left|\dfrac{2\pi}{b}\right| = \dfrac{2\pi}{\frac{1}{2}} = 4\pi$, the amplitude is $|3| = 3$, and the phase shift is $\dfrac{\pi}{4}$.

(b) We begin by sketching the graph of $y = 3 \cos\left(\dfrac{1}{2}x\right)$ as a dashed curve, then translate this graph $\dfrac{\pi}{4}$ units to the right and one unit downward to produce the graph of $y = 3 \cos \dfrac{1}{2}\left(x - \dfrac{\pi}{4}\right) - 1$.

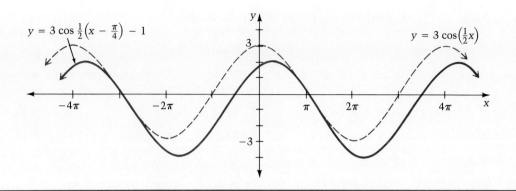

Goal C to graph by the addition of ordinates

Suppose we wish to sketch the graph of the function $f(x) = \frac{1}{2}x + \sin x$. Taken individually, the two component functions $y = \frac{1}{2}x$ and $y = \sin x$ are familiar to us (a straight line and the sine curve, respectively). One method for graphing the function $f(x) = \frac{1}{2}x + \sin x$ is to shift each point on the graph of $y = \sin x$ by an amount equal to $\frac{1}{2}x$. Example 5 uses this method, which is called graphing by **addition of ordinates.**

Example 5 Sketch the graph of $y = \frac{1}{2}x + \sin x$.

Solution

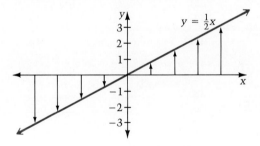

■ Sketch the graph of $y = \frac{1}{2}x$ and use arrows to represent y-coordinates for various values of x.

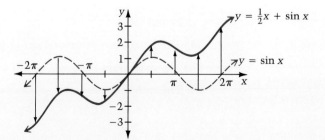

■ Sketch the graph of $y = \sin x$ as a dashed curve and use the arrows from the figure above to indicate the result of adding $\frac{1}{2}x$. Draw a smooth curve through the arrowheads.

Exercise Set 8.4 ▣ Calculator Exercises in the Appendix

Goal A

In exercises 1–16, sketch the graph of the given function.

1. $y = \frac{1}{3} \cos x$
2. $y = 4 \sin x$
3. $y = -\frac{1}{2} \sin x$
4. $y = -\frac{3}{2} \cos x$
5. $y = -3 \sin x$
6. $y = \sqrt{2} \sin x$
7. $y = \cos \frac{1}{2}x$
8. $y = \sin(-\frac{1}{3}x)$
9. $y = \cos(-3x)$
10. $y = \sin \frac{3}{2}x$
11. $y = 2 \sin \frac{1}{2}x$
12. $y = \frac{1}{2} \cos 2x$
13. $y = -3 \sin \frac{1}{3}x$
14. $y = -2 \cos 3x$
15. $y = \sqrt{2} \cos(-2x)$
16. $y = \sqrt{3} \cos(-\frac{1}{2}x)$

Goal B

In exercises 17–30, determine the amplitude, period, and phase shift, and sketch the graph of the given function.

17. $y = 2 \cos\left(x + \frac{\pi}{2}\right)$
18. $y = 3 \cos\left(x - \frac{\pi}{2}\right)$
19. $y = 2 \cos \frac{1}{2}\left(x - \frac{\pi}{3}\right)$
20. $y = \frac{1}{2} \sin 2(x - \pi)$
21. $y = 2 \sin\left(\frac{1}{2}x + \frac{\pi}{4}\right)$
22. $y = 2 \cos(2x + \pi)$
23. $y = \sqrt{3} \sin\left(-\frac{x}{2} - \frac{\pi}{2}\right)$
24. $y = 3 \sin\left(-x + \frac{\pi}{2}\right)$
25. $y = -3 \sin(2x - \pi)$
26. $y = -2 \cos\left(\frac{x}{2} + \frac{\pi}{2}\right)$
27. $y = 3 \cos\left(\frac{1}{2}x + \frac{\pi}{2}\right) - 2$
28. $y = -3 \sin\left(2x + \frac{\pi}{2}\right) + 1$
29. $y = 2 \sin\left(x - \frac{\pi}{2}\right) + 1$
30. $y = -2 \cos\left(\frac{x}{2} + \frac{\pi}{2}\right) - 1$

Goal C

In exercises 31–38, sketch the graph of the given function.

31. $y = \sin x - 2x$
32. $y = 2x - \sin x$
33. $y = \frac{1}{2}x + \sin 2x$
34. $y = \frac{1}{2}x - \cos 2x$
35. $y = -3 \cos 2x + x$
36. $y = -2 \sin \frac{1}{2}x + 3x$
37. $y = -2 \sin \frac{1}{2}\left(x - \frac{\pi}{2}\right) - 2x$
38. $y = -\frac{1}{2} \cos\left(x - \frac{\pi}{2}\right) + 4x$

Superset

39. Find b such that the period of the function $y = \frac{1}{2} \sin bx$ is $\frac{\pi}{4}$.

40. Find b such that the period of the function $y = 3 \cos b(x - \pi)$ is 4π.

In exercises 41–48, sketch the graph of the given function.

41. $y = |\sin 2x|$
42. $y = |3 \sin x|$
43. $y = |-2 \cos x|$
44. $y = \cos|-2x|$
45. $y = \sin x + \cos x$
46. $y = \sin x - \cos x$
47. $y = \sqrt{3} \sin x - \cos x$
48. $y = \frac{\sqrt{3}}{2} \sin x + \frac{1}{2} \cos x$

In exercises 49–52, determine whether the statement is True or False.

49. The function $y = \sin x + \cos x$ is periodic with period 4π.
50. The maximum value of $4 \sin \frac{1}{2}x$ is 2.
51. The maximum value of $-2 \cos \frac{1}{2}x$ is 2.
52. The minimum value of $-\frac{3}{2} \cos \frac{2}{3}x$ is $\frac{3}{2}$.

8.5 The Inverse Trigonometric Functions

Goal A *to use special angles to evaluate inverse trigonometric functions*

By the definition of a function, we are assured that to each value in the domain of a trigonometric function, there is assigned exactly one range value. For example, the sine function assigns to the domain value $\frac{\pi}{6}$ exactly one range value, namely $\frac{1}{2}$. Thus, if we know that $\sin \frac{\pi}{6} = y$, then y must be $\frac{1}{2}$. However, it is not true that if we know a particular y-value, say $\frac{1}{2}$, the corresponding x-value is unique. For example, if $\sin x = \frac{1}{2}$, then x can be any number in the set $\{\cdots, -\frac{11\pi}{6}, -\frac{7\pi}{6}, \frac{\pi}{6}, \frac{5\pi}{6}, \frac{13\pi}{6}, \cdots\}$.

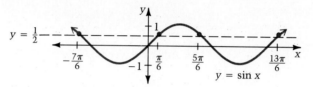

Figure 8.25 There are an infinite number of values of x such that $\sin x = \frac{1}{2}$.

In this section we would like to define an inverse sine function, that is a function which "undoes" what the sine function "does." To be a function, this "inverse sine" must take a value like $\frac{1}{2}$ and produce *exactly one number* whose sine is $\frac{1}{2}$, for example, $\frac{\pi}{6}$. (If it produced more than one, it would not be a function!) The problem is that the sine function is not one-to-one, and thus cannot have an inverse function. To resolve this problem, we look at the sine function and restrict its domain to some interval on which it *is* one-to-one. In Figure 8.26(a), we have graphed the sine function for domain values in the interval $[-\frac{\pi}{2}, \frac{\pi}{2}]$. For these values, the sine function is one-to-one and thus has an inverse function. Recall that the graph of an inverse can be found by reflecting the original graph across the line $y = x$. The graph of the inverse sine function is shown below. We describe this inverse with the equation $y = \sin^{-1} x$.

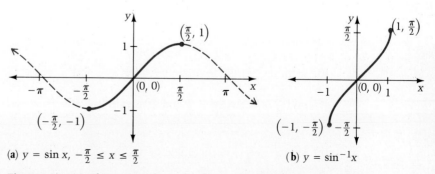

(a) $y = \sin x,\ -\frac{\pi}{2} \leq x \leq \frac{\pi}{2}$ **(b)** $y = \sin^{-1} x$

Figure 8.26 (a) The restricted sine function. (b) The inverse sine function.

Careful! The symbol $\sin^{-1} x$ does not mean $\dfrac{1}{\sin x}$; as you know, $\dfrac{1}{\sin x} = \csc x$.)

Definition
The inverse sine function is defined as follows:

$$y = \sin^{-1} x \quad \text{with domain } -1 \le x \le 1$$

if and only if $\sin y = x$ and $-\dfrac{\pi}{2} \le y \le \dfrac{\pi}{2}$.

As a consequence of the definition of $y = \sin^{-1} x$, it is often useful to think of the expression $\sin^{-1} x$ as

the angle between $-\dfrac{\pi}{2}$ and $\dfrac{\pi}{2}$ inclusive whose sine is x.

The function name "arcsin" is sometimes used instead of $\sin^{-1}$ to refer to the inverse sine function. Thus, arcsin and $\sin^{-1}$ mean the same thing and can be used interchangeably.

Example 1 Determine the value of each expression without using Table 4 or a calculator.

(a) $\sin^{-1}\left(\dfrac{1}{2}\right)$ (b) $\arcsin\left(\dfrac{\sqrt{2}}{2}\right)$ (c) $\sin^{-1}\left(-\dfrac{\sqrt{3}}{2}\right)$ (d) $\tan\left(\arcsin\left(-\dfrac{1}{2}\right)\right)$

(e) $\sin^{-1}\left(\sin\dfrac{5\pi}{4}\right)$

Solution (a) Think of $\sin^{-1}\left(\dfrac{1}{2}\right)$ as the angle between $-\dfrac{\pi}{2}$ and $\dfrac{\pi}{2}$ whose sine is $\dfrac{1}{2}$. Since $\sin\left(\dfrac{\pi}{6}\right) = \dfrac{1}{2}$, and $\dfrac{\pi}{6}$ is between $-\dfrac{\pi}{2}$ and $\dfrac{\pi}{2}$, we conclude that $\sin^{-1}\left(\dfrac{1}{2}\right) = \dfrac{\pi}{6}$.

(b) $\arcsin\left(\dfrac{\sqrt{2}}{2}\right)$ is the angle between $-\dfrac{\pi}{2}$ and $\dfrac{\pi}{2}$ whose sine is $\dfrac{\sqrt{2}}{2}$. Since $\sin\left(\dfrac{\pi}{4}\right) = \dfrac{\sqrt{2}}{2}$, and $\dfrac{\pi}{4}$ is between $-\dfrac{\pi}{2}$ and $\dfrac{\pi}{2}$, $\arcsin\left(\dfrac{\sqrt{2}}{2}\right) = \dfrac{\pi}{4}$.

(c) $\sin^{-1}\left(-\dfrac{\sqrt{3}}{2}\right)$ is the angle between $-\dfrac{\pi}{2}$ and $\dfrac{\pi}{2}$ whose sine is $-\dfrac{\sqrt{3}}{2}$. Since $\sin\left(-\dfrac{\pi}{3}\right) = -\dfrac{\sqrt{3}}{2}$, and $-\dfrac{\pi}{3}$ is between $-\dfrac{\pi}{2}$ and $\dfrac{\pi}{2}$, $\sin^{-1}\left(-\dfrac{\sqrt{3}}{2}\right) = -\dfrac{\pi}{3}$.

(d) First determine $\arcsin\left(-\frac{1}{2}\right)$: $\arcsin\left(-\frac{1}{2}\right)$ is the angle between $-\frac{\pi}{2}$ and $\frac{\pi}{2}$ whose sine is $-\frac{1}{2}$. Since $\sin\left(-\frac{\pi}{6}\right) = -\frac{1}{2}$,

$$\arcsin\left(-\frac{1}{2}\right) = -\frac{\pi}{6}.$$

Thus,

$$\tan\left(\arcsin\left(-\frac{1}{2}\right)\right) = \tan\left(-\frac{\pi}{6}\right)$$

$$= -\tan\left(\frac{\pi}{6}\right) \quad \blacksquare \text{ Tangent is an odd function.}$$

$$= -\frac{\sqrt{3}}{3}$$

$$\tan\left(\arcsin\left(-\frac{1}{2}\right)\right) = -\frac{\sqrt{3}}{3}$$

(e) First determine $\sin\frac{5\pi}{4}$.

$$\sin\left(\frac{5\pi}{4}\right) = -\sin\left(\frac{\pi}{4}\right) = -\frac{\sqrt{2}}{2} \quad \blacksquare \text{ Use the reference angle } \frac{\pi}{4}.$$

Thus,

$$\sin^{-1}\left(\sin\frac{5\pi}{4}\right) = \sin^{-1}\left(-\frac{\sqrt{2}}{2}\right)$$

$$= -\frac{\pi}{4} \quad \blacksquare \; \sin^{-1}\left(-\frac{\sqrt{2}}{2}\right) \text{ is the angle between } -\frac{\pi}{2} \text{ and } \frac{\pi}{2}$$
$$\text{whose sine is } -\frac{\sqrt{2}}{2}, \text{ namely, } -\frac{\pi}{4}.$$

$$\sin^{-1}\left(\sin\frac{5\pi}{4}\right) = -\frac{\pi}{4}$$

Recall that if f and f^{-1} are inverses of one another, then

$$f(f^{-1}(x)) = x \quad \text{for all } x \text{ in the domain of } f^{-1}, \text{ and}$$
$$f^{-1}(f(x)) = x \quad \text{for all } x \text{ in the domain of } f.$$

If we let $f(x) = \sin x$ and $f^{-1}(x) = \sin^{-1} x$, then those statements become

$$\sin(\sin^{-1} x) = x \quad \text{for all } x \text{ such that } -1 \leq x \leq 1, \text{ and}$$
$$\sin^{-1}(\sin x) = x \quad \text{for all } x \text{ such that } -\frac{\pi}{2} \leq x \leq \frac{\pi}{2}.$$

By virtue of the last equation, it is *not* true that $\sin^{-1}(\sin x) = x$ for all x in the domain of the sine function—Example 1(e) is a case in point. However, for all x between $-\frac{\pi}{2}$ and $\frac{\pi}{2}$ (the domain of the restricted sine function), it is true that $\sin^{-1}(\sin x) = x$.

In a manner similar to the case of $\sin^{-1} x$, we can restrict the domains of the other trigonometric functions so that inverse functions can be defined. By restricting $y = \cos x$ to values of x between 0 and π inclusive, the function becomes one-to-one and has an inverse cosine function.

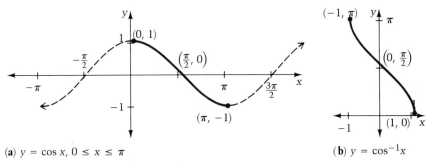

(a) $y = \cos x$, $0 \leq x \leq \pi$ (b) $y = \cos^{-1} x$

Figure 8.27 (a) The restricted cosine function. (b) The inverse cosine function.

Definition
The inverse cosine function is defined as follows:
$$y = \cos^{-1} x \quad \text{with domain } -1 \leq x \leq 1$$
if and only if $\cos y = x$ and $0 \leq y \leq \pi$.

By restricting $y = \tan x$ to values of x between $-\frac{\pi}{2}$ and $\frac{\pi}{2}$, we can define the inverse function $y = \tan^{-1} x$.

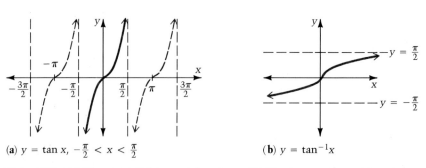

(a) $y = \tan x$, $-\frac{\pi}{2} < x < \frac{\pi}{2}$ (b) $y = \tan^{-1} x$

Figure 8.28 (a) The restricted tangent function. (b) The inverse tangent function.

Definition

The inverse tangent function is defined as follows:

$$y = \tan^{-1} x \quad \text{with domain } -\infty < x < +\infty,$$

if and only if $\tan y = x$ and $-\dfrac{\pi}{2} < y < \dfrac{\pi}{2}$.

By virtue of the definitions of $\cos^{-1} x$ and $\tan^{-1} x$, we can think of $\cos^{-1} x$ as "the angle between 0 and π whose cosine is equal to x," and $\tan^{-1} x$ as "the angle between $-\dfrac{\pi}{2}$ and $\dfrac{\pi}{2}$ whose tangent is equal to x." The expressions arccos x and arctan x are used interchangeably with $\cos^{-1} x$ and $\tan^{-1} x$.

Example 2 Determine the value of each expression, without using Table 4.

(a) $\arctan(1)$ (b) $\cos^{-1}\left(-\dfrac{\sqrt{3}}{2}\right)$ (c) $\tan^{-1}\left(\tan \dfrac{2\pi}{3}\right)$

Solution (a) $\arctan(1)$ is the angle between $-\dfrac{\pi}{2}$ and $\dfrac{\pi}{2}$ whose tangent is 1. Since $\tan\left(\dfrac{\pi}{4}\right) = 1$, and $\dfrac{\pi}{4}$ is between $-\dfrac{\pi}{2}$ and $\dfrac{\pi}{2}$, $\arctan(1) = \dfrac{\pi}{4}$.

(b) $\cos^{-1}\left(-\dfrac{\sqrt{3}}{2}\right)$ is the angle between 0 and π inclusive whose cosine is $-\dfrac{\sqrt{3}}{2}$. Since $\cos\left(\dfrac{\pi}{6}\right)$ is $\dfrac{\sqrt{3}}{2}$, we are looking for an angle that has $\dfrac{\pi}{6}$ as a reference angle, and whose cosine is negative. Since the angle must be between 0 and π, our only choice is $\dfrac{5\pi}{6}$. Thus, $\cos^{-1}\left(-\dfrac{\sqrt{3}}{2}\right) = \dfrac{5\pi}{6}$.

(c) We begin by determining $\tan\left(\dfrac{2\pi}{3}\right)$:

$$\tan\left(\dfrac{2\pi}{3}\right) = -\tan\left(\dfrac{\pi}{3}\right) = -\sqrt{3}.$$

Now determine $\tan^{-1}(-\sqrt{3})$: $\tan^{-1}(-\sqrt{3})$ is the angle between $-\dfrac{\pi}{2}$ and $\dfrac{\pi}{2}$ whose tangent is $-\sqrt{3}$. Since the tangent is negative, the angle is between $-\dfrac{\pi}{2}$ and 0, and since the value of the tangent is $-\sqrt{3}$, the reference angle is $\dfrac{\pi}{3}$. Thus $\tan^{-1}(-\sqrt{3}) = -\dfrac{\pi}{3}$, and we have $\tan^{-1}\left(\tan\left(\dfrac{2\pi}{3}\right)\right) = -\dfrac{\pi}{3}$.

Inverses can be defined for the cotangent, secant, and cosecant functions by suitably restricting the domains of these three trigonometric functions. We leave these problems for the exercise set.

Goal B to use right triangles to evaluate trigonometric and inverse trigonometric functions

Sometimes it is necessary to evaluate expressions such as $\tan(\arcsin(-\frac{3}{5}))$ or $\cos(\tan^{-1}(\frac{5}{12}))$. Such composite expressions can be evaluated without resorting to calculators, tables, or facts about special angles. These problems require that you recall three things:

- The ranges of the inverse trigonometric functions:

$$-\frac{\pi}{2} \leq \sin^{-1} x \leq \frac{\pi}{2}, \quad 0 \leq \cos^{-1} x \leq \pi, \quad -\frac{\pi}{2} < \tan^{-1} x < \frac{\pi}{2}.$$

- The ASTC memory device for determining the quadrant in which an angle in standard position will terminate.
- The trigonometric function values of an angle in standard position can be determined by the coordinates of any point on the terminal side of the angle.

Example 3   Determine the exact values of the following:

(a) $\tan\left(\arcsin\left(-\frac{3}{5}\right)\right)$ (b) $\cos\left(\tan^{-1}\left(\frac{5}{12}\right)\right)$

Solution   (a) Begin by letting $t = \arcsin\left(-\frac{3}{5}\right)$. Then $\sin t = \sin\left(\arcsin\left(-\frac{3}{5}\right)\right) = -\frac{3}{5}$. Since $\sin t = -\frac{3}{5}$, a negative number, t must terminate in either the third or fourth quadrant (use ASTC). But, because t is in the range of the arcsin function, $-\frac{\pi}{2} \leq t \leq \frac{\pi}{2}$, and thus t cannot terminate in the third quadrant. Thus, t must be a negative angle terminating in the fourth quadrant.

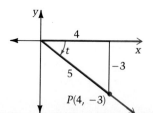

- Draw angle t and label point P on its terminal side.
- Since $\sin t = \frac{y}{r} = \frac{-3}{5}$, let P have y-coordinate -3, and lie 5 units from the origin.

Find x by applying the Pythagorean Theorem to the right triangle with legs x and 3 and hypotenuse 5:

$$x^2 + y^2 = r^2$$
$$x^2 + 3^2 = 5^2$$
$$x^2 = 16$$

Since $x > 0$, we have $x = 4$. Thus, $\tan\left(\arcsin\left(-\frac{3}{5}\right)\right) = \tan t = \frac{y}{x} = -\frac{3}{4}$.

(b) Begin by letting $t = \tan^{-1}\left(\frac{5}{12}\right)$. Then $\tan t = \tan\left(\tan^{-1}\left(\frac{5}{12}\right)\right) = \frac{5}{12}$. Since $\tan t$ is a positive number, t must terminate in either the first or third quadrant. The range of the inverse tangent function requires that $-\frac{\pi}{2} < t < \frac{\pi}{2}$. Thus, t is a first quadrant angle.

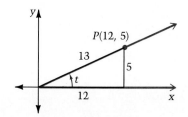

- Draw t and a point P on its terminal side. Since $\tan t = \frac{y}{x} = \frac{5}{12}$, let P have coordinates $(12, 5)$. By the Pythagorean Theorem, we have
$$r^2 = 12^2 + 5^2 = 169$$
$$r = 13.$$

Thus, $\cos\left(\tan^{-1}\left(\frac{5}{12}\right)\right) = \cos t = \frac{x}{r} = \frac{12}{13}$.

Example 4 Write $\cos^2(\tan^{-1} z)$, for $z \geq 0$, as an expression involving no trigonometric functions.

Solution Let $t = \tan^{-1} z$. Then $\tan t = z$. Since $z \geq 0$, t must be between 0 and $\frac{\pi}{2}$.

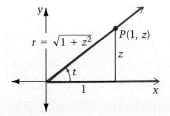

- Since $\tan t = \frac{y}{x} = \frac{z}{1}$, draw angle t with $P(1, z)$ on its terminal side. By the Pythagorean Theorem,
$$r^2 = 1^2 + z^2$$
$$r = \sqrt{1 + z^2}.$$

Thus, $\cos^2(\tan^{-1} z) = (\cos t)^2 = \left(\frac{x}{r}\right)^2 = \left(\frac{1}{\sqrt{1+z^2}}\right)^2 = \frac{1}{1+z^2}$.

Exercise Set 8.5 ▣ Calculator Exercises in the Appendix

Goal A

In exercises 1–24, evaluate the expression without using Table 4 or a calculator.

1. $\sin^{-1}\frac{\sqrt{3}}{2}$
2. $\sin^{-1} 1$
3. $\cos^{-1}(-1)$
4. $\tan^{-1}(\sqrt{3})$
5. $\tan^{-1}\left(-\frac{\sqrt{3}}{3}\right)$
6. $\cos^{-1}\left(-\frac{\sqrt{2}}{2}\right)$
7. $\arctan(-1)$
8. $\tan^{-1}(-\sqrt{3})$
9. $\arcsin 0$
10. $\arccos 0$
11. $\sin\left(\cos^{-1}\left(\frac{\sqrt{3}}{2}\right)\right)$
12. $\tan\left(\cos^{-1}\left(\frac{\sqrt{3}}{2}\right)\right)$
13. $\tan\left(\cos^{-1}\left(-\frac{1}{2}\right)\right)$
14. $\sin\left(\cos^{-1}\left(-\frac{1}{2}\right)\right)$
15. $\cos(\sin^{-1}(-1))$
16. $\cos\left(\sin^{-1}\left(-\frac{\sqrt{2}}{2}\right)\right)$
17. $\sin(\tan^{-1}(-\sqrt{3}))$
18. $\cos(\tan^{-1}(-1))$
19. $\cos\left(\cos^{-1}\left(-\frac{\sqrt{2}}{2}\right)\right)$
20. $\cos\left(\sin^{-1}\left(-\frac{1}{2}\right)\right)$
21. $\tan^{-1}\left(\tan\frac{3\pi}{4}\right)$
22. $\cos^{-1}\left(\cos\frac{\pi}{3}\right)$
23. $\sin^{-1}\left(\cos\left(-\frac{\pi}{3}\right)\right)$
24. $\tan(\tan^{-1}(-\sqrt{3}))$

Goal B

In exercises 25–34, use a right triangle to evaluate the expression.

25. $\sin\left(\sin^{-1}\frac{4}{5}\right)$
26. $\cos\left(\sin^{-1}\frac{4}{5}\right)$
27. $\cos\left(\sin^{-1}\left(-\frac{3}{5}\right)\right)$
28. $\tan\left(\arccos\frac{4}{5}\right)$
29. $\cos\left(\tan^{-1}\left(-\frac{3}{4}\right)\right)$
30. $\sin\left(\tan^{-1}\frac{7}{24}\right)$
31. $\sin\left(\arccos\left(-\frac{12}{13}\right)\right)$
32. $\tan\left(\sin^{-1}\left(\frac{5}{13}\right)\right)$
33. $\tan\left(\arcsin\left(-\frac{2}{7}\right)\right)$
34. $\sin(\arctan 2)$

In exercises 35–40, rewrite each expression without using trigonometric functions. Assume $z \geq 0$.

35. $\sin^2(\arctan z)$
36. $\cos^2(\tan^{-1} z)$
37. $\cos^2(\sin^{-1} z)$
38. $\tan^2(\sin^{-1} z)$
39. $\tan(\arcsin(-z))$
40. $\sin(\arcsin(-z))$

Superset

Inverse cotangent, secant, and cosecant functions can be defined as follows:

$y = \cot^{-1} x$ if and only if $\cot y = x$, where $-\infty < x < +\infty$, and $0 < y < \pi$,

$y = \sec^{-1} x$ if and only if $\sec y = x$, where $x \leq -1$ or $x \geq 1$, and $0 \leq y \leq \pi$, $y \neq \frac{\pi}{2}$,

$y = \csc^{-1} x$ if and only if $\csc y = x$, where $x \leq -1$ or $x \geq 1$, and $-\frac{\pi}{2} \leq y \leq \frac{\pi}{2}$, $y \neq 0$.

Use these definitions in evaluating the expressions in exercises 41–56.

41. $\operatorname{arcsec}(-2)$
42. $\operatorname{arccsc}(-\sqrt{2})$
43. $\sin\left(\csc^{-1}\left(-\frac{13}{5}\right)\right)$
44. $\cos\left(\sec^{-1}\left(-\frac{13}{12}\right)\right)$
45. $\cos\left(\sec^{-1}\frac{3}{\sqrt{5}}\right)$
46. $\cot\left(\sec^{-1}\frac{7}{3}\right)$
47. $\cot\left(\sin^{-1}\frac{2}{3}\right)$
48. $\sec\left(\sin^{-1}\left(-\frac{1}{2}\right)\right)$
49. $\sec\left(\cot^{-1}\left(-\frac{\sqrt{11}}{2}\right)\right)$
50. $\cot\left(\tan^{-1}\left(\frac{\sqrt{6}}{4}\right)\right)$
51. $\arccos(0.9483)$
52. $\sin^{-1}(0.4226)$
53. $\tan^{-1}(7.115)$
54. $\operatorname{arccot}(-28.64)$
55. $\sin(\cot^{-1}(12.25))$
56. $\sin(\cos^{-1}(-0.2221))$

In exercises 57–60, sketch the graph.

57. $y = 3\cos^{-1} x$
58. $y = \cos^{-1}(\frac{1}{3}x)$
59. $y = \sin^{-1}(2x)$
60. $y = \frac{1}{2}\sin^{-1} x$

Chapter Review & Test

Chapter Review

8.1 Radian Measure and the Unit Circle (pp. 318–324)

The unit circle is the circle of radius 1 with center at the origin. A point $P(u, v)$ is on the unit circle if and only if $u^2 + v^2 = 1$ (p. 318) A *standard arc s* is an arc on the unit circle which starts at $(1, 0)$, has length $|s|$, and travels counterclockwise if s is positive, and clockwise if s is negative. (p. 319)

A *central angle* θ is an angle whose vertex is at the center of a circle. (p. 320) The *radian measure* of a positive central angle θ is defined as a ratio:

$$\theta = \frac{s}{r},$$

where s is the arclength of the arc cut by θ, and r is the radius of the circle. (p. 321)

To convert the degree measure d of an angle to its radian measure α, and vice versa, use the following proportion. (p. 322)

$$\frac{d}{180°} = \frac{\alpha}{\pi}.$$

8.2 Trigonometric Functions of Real Numbers (pp. 325–332)

Every real number x can be associated uniquely with a central angle θ that cuts standard arc x on the unit circle. (p. 325) To define the trigonometric functions of any real number x, let $T(u, v)$ be the terminal point on the standard arc x. (p. 326)

$$\sin x = v \qquad \cos x = u \qquad \tan x = \frac{v}{u}$$

$$\csc x = \frac{1}{v} \qquad \sec x = \frac{1}{u} \qquad \cot x = \frac{u}{v}$$

To find the trigonometric function value of any real number x, we determine the sign by means of the ASTC chart, and then find the trigonometric function value of the reference value for x:

x between	0 and $\frac{\pi}{2}$	$\frac{\pi}{2}$ and π	π and $\frac{3\pi}{2}$	$\frac{3\pi}{2}$ and 2π
reference value	x	$\pi - x$	$x - \pi$	$2\pi - x$

If x is outside the interval $[0, 2\pi)$, begin by adding an appropriate integer multiple of 2π to produce a value between 0 and 2π, then use the procedure described above. (p. 329)

8.3 The Trigonometric Functions: Basic Graphs and Properties (pp. 333–343)

A function f is called a *periodic function* if there is a positive real number p such that $f(x) = f(x + p)$ for all x in the domain of f. The smallest such positive number p is called the *period* of the function. (p. 333)

8.4 Transformations of the Trigonometric Functions (pp. 344–350)

For functions described by equations of the form $y = k + a \sin b(x - h)$ or $y = k + a \cos b(x - h)$, the number $|a|$ is the *amplitude*, the number $\left|\dfrac{2\pi}{b}\right|$ is the *period*, and the number h is the *phase shift*. (pp. 344–348)

To graph functions of the form $y = a \sin bx + g(x)$ or $y = a \cos bx + g(x)$, use the method of *addition of ordinates*. (p. 349)

8.5 The Inverse Trigonometric Functions (pp. 351–358)

If each of the trigonometric functions is restricted to a portion of its domain, the resulting function is one-to-one, and thus has an inverse function. The inverses of the sine, cosine, and tangent functions are defined as follows:

$y = \sin^{-1} x$ with domain $-1 \leq x \leq 1$ if and only if $\sin y = x$ and $-\dfrac{\pi}{2} \leq y \leq \dfrac{\pi}{2}$,

$y = \cos^{-1} x$ with domain $-1 \leq x \leq 1$ if and only if $\cos y = x$ and $0 \leq y \leq \pi$,

$y = \tan^{-1} x$ with domain $-\infty < x < +\infty$ if and only if $\tan y = x$ and $-\dfrac{\pi}{2} < x < \dfrac{\pi}{2}$.

Chapter Test

8.1A Determine whether the given point is on the unit circle.

1. $\left(-\dfrac{5}{13}, \dfrac{12}{13}\right)$
2. $\left(\dfrac{\sqrt{2}}{6}, \dfrac{4}{6}\right)$

Represent the given real number as a standard arc.

3. $\dfrac{10\pi}{3}$
4. $\dfrac{7\pi}{2}$
5. $\dfrac{-25\pi}{4}$

8 / Trigonometric Functions

8.1B Convert the measure of the given angle from degrees to radians.

    6. $-135°$ **7.** $240°$ **8.** $-735°$

Convert the measure of the given angle from radians to degrees.

    9. $\dfrac{9\pi}{2}$ **10.** $\dfrac{16\pi}{3}$

8.1C Determine the measure in radians and in degrees of the central angle which cuts an arc of length s in a circle with radius r.

    11. $s = 12\pi$ cm, $r = 8$ cm   **12.** $s = \dfrac{3\pi}{2}$ cm, $r = 6$ cm

    13. On a steeple clock the hour hand is 3 ft long and the minute hand is 4 ft long. How far does the tip of the hour hand move in 4 hr 30 min?

8.2A Determine the given trigonometric function value.

    14. $\sin\left(\dfrac{5\pi}{4}\right)$ **15.** $\cos\left(-\dfrac{4\pi}{3}\right)$ **16.** $\sec\left(\dfrac{19\pi}{2}\right)$ **17.** $\tan\left(-\dfrac{17\pi}{3}\right)$

Use Table 4 or a calculator to determine the trigonometric function value.

    18. $\sin 4.3$ **19.** $\cos 5.6$

8.3A Graph the following pair of functions on the same axes.

    20. $y = \sin x,\ y = -\sin\left(x + \dfrac{\pi}{2}\right)$

8.3B Graph the following pair of functions on the same axes.

    21. $y = \tan x,\ y = \tan\left(x - \dfrac{\pi}{4}\right)$

Determine whether each statement is true or false.

    22. $\tan x$ is not defined for $x = -3\pi$.

    23. $\cot x$ is not defined for $x = -\dfrac{3\pi}{2}$.

    24. $\csc(-x) = -\dfrac{1}{\sin x}$ is an identity.

    25. $\cos(-x) = -\cos x$ is an identity.

8.4A Determine the amplitude and period, and sketch the graph.

26. $y = 3\cos\frac{1}{2}x$
27. $y = -3\sin 2x$

8.4B Determine the amplitude, period, and phase shift, and sketch the graph.

28. $y = 2\cos\left(2x - \frac{\pi}{2}\right)$
29. $y = -4\sin(2x + \pi)$

8.4C Sketch the graph of each of the following trigonometric functions.

30. $y = -3\sin\frac{1}{2}x + x$
31. $y = \frac{1}{2}\cos 2x - 2x$

8.5A Evaluate each expression without using Table 4 or a calculator.

32. $\arcsin\left(-\frac{1}{2}\right)$
33. $\cos^{-1}\left(\frac{\sqrt{3}}{2}\right)$
34. $\sin^{-1}\left(\cot\frac{3\pi}{4}\right)$

8.5B Use a right triangle to evaluate the expression.

35. $\cos\left(\tan^{-1}\frac{3}{4}\right)$
36. $\sin\left(\text{arcsec}\,\frac{13}{5}\right)$

Superset

In exercises 37–42, determine whether the statement is true or false.

37. $2\sin\frac{\pi}{6}\cos\frac{\pi}{6} = \sin\frac{\pi}{3}$
38. $1 + \tan^2\frac{\pi}{4} = \sec^2\frac{\pi}{2}$

39. $\sin\frac{\pi}{6} = \sqrt{\dfrac{1 + \cos\frac{\pi}{3}}{2}}$
40. $\sin\frac{\pi}{4} = \sqrt{\dfrac{1 - \sin\frac{\pi}{2}}{2}}$

In exercises 41–44, find a value of b such that the period p of the function is the given value.

41. $y = \sin bx,\ p = \frac{\pi}{3}$
42. $y = \cos bx,\ p = \frac{\pi}{2}$

43. $y = \cot bx,\ p = 2\pi$
44. $y = \tan bx,\ p = \frac{\pi}{6}$

45. An equilateral triangle is inscribed in a circle with a 6-inch radius. Find the length of the arc cut by one side of the triangle.

46. An automobile has tires with a 30-inch diameter. If the tires are revolving at 264 rpm, find the speed of the car in miles per hour.

Cumulative Test: *Chapters 6–8*

In exercises 1–2, determine the distance between the given points.

1. $(-7, 2), (-1, -6)$ **2.** $(-4, 3), (8, -2)$

In exercises 3–4, determine $f(3), f(-2), f(h^2), f(h-4)$ for each function.

3. $f(x) = x^2 - 2x + 3$ **4.** $f(x) = 9 - x^2$

In exercises 5–6, determine whether each of the functions is one-to-one.

5. $f(x) = x^3 - 3$ **6.** $f(x) = -3x + 5$

In exercises 7–8, find an equation that describes $(f \circ g)(x)$ and state the domain of the composite function.

7. $f(x) = x^2 - 4; g(x) = 2x - 5$ **8.** $f(x) = \sqrt{x - 2}; g(x) = 4x^2 + 1$

In exercises 9–10, the given function $y = f(x)$ is one-to-one. Find the equation describing the inverse function, sketch the inverse, and identify its domain and range.

9. $f(x) = 2x - 5; 0 \leq x \leq 3$ **10.** $f(x) = \sqrt{x - 1}, 1 \leq x \leq 5$

In exercises 11–12, sketch the graph of the second equation by reflecting the graph of the first equation.

11. $y = |x - 1|; \ y = -|x - 1|$ **12.** $y = \sqrt{x + 2}; \ y = \sqrt{2 - x}$

In exercises 13–14, use transformations to sketch the following.

13. $y = -(x - 2)^2 + 2$ **14.** $y = -2|x + 1|$

In exercises 15–16, determine an equation of the line satisfying the stated conditions. Write your answer in (a) point-slope form, (b) slope-intercept form, and (c) general form.

15. The line is perpendicular to the graph of $2x - 3y + 5 = 0$ with y-intercept -2.

16. The line is parallel to the graph of $2y - 6x + 3 = 0$ and passes through $(-3, 0)$.

In exercises 17–18, sketch the graph of each function. Label the vertex, the x-intercepts, the axis of symmetry, and two additional points.

17. $f(x) = x^2 - 4x + 4$ **18.** $f(x) = x^2 - 2x + 5$

In exercises 19–20, a right triangle ABC is given with c the length of the hypotenuse, and a and b the lengths of the legs. Find the length of the third side given the lengths of two sides.

19. $a = 8, b = 15$ **20.** $c = 25, b = 7$

In exercises 21–22, $\triangle ABC$ has right angle at C and sides of length a, b, and c. Given the following information, find the indicated parts.

21. $m(\angle A) = 60°$, $a = 20$, find c **22.** $m(\angle A) = 60°$, $c = 14$, find b.

In exercises 23–24, the given point is located on the terminal side of angle φ. Find $\sin \varphi$, $\cos \varphi$ and $\tan \varphi$.

23. $(-8, -20)$ **24.** $(22, -12)$

In exercises 25–28, evaluate each of the following.

25. $\sin 225°$ **26.** $\cos(-585°)$ **27.** $\tan(-480°)$ **28.** $\cot 330°$

In exercises 29–32, convert the measure of the given angle from degrees to radians or vice versa.

29. $-630°$ **30.** $-315°$ **31.** $\dfrac{8\pi}{3}$ **32.** $\dfrac{9\pi}{4}$

In exercises 33–34, use Table 4 or a calculator to determine the trigonometric function value.

33. $\cos(-3.8)$ **34.** $\sin(7.2)$

In exercises 35–36, determine the amplitude, period and phase shift, and sketch the graph.

35. $y = 3 \sin(2x - \pi)$ **36.** $y = 2 \cos\left(\dfrac{1}{2}x + \pi\right)$

Superset

37. For the function $f(x) = 3x^2 + 2x$, find the following: $\dfrac{f(2+h) - f(2)}{h}$.

38. Sketch the graph of the following function: $f(x) = \begin{cases} x, & \text{if } x < -3, \\ x^2 + 1, & \text{if } x > 0. \end{cases}$

39. The value of one trigonometric function of θ is given. If the terminal side of θ lies in the given quadrant, find the values of the other five trigonometric functions.

(a) fourth quadrant, $\cos \theta = \dfrac{9}{41}$ (b) third quadrant, $\sin \theta = -\dfrac{7}{25}$

40. A jogger runs six laps in five minutes on a circular track that has a diameter of 60 yd. What is the jogger's average speed?

9 Trigonometric Identities and Equations

9.1 Basic Trigonometric Identities

Goal A *to simplify an expression by using identities*

We refer to equations such as

$$(x + 1)^2 = x^2 + 2x + 1 \quad \text{or} \quad \frac{x^2 - 1}{x - 1} = x + 1$$

as identities. An **identity** is an equation that is true for all values of x for which the expressions are defined. The equation

$$(x + 1)^2 = x^2 + 2x + 1$$

is true for all real numbers x. The equation

$$\frac{x^2 - 1}{x - 1} = x + 1$$

is true for $x \neq 1$. As long as $x \neq 1$, the expression on the left-hand side of the equation is defined, and may be simplified by factoring and canceling:

$$\frac{x^2 - 1}{x - 1} = \frac{(x - 1)(x + 1)}{x - 1} = x + 1.$$

We have already seen trigonometric identities such as

$$\tan \theta = \frac{\sin \theta}{\cos \theta}.$$

This identity is true for all θ such that $\cos \theta \neq 0$. It is one of the basic identities we derived earlier. Recall that we derived eleven basic identities using the definitions of the trigonometric functions and the properties of the unit circle. They are called the **basic identities** because only the trigonometric definitions are used to derive them.

We classified the eleven basic identities into three categories:

Basic Trigonometric Identities

I. Reciprocal Identities

(1) $\qquad \sin \theta = \dfrac{1}{\csc \theta}$

(2) $\qquad \cos \theta = \dfrac{1}{\sec \theta}$

(3) $\qquad \tan \theta = \dfrac{1}{\cot \theta}$

(4) $\qquad \cot \theta = \dfrac{1}{\tan \theta}$

(5) $\qquad \sec \theta = \dfrac{1}{\cos \theta}$

(6) $\qquad \csc \theta = \dfrac{1}{\sin \theta}$

II. Quotient Identities

(7) $\qquad \dfrac{\sin \theta}{\cos \theta} = \tan \theta$

(8) $\qquad \dfrac{\cos \theta}{\sin \theta} = \cot \theta$

III. Pythagorean Identities

(9) $\qquad \sin^2 \theta + \cos^2 \theta = 1$

(10) $\qquad \tan^2 \theta + 1 = \sec^2 \theta$

(11) $\qquad 1 + \cot^2 \theta = \csc^2 \theta$

You should memorize the basic identities listed above. We now consider an example in which knowledge of the basic identities is essential in simplifying the given expression.

Example 1 Show that the expression $\dfrac{1 + \tan^2 \theta}{\csc^2 \theta}$ may be simplified to $\tan^2 \theta$.

Solution
$$\dfrac{1 + \tan^2 \theta}{\csc^2 \theta} = \dfrac{\sec^2 \theta}{\csc^2 \theta} \qquad \blacksquare \text{ By identity (10), } \tan^2 \theta + 1 = \sec^2 \theta.$$

$$= \dfrac{\dfrac{1}{\cos^2 \theta}}{\dfrac{1}{\sin^2 \theta}} \qquad \begin{array}{l} \blacksquare \text{ By (5), } \sec^2 \theta = \dfrac{1}{\cos^2 \theta}. \\[6pt] \blacksquare \text{ By (6), } \csc^2 \theta = \dfrac{1}{\sin^2 \theta}. \end{array}$$

$$= \dfrac{\sin^2 \theta}{\cos^2 \theta} \qquad \blacksquare \text{ Divide: multiply by the reciprocal of the denominator.}$$

$$= \tan^2 \theta \qquad \blacksquare \text{ By (7), } \dfrac{\sin \theta}{\cos \theta} = \tan \theta.$$

There are no standard steps to take to simplify a trigonometric expression. Simplifying trigonometric expressions is similar to factoring polynomials: by trial and error and by experience, you learn what will work in which situations. One useful technique is to begin by rewriting the entire expression in terms of sines and cosines.

Example 2 Express $\left(1 - \dfrac{1}{\csc \theta}\right)^2 + \cos^2 \theta$ in terms of $\sin \theta$.

Solution
$$\left(1 - \dfrac{1}{\csc \theta}\right)^2 + \cos^2 \theta = (1 - \sin \theta)^2 + \cos^2 \theta \qquad \blacksquare \text{ By (1)}$$
$$= 1 - 2 \sin \theta + \sin^2 \theta + \cos^2 \theta \qquad \blacksquare \ (1 - \sin \theta)^2 \text{ is expanded.}$$
$$= 1 - 2 \sin \theta + 1 \qquad \blacksquare \text{ By (9)}$$
$$= 2 - 2 \sin \theta$$

Goal B *to verify simple identities*

We are now in a position to begin verifying new trigonometric identities. The primary strategy that we will use is to choose the expression on one side of the identity and to transform it so that it is identical to the expression on the other side. This "transformation" will come about by using the rules of algebra and the Basic Identities. Before we begin with some examples, let us offer a few words of advice.

1. Memorize the Basic Identities. Verifying a new identity often means that you must run through a mental checklist of options that might prove effective in rewriting one of the expressions.
2. It is often best to proceed by first writing down the more complicated side of the identity, and then transforming it in a step-by-step fashion until it looks exactly like the other side of the identity.
3. As an aid in rewriting the more complicated side of the identity, it is often useful to rewrite this expression in terms of sines and cosines only.

Careful! One approach *not* to be used is to write down the entire identity as your first step, and "do the same thing to both sides," as you would in solving an equation. Working on both sides of an equation in this way can be done only when the equation is assumed to be true. In verifying an identity, you are trying to prove that the equation is true. Thus, you may not assume that it is already true.

Example 3 Verify the identity: $\dfrac{1 - \cos^2 \theta}{\tan \theta} = \sin \theta \cos \theta$.

Solution
$$\dfrac{1 - \cos^2 \theta}{\tan \theta} = \dfrac{\sin^2 \theta}{\tan \theta} \qquad \blacksquare \text{ By (9), } 1 - \cos^2 \theta = \sin^2 \theta.$$

$$= \dfrac{\sin^2 \theta}{\dfrac{\sin \theta}{\cos \theta}} \qquad \blacksquare \text{ By (7)}$$

$$= \dfrac{\sin^2 \theta}{1} \cdot \dfrac{\cos \theta}{\sin \theta}$$

$$= \sin \theta \cos \theta$$

Example 4 Verify the identity: $\dfrac{\tan x}{1 - \sec x} = -\dfrac{1 + \sec x}{\tan x}$.

Solution
$$\dfrac{\tan x}{1 - \sec x} = \dfrac{\tan x}{1 - \sec x} \cdot \dfrac{1 + \sec x}{1 + \sec x} \qquad \blacksquare \text{ Notice that multiplying by } \dfrac{1 + \sec x}{1 + \sec x} \text{ transforms the denominator into } 1 - \sec^2 x.$$

$$= \dfrac{(\tan x)(1 + \sec x)}{1 - \sec^2 x}$$

$$= \dfrac{(\tan x)(1 + \sec x)}{-\tan^2 x} \qquad \blacksquare \text{ By (10), } 1 - \sec^2 x = -\tan^2 x.$$

$$= -\dfrac{1 + \sec x}{\tan x} \qquad \blacksquare \text{ tan } x \text{ has been canceled from the numerator and denominator.}$$

Exercise Set 9.1

Goal A

In exercises 1–8, show that the first expression may be rewritten as the second by using the Reciprocal and Quotient Identities.

1. $\sin \alpha \sec \alpha$; $\tan \alpha$
2. $\dfrac{\csc \alpha}{\sec \alpha}$; $\cot \alpha$
3. $\dfrac{\sin \varphi}{\csc \varphi}$; $\sin^2 \varphi$
4. $\dfrac{\cos \varphi}{\sec \varphi}$; $\cos^2 \varphi$
5. $\csc \beta$; $\sec \beta \cot \beta$
6. $\sin \beta$; $\dfrac{\tan \beta}{\sec \beta}$
7. $\csc \theta \cot \theta$; $\cos \theta \csc^2 \theta$
8. $\csc^2 \theta \tan \theta$; $\sec \theta \csc \theta$

In exercises 9–18, rewrite the expression in terms of only the sine or only the tangent.

9. $\dfrac{\csc \alpha}{1 + \cot^2 \alpha}$
10. $\dfrac{\tan \alpha \sec \alpha}{1 + \tan^2 \alpha}$
11. $\cot \beta (\sec^2 \beta - 1)$
12. $\sec^2 \alpha - 2$
13. $\csc \beta - \cos \beta \cot \beta$
14. $\dfrac{\sec \beta - \cos \beta}{\tan \beta}$
15. $(1 + \tan^2 \varphi)\sin^2 \varphi$
16. $\dfrac{\sec^2 \varphi - 1}{\sec^2 \varphi}$
17. $\dfrac{\sin \alpha + \cos \alpha}{\cos \alpha}$
18. $\dfrac{\csc \alpha + \sec \alpha}{\csc \alpha}$

Goal B

In exercises 19–38, verify each identity by using the Basic Identities.

19. $\tan \alpha + \cot \alpha = \csc \alpha \sec \alpha$
20. $\dfrac{\sec \alpha}{\tan \alpha} - \dfrac{\tan \alpha}{\sec \alpha} = \cos \alpha \cot \alpha$
21. $(1 + \cot \beta)^2 - \csc^2 \beta = 2 \cot \beta$
22. $(1 + \csc^2 \beta) - \cot^2 \beta = 2$
23. $\cos \alpha (\tan \alpha + \sec \alpha) = 1 + \sin \alpha$
24. $\sin \alpha (\cot \alpha + \csc \alpha) = \cos \alpha + 1$
25. $\sec \beta + \csc \beta \cot \beta = \csc^2 \beta \sec \beta$
26. $\sec \beta - \sin \beta \cot \beta = \tan \beta \sin \beta$
27. $\dfrac{\sin \gamma}{\cos \gamma + 1} + \dfrac{\cos \gamma - 1}{\sin \gamma} = 0$
28. $\dfrac{\sin \gamma}{\csc \gamma} + \dfrac{\cos \gamma}{\sec \gamma} = 1$
29. $\dfrac{(\sin \gamma + \cos \gamma)^2}{\sin \gamma \cos \gamma} = \csc \gamma \sec \gamma + 2$
30. $\left(\dfrac{\sin \gamma}{\cos \gamma} + \dfrac{\cos \gamma}{\sin \gamma}\right)^2 = \csc^2 \gamma \sec^2 \gamma$
31. $(\tan^2 \alpha + 1)(1 + \cos^2 \alpha) = \tan^2 \alpha + 2$
32. $(1 + \cot^2 \alpha)(1 + \sin^2 \alpha) = 2 + \cot^2 \alpha$
33. $\csc x \sec x = \tan x + \cot x$
34. $\cos x = (\csc x - \sin x)\tan x$
35. $\dfrac{\tan \beta - \sin \beta}{\sin^3 \beta} = \dfrac{\sec \beta}{1 + \cos \beta}$
36. $\dfrac{\cot \beta - \cos \beta}{\cos^3 \beta} = \dfrac{\csc \beta}{1 + \sin \beta}$
37. $(\sin \alpha + \cos \alpha)^2 - (\sin \beta + \cos \beta)^2$
$= 2(\sin \alpha \cos \alpha - \sin \beta \cos \beta)$
38. $(\sin \alpha - \sin \beta)^2 + (\cos \alpha - \cos \beta)^2$
$= 2(1 - \sin \alpha \sin \beta - \cos \alpha \cos \beta)$

Superset

In exercises 39–43, the quadrantal angle is given. Express the first trigonometric function in terms of the second.

39. $\sin \alpha$ in terms of $\cot \alpha$; α in Quadrant I
40. $\cos \alpha$ in terms of $\cot \alpha$; α in Quadrant III
41. $\sec \beta$ in terms of $\sin \beta$; β in Quadrant II
42. $\tan \varphi$ in terms of $\cos \varphi$; φ in Quadrant IV
43. $\csc \gamma$ in terms of $\sec \gamma$; γ in Quadrant III
44. Express the other 5 trigonometric functions in terms of (a) $\sin \alpha$ for α in Quadrant II; (b) $\cos \alpha$ for α in the Quadrant III.

9.2 Sum and Difference Identities

Goal A *to apply the sum and difference identities*

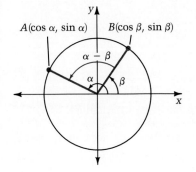

Figure 9.1

Suppose you wish to determine $\sin(\alpha + \beta)$ and you already know the trigonometric function values of α and β. Although it might be tempting to write $\sin \alpha + \sin \beta$ as your answer, it would be incorrect to do so. In this section we will derive identities for the trigonometric functions of $\alpha + \beta$ and $\alpha - \beta$, where α and β represent any real numbers or angle measurements. We begin by determining a formula for $\cos(\alpha + \beta)$, from which the other sum and difference formulas can be easily derived.

Suppose α and β are positive real numbers and $\alpha > \beta$. To visualize what we are about to do, we can think of α and β as the lengths of standard arcs, or the measures of angles in standard position on the unit circle. Interpreted this way, α and β determine points A and B on the unit circle, shown in Figure 9.1. Note that A has coordinates $(\cos \alpha, \sin \alpha)$, and B has coordinates $(\cos \beta, \sin \beta)$.

In Figure 9.2(a) we show $\alpha - \beta$ and a line segment connecting the points A and B. The line segment joining the endpoints of an arc is called a **chord**. We let d be the length of chord AB. In Figure 9.2(b), the angle $\alpha - \beta$ has been redrawn in standard position, and the chord of length d now has endpoints $(\cos(\alpha - \beta), \sin(\alpha - \beta))$ and $(1, 0)$, labeled A' and B' respectively.

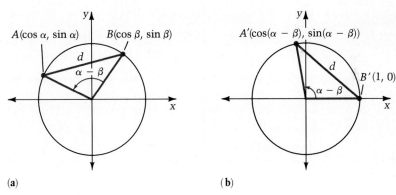

(a)   (b)

Figure 9.2

By means of the distance formula, we express d^2 in two different ways: first by using the coordinates of A and B (Figure 9.2(a)), and then by using the coordinates of A' and B' (Figure 9.2(b)). Using A and B, we have

$$d^2 = \text{(distance between } A \text{ and } B)^2 = (\cos \alpha - \cos \beta)^2 + (\sin \alpha - \sin \beta)^2$$
$$= (\cos^2 \alpha - 2 \cos \alpha \cos \beta + \cos^2 \beta) + (\sin^2 \alpha - 2 \sin \alpha \sin \beta + \sin^2 \beta)$$
$$= (\sin^2 \alpha + \cos^2 \alpha) + (\sin^2 \beta + \cos^2 \beta) - 2 \cos \alpha \cos \beta - 2 \sin \alpha \sin \beta$$
$$= \quad 1 \quad + \quad 1 \quad - 2(\cos \alpha \cos \beta + \sin \alpha \sin \beta)$$
$$= 2 - 2(\cos \alpha \cos \beta + \sin \alpha \sin \beta)$$

Using A' and B', we have

$$\begin{aligned} d^2 &= (\text{distance between } A' \text{ and } B')^2 \\ &= (\cos(\alpha - \beta) - 1)^2 + (\sin(\alpha - \beta) - 0)^2 \\ &= \cos^2(\alpha - \beta) - 2\cos(\alpha - \beta) + 1 + \sin^2(\alpha - \beta) \\ &= \underbrace{\sin^2(\alpha - \beta) + \cos^2(\alpha - \beta)}_{1} + 1 - 2\cos(\alpha - \beta) \\ &= 2 - 2\cos(\alpha - \beta) \end{aligned}$$

Equating the two expressions for d^2, we have

$$2 - 2\cos(\alpha - \beta) = 2 - 2(\cos\alpha\cos\beta + \sin\alpha\sin\beta).$$

Then, by adding (-2) to both sides and dividing by (-2), we conclude that

$$(12) \qquad \cos(\alpha - \beta) = \cos\alpha\cos\beta + \sin\alpha\sin\beta.$$

Although we have developed identity (12) for α and β that are positive with $\alpha > \beta$, a similar argument establishes (12) for all α and β.

Example 1 Determine $\cos 15°$ without using Table 4 or a calculator.

Solution
$$\begin{aligned} \cos 15° &= \cos(45° - 30°) \\ &= \cos 45° \cos 30° + \sin 45° \sin 30° \qquad \blacksquare \text{ By identity (12)} \\ &= \left(\frac{\sqrt{2}}{2}\right)\left(\frac{\sqrt{3}}{2}\right) + \left(\frac{\sqrt{2}}{2}\right)\left(\frac{1}{2}\right) \\ &= \frac{\sqrt{6}}{4} + \frac{\sqrt{2}}{4} \end{aligned}$$

Thus, $\cos 15° = \dfrac{\sqrt{6} + \sqrt{2}}{4}$.

Recall that sine is an odd function and cosine is an even function. These two facts are represented as follows: for all real numbers x,

$$\sin(-x) = -\sin x \quad \text{and} \quad \cos(-x) = \cos x.$$

We use these facts to establish another identity:

$$\begin{aligned} \cos(\alpha + \beta) &= \cos(\alpha - (-\beta)) \\ &= \cos\alpha\cos(-\beta) + \sin\alpha \cdot (\sin(-\beta)) \qquad &&\blacksquare \text{ By (12)} \\ &= \cos\alpha\cos\beta + \sin\alpha\,(-\sin\beta) \qquad &&\blacksquare \text{ cosine is even;} \\ &= \cos\alpha\cos\beta - \sin\alpha\sin\beta &&\text{ sine is odd} \end{aligned}$$

Thus,

(13) $$\cos(\alpha + \beta) = \cos\alpha\cos\beta - \sin\alpha\sin\beta$$

Recall that a trigonometric function value of an acute angle θ is equal to the co-trigonometric function value of the complement of θ. For example, for any acute angle θ,

$$\sin\theta = \cos\left(\frac{\pi}{2} - \theta\right).$$

By virtue of (12) we can now prove such a statement true for any θ.

$$\cos\left(\frac{\pi}{2} - \theta\right) = \cos\frac{\pi}{2}\cos\theta + \sin\frac{\pi}{2}\sin\theta$$
$$= (0)\cdot\cos\theta + (1)\cdot\sin\theta$$
$$= \sin\theta$$

In a similar manner, we can show that $\sin\left(\frac{\pi}{2} - \theta\right) = \cos\theta$.

We will use these results in deriving an identity for $\sin(\alpha + \beta)$.

$$\sin(\alpha + \beta) = \cos\left(\frac{\pi}{2} - (\alpha + \beta)\right) \qquad \blacksquare \; \sin\theta = \cos\left(\frac{\pi}{2} - \theta\right).$$
$$= \cos\left(\left(\frac{\pi}{2} - \alpha\right) - \beta\right)$$
$$= \cos\left(\frac{\pi}{2} - \alpha\right)\cos\beta + \sin\left(\frac{\pi}{2} - \alpha\right)\sin\beta \qquad \blacksquare \; \text{By (12)}$$
$$= \sin\alpha\cos\beta + \cos\alpha\sin\beta$$

Thus,

(14) $$\sin(\alpha + \beta) = \sin\alpha\cos\beta + \cos\alpha\sin\beta.$$

Replacing β with $(-\beta)$ in (14), we have

$$\sin(\alpha - \beta) = \sin(\alpha + (-\beta))$$
$$= \sin\alpha\cos(-\beta) + \cos\alpha\sin(-\beta) \qquad \blacksquare \; \text{By (14)}$$
$$= \sin\alpha\cos\beta + \cos\alpha(-\sin\beta) \qquad \blacksquare \; \text{Cosine is even;}$$
$$\qquad\qquad\qquad\qquad\qquad\qquad\qquad\qquad\qquad\text{sine is odd.}$$
$$= \sin\alpha\cos\beta - \cos\alpha\sin\beta$$

Thus,

(15) $$\sin(\alpha - \beta) = \sin\alpha\cos\beta - \cos\alpha\sin\beta.$$

Example 2 Determine $\sin \dfrac{5\pi}{12}$ without using Table 4 or a calculator.

Solution
$$\sin \dfrac{5\pi}{12} = \sin\left(\dfrac{\pi}{4} + \dfrac{\pi}{6}\right)$$
$$= \sin \dfrac{\pi}{4} \cos \dfrac{\pi}{6} + \cos \dfrac{\pi}{4} \sin \dfrac{\pi}{6}$$
$$= \dfrac{\sqrt{2}}{2} \cdot \dfrac{\sqrt{3}}{2} + \dfrac{\sqrt{2}}{2} \cdot \dfrac{1}{2}$$

Thus,
$$\sin \dfrac{5\pi}{12} = \dfrac{\sqrt{6}}{4} + \dfrac{\sqrt{2}}{4} = \dfrac{\sqrt{6} + \sqrt{2}}{4}.$$

Notice that in Example 1 we found that $\cos 15° = \dfrac{\sqrt{6} + \sqrt{2}}{4}$, and in Example 2 we found that $\sin \dfrac{5\pi}{12}$ also equals $\dfrac{\sqrt{6} + \sqrt{2}}{4}$. Since $\dfrac{5\pi}{12} = 75°$, this result is expected because of the cofunction identity:
$$\cos 15° = \sin(90° - 15°) = \sin 75°.$$

Example 3 If $\sin \alpha = \tfrac{12}{13}$ for some second quadrant angle α, and $\sin \beta = \tfrac{3}{5}$ for some first quadrant angle β, determine (a) $\sin(\alpha - \beta)$ and (b) $\sin(\alpha + \beta)$.

Solution We begin by using the Pythagorean Identity (9) to find $\cos \alpha$ and $\cos \beta$.

$\cos \alpha = \pm\sqrt{1 - \sin^2 \alpha}$

$\cos \alpha = -\sqrt{1 - \left(\dfrac{12}{13}\right)^2}$ ■ Since α is a second quadrant angle, $\cos \alpha$ is negative.

$\cos \alpha = -\sqrt{1 - \dfrac{144}{169}}$

$\cos \alpha = -\sqrt{\dfrac{25}{169}} = -\dfrac{5}{13}$

$\cos \beta = \sqrt{1 - \left(\dfrac{3}{5}\right)^2}$ ■ Since β is a first quadrant angle, $\cos \beta$ is positive.

$\cos \beta = \sqrt{\dfrac{16}{25}} = \dfrac{4}{5}$

(a) $\sin(\alpha - \beta) = \sin\alpha\cos\beta - \cos\alpha\sin\beta = \left(\frac{12}{13}\right)\left(\frac{4}{5}\right) - \left(\frac{-5}{13}\right)\left(\frac{3}{5}\right) = \frac{63}{65}$

(b) $\sin(\alpha + \beta) = \sin\alpha\cos\beta + \cos\alpha\sin\beta = \left(\frac{12}{13}\right)\left(\frac{4}{5}\right) + \left(\frac{-5}{13}\right)\left(\frac{3}{5}\right) = \frac{33}{65}$

The identity for $\tan(\alpha + \beta)$ may be derived by using the quotient identity (7) and the identities for the sine and cosine of a sum.

$$\tan(\alpha + \beta) = \frac{\sin(\alpha + \beta)}{\cos(\alpha + \beta)} \quad \blacksquare \text{ By (7)}$$

$$= \frac{\sin\alpha\cos\beta + \cos\alpha\sin\beta}{\cos\alpha\cos\beta - \sin\alpha\sin\beta} \quad \blacksquare \text{ By (14) and (13)}$$

$$= \frac{\dfrac{\sin\alpha\cos\beta}{\cos\alpha\cos\beta} + \dfrac{\cos\alpha\sin\beta}{\cos\alpha\cos\beta}}{\dfrac{\cos\alpha\cos\beta}{\cos\alpha\cos\beta} - \dfrac{\sin\alpha\sin\beta}{\cos\alpha\cos\beta}} \quad \blacksquare \text{ Divide each term by } \cos\alpha\cos\beta.$$

$$= \frac{\tan\alpha + \tan\beta}{1 - \tan\alpha\tan\beta} \quad \blacksquare \text{ (7) is used four times.}$$

Therefore,

(16) $$\tan(\alpha + \beta) = \frac{\tan\alpha + \tan\beta}{1 - \tan\alpha\tan\beta}$$

In a similar manner,

(17) $$\tan(\alpha - \beta) = \frac{\tan\alpha - \tan\beta}{1 + \tan\alpha\tan\beta}$$

The derivation of identity (17) is left as an exercise.

Example 4 Without using Table 4 or a calculator, find $\tan\frac{7\pi}{12}$.

Solution

$$\tan\left(\frac{7\pi}{12}\right) = \tan\left(\frac{\pi}{4} + \frac{\pi}{3}\right) = \frac{\tan\dfrac{\pi}{4} + \tan\dfrac{\pi}{3}}{1 - \tan\dfrac{\pi}{4}\tan\dfrac{\pi}{3}} \quad \blacksquare \text{ By (16)}$$

$$= \frac{1 + \sqrt{3}}{1 - \sqrt{3}}$$

When an answer has a radical in both the numerator and the denominator, it is often easier to approximate the answer if we rationalize the denominator first. In the case of Example 4, we have,

$$\frac{1+\sqrt{3}}{1-\sqrt{3}} = \frac{(1+\sqrt{3})(1+\sqrt{3})}{(1-\sqrt{3})(1+\sqrt{3})} = \frac{1+2\sqrt{3}+3}{1-3} = \frac{4+2\sqrt{3}}{-2} = -(2+\sqrt{3}).$$

Thus, we see that $\tan \frac{7\pi}{12}$ is approximately -3.73. ($\sqrt{3} \approx 1.73$.)

Example 5 Simplify the following trigonometric expression as a single function of some angle:

$$\frac{\sin(\alpha+\beta) + \sin(\alpha-\beta)}{\cos(\alpha+\beta) + \cos(\alpha-\beta)}.$$

Solution

$$\frac{\sin(\alpha+\beta) + \sin(\alpha-\beta)}{\cos(\alpha+\beta) + \cos(\alpha-\beta)}$$

$$= \frac{(\sin\alpha\cos\beta + \cos\alpha\sin\beta) + (\sin\alpha\cos\beta - \cos\alpha\sin\beta)}{(\cos\alpha\cos\beta - \sin\alpha\sin\beta) + (\cos\alpha\cos\beta + \sin\alpha\sin\beta)}$$

$$= \frac{2\sin\alpha\cos\beta}{2\cos\alpha\cos\beta}$$

$$= \frac{\sin\alpha}{\cos\alpha} = \tan\alpha$$

The sum and difference identities derived in this section are listed below for easy reference.

Sum and Difference Identities

(12) $\cos(\alpha-\beta) = \cos\alpha\cos\beta + \sin\alpha\sin\beta$
(13) $\cos(\alpha+\beta) = \cos\alpha\cos\beta - \sin\alpha\sin\beta$
(14) $\sin(\alpha+\beta) = \sin\alpha\cos\beta + \cos\alpha\sin\beta$
(15) $\sin(\alpha-\beta) = \sin\alpha\cos\beta - \cos\alpha\sin\beta$
(16) $\tan(\alpha+\beta) = \dfrac{\tan\alpha + \tan\beta}{1 - \tan\alpha\tan\beta}$
(17) $\tan(\alpha-\beta) = \dfrac{\tan\alpha - \tan\beta}{1 + \tan\alpha\tan\beta}$

Exercise Set 9.2   ≡ Calculator Exercises in the Appendix

Goal A

In exercises 1–12, without using Table 4 or a calculator, find the following trigonometric values by using the sum and difference identities.

1. $\sin 120°$
2. $\cos 120°$
3. $\sin 15°$
4. $\cos 15°$
5. $\cos 75°$
6. $\sin 75°$
7. $\sin 195°$
8. $\cos 105°$
9. $\cos(-45°)$
10. $\sin(-45°)$
11. $\sin(-135°)$
12. $\cos(-135°)$

In exercises 13–20, without using Table 4 or a calculator, find the following trigonometric values.

13. $\cos \frac{5\pi}{12}$
14. $\sin \frac{5\pi}{12}$
15. $\sin \frac{7\pi}{12}$
16. $\cos \frac{7\pi}{12}$
17. $\cos\left(-\frac{5\pi}{12}\right)$
18. $\sin\left(-\frac{5\pi}{12}\right)$
19. $\cos\left(-\frac{11\pi}{12}\right)$
20. $\sin\left(-\frac{11\pi}{12}\right)$

In exercises 21–26, find the following trigonometric values without using Table 4 or a calculator.

21. $\tan 75°$
22. $\tan 15°$
23. $\tan \frac{19\pi}{12}$
24. $\tan \frac{\pi}{12}$
25. $\tan\left(-\frac{\pi}{12}\right)$
26. $\tan\left(-\frac{5\pi}{12}\right)$

In exercises 27–30, use the given information to determine the following: (a) $\sin(\alpha + \beta)$, (b) $\cos(\alpha + \beta)$, and (c) $\tan(\alpha + \beta)$.

27. $\sin \alpha = \frac{4}{5}$, α is a first quadrant angle, β is a third quadrant angle, and $\sin \beta = -\frac{12}{13}$.
28. $\sin \alpha = -\frac{4}{5}$, α is a fourth quadrant angle, β is a second quadrant angle, and $\sin \beta = \frac{3}{5}$.
29. $\cos \alpha = \frac{5}{13}$, α is a first quadrant angle, β is a second quadrant angle, and $\tan \beta = -\frac{3}{2}$.
30. $\cos \alpha = -\frac{5}{12}$, α is a second quadrant angle, β is a third quadrant angle, and $\tan \beta = \frac{3}{2}$.

In exercises 31–34, given the information in exercises 27–30, find the following:
(a) $\sin(\alpha - \beta)$, (b) $\cos(\alpha - \beta)$, and (c) $\tan(\alpha - \beta)$.

In exercises 35–44, verify each identity.

35. $\sin(\alpha - \beta)\cos \beta + \cos(\alpha - \beta)\sin \beta = \sin \alpha$
36. $\sin(\alpha - \beta)\sin(\alpha + \beta) = \sin^2 \alpha - \sin^2 \beta$
37. $\cos(\alpha - \beta)\cos(\alpha + \beta) = \cos^2 \alpha - \sin^2 \beta$
38. $\cos(\alpha - \beta) + \cos(\alpha + \beta) = 2 \cos \alpha \cos \beta$
39. $\tan\left(\alpha + \frac{\pi}{4}\right) = \frac{\cos \alpha + \sin \alpha}{\cos \alpha - \sin \alpha}$
40. $\tan\left(\frac{\pi}{4} - \alpha\right) = \frac{\cos \alpha - \sin \alpha}{\cos \alpha + \sin \alpha}$
41. $\tan\left(\frac{\pi}{4} + \alpha\right)\tan\left(\frac{\pi}{4} - \alpha\right) = \tan \frac{\pi}{4}$
42. $\frac{\tan(\alpha - \beta)}{\tan(\beta - \alpha)} = -1$
43. $\frac{\sin(x + y)}{\sin(x - y)} = \frac{\tan x + \tan y}{\tan x - \tan y}$
44. $\frac{\sin(x + y)}{\sin x \sin y} = \frac{1}{\tan x} + \frac{1}{\tan y}$

Superset

In exercises 45–52, simplify the given sum or difference.

45. $\cos\left(\alpha + \frac{\pi}{6}\right)$
46. $\sin\left(\alpha + \frac{\pi}{6}\right)$
47. $\tan\left(\alpha + \frac{\pi}{4}\right)$
48. $\cos\left(\beta + \frac{\pi}{4}\right)$
49. $\sin\left(\frac{\pi}{2} - \varphi\right)$
50. $\tan\left(\frac{\pi}{2} - \beta\right)$
51. $\sin\left(\frac{\pi}{4} - \alpha\right)$
52. $\cos\left(\frac{\pi}{4} - \beta\right)$

In exercises 53–58, derive an identity involving the given expression.

53. $\tan(\alpha - \beta)$
54. $\cot(\alpha + \beta)$
55. $\sec(\alpha + \beta)$
56. $\csc(\alpha + \beta)$
57. $\sin 2\alpha$ (Hint: $2\alpha = \alpha + \alpha$.)
58. $\cos 2\alpha$

9.3 The Double-Angle and Half-Angle Identities

Goal A *to apply the double-angle identities*

In this section we will derive identities for $\sin 2\theta$, $\cos 2\theta$, $\tan 2\theta$, $\sin \frac{\theta}{2}$, $\cos \frac{\theta}{2}$, and $\tan \frac{\theta}{2}$. The first three identities are called **double-angle identities**; the other three are called **half-angle identities**.

The double-angle identities follow easily from the sum identities. For example, consider the identity $\sin(\alpha + \beta) = \sin \alpha \cos \beta + \cos \alpha \sin \beta$. If $\alpha = \beta$, we have $\sin(2\alpha) = \sin(\alpha + \alpha) = \sin \alpha \cos \alpha + \cos \alpha \sin \alpha$. Thus, for any angle θ

(18) $$\sin 2\theta = 2 \sin \theta \cos \theta.$$

Similarly, $\cos 2\theta = \cos(\theta + \theta) = \cos \theta \cos \theta - \sin \theta \sin \theta$, that is,

(19) $$\cos 2\theta = \cos^2 \theta - \sin^2 \theta.$$

Using the Pythagorean Identity, $\sin^2 \theta + \cos^2 \theta = 1$, we can use (19) to derive two other identities for $\cos 2\theta$:

$$\cos 2\theta = \cos^2 \theta - \sin^2 \theta$$
$$= (1 - \sin^2 \theta) - \sin^2 \theta$$

(20) $$\cos 2\theta = 1 - 2 \sin^2 \theta.$$

Also,

$$\cos 2\theta = \cos^2 \theta - \sin^2 \theta$$
$$= \cos^2 \theta - (1 - \cos^2 \theta)$$

(21) $$\cos 2\theta = 2 \cos^2 \theta - 1.$$

We leave as an exercise to show that

(22) $$\tan 2\theta = \frac{2 \tan \theta}{1 - \tan^2 \theta}.$$

Example 1 If $\sin \theta = 0.6$ and $\cos \theta = 0.8$, find (a) $\sin 2\theta$, and (b) $\cos 2\theta$.

Solution (a) $\sin 2\theta = 2 \sin \theta \cos \theta = 2(0.6)(0.8) = 0.96$ ■ By (18)

(b) $\cos 2\theta = \cos^2 \theta - \sin^2 \theta = (0.8)^2 - (0.6)^2 = 0.28$ ■ By (19)

Example 2 Use double-angle identities and the values of sin 30°, cos 30°, and tan 30° from the Table of Special Angles to determine the following: (a) sin 60°, (b) cos 60°, (c) tan 60°.

Solution (a) $\sin 60° = \sin 2(30°) = 2 \sin 30° \cos 30° = 2\left(\frac{1}{2}\right)\left(\frac{\sqrt{3}}{2}\right) = \frac{\sqrt{3}}{2}$

(b) $\cos 60° = \cos 2(30°) = \cos^2 30° - \sin^2 30° = \left(\frac{\sqrt{3}}{2}\right)^2 - \left(\frac{1}{2}\right)^2 = \frac{1}{2}$

(c) $\tan 60° = \tan 2(30°) = \dfrac{2 \tan 30°}{1 - \tan^2 30°} = \dfrac{2\left(\frac{\sqrt{3}}{3}\right)}{1 - \left(\frac{\sqrt{3}}{3}\right)^2}$

$= \dfrac{\frac{2\sqrt{3}}{3}}{1 - \frac{3}{9}} = \dfrac{2\sqrt{3}}{3} \cdot \dfrac{9}{6} = \sqrt{3}$

Example 3 Given that $\sin \frac{\pi}{9} \approx 0.3420$, find $\tan \frac{2\pi}{9}$.

Solution
$\tan \dfrac{2\pi}{9} = \dfrac{\sin \frac{2\pi}{9}}{\cos \frac{2\pi}{9}} = \dfrac{\sin 2\left(\frac{\pi}{9}\right)}{\cos 2\left(\frac{\pi}{9}\right)}$ ■ By (7)

$= \dfrac{2 \sin \frac{\pi}{9} \cos \frac{\pi}{9}}{1 - 2 \sin^2\left(\frac{\pi}{9}\right)}$ ■ By (18)

■ By (20)

To proceed we need to determine $\cos \frac{\pi}{9}$.

$\sin^2\left(\frac{\pi}{9}\right) + \cos^2\left(\frac{\pi}{9}\right) = 1$ ■ By (9)

$\cos \frac{\pi}{9} = \sqrt{1 - \sin^2\left(\frac{\pi}{9}\right)}$ ■ Since $\frac{\pi}{9}$ is a first quadrant angle, $\cos \frac{\pi}{9}$ is positive.

$\approx \sqrt{1 - (0.3420)^2} = 0.9397$

Thus, $\tan \dfrac{2\pi}{9} = \dfrac{2 \sin \frac{\pi}{9} \cos \frac{\pi}{9}}{1 - 2 \sin^2 \frac{\pi}{9}} \approx \dfrac{2(0.3420)(0.9397)}{1 - 2(0.3420)^2} \approx 0.8390.$

Example 4 Given $\sin\theta = -\frac{4}{5}$ and that θ is a third quadrant angle, determine (a) $\sin 2\theta$ (b) $\cos 2\theta$ (c) $\tan 2\theta$.

Solution (a) To apply the identity $\sin 2\theta = 2\sin\theta\cos\theta$, we need to determine $\cos\theta$.

$$\cos\theta = -\sqrt{1 - \sin^2\theta} \qquad \blacksquare \text{ By (9). Since } \theta \text{ is a third quadrant angle, } \cos\theta \text{ is negative.}$$

$$= -\sqrt{1 - \left(-\frac{4}{5}\right)^2} = -\sqrt{1 - \frac{16}{25}} = -\frac{3}{5}$$

$$\sin 2\theta = 2\left(-\frac{4}{5}\right)\left(-\frac{3}{5}\right) = \frac{24}{25}. \qquad \blacksquare \text{ By (18)}$$

(b) $\cos 2\theta = 1 - 2\sin^2\theta$ $\qquad\blacksquare$ Identity (20)

$$= 1 - 2\left(-\frac{4}{5}\right)^2 = 1 - \frac{32}{25} = -\frac{7}{25}$$

(c) $\tan 2\theta = \dfrac{\sin 2\theta}{\cos 2\theta} = \dfrac{\frac{24}{25}}{-\frac{7}{25}} = -\dfrac{24}{7}$

In Example 4(c), we used the quotient identity (7) to determine $\tan 2\theta$, since we already knew the values of $\sin 2\theta$ and $\cos 2\theta$ from parts (a) and (b). We could also have used the double-angle identity (22),

$$\tan 2\theta = \frac{2\tan\theta}{1 - \tan^2\theta}.$$

In that case, we would first need to find the value of $\tan\theta$ by computing $\dfrac{\sin\theta}{\cos\theta}$.

Goal B *to use the half-angle identities*

To derive the first half-angle identity, we solve the double-angle identity $\cos 2\theta = 1 - 2\sin^2\theta$ for $\sin^2\theta$:

$$\sin^2\theta = \frac{1 - \cos 2\theta}{2}.$$

If we replace θ with $\dfrac{\theta}{2}$, we have

$$\sin^2\frac{\theta}{2} = \frac{1 - \cos\theta}{2}.$$

9 / Trigonometric Identities and Equations

Thus, we can state our first half-angle identity as follows:

(23) $$\sin \frac{\theta}{2} = \pm \sqrt{\frac{1 - \cos \theta}{2}}$$

■ The positive or negative sign is used depending on the quadrant in which $\frac{\theta}{2}$ is located.

Example 5 Use identity (23) to determine $\sin 15°$.

Solution
$$\sin 15° = \sin \frac{30°}{2} = \sqrt{\frac{1 - \cos 30°}{2}}$$
$$= \sqrt{\frac{1 - \frac{\sqrt{3}}{2}}{2}} = \sqrt{\frac{1}{2} - \frac{\sqrt{3}}{4}} = \sqrt{\frac{2}{4} - \frac{\sqrt{3}}{4}} = \frac{\sqrt{2 - \sqrt{3}}}{2}$$

■ We choose the positive sign since $15°$ is a first quadrant angle.

To derive a half-angle identity for the cosine function, we solve the identity $\cos 2\theta = 2\cos^2 \theta - 1$ for $\cos \theta$:

$$\cos \theta = \pm \sqrt{\frac{1 + \cos 2\theta}{2}}.$$

Replacing θ with $\frac{\theta}{2}$, we conclude that

(24) $$\cos \frac{\theta}{2} = \pm \sqrt{\frac{1 + \cos \theta}{2}}.$$

Example 6 Use identity (24) to determine $\cos \frac{7\pi}{12}$.

Solution
$$\cos \frac{7\pi}{12} = \cos\left(\frac{1}{2}\left(\frac{7\pi}{6}\right)\right) = -\sqrt{\frac{1 + \cos \frac{7\pi}{6}}{2}}$$
$$= -\sqrt{\frac{1 - \frac{\sqrt{3}}{2}}{2}} = -\frac{\sqrt{2 - \sqrt{3}}}{2}$$

■ We choose the negative sign since $\frac{7\pi}{12}$ is a second quadrant angle and cosine is negative there.

■ $\cos \frac{7\pi}{6} = -\cos \frac{\pi}{6} = -\frac{\sqrt{3}}{2}$.

Using identities (7), (23), and (24), we can establish the following:

(25) $$\tan\frac{\theta}{2} = \pm\sqrt{\frac{1-\cos\theta}{1+\cos\theta}}.$$

There are two other ways of describing $\tan\frac{\theta}{2}$ that are often useful.

$$\tan\frac{\theta}{2} = \frac{\sin\frac{\theta}{2}}{\cos\frac{\theta}{2}} = \frac{\sin\frac{\theta}{2}}{\cos\frac{\theta}{2}} \cdot \frac{2\cos\frac{\theta}{2}}{2\cos\frac{\theta}{2}} = \frac{2\sin\frac{\theta}{2}\cos\frac{\theta}{2}}{2\cos^2\left(\frac{\theta}{2}\right)}$$

$$= \frac{\sin\left(2\cdot\frac{\theta}{2}\right)}{1+\left(2\cos^2\left(\frac{\theta}{2}\right)-1\right)} \quad \blacksquare \text{ By (18)}$$
$$\quad\quad\quad\quad\quad\quad\quad\quad\quad \blacksquare \text{ Add 1 and subtract 1.}$$

$$= \frac{\sin\theta}{1+\cos\theta} \quad \blacksquare \text{ By (21)}$$

Thus,

(26) $$\tan\frac{\theta}{2} = \frac{\sin\theta}{1+\cos\theta}.$$

Multiplying the right-hand side of (26) by $\frac{1-\cos\theta}{1-\cos\theta}$, yields still another identity for $\tan\frac{\theta}{2}$.

(27) $$\tan\frac{\theta}{2} = \frac{1-\cos\theta}{\sin\theta}.$$

Identities (26) and (27) have an advantage over identity (25) in that they do not contain a radical sign.

Example 7 Use a half-angle identity to determine $\tan(-202.5°)$.

Solution Solving this problem requires that we recognize that $-202.5°$ is $\frac{1}{2}(-405°)$ and that $-405°$ is a fourth quadrant angle which has $45°$ as its reference angle.

$$\tan(-202.5°) = \tan\left(\frac{-405°}{2}\right) = \frac{\sin(-405°)}{1+\cos(-405°)} \quad \blacksquare \text{ By (26)}$$

$$= \frac{-\frac{\sqrt{2}}{2}}{1+\frac{\sqrt{2}}{2}} = -\frac{\sqrt{2}}{2+\sqrt{2}} \quad \blacksquare \sin(-405°) = -\sin 45° = -\frac{\sqrt{2}}{2}$$
$$\quad\quad\quad\quad\quad\quad\quad\quad\quad \blacksquare \cos(-405°) = +\cos 45° = \frac{\sqrt{2}}{2}$$

In Example 7 we can estimate the value of tan $(-202.5°)$ most easily if we rationalize the denominator first.

$$\frac{-\sqrt{2}}{2+\sqrt{2}} = \frac{-\sqrt{2}}{2+\sqrt{2}} \cdot \frac{2-\sqrt{2}}{2-\sqrt{2}} = \frac{-2\sqrt{2}+2}{2}$$
$$= -\sqrt{2} + 1 \approx 1 - 1.414 = -0.414$$

That is, tan$(-202.5°)$ is approximately -0.414.

For easy reference, we summarize the identities presented in this section.

Double-Angle Identities

(18) $\qquad \sin 2\theta = 2 \sin \theta \cos \theta$

(19) $\qquad \cos 2\theta = \cos^2 \theta - \sin^2 \theta$

(20) $\qquad \cos 2\theta = 1 - 2 \sin^2 \theta$

(21) $\qquad \cos 2\theta = 2 \cos^2 \theta - 1$

(22) $\qquad \tan 2\theta = \dfrac{2 \tan \theta}{1 - \tan^2 \theta}$

Half-Angle Identities

(23) $\qquad \sin \dfrac{\theta}{2} = \pm \sqrt{\dfrac{1 - \cos \theta}{2}}$

(24) $\qquad \cos \dfrac{\theta}{2} = \pm \sqrt{\dfrac{1 + \cos \theta}{2}}$

(25) $\qquad \tan \dfrac{\theta}{2} = \pm \sqrt{\dfrac{1 - \cos \theta}{1 + \cos \theta}}$

(26) $\qquad \tan \dfrac{\theta}{2} = \dfrac{\sin \theta}{1 + \cos \theta}$

(27) $\qquad \tan \dfrac{\theta}{2} = \dfrac{1 - \cos \theta}{\sin \theta}$

Exercise Set 9.3 ⊟ Calculator Exercises in the Appendix

Goal A

In exercises 1–18, given α, find $\sin 2\alpha$, $\cos 2\alpha$, and $\tan 2\alpha$ using the double-angle identities.

1. 60°
2. 45°
3. 120°
4. 180°
5. 135°
6. 150°
7. $-300°$
8. $-225°$
9. 585°
10. 765°
11. $\dfrac{5\pi}{6}$
12. $\dfrac{3\pi}{4}$
13. $-\dfrac{2\pi}{3}$
14. $-\dfrac{7\pi}{6}$
15. $-\dfrac{3\pi}{4}$

16. -3π 17. $\frac{22\pi}{8}$ 18. $\frac{26\pi}{12}$

In exercises 19–26, from the given information, determine (a) $\sin 2\alpha$, (b) $\cos 2\alpha$, (c) $\tan 2\alpha$.

19. $\cos \alpha = -\frac{4}{5}$, α in Quadrant II
20. $\cos \alpha = \frac{4}{5}$, α in Quadrant IV
21. $\sin \alpha = -\frac{12}{13}$, α in Quadrant IV
22. $\tan \alpha = \frac{4}{3}$, α in Quadrant III
23. $\tan \alpha = \frac{9}{40}$, α in Quadrant I
24. $\sin \alpha = \frac{5}{13}$, α in Quadrant I
25. $\cot \alpha = 2$, α in Quadrant III
26. $\sec \alpha = -\frac{25}{24}$, α in Quadrant II

Goal B

In exercises 27–46, given α, find $\sin \frac{\alpha}{2}$, $\cos \frac{\alpha}{2}$, and $\tan \frac{\alpha}{2}$ using the half-angle identities.

27. $60°$ 28. $180°$ 29. $90°$
30. $120°$ 31. $210°$ 32. $315°$
33. $-225°$ 34. $-495°$ 35. $675°$
36. $855°$ 37. 4π 38. $\frac{2\pi}{3}$
39. $\frac{4\pi}{3}$ 40. $\frac{5\pi}{2}$ 41. $\frac{\pi}{6}$
42. $\frac{5\pi}{4}$ 43. $-\frac{11\pi}{6}$ 44. $-\frac{3\pi}{4}$
45. $\frac{19\pi}{6}$ 46. $\frac{16\pi}{3}$

In exercises 47–54, use the information given in exercises 19–26 to determine (a) $\sin \frac{\alpha}{2}$, (b) $\cos \frac{\alpha}{2}$, and (c) $\tan \frac{\alpha}{2}$.

Superset

55. Derive identity (22) for $\tan 2\alpha$.
56. Derive identity (25) for $\tan \frac{\alpha}{2}$.
57. Work out the details to show identity (27).

58. Express $\sin 3\alpha$ in terms of $\sin \alpha$ only.
59. Express $\cos 3\alpha$ in terms of $\cos \alpha$ only.
60. Express $\cos 4\alpha$ in terms of $\cos \alpha$ only.
61. Express $\sin 4\alpha$ in terms of $\sin \alpha$ and $\cos \alpha$.
62. Express $\tan 4\alpha$ in terms of $\tan \alpha$ only.

In exercises 63–72, express the first trigonometric function as specified.

63. $\cos 12\alpha$ in terms of $\cos 6\alpha$
64. $\cos 12\alpha$ in terms of $\sin 6\alpha$
65. $\sin 8\alpha$ in terms of $\sin 4\alpha$ and $\cos 4\alpha$
66. $\cos 8\alpha$ in terms of $\sin 4\alpha$ and $\cos 4\alpha$
67. $\tan 6\alpha$ in terms of $\tan 2\alpha$
68. $\cot 6\alpha$ in terms of $\tan 2\alpha$
69. $\cot 6\alpha$ in terms of $\cot 12\alpha$ and $\csc 12\alpha$
70. $\tan 6\alpha$ in terms of $\tan 12\alpha$ and $\sec 12\alpha$
71. $\sin 8\alpha$ in terms of $\sin 4\alpha$
72. $\sin 2\alpha$ in terms of $\cot 4\alpha$ and $\csc 4\alpha$

In exercises 73–84, simplify each of the trigonometric expressions.

73. $\dfrac{\sin 2\alpha}{2 \sin \alpha}$ 74. $\dfrac{\sin 2\alpha}{2 \cos \alpha}$

75. $2 \sin \frac{\alpha}{2} \cos \frac{\alpha}{2}$ 76. $2 \sin 2\alpha \cos 2\alpha$

77. $1 - 2 \sin^2 \frac{\alpha}{2}$ 78. $2 \cos^2 \frac{\alpha}{2} - 1$

79. $(\sin \alpha + \cos \alpha)^2 - \sin 2\alpha$
80. $(\sin \alpha - \cos \alpha)^2 + \sin 2\alpha$

81. $\dfrac{\cos^2 \alpha}{1 + \sin \alpha}$

82. $\dfrac{\cos^3 \alpha - \sin^3 \alpha}{\cos \alpha - \sin \alpha}$

83. $\cos^4 \alpha - \sin^4 \alpha$

84. $\dfrac{\cos^4 \alpha - \sin^4 \alpha}{\sin^2 \alpha - \cos^2 \alpha}$

9.4 Identities Revisited

Goal A *to simplify a trigonometric expression using identities*

As we have already seen, the primary strategy used to verify a trigonometric identity is to transform the expression on one side of the identity so that it is identical to the expression on the other side. Frequently this means that you must recognize that a form present in the problem, is a part of a known identity.

Example 1 Simplify the expression $\cos 2x \cos x + \sin 2x \sin x$.

Solution Observe that $\cos 2x \cos x + \sin 2x \sin x$ is simply the right-hand side of identity (12): $\cos(\alpha - \beta) = \cos \alpha \cos \beta + \sin \alpha \sin \beta$, with α replaced by $2x$ and β replaced by x. Thus,

$$\cos 2x \cos x + \sin 2x \sin x = \cos(2x - x) = \cos x.$$

Example 2 Simplify the expression $\dfrac{\sin 2x}{1 + \cos 2x}$.

Solution The given expression is the right-hand side of identity (26): $\tan \dfrac{\theta}{2} = \dfrac{\sin \theta}{1 + \cos \theta}$, with θ replaced by $2x$. Thus,

$$\frac{\sin 2x}{1 + \cos 2x} = \tan \frac{2x}{2} = \tan x.$$

Example 3 Simplify the expression $1 - 2 \sin^2 3A$.

Solution This expression is the right-hand side of identity (20): $\cos 2\theta = 1 - 2 \sin^2 \theta$, with θ replaced by $3A$. Thus

$$1 - 2 \sin^2 3A = \cos 2(3A) = \cos 6A.$$

Goal B *to verify trigonometric identities*

In section 1 of this chapter, we introduced techniques for verifying identities, but had only the Basic Identities (1)–(11) at our disposal. We now have, in addition, identities (12)–(27) to assist us in verifying new identities. We now consider several examples that draw upon the knowledge of all the identities established thus far.

Example 4 Verify the identity: $\dfrac{\sin 2\theta}{1 + \cos 2\theta} = \tan \theta$.

Solution
$$\dfrac{\sin 2\theta}{1 + \cos 2\theta} = \dfrac{2 \sin \theta \cos \theta}{1 + (2 \cos^2 \theta - 1)}$$

- (18) is used in the numerator. (21) is used in the denominator to produce an expression that will have only one term.

$$= \dfrac{2 \sin \theta \cos \theta}{2 \cos^2 \theta}$$

- Denominator is simplified.

$$= \dfrac{\sin \theta}{\cos \theta} = \tan \theta$$

- Common factors are canceled.

Example 5 Verify the identity: $\dfrac{\tan \theta + \sin \theta}{2 \tan \theta} = \cos^2 \dfrac{\theta}{2}$.

Solution
$$\dfrac{\tan \theta + \sin \theta}{2 \tan \theta} = \dfrac{\tan \theta}{2 \tan \theta} + \dfrac{\sin \theta}{2 \tan \theta}$$

- We work with the left-hand side of the identity since it is more complicated and offers the best chance for simplification.

$$= \dfrac{\tan \theta}{2 \tan \theta} + \dfrac{\sin \theta}{2} \cdot \dfrac{\cos \theta}{\sin \theta}$$

- Since $\tan \theta = \dfrac{\sin \theta}{\cos \theta}$, we can divide by $\tan \theta$ by multiplying by its reciprocal $\dfrac{\cos \theta}{\sin \theta}$.

$$= \dfrac{1}{2} + \dfrac{\cos \theta}{2} = \dfrac{1 + \cos \theta}{2}$$

$$= \cos^2 \dfrac{\theta}{2}$$

- By (24)

Example 6 Verify the identity: $\sin(\alpha + \beta) \sin(\alpha - \beta) = \sin^2 \alpha - \sin^2 \beta$.

Solution
$$\sin(\alpha + \beta) \sin(\alpha - \beta)$$
$$= (\sin \alpha \cos \beta + \cos \alpha \sin \beta)(\sin \alpha \cos \beta - \cos \alpha \sin \beta)$$

- By (14) and (15)

$$= (\sin \alpha \cos \beta)^2 - (\cos \alpha \sin \beta)^2$$

- $(A + B)(A - B) = A^2 - B^2$.

$$= \sin^2 \alpha \cos^2 \beta - \cos^2 \alpha \sin^2 \beta$$
$$= \sin^2 \alpha (1 - \sin^2 \beta) - (1 - \sin^2 \alpha)(\sin^2 \beta)$$

- We want an expression that involves only $\sin \alpha$ and $\sin \beta$.

$$= \sin^2 \alpha - \sin^2 \alpha \sin^2 \beta - \sin^2 \beta + \sin^2 \alpha \sin^2 \beta$$
$$= \sin^2 \alpha - \sin^2 \beta$$

Example 7 Verify the identity: $\dfrac{\sec \beta}{1 + \cos \beta} = \dfrac{\tan \beta - \sin \beta}{\sin^3 \beta}$.

Solution
$$\dfrac{\sec \beta}{1 + \cos \beta} = \dfrac{\sec \beta}{1 + \cos \beta} \cdot \dfrac{1 - \cos \beta}{1 - \cos \beta}$$

- We will work with the left-hand side of the identity. The denominator is multiplied by $1 - \cos \beta$ in order to get an expression involving only $\sin \beta$.

$$= \frac{\sec \beta (1 - \cos \beta)}{1 - \cos^2 \beta}$$

$$= \frac{\sec \beta - \sec \beta (\cos \beta)}{\sin^2 \beta} \qquad \blacksquare \text{ By (9)}$$

$$= \frac{\sec \beta - 1}{\sin^2 \beta} \qquad \blacksquare \sec \beta (\cos \beta) = \sec \beta \left(\frac{1}{\sec \beta}\right) = 1.$$

$$= \frac{\sec \beta - 1}{\sin^2 \beta} \cdot \frac{\sin \beta}{\sin \beta} \qquad \blacksquare \text{ We know that we want } \sin^3 \beta \text{ in the denominator, so we multiply by } \frac{\sin \beta}{\sin \beta}.$$

$$= \frac{\sec \beta \sin \beta - \sin \beta}{\sin^3 \beta}$$

$$= \frac{\tan \beta - \sin \beta}{\sin^3 \beta} \qquad \blacksquare \sec \beta \sin \beta = \frac{1}{\cos \beta} \sin \beta = \tan \beta.$$

Goal C *to use conversion identities*

We now consider a group of identities that allow us to restate a product of two trigonometric function values as a sum, and a sum of two trigonometric function values as a product. These new identities are derived from the sum and difference identities for the sine and cosine (identities (12)–(15)). For example, based on identities (14) and (15), we have

$$\sin(\alpha + \beta) + \sin(\alpha - \beta) = (\sin \alpha \cos \beta + \cos \alpha \sin \beta)$$
$$+ (\sin \alpha \cos \beta - \cos \alpha \sin \beta)$$
$$= 2 \sin \alpha \cos \beta.$$

Thus, $\sin(\alpha + \beta) + \sin(\alpha - \beta) = 2 \sin \alpha \cos \beta$, or

(28) $$\sin \alpha \cos \beta = \frac{1}{2}[\sin(\alpha + \beta) + \sin(\alpha - \beta)]$$

Example 8 Determine $\sin 15° \cos 75°$ by using identity (28).

Solution
$$\sin 15° \cos 75° = \frac{1}{2}[\sin(15° + 75°) + \sin(15° - 75°)]$$

$$= \frac{1}{2}[\sin 90° + \sin(-60°)]$$

$$= \frac{1}{2}[\sin 90° - \sin 60°] \qquad \blacksquare \text{ Recall that } \sin(-\theta) = -\sin \theta \text{ because sine is an odd function.}$$

$$= \frac{1}{2}\left(1 - \frac{\sqrt{3}}{2}\right) = \frac{2 - \sqrt{3}}{4}$$

In identity (28), let us replace $\alpha + \beta$ with x and $\alpha - \beta$ with y. Since

$$x + y = (\alpha + \beta) + (\alpha - \beta) = 2\alpha, \quad \text{and}$$
$$x - y = (\alpha + \beta) - (\alpha - \beta) = 2\beta,$$

we have $\alpha = \dfrac{x + y}{2}$, and $\beta = \dfrac{x - y}{2}$. Thus identity (28) can be rewritten as

$$\sin\left(\frac{x + y}{2}\right) \cos\left(\frac{x - y}{2}\right) = \frac{1}{2}(\sin x + \sin y),$$

or equivalently as

(31) $$\sin x + \sin y = 2\left[\sin\left(\frac{x + y}{2}\right) \cos\left(\frac{x - y}{2}\right)\right].$$

Notice that identity (31) expresses the sum of two trigonometric function values as a product of trigonometric function values.

Employing arguments similar to those used to derive identities (28) and (31), we can develop five other identities. These identities are called **conversion identities** because they allow us to convert products of trigonometric function values into sums, and vice versa.

Conversion Identities

(28) $$\sin \alpha \cos \beta = \frac{1}{2}[\sin(\alpha + \beta) + \sin(\alpha - \beta)]$$

(29) $$\sin \alpha \sin \beta = \frac{1}{2}[\cos(\alpha - \beta) - \cos(\alpha + \beta)]$$

(30) $$\cos \alpha \cos \beta = \frac{1}{2}[\cos(\alpha + \beta) + \cos(\alpha - \beta)]$$

(31) $$\sin x + \sin y = 2 \sin\left(\frac{x + y}{2}\right) \cos\left(\frac{x - y}{2}\right)$$

(32) $$\sin x - \sin y = 2 \cos\left(\frac{x + y}{2}\right) \sin\left(\frac{x - y}{2}\right)$$

(33) $$\cos x + \cos y = 2 \cos\left(\frac{x + y}{2}\right) \cos\left(\frac{x - y}{2}\right)$$

(34) $$\cos x - \cos y = -2 \sin\left(\frac{x + y}{2}\right) \sin\left(\frac{x - y}{2}\right)$$

Example 9 Express (a) $\cos 8\alpha + \cos 6\alpha$ as a product (b) $\cos 2\theta \sin \theta$ as a sum or difference.

Solution (a) $\cos 8\alpha + \cos 6\alpha = 2 \cos\left(\dfrac{8\alpha + 6\alpha}{2}\right) \cos\left(\dfrac{8\alpha - 6\alpha}{2}\right)$ ■ Identity (33) is used with $x = 8\alpha$ and $y = 6\alpha$.

$= 2 \cos 7\alpha \cos \alpha.$

(b) $\cos 2\theta \sin \theta = \dfrac{1}{2}[\sin(3\theta) + \sin(-\theta)]$ ■ Identity (28) is used with $\alpha = \theta$ and $\beta = 2\theta$.

$= \dfrac{1}{2}[\sin 3\theta - \sin \theta].$

Example 10 Verify the following identity: $\dfrac{\sin 4x - \sin 3x}{\cos 4x + \cos 3x} = \dfrac{1 - \cos x}{\sin x}.$

Solution

$\dfrac{\sin 4x - \sin 3x}{\cos 4x + \cos 3x} = \dfrac{2 \cos\left(\dfrac{4x + 3x}{2}\right) \sin\left(\dfrac{4x - 3x}{2}\right)}{2 \cos\left(\dfrac{4x + 3x}{2}\right) \cos\left(\dfrac{4x - 3x}{2}\right)}$ ■ By (32)

■ By (33)

$= \dfrac{\sin \dfrac{x}{2}}{\cos \dfrac{x}{2}} = \tan \dfrac{x}{2} = \dfrac{1 - \cos x}{\sin x}$ ■ By (27)

Exercise Set 9.4

Goal A

In exercises 1–8, simplify the expression.

1. $\cos 4\alpha \cos \alpha + \sin \alpha \sin 4\alpha$
2. $\cos 4\alpha \cos \alpha - \sin \alpha \sin 4\alpha$
3. $\sin 4\alpha \cos \alpha - \sin \alpha \cos 4\alpha$
4. $\sin 4\alpha \cos \alpha + \sin \alpha \cos 4\alpha$
5. $\dfrac{2 \tan 2\varphi}{1 - \tan^2 2\varphi}$
6. $\dfrac{\sin 2\alpha}{1 + \cos 2\alpha}$
7. $\dfrac{\sec \alpha - 1}{\sec \alpha}$
8. $\dfrac{\tan \beta}{\sec^2 \beta - 2}$

Goal B

In exercises 9–20, verify the identity.

9. $\tan x = \dfrac{2 \tan \dfrac{1}{2}x}{1 - \tan^2 \dfrac{1}{2}x}$

10. $\cot x = \dfrac{\cot^2 \dfrac{1}{2}x - 1}{2 \cot \dfrac{1}{2}x}$

11. $\dfrac{\cos(x + y)}{\cos x \sin y} = \cot y - \tan x$

12. $\dfrac{\sin(x + y)}{\sin(x - y)} = \dfrac{\tan x + \tan y}{\tan x - \tan y}$

13. $\dfrac{1 - \cos 2\alpha}{1 + \cos 2\alpha} = \tan^2 \alpha$

14. $\dfrac{\sin(\alpha + \beta)}{\sin \alpha \sin \beta} = \cot \alpha + \cot \beta$

15. $\dfrac{\sin \alpha + \sin \beta}{\cos \alpha - \cos \beta} = -\cot \dfrac{1}{2}(\alpha - \beta)$

16. $\dfrac{\cos(\alpha + \beta)}{\cos(\alpha - \beta)} = -\dfrac{\tan \beta - \cot \alpha}{\tan \beta + \cot \alpha}$

17. $\csc x - \sin x = \sin x \cot^2 x$

18. $\cos x \csc x \tan x = 1$

19. $\dfrac{1 + \cos \varphi}{\sin \varphi} + \dfrac{\sin \varphi}{1 + \cos \varphi} = 2 \csc \varphi$

20. $\dfrac{\sin \varphi}{1 - \sin \varphi} + \dfrac{1 + \sin \varphi}{\sin \varphi} = \dfrac{\csc \varphi}{1 - \sin \varphi}$

In exercises 21–26, determine whether the given equation is or is not an identity.

21. $\tan s = \dfrac{\tan s + 1}{\cot s + 1}$

22. $\dfrac{\sin s}{1 + \cos s} = \dfrac{1 - \cos s}{\sin s}$

23. $\tan 3\alpha = \dfrac{1 + \cos 2\alpha}{\sin 2\alpha}$

24. $\dfrac{\sin 3\alpha}{\sin \alpha} + \dfrac{\cos 3\alpha}{\cos \alpha} = 1$

25. $\dfrac{1 + \sin \theta - \cos \theta}{1 + \sin \theta + \cos \theta} = \tan \dfrac{1}{2}\theta$

26. $\dfrac{1 + \sin \theta}{1 - \sin \theta} - \dfrac{1 - \sin \theta}{1 + \sin \theta} = \dfrac{4 \tan \theta}{\cos \theta}$

Goal C

In exercises 27–34, evaluate by using the conversion identities.

27. $\sin 75° + \sin 15°$

28. $\cos 75° - \cos 15°$

29. $\cos 105° + \cos 15°$

30. $\sin 105° - \sin 15°$

31. $\sin \dfrac{\pi}{12} - \sin \dfrac{5\pi}{12}$

32. $\cos \dfrac{11\pi}{12} + \cos \dfrac{5\pi}{12}$

33. $\cos \dfrac{7\pi}{12} - \cos \dfrac{\pi}{12}$

34. $\sin \dfrac{13\pi}{12} + \sin \dfrac{7\pi}{12}$

For exercises 35–38, derive the identities (29), (30), (32), and (33).

In exercises 39–44, express the trigonometric expression as a product.

39. $\sin 3x + \sin x$

40. $\sin 3x - \sin x$

41. $\sin 6x + \sin 2x$

42. $\sin 6x - \sin 2x$

43. $\cos 6x - \cos 2x$

44. $\cos 6x + \cos 2x$

In exercises 45–48, write the expression as a sum or difference involving sines and cosines.

45. $\sin 2x \cos 3x$

46. $\sin 3x \cos 2x$

47. $\sin 3x \sin 2x$

48. $\cos 3x \cos 2x$

Superset

49. Find an expression for $\tan 2x + \sec 2x$ in terms only of $\sin x$ and $\cos x$.

50. If $0 < \alpha < \dfrac{\pi}{2}$ and $\beta = \dfrac{\pi}{2} - \alpha$, show that
    (a) $\sin \alpha \cos \beta + \cos \alpha \sin \beta = 1$
    (b) $\cos \alpha \cos \beta - \sin \alpha \sin \beta = 0$
    (c) $\sin \alpha + \sin \beta = \cos \alpha + \cos \beta$

51. Verify each of the following.
    (a) $2 \arctan(\tfrac{1}{3}) + \arctan(\tfrac{1}{7}) = \dfrac{\pi}{4}$
    (b) $\arctan(\tfrac{1}{2}) + \arctan(\tfrac{1}{3}) = \dfrac{\pi}{4}$
    (c) $\tan(\arctan(a) - \arctan(b)) = \dfrac{a - b}{1 + ab}$

52. For each of the following, find two nonzero numbers M and N such that
    (a) $M + N \cos^2 x = \sin^4 x - \cos^4 x$
    (b) $\cos^4 x - \sin^4 x = M - N \sin^2 x$
    (c) $M + N \cos x = \dfrac{\sin^2 x}{1 - \cos x}$
    (d) $\tan x \csc x \cos x + \cot x \sec x \sin x = M$
    (e) $(\tan x + \cot x)^2 \sin^2 x = M + N \tan^2 x$

9.5 Trigonometric Equations

Goal A *to solve trigonometric equations using knowledge of the special angles*

Much of algebra is concerned with techniques for solving equations like

$$2x + 1 = 0, \quad \text{or} \quad 2x^2 - x = 1,$$

where x represents a real number. We now wish to consider equations that involve trigonometric functions, such as

$$2 \cos \alpha + 1 = 0, \quad \text{or} \quad 2 \cos^2 \alpha - \cos \alpha = 1.$$

To solve such equations, we first solve for $\cos \alpha$ using algebraic techniques, and then we use our knowledge of trigonometry to solve for α. For example, to solve the trigonometric equation

$$2 \cos \alpha + 1 = 0,$$

we begin by solving for $\cos \alpha$. This requires the same algebraic steps as solving the equation $2x + 1 = 0$ for x.

Example 1 Solve $2 \cos \alpha + 1 = 0$ for α. Express the solution in radians.

Solution

$2 \cos \alpha + 1 = 0$ ■ Begin by solving for $\cos \alpha$.

$2 \cos \alpha = -1$

$\cos \alpha = -\dfrac{1}{2}$ ■ Having solved for $\cos \alpha$, we now solve for α.

Recall that $\cos \frac{\pi}{3} = \frac{1}{2}$, and that cosine is negative for second and third quadrant angles. Thus, α must be a second or third quadrant angle whose reference angle is $\frac{\pi}{3}$, namely $\frac{2\pi}{3}$ or $\frac{4\pi}{3}$. Since adding any multiple of 2π to either of these values produces an angle whose cosine is also $-\frac{1}{2}$, we conclude

$$\alpha = \frac{2\pi}{3} + 2n\pi \quad \text{or} \quad \alpha = \frac{4\pi}{3} + 2n\pi \quad \text{for any integer } n.$$

When solving trigonometric equations, we often restrict the solutions to values in the interval $[0, 2\pi)$ or angles in the interval $[0°, 360°)$.

Example 2 Determine all solutions of $2\sin^2\alpha - \sin\alpha = 1$ in the interval $[0, 2\pi)$.

Solution

$$2\sin^2\alpha - \sin\alpha = 1$$ ■ Begin by solving for $\sin\alpha$.

$$2\sin^2\alpha - \sin\alpha - 1 = 0$$ ■ Now factor the left side; treat $\sin\alpha$ as the variable.

$$(2\sin\alpha + 1)(\sin\alpha - 1) = 0$$

$2\sin\alpha + 1 = 0$	$\sin\alpha - 1 = 0$	■ Use the Principle of Zero Products.
$\sin\alpha = -\dfrac{1}{2}$	$\sin\alpha = 1$	

To solve $\sin\alpha = -\frac{1}{2}$, recall that $\sin\frac{\pi}{6} = \frac{1}{2}$ and sine is negative in the third and fourth quadrants. Thus, α is a third or fourth quadrant angle whose reference angle is $\frac{\pi}{6}$. So $\alpha = \frac{7\pi}{6}$ or $\frac{11\pi}{6}$. To solve $\sin\alpha = 1$, recall that the only value in the interval $[0, 2\pi)$ whose sine is 1 is $\alpha = \frac{\pi}{2}$.

The solutions of the given equation in the interval $[0, 2\pi)$ are $\frac{\pi}{2}, \frac{7\pi}{6}$, and $\frac{11\pi}{6}$.

If the domain in the Example 2 had not been restricted to the interval $[0, 2\pi)$, the solutions would be $\frac{7\pi}{6} + 2n\pi$, $\frac{11\pi}{6} + 2n\pi$, and $\frac{\pi}{2} + 2n\pi$, for any integer n.

When solving trigonometric equations, it is important to remember the domains and ranges of the trigonometric functions. Example 3 illustrates how knowledge of the range can be crucial.

Example 3 Determine the solutions of $\cos^2\varphi = 2 - \cos\varphi$ in the interval $[0°, 360°)$.

Solution

$$\cos^2\varphi = 2 - \cos\varphi$$ ■ First solve for $\cos\varphi$.

$$\cos^2\varphi + \cos\varphi - 2 = 0$$

$$(\cos\varphi + 2)(\cos\varphi - 1) = 0$$ ■ Now use the Principle of Zero Products.

$\cos\varphi + 2 = 0$	$\cos\varphi - 1 = 0$
$\cos\varphi = -2$	$\cos\varphi = 1$

There is no solution to the equation $\cos\varphi = -2$ because the range of the cosine function is the interval $[-1, 1]$, therefore there is no angle having a cosine of -2. To solve $\cos\varphi = 1$, recall that the only value in $[0°, 360°)$ whose cosine is 1 is $\varphi = 0°$. The only solution of the given equation in the interval $[0°, 360°)$ is $0°$.

9 / Trigonometric Identities and Equations

In the next two examples, we solve equations involving two different trigonometric functions. Given such an equation, you should try to use algebraic techniques and trigonometric identities to rewrite the equation in a form involving only one trigonometric function.

Example 4 Determine all solutions of the equation $\sin x + 1 = \cos x$ in the interval $[0, 2\pi)$.

Solution

$$\sin x + 1 = \cos x$$
$$(\sin x + 1)^2 = (\cos x)^2$$ ■ Square both sides of the equation to produce $\cos^2 x$ on the right, which can be rewritten in terms of $\sin x$.
$$\sin^2 x + 2 \sin x + 1 = \cos^2 x$$
$$\sin^2 x + 2 \sin x + 1 = 1 - \sin^2 x$$ ■ $\cos^2 x$ is replaced with $1 - \sin^2 x$ (identity (9)).
$$\sin^2 x + 2 \sin x = -\sin^2 x$$
$$2 \sin^2 x + 2 \sin x = 0$$
$$2 \sin x (\sin x + 1) = 0$$

$2 \sin x = 0$	$\sin x + 1 = 0$
$\sin x = 0$	$\sin x = -1$
$x = 0$ or π	$x = \dfrac{3\pi}{2}$

■ Because we squared both sides of the original equation, we must check for extraneous solutions.

Must check:

$x = 0$ $x = \pi$ $x = \dfrac{3\pi}{2}$

$\sin 0 + 1 = \cos 0$ $\sin \pi + 1 = \cos \pi$ $\sin \dfrac{3\pi}{2} + 1 = \cos \dfrac{3\pi}{2}$

$0 + 1 = 1$ ■ True $0 + 1 = -1$ ■ False $-1 + 1 = 0$ ■ True

The solutions of the given equation in the interval $[0, 2\pi)$ are 0 and $\dfrac{3\pi}{2}$.

Example 5 Solve the equation $\cos 2x + \sin x = 0$ for x. Express solutions in radians.

Solution

$$\cos 2x + \sin x = 0$$ ■ Use identity (20) to rewrite $\cos 2x$ in terms of $\sin x$.
$$1 - 2 \sin^2 x + \sin x = 0$$
$$2 \sin^2 x - \sin x - 1 = 0$$ ■ The equation was multiplied by -1.

This equation was solved over the interval $[0, 2\pi)$ in Example 2. The solutions were found to be $\dfrac{\pi}{2}$, $\dfrac{7\pi}{6}$, and $\dfrac{11\pi}{6}$. Thus the solutions of the given equation are

$$\dfrac{\pi}{2} + 2n\pi, \quad \dfrac{7\pi}{6} + 2n\pi, \quad \text{and} \quad \dfrac{11\pi}{6} + 2n\pi \quad \text{for any integer } n.$$

Example 6 Solve the equation $\sin \theta + 2 \sin \dfrac{\theta}{2} = \cos \dfrac{\theta}{2} + 1$ over the interval $[0°, 360°)$.

Solution We begin by making the substitution $\alpha = \dfrac{\theta}{2}$ in order to simplify the equation.

$$\sin 2\alpha + 2 \sin \alpha = \cos \alpha + 1 \qquad \blacksquare \text{ Since } \alpha = \dfrac{\theta}{2}, \theta = 2\alpha.$$
$$2 \sin \alpha \cos \alpha + 2 \sin \alpha = \cos \alpha + 1 \qquad \blacksquare \text{ By identity (18)}$$
$$2 \sin \alpha \cos \alpha + 2 \sin \alpha - \cos \alpha - 1 = 0$$
$$2 \sin \alpha (\cos \alpha + 1) - 1(\cos \alpha + 1) = 0 \qquad \blacksquare \text{ Factor.}$$
$$(2 \sin \alpha - 1)(\cos \alpha + 1) = 0 \qquad \blacksquare \text{ Factor again, and use the Principle of Zero Products.}$$

$$\begin{array}{c|c} 2 \sin \alpha - 1 = 0 & \cos \alpha + 1 = 0 \\ \sin \alpha = \dfrac{1}{2} & \cos \alpha = -1 \\ \alpha = 30° \text{ or } 150° & \alpha = 180° \end{array}$$

By virtue of our substitution, $\theta = 2\alpha$, and so the values of θ corresponding to $\alpha = 30°$, $\alpha = 150°$, and $\alpha = 180°$, are $\theta = 60°$, $\theta = 300°$, and $\theta = 360°$. Since the last value is not in the interval $[0°, 360°)$, we conclude that the solutions of the original equation in the given interval are

$$\theta = 60° \quad \text{and} \quad \theta = 300°.$$

Example 7 Solve the equation $\sin 3x + \sin x + \cos x = 0$ for x. Express solutions in degrees.

Solution
$$\sin 3x + \sin x + \cos x = 0$$
$$\sin(2x + x) + \sin(2x - x) + \cos x = 0 \qquad \blacksquare \text{ 3x and x are rewritten to allow the use of identity (28).}$$
$$2 \sin 2x \cos x + \cos x = 0 \qquad \blacksquare \text{ By (28) with } \alpha = 2x \text{ and } \beta = x\text{:}$$
$$\cos x(2 \sin 2x + 1) = 0 \qquad 2 \sin \alpha \cos \beta = \sin(\alpha + \beta) + \sin(\alpha - \beta).$$

$$\begin{array}{c|c} \cos x = 0 & 2 \sin 2x + 1 = 0 \\ x = 90° \text{ or } 270° & \sin 2x = -\dfrac{1}{2} \\ & 2x = 210° \text{ or } 330° \\ & x = 105° \text{ or } 165° \end{array}$$

The solutions of the original equation are all angles of the form

$$90° + n \cdot 360°, \quad 105° + n \cdot 360°, \quad 165° + n \cdot 360°, \quad \text{and} \quad 270° + n \cdot 360° \quad \text{for any integer } n.$$

Goal B *to solve trigonometric equations by using tables or a calculator*

In the previous section, the solutions of the trigonometric equations involved special angles $(0, \frac{\pi}{6}, \frac{\pi}{4}, $ etc$)$. An equation such as $2 - 7 \cos x = 0$ has a solution which is not a special angle. We must use a table or a calculator to determine such a solution.

Example 8 Solve the equation $2 - 7 \cos x = 0$ over the interval $[0°, 360°)$.

Solution $2 - 7 \cos x = 0$

$$\cos x = \frac{2}{7} \approx 0.2857$$

Using a calculator or Table 4 and rounding to the nearest ten minutes, we find that $x \approx 73°20'$ (the interpolated value is $73°24'$). Since we are given that $\cos x$ is positive, x could be a first or fourth quadrant angle. The fourth quadrant angle having $73°20'$ as a reference angle is $286°40'$.

The solutions rounded to the nearest ten minutes are $73°20'$ and $286°40'$.

Example 9 Solve the equation $\sin^2 \theta + 2 \sin \theta - 1 = 0$ over the interval $[0°, 360°)$.

Solution $\sin^2 \theta + 2 \sin \theta - 1 = 0$

$x^2 + 2x - 1 = 0$

■ Replace $\sin \theta$ with x, and solve the quadratic equation for x.

$$x = \frac{-2 \pm \sqrt{2^2 - 4(1)(-1)}}{2(1)}$$

■ Use the quadratic formula with $a = 1$, $b = 2$, and $c = -1$.

$$x = \frac{-2 \pm \sqrt{8}}{2} = \frac{-2 \pm 2\sqrt{2}}{2} = -1 \pm \sqrt{2}$$

Using $\sqrt{2} \approx 1.41$, and recalling that x represents $\sin \theta$, we have

$$\sin \theta \approx -1 + 1.41 = 0.41 \quad \text{or} \quad \sin \theta \approx -1 - 1.41 = -2.41.$$

Since the range of the sine function is the interval $[-1, 1]$, there is no θ such that $\sin \theta = -2.41$. Thus, our only solutions result from the equation $\sin \theta = 0.41$. Using a calculator or Table 4 and rounding to the nearest ten minutes, we obtain $\theta \approx 24°10'$ (the interpolated value is $24°12'$). Since 0.41 is positive, θ could also be the second quadrant angle with $24°10'$ as its reference angle, namely $155°50'$. Thus, the solutions to the nearest ten minutes are

$$\theta = 24°10' \quad \text{and} \quad \theta = 155°50'.$$

Goal C *to solve equations involving inverse trigonometric functions*

The following examples illustrate techniques for solving equations that involve inverse trigonometric functions.

Example 10 Solve $\sin^{-1}\left(\frac{5}{13}\right) + \cos^{-1}\left(\frac{3}{5}\right) = \sin^{-1} x$ for x.

Solution By the definition of $\sin^{-1}(x)$, we know that $\sin(\sin^{-1}(x)) = x$. Taking the sine of both sides of the given equation we get

$$\sin\left(\sin^{-1}\left(\frac{5}{13}\right) + \cos^{-1}\left(\frac{3}{5}\right)\right) = x.$$

Let $\alpha = \sin^{-1}\left(\frac{5}{13}\right)$, and $\beta = \cos^{-1}\left(\frac{3}{5}\right)$. Draw α and β as acute angles in two different right triangles.

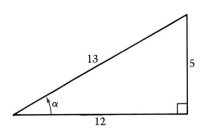

- Since $\alpha = \sin^{-1}(\frac{5}{13})$, $\sin \alpha = \sin(\sin^{-1}(\frac{5}{13})) = \frac{5}{13}$. We label the sides of the triangle accordingly.

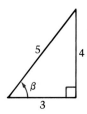

- Since $\beta = \cos^{-1}(\frac{3}{5})$, $\cos \beta = \cos(\cos^{-1}(\frac{3}{5})) = \frac{3}{5}$. We label the triangle accordingly.

$$x = \sin\left(\sin^{-1}\left(\frac{5}{13}\right) + \cos^{-1}\left(\frac{3}{5}\right)\right)$$
$$= \sin(\alpha + \beta) = \sin \alpha \cos \beta + \cos \alpha \sin \beta \quad \blacksquare \text{ By identity (14).}$$
$$= \frac{5}{13} \cdot \frac{3}{5} + \frac{12}{13} \cdot \frac{4}{5} = \frac{15}{65} + \frac{48}{65} = \frac{63}{65} \quad \blacksquare \text{ From the two triangles.}$$

Example 11 Find the smallest positive angle x, in degrees, such that

$$\tan^{-1}\left(\frac{1}{2}\right) + \tan^{-1}\left(\frac{1}{3}\right) = x.$$

Solution
$$\tan^{-1}\left(\frac{1}{2}\right) + \tan^{-1}\left(\frac{1}{3}\right) = x$$

Let $\alpha = \tan^{-1}(\frac{1}{2})$ and $\beta = \tan^{-1}(\frac{1}{3})$; the equation becomes $\alpha + \beta = x$. Then, taking the tangent of both sides of this equation, we have

$$\tan x = \tan(\alpha + \beta) = \frac{\tan \alpha + \tan \beta}{1 - \tan \alpha \tan \beta} \quad \blacksquare \text{ Identity (16)}$$

Therefore, $\tan x = \dfrac{\frac{1}{2} + \frac{1}{3}}{1 - \left(\frac{1}{2}\right)\left(\frac{1}{3}\right)} = \dfrac{\frac{5}{6}}{\frac{5}{6}} = 1 \quad \blacksquare \begin{array}{l} \tan \alpha = \tan(\tan^{-1}(\frac{1}{2})) = \frac{1}{2}. \\ \tan \beta = \tan(\tan^{-1}(\frac{1}{3})) = \frac{1}{3}. \end{array}$

The smallest positive angle having a tangent function value of 1 is 45°.

Exercise Set 9.5   ▢ Calculator Exercises in the Appendix

Goal A

In exercises 1–10, solve the equation for x. Express your answer in radians.

1. $2 \sin x + \sqrt{3} = 0$
2. $3 \cot x - \sqrt{3} = 0$
3. $2 \sin^2 x - \sin x = 0$
4. $2 \cos^2 x - \cos x - 1 = 0$
5. $4 \cos^2 x - 3 = 0$
6. $\tan^2 x - 1 = 0$
7. $2 \cos^2 x + 5 \cos x - 3 = 0$
8. $\sin x \cos x - \cos x = 0$
9. $\sin x \cos x - \sin x + \cos x - 1 = 0$
10. $2 \sin^2 x - \sqrt{2} \sin x - 4 \sin x + 2\sqrt{2} = 0$

In exercises 11–18, determine all the solutions of the equation in the interval $[0, 2\pi)$.

11. $2 \sin^2 x + \sin x - 1 = 0$
12. $2 \cos^2 x - \cos x - 1 = 0$
13. $2 \sin^2 x + 3 \sin x - 2 = 0$
14. $\sqrt{2} \cos^2 x + (1 - \sqrt{2}) \cos x - 1 = 0$
15. $2 \cos^2 x + 3 \sin x - 2 = 0$
16. $3 \cot^2 x - 1 = 0$
17. $2 \sin^2 3\theta - 4 \sin 3\theta - 6 = 0$
18. $2 \sin^2 \frac{1}{2}\theta + 3 \sin \frac{1}{2}\theta - 2 = 0$

In exercises 19–32, solve the equation for x in the indicated domain.

19. $\sin 2x = \cos 2x$; x in $[0, 2\pi)$
20. $\cos x \tan x = 0$; x in $[0, 360°)$
21. $3\cos^2 x = \sin^2 x$; x in $[-90°, 90°]$
22. $\sin x \cos x = \cos x$; x in $[0, 180°]$
23. $\dfrac{\sqrt{2}\sin^2 x}{\cos x} - \tan x = 0$; x in $[0, 2\pi)$
24. $\cos^2 x - \sin x = 1$; x in $[0, \pi]$
25. $2\sin x - \tan x = 0$; x in $[0°, 180°]$
26. $3\sin^2 x - \cos x = 3$; x in $[0, \pi]$
27. $\cos 5x + \cos x = 0$; x in $[0°, 360°)$
28. $\cos \dfrac{7}{2}x + \cos \dfrac{1}{2}x = 0$; x in $[0, \pi)$
29. $\sin 2x + \cos x = 0$; x in $[0, 2\pi)$
30. $\cos 2x + \cos x + 1 = 0$; x in $[0, 2\pi)$
31. $\cos 2x + \sin x = 1$; x in $[0, 2\pi)$
32. $\cos 2x - 1 - \tan x = 0$; x in $[0, 2\pi)$

Goal B

In exercises 33–44, solve the equation for x in $[0°, 360°)$ using Table 4 or a calculator to the nearest ten minutes.

33. $4\sin^2 x - 3\sin x = 0$
34. $5\cos^2 x - 2\cos x = 0$
35. $3\sin^2 x + 2\sin x - 1 = 0$
36. $3\cos^2 x - \cos x - 2 = 0$
37. $6\sin^2 x - \sin x - 2 = 0$
38. $2\sin^2 x + \dfrac{1}{2}\sin x - 2 = 0$
39. $(4\sin x + 1)^2 - 3 = 0$
40. $(3\cos x - 5)^2 - 8 = 0$
41. $2\tan^2 x + 2\tan x + 3 = 0$
42. $3\tan^2 x + 4\tan x - 4 = 0$
43. $3\sin^2 2x - 2 = 0$
44. $(2\cos^2 2x + 3)^2 = 1$

Goal C

In exercises 45–56, solve the equation for x.

45. $\arcsin\left(-\dfrac{1}{2}\right) = x$
46. $\arccos\left(-\dfrac{1}{2}\right) = x$
47. $\tan^{-1}\sqrt{3} = x$
48. $\dfrac{1}{2}\pi - \tan^{-1}(-1) = x$
49. $\sin\left(2\arcsin\dfrac{3}{5}\right) = x$
50. $\sin\left(2\arccos\dfrac{3}{5}\right) = x$
51. $\cos^{-1}\dfrac{3}{5} + \sin^{-1}\dfrac{4}{5} = \cos^{-1}x$
52. $\cos^{-1}\dfrac{3}{5} - \sin^{-1}\dfrac{4}{5} = \cos^{-1}x$
53. $\arcsin\dfrac{3}{5} - \arccos\dfrac{5}{13} = \arcsin x$
54. $\arcsin\dfrac{3}{5} + \arccos\dfrac{12}{13} = \arccos x$
55. $\tan^{-1}\dfrac{4}{3} + \tan^{-1}\dfrac{1}{7} = x$
56. $\arctan\left(-\dfrac{1}{3}\right) - \arctan\left(\dfrac{2}{3}\right) = \arctan x$

Superset

57. Solve for x: $\arctan x + \arctan \dfrac{1}{3} = \dfrac{\pi}{4}$.
58. Solve the following system of equations:
$$\begin{cases} \sin x + \cos x = 1 \\ 2\sin x = \sin 2x \end{cases}$$
59. Solve simultaneously:
$$\begin{cases} r = 1 - \cos\theta \\ r = \cos\theta \end{cases}$$
where $r > 0$ and $0 \le \theta \le 2\pi$
60. Solve for x: $\tan x + \sqrt{\tan x + 5} = 7$.
61. Solve for x: $\sqrt{2\cot x - 1} + \sqrt{\cot x - 4} = 4$.

Chapter Review and Test

Chapter Review

9.1 Basic Trigonometric Identities (pp. 366–370)

Reciprocal Identities

(1) $\sin \theta = \dfrac{1}{\csc \theta}$

(2) $\cos \theta = \dfrac{1}{\sec \theta}$

(3) $\tan \theta = \dfrac{1}{\cot \theta}$

(4) $\cot \theta = \dfrac{1}{\tan \theta}$

(5) $\sec \theta = \dfrac{1}{\cos \theta}$

(6) $\csc \theta = \dfrac{1}{\sin \theta}$

Quotient Identities

(7) $\tan \theta = \dfrac{\sin \theta}{\cos \theta}$

(8) $\cot \theta = \dfrac{\cos \theta}{\sin \theta}$

Pythagorean Identities

(9) $\sin^2 \theta + \cos^2 \theta = 1$

(10) $\tan^2 \theta + 1 = \sec^2 \theta$

(11) $1 + \cot^2 \theta = \csc^2 \theta$

9.2 Sum and Difference Identities (pp. 371–377)

(12) $\cos(\alpha - \beta) = \cos \alpha \cos \beta + \sin \alpha \sin \beta$

(13) $\cos(\alpha + \beta) = \cos \alpha \cos \beta - \sin \alpha \sin \beta$

(14) $\sin(\alpha + \beta) = \sin \alpha \cos \beta + \cos \alpha \sin \beta$

(15) $\sin(\alpha - \beta) = \sin \alpha \cos \beta - \cos \alpha \sin \beta$

(16) $\tan(\alpha + \beta) = \dfrac{\tan \alpha + \tan \beta}{1 - \tan \alpha \tan \beta}$

(17) $\tan(\alpha - \beta) = \dfrac{\tan \alpha - \tan \beta}{1 + \tan \alpha \tan \beta}$

9.3 The Double-Angle and Half-Angle Identities (pp. 378–384)

(18) $\sin 2\theta = 2 \sin \theta \cos \theta$

(19) $\cos 2\theta = \cos^2 \theta - \sin^2 \theta$

(20) $\cos 2\theta = 1 - 2 \sin^2 \theta$

(21) $\cos 2\theta = 2 \cos^2 \theta - 1$

(22) $\tan 2\theta = \dfrac{2 \tan \theta}{1 - \tan^2 \theta}$

(23) $\sin \dfrac{\theta}{2} = \pm \sqrt{\dfrac{1 - \cos \theta}{2}}$

(24) $\cos \dfrac{\theta}{2} = \pm \sqrt{\dfrac{1 + \cos \theta}{2}}$

(25) $\tan \dfrac{\theta}{2} = \pm \sqrt{\dfrac{1 - \cos \theta}{1 + \cos \theta}}$

(26) $\tan\dfrac{\theta}{2} = \dfrac{\sin\theta}{1 + \cos\theta}$ (27) $\tan\dfrac{\theta}{2} = \dfrac{1 - \cos\theta}{\sin\theta}$

9.4 The Conversion Identities (pp. 385–390)

(28) $\sin\alpha \cos\beta = \dfrac{1}{2}[\sin(\alpha + \beta) + \sin(\alpha - \beta)]$

(29) $\sin\alpha \sin\beta = \dfrac{1}{2}[\cos(\alpha - \beta) - \cos(\alpha + \beta)]$

(30) $\cos\alpha \cos\beta = \dfrac{1}{2}[\cos(\alpha + \beta) + \cos(\alpha - \beta)]$

(31) $\sin x + \sin y = 2\sin\left(\dfrac{x+y}{2}\right)\cos\left(\dfrac{x-y}{2}\right)$

(32) $\sin x - \sin y = 2\cos\left(\dfrac{x+y}{2}\right)\sin\left(\dfrac{x-y}{2}\right)$

(33) $\cos x + \cos y = 2\cos\left(\dfrac{x+y}{2}\right)\cos\left(\dfrac{x-y}{2}\right)$

(34) $\cos x - \cos y = -2\sin\left(\dfrac{x+y}{2}\right)\sin\left(\dfrac{x-y}{2}\right)$

9.5 Trigonometric Equations (pp. 391–398)

A trigonometric equation is an equation that involves trigonometric functions. For example, $2\cos^2 x - \cos x = 1$ is a trigonometric equation. To solve such equations usually involves a two step process, first solve for $\cos x$ and then, using trigonometry, solve for x. We often restrict the solutions to real numbers in the interval $[0, 2\pi)$ or angles in the interval $[0°, 360°)$. p. 143

Chapter Test

9.1A Show that the first expression may be rewritten as the second by using the Basic Trigonometric Identities.

1. $\dfrac{1 - \tan x}{\sec x} + \sin x$; $\cos x$

2. $\dfrac{1}{2}\left(\dfrac{1 + \cos x}{\sin x} + \dfrac{\sin x}{1 + \cos x}\right)$; $\csc x$

9.1B Verify the identity.

3. $\dfrac{\tan^2 x + 1}{\tan^2 x + 2} = \dfrac{1}{1 + \cos^2 x}$

4. $\dfrac{1 + \cot^2 x}{2 + \cot^2 x} = \dfrac{1}{1 + \sin^2 x}$

9.2A Find the following trigonometric values without using Table 4 or a calculator.

5. $\cot \dfrac{5\pi}{12}$
6. $\tan\left(-\dfrac{11\pi}{12}\right)$

Verify each identity.

7. $\dfrac{\sin(a+b)}{\cos a \cos b} = \tan a + \tan b$
8. $\dfrac{\tan a - \tan b}{\tan a + \tan b} = \dfrac{\sin(a-b)}{\sin(a+b)}$

9.3A Given α, find $\sin 2\alpha$, $\cos 2\alpha$, and $\tan 2\alpha$ using the double-angle identities.

9. $315°$
10. $-\dfrac{7\pi}{4}$

9.3B From the given information, determine $\cos \dfrac{\alpha}{2}$.

11. $\sin \alpha = -\dfrac{2}{3}$, α in Quadrant III
12. $\tan \alpha = -\dfrac{3}{2}$, α in Quadrant II
13. $\tan \alpha = \dfrac{3}{4}$, α in Quadrant III
14. $\cot \alpha = -\dfrac{3}{2}$, α in Quadrant II

9.4A Simplify each expression.

15. $\cos x (\tan x + \cot x)$
16. $\dfrac{1 + \cot x}{\sin x + \cos x}$
17. $\dfrac{\tan x + 1}{\sec x + \csc x}$
18. $\dfrac{\sec^2 \theta - \tan^2 \theta}{\csc \theta}$
19. $\dfrac{\cot \varphi - \tan \varphi}{\cot \varphi + \tan \varphi}$
20. $\dfrac{\sin 2\alpha}{1 - \cos 2\alpha}$

9.4B Verify each identity.

21. $\cot \theta \sin^2 \theta = \tan \theta \cos^2 \theta$
22. $\dfrac{1}{\sin x} - \sin x = \dfrac{\sin x}{\tan^2 x}$
23. $\sin\left(\dfrac{\pi}{4} - \alpha\right) = \cos\left(\alpha + \dfrac{\pi}{4}\right)$
24. $\sin \alpha \cos \beta \cot \alpha \tan \beta = \cos \alpha \sin \beta$

9.4C Evaluate by using the conversion identities.

25. $\cos \dfrac{11\pi}{12} + \cos \dfrac{5\pi}{12}$
26. $\cos \dfrac{13\pi}{12} - \cos \dfrac{5\pi}{12}$

Write each expression as a sum or difference.

27. $\sin 4x \cos 3x$
28. $\cos 4x \cos 2x$

9.5A Solve for x. Express the answer in radians.

29. $2 \cos^2 x - \cos x = 1$
30. $2 \sin^2 x + \sin x - 1 = 0$
31. $\sin^2 x + 3 \sin x = 4$
32. $\cos^2 x + \cos x - 2 = 0$

9.5B Solve each equation for x in $[0°, 360°)$ using tables or a calculator to the nearest ten minutes.

33. $3 \sin^2 x - 2 \sin x - 1 = 0$
34. $\tan^2 x - 2 \tan x - 3 = 0$
35. $7 \cos^2 x - 6 \cos x - 1 = 0$
36. $12 \sin^2 x + 12 \sin x - 6 = 0$

9.5C Solve each equation for x.

37. $\arctan(1) = x - \dfrac{\pi}{3}$
38. $\arccos\left(-\dfrac{\sqrt{3}}{2}\right) = x$
39. $\arcsin\left(\dfrac{3}{5}\right) - \arccos\left(\dfrac{4}{5}\right) = \sin^{-1} x$
40. $\tan^{-1}(1) + \tan^{-1}\left(\dfrac{12}{5}\right) = x$

Superset

41. Express $\sin \alpha$ in terms of $\cot \alpha$ for α in Quadrant II.
42. Express $\sec \alpha$ in terms of $\sin \alpha$ for α in Quadrant I.
43. Simplify each expression.
    (a) $\cos\left(\dfrac{3\pi}{2} - \alpha\right)$
    (b) $\sin(\pi + \beta)$
44. Simplify each expression.
    (a) $\dfrac{\sin 2x}{\sin x} - \dfrac{\cos 2x}{\cos x}$
    (b) $\dfrac{1}{\csc \theta - \cot \theta} - \dfrac{1}{\csc \theta + \cot \theta}$
45. Express $\tan 6x$ in terms of $\tan 2x$.

10 Applications in Trigonometry

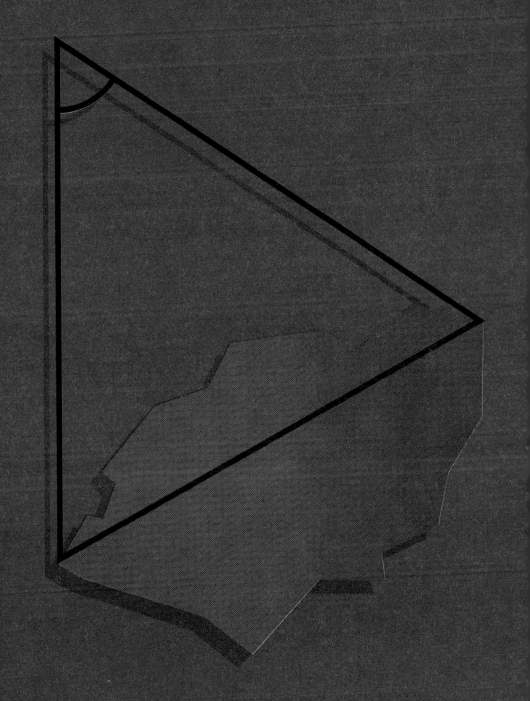

10.1 Applications Involving Right Triangles

Goal A *to solve a right triangle*

One important use of trigonometry is to solve problems that can be modeled by a triangle. Problems involving right triangles are the simplest, and we shall consider them first. We usually must determine the measure of one or more of the sides or angles of the triangle. Determining the measures of all sides and angles of a triangle is referred to as **solving the triangle.**

To simplify our discussion, we will agree that in $\triangle ABC$, the vertices are A, B, and C, and the sides opposite these vertices are a, b, and c, respectively (Figure 10.1). Also, we will agree that $A = 42°$ will mean "the measure of the angle at vertex A is $42°$."

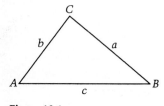

Figure 10.1

Example 1 Solve the right triangle ABC, given that $C = 90°$, $B = 40°$, and $a = 10.25$.

Solution

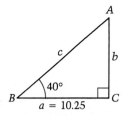

■ Begin by drawing a triangle and labeling the parts.

404

Find c: $\cos 40° = \dfrac{10.25}{c}$ ■ $\cos B = \dfrac{a}{c}$.

$$c = \dfrac{10.25}{\cos 40°} = \dfrac{10.25}{0.7660} \approx 13.38$$

We chose $\cos B = \dfrac{a}{c}$ because this equation involved the unknown c that we wished to find and two variables, B and a, that were known. Trying $\sin B = \dfrac{b}{c}$ would have been pointless because two variables in that equation were then unknown.

Find b: $\tan 40° = \dfrac{b}{10.25}$ ■ $\tan B = \dfrac{b}{a}$ is used to find b because B and a are known.

$b = (10.25)(\tan 40°)$
$b = (10.25)(0.8391) \approx 8.60$

Find a: Since A and B are complementary angles, $A = 50°$.

We must digress for a moment to discuss accuracy of answers to applied problems. The trigonometric function values in Table 4 are approximations, and measuring instruments are subject to some measurement error and thus produce approximations. Thus, when we solve problems involving lengths and angles, our solutions are approximations. The question then is how should such solutions be "rounded"? To answer this we must discuss the notion of *significant digits*.

The length 10.25 in Example 1 is said to have four significant digits. The number 0.061 has two significant digits, and the number 72,000, if it has been rounded to the nearest thousand, also has two significant digits. We apply the following rule of thumb:

> If a number N can be written in scientific notation as $N = A \cdot 10^n$ where $|A|$ is greater than or equal to 1 and less than 10, then the number of **significant digits** in N is the number of digits in A.

When solving problems involving angles and sides of triangles, you should round your calculations according to these standard rules.

Number of significant digits of the length	Measure of angle in degrees should be rounded to
1	the nearest multiple of 10°
2	the nearest degree
3	the nearest multiple of 10'
4	the nearest minute

When measurements are added or subtracted, the answer is rounded to *the least number of decimal places* of any of the measurements. When measurements are multiplied or divided, the answer is rounded to the *least number of significant digits* of any of the measurements. The power or root of a measurement is rounded to the same number of significant digits as is the measurement itself.

Example 2 Solve $\triangle ABC$ given that it is a right triangle with $C = 90°$, $a = 16.5$, and $c = 30.2$.

Solution

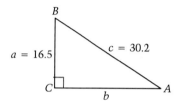

■ Draw a triangle and label the parts.

Find b: $(30.2)^2 = (16.5)^2 + b^2$
$b^2 \approx 912 - 272 = 640$
$b \approx 25.3$

Find A: $\sin A = \dfrac{16.5}{30.2} \approx 0.546$
$A \approx 33°10'$

Find B: $B \approx 56°50'$

Note that A and B are rounded to the nearest $10'$.

Goal B *to solve applied problems by using right triangles*

It should be clear from Examples 1 and 2 that a right triangle can be solved provided either one side and one of the acute angles are known, or two sides are known. Suppose you are given both acute angles. Is that enough information to solve the triangle? (The answer is "No.")

Some practical applications involve angles formed by the horizontal and the line of sight of an observer. If the line of sight is above the horizontal, the angle is called an **angle of elevation.** If the line of sight is below the horizontal, the angle is called an **angle of depression.**

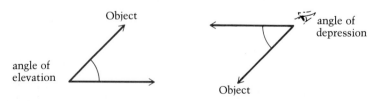

Figure 10.2

Example 3 A hot air balloon is rising vertically in still air. An observer is standing on level ground, 100 feet away from the point of launch. At one instant the observer measures the angle of elevation of the balloon as 30°00′. One minute later, the angle of elevation is 76°10′. How far did the balloon travel during that minute?

Solution

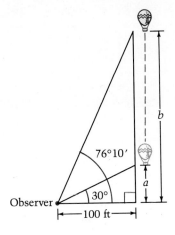

■ Draw a triangle that models the problem. We wish to determine $b - a$, the distance traveled during the one-minute period.

Find a: $\tan 30° = \dfrac{a}{100}$

$a = 100(\tan 30°)$

$= 100\left(\dfrac{\sqrt{3}}{3}\right) \approx 57.7$

Find b: $\tan 76°10′ = \dfrac{b}{100}$

$b = 100(\tan 76°10′)$

$= 100(4.061) \approx 406$

The balloon traveled $406 - 57.7 \approx 348$ feet during the minute.

Surveyors and navigators measure angles in terms of the north-south line. Two methods of measure are used: azimuth and bearing. The **azimuth** is the measure of an angle from due north in the clockwise direction.

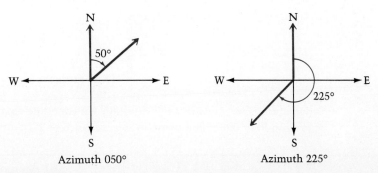

Figure 10.3

The **bearing** is the measure of the acute angle between the north-south line and the line representing the direction. It is described in terms of compass directions.

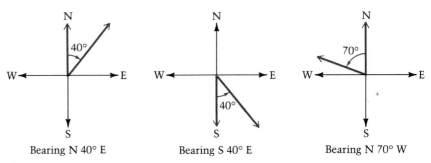

Bearing N 40° E   Bearing S 40° E   Bearing N 70° W

Figure 10.4

Example 4 A plane flies from County Airport on a bearing of N 45° E for three hours, and then flies on a bearing of S 45° E for four hours. If the speed of the plane is 400 mph, and we ignore the effects of wind, what is the plane's distance and azimuth from the County Airport after the seven hours?

Solution

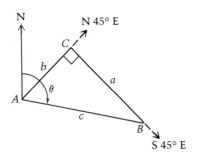

■ We begin by drawing a figure. Note that $C = 90°$ so $\triangle ABC$ is a right triangle. We wish to determine side c and angle (azimuth) θ. Note that $\theta = A + 45°$.

$$b = \text{(speed)(time on N 45° E bearing)}$$
$$= (400 \text{ mph})(3 \text{ hr}) = 1200 \text{ mi}$$

$$a = \text{(speed)(time on S 45° E bearing)}$$
$$= (400)(4 \text{ hr}) = 1600 \text{ mi}$$

$$c = \sqrt{1200^2 + 1600^2} = 2000$$

$$\tan A = \frac{\text{opposite}}{\text{adjacent}} = \frac{a}{b} = \frac{1600}{1200} = \frac{4}{3}$$

$$A \approx 53°$$ ■ A is rounded to the nearest degree.

Since $\theta = 45° + A = 45° + 53° = 98°$, we conclude that the azimuth of the plane is approximately 098°. The plane is at a distance of 2000 mi from County Airport.

Exercise Set 10.1 = Calculator Exercises in the Appendix

Goal A

In exercises 1–10, some information about the right triangle ABC is given. Use only the Table of Special Angles to solve △ABC.

1. $B = 45°, c = 64$
2. $A = 30°, c = 48$
3. $A = 60°, b = 44$
4. $B = 60°, b = 12$
5. $a = 15, b = 15\sqrt{3}$
6. $b = 40, c = 80$
7. $B = 30°, a = 24$
8. $A = 45°, a = 44$
9. $a = 25\sqrt{2}, c = 50$
10. $b = 25, a = 25$

In exercises 11–20, use the given information to solve the right triangle ABC. Be sure to adjust the number of significant digits in the answer.

11. $A = 36°, a = 70$
12. $B = 72°, a = 40$
13. $B = 12°, a = 9$
14. $A = 53°, c = 20$
15. $c = 12.3, A = 38°$
16. $a = 42.5, A = 67°$
17. $b = 0.244, A = 28°20'$
18. $A = 51°40', b = 7.85$
19. $c = 142.5, B = 71°18'$
20. $c = 6.782, B = 23°37'$

Goal B

21. A 24 ft ladder is leaning against a building. If the ladder makes an angle of 60° with the ground, how far up the building does the ladder reach?

22. If an 18 ft ladder is placed against a building so that it reaches a window sill 9 ft off the ground, what is the acute angle between the ladder and the ground?

23. A jet is flying at an altitude of 2000 ft. The angle of depression to an aircraft carrier is 20°. How far is the jet from the carrier?

24. When the angle of elevation of the sun is 60°, a certain flagpole casts a shadow 30 ft long. How tall is this flagpole?

25. A 60 ft long ramp is inclined at an angle of 5° with the level ground. How high does the ramp rise above the ground?

26. A ramp for wheelchairs is to be built beside the main steps of the library. The total vertical rise of the steps is 3 ft and the ramp will be inclined at an angle of 12°. How long a ramp is needed?

27. A television crew 2600 ft from a launch pad is filming the launch of a space shuttle. What is the angle of elevation of the camera when the shuttle is 4000 ft directly above the pad?

28. From an 80 ft lighthouse on the coast, an overturned sailboat is sighted. If the angle of depression is 9°, how far is the boat from the lighthouse?

29. A 50 ft tall flagpole casts a shadow on level ground. What is the angle of elevation of the sun when the shadow is (a) 29 ft long? (b) 60.0 ft long? (c) 12.2 ft long? (d) 125.00 ft long?

30. The Charleston Light in Charleston, SC is one of the most powerful lighthouses in the Western Hemisphere. It is 163 ft high, and its light can be seen 19 mi out at sea. What is the distance of a small boat from the foot of the tower if the angle of depression of the boat from the tower is (a) 30°? (b) 9°32'?

Superset

31. From level ground, the angle of elevation to a distant cliff is 30°. By walking a distance of 2000 ft directly toward the foot of the cliff, the angle of elevation becomes 45°. What is the height of the cliff?

32. A 20 ft flagpole is mounted on the edge of the roof of a building. A person standing level with the base of the building measures the angle of elevation to the top of the flagpole to be 65°. From the same spot, the angle of elevation of the foot of the flagpole is 60°. What is the height of the building?

33. A video camera is to be installed in a bank to monitor the bank teller's counter, which is 4 ft high. The camera will be mounted on the wall at a height of 10 ft. The counter is 20 ft from the wall on which the camera is to be mounted. To aim the camera, what should the angle of depression of the camera be?

10.2 Law of Sines

Goal A   *to solve a triangle, given one side and two angles*

In the last section we used trigonometric functions to solve right triangles. In this section and the next, we shall derive the Law of Sines and the Law of Cosines, which allow us to solve triangles which are *not* right triangles. Triangles that are not right triangles are called **oblique.**

First, we shall consider the Law of Sines. We begin by considering any oblique triangle *ABC*. There are two possibilities. **Case 1:** $\triangle ABC$ is an *acute* triangle (all angles are less than 90°) as shown in Figure 10.5(a). **Case 2:** $\triangle ABC$ is an *obtuse* triangle (one of its angles is greater than 90°) as shown in Figure 10.5(b). In each case, we have drawn the altitude *h* from vertex *B* to side *b*, or the line containing side *b* (in Case 2).

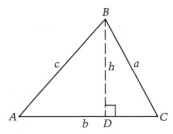
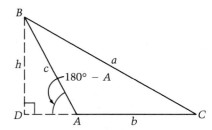

(a) Case 1: acute triangle   **(b)** Case 2: obtuse triangle

Figure 10.5

Note that in Case 2 we have

$$\sin(180° - A) = \sin 180° \cos A - \cos 180° \sin A$$
$$= (0)(\cos A) - (-1)(\sin A)$$
$$= \sin A.$$

In both cases,

$$h = a \sin C, \quad \text{and} \quad h = c \sin A.$$

Equating the two expressions for *h*, we get $a \sin C = c \sin A$, or

$$\frac{\sin A}{a} = \frac{\sin C}{c}.$$

Had we drawn the altitude from vertex *C* to side *c*, we would have concluded

$$\frac{\sin A}{a} = \frac{\sin B}{b}.$$

10 / Applications in Trigonometry

We summarize our results as follows:

Law of Sines
In any triangle with angles A, B, and C, and opposite sides a, b, and c, respectively,

$$\frac{\sin A}{a} = \frac{\sin B}{b} = \frac{\sin C}{c}.$$

Example 1 Given $\triangle ABC$ with $A = 50°10'$, $B = 70°40'$, and $c = 10.5$, solve the triangle.

Solution

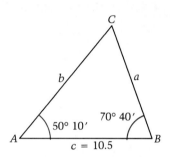

■ Begin by drawing a triangle and labeling the parts.

Find C: $C = 180° - 50°10' - 70°40' = 59°10'$

Find a: $\dfrac{\sin 50°10'}{a} = \dfrac{\sin 59°10'}{10.5}$ ■ $\dfrac{\sin A}{a} = \dfrac{\sin C}{c}.$

$a = \dfrac{(10.5)(\sin 50°10')}{\sin 59°10'}$

$= \dfrac{(10.5)(0.7679)}{0.8587} \approx 9.39$ ■ Round to three significant digits.

Find b: $\dfrac{\sin 70°40'}{b} = \dfrac{\sin 59°10'}{10.5}$ ■ $\dfrac{\sin B}{b} = \dfrac{\sin C}{c}.$

$b = \dfrac{(10.5)(0.9436)}{0.8587} \approx 11.5$ ■ Round to three significant digits.

Goal B *to solve a triangle given two sides and an angle opposite one of them*

We use the Law of Sines to solve oblique triangles when we are given (1) one side and two angles, or (2) two sides and the angle opposite one of them. The first type of problem was treated in Example 1. The second type is more complicated.

Consider what might happen if we are given two sides, say a and b, and acute angle A. There are four situations that can occur. (In the figures below, h is the altitude from vertex C.)

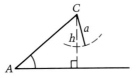

(a) $a < h$: no triangle possible

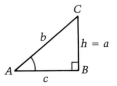

(b) $a = h$: one triangle possible

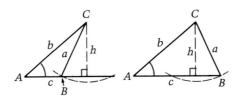

(c) $h < a < b$: two triangles possible

Figure 10.6

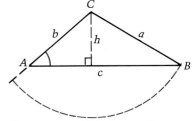

(d) $a \geq b$: one triangle possible

Example 2 illustrates case (c) above, namely the case where two triangles are possible. This is generally called *the ambiguous case*.

Example 2 Solve $\triangle ABC$, given that $A = 24°30'$, $a = 6.00$, and $b = 12.2$.

Solution Before we can draw the triangle, we must determine another angle.

Find B: $\dfrac{\sin B}{12.2} = \dfrac{\sin 24°30'}{6.00}$ ∎ $\dfrac{\sin B}{b} = \dfrac{\sin A}{a}$.

$\sin B = \dfrac{0.4147}{6.00}(12.2)$

Recall that the sine function is positive for angles in the first or second quadrant. Since $\sin 57°30' \approx 0.8432$, the second quadrant angle having the same sine function value is $180° - 57°30' = 122°30'$. Thus, we have two cases:

$$B = 57°30' \quad \text{or} \quad 122°30'.$$

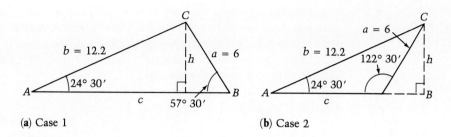

(a) Case 1 **(b)** Case 2

Case 1: $B = 57°30'$

Find C: $C = 180° - 24°30' - 57°30' = 98°00'$

Find c: $\dfrac{\sin 98°00'}{c} = \dfrac{\sin 24°30'}{6.00}$ ■ $\dfrac{\sin C}{c} = \dfrac{\sin A}{a}$.

$\dfrac{0.9903}{c} = \dfrac{0.4147}{6.00}$ ■ $\sin 98°00' = \sin(180° - 98°00')$
$\phantom{\dfrac{0.9903}{c} = \dfrac{0.4147}{6.00}\ \ \ }= \sin 82°00' = 0.9903$

$c = \left(\dfrac{6.00}{0.4147}\right)(0.9903) \approx 14.3$

Case 2: $B = 122°30'$

Find C: $C = 180° - 24°30' - 122°30' = 33°00'$

Find c: $\dfrac{\sin 33°00'}{c} = \dfrac{0.4147}{6.00}$ ■ $\dfrac{\sin C}{c} = \dfrac{\sin A}{a}$.

$c = \dfrac{6.00}{0.4147}(0.5446) \approx 7.9$

In discussing the case where two sides and an opposite angle are given, we have not considered the case where the given angle is obtuse.

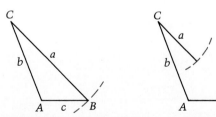

Figure 10.7

Clearly in this situation, a triangle can be formed only if $a > b$.

In practice, applying the Law of Sines will supply you with the information you need in order to decide which of the many possible cases can occur for the given data. For example, in the case of "no possible triangle," one step in the computation might yield that the sine of an unknown angle is greater than 1, which is impossible.

Goal C to solve applied problems by using the Law of Sines

Example 3 A forest fire is spotted by observers in two fire towers 12 miles apart. Tower B is on a bearing of S 12°10′ E from Tower A. If the bearing of the fire from Tower A is S 45°40′ W and from Tower B is N 75°20′ W, how far is the fire from Tower B.

Solution

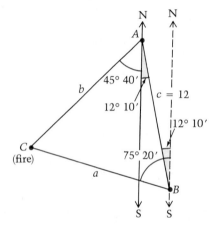

- Draw a figure to represent the given data.
- To determine B, use the fact that when parallel lines are cut by a transversal, alternate interior angles (shaded angles) are equal. Thus $B = 75°20′ − 12°10′ = 63°10′$.
- Note that $A = 45°40′ + 12°10′ = 57°50′$.

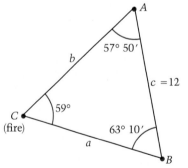

- Let us now simplify our diagram, showing the known parts of the triangle.
- $C = 180° − A − B$
  $= 180° − 57°50′ − 63°10′$
  $= 59°00′$.

- $\dfrac{\sin A}{a} = \dfrac{\sin C}{c}$.

Find a: $\dfrac{\sin 57°50′}{a} = \dfrac{\sin 59°00′}{12.0}$

$$a = \dfrac{12.0}{0.8572}(0.8465) \approx 11.9 \text{ mi}$$

The fire is approximately 11.9 mi from Tower B.

Exercise Set 10.2   =   Calculator Exercises in the Appendix

Goal A

In exercises 1–10, the measures of the two angles and one side of $\triangle ABC$ are given. Solve the triangle.

1. $A = 30°$, $B = 45°$, $a = 10\sqrt{2}$
2. $A = 60°$, $B = 45°$, $b = 8\sqrt{6}$
3. $B = 30°$, $C = 135°$, $b = 4\sqrt{2}$
4. $B = 45°$, $C = 120°$, $c = 4\sqrt{6}$
5. $a = 24.0$, $A = 100°$, $B = 24°$
6. $b = 54.0$, $A = 76°$, $B = 41°$
7. $c = 42.0$, $B = 36°$, $C = 64°$
8. $a = 96.0$, $A = 105°$, $C = 21°$
9. $a = 0.7280$, $B = 71°36'$, $C = 66°54'$
10. $b = 0.4980$, $A = 37°18'$, $C = 92°06'$

Goal B

In exercises 11–18, you are given measures for A, a, and b. State whether it is possible to have $\triangle ABC$. If it is, solve the triangle. (Determine both solutions if there is more than one.)

11. $A = 30°$, $a = 12$, $b = 24$
12. $A = 60°$, $a = 36$, $b = 24$
13. $A = 120°$, $a = 18$, $b = 24$
14. $A = 135°$, $a = 24$, $b = 24\sqrt{2}$
15. $A = 120°$, $a = 54.3$, $b = 48.8$
16. $A = 135°$, $b = 36.4$, $a = 44.7$
17. $a = 32.2$, $b = 36.4$, $A = 52°30'$
18. $b = 24.8$, $a = 20.6$, $A = 38°30'$

Goal C

19. Points A and B are on opposite sides of the Grand Canyon. Point C is 200 yd from A, $m(\angle BAC) = 87°30'$, and $m(\angle ACB) = 67°12'$. What is the distance between A and B?

20. Two observers standing on shore $\frac{1}{2}$ mi apart at points A and B measure the angle to a sailboat at point C at the same time. If $m(\angle CAB) = 63°24'$ and $m(\angle CBA) = 56°36'$, find the distance from each observer to the boat.

21. Ship A is 485 m due east of ship B. A lighthouse 1600 m from ship A is on a bearing of N 17°18' E from ship B. What is the bearing of the lighthouse from ship A?

22. Two observers 2 mi apart on level ground are in line with a spot directly below a hot air balloon. If the angle of elevation of the balloon for one observer is 68°54' and, at the same time, 26°24' for the other, what is the altitude of the balloon to the nearest tenth of a mile?

23. An observer on a ship spots a life raft at a bearing of N 75°24' E while, at the same instant, a second observer on another ship takes a bearing of the life raft of N 15°54' E. If the second ship is at a distance of 5.5 miles and a bearing of S 27°54' E from the first ship, find the distance from each ship to the life raft.

24. An observer on a ship spots a life raft at a bearing of N 35°18' W while, at the same instant, a second observer on another ship takes a bearing of the life raft of N 10°36' E. If the second ship is 4.8 mi and on a bearing of S 55°42' W from the first ship, how far is each ship from the life raft.

Superset

25. Express x in terms of θ, φ and m.

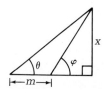

In exercises 26–27, prove each of the following.

26. $\dfrac{a + b}{b} = \dfrac{\sin A + \sin B}{\sin B}$

27. $\dfrac{a - b}{a + b} = \dfrac{\sin A - \sin B}{\sin A + \sin B}$

10.3 Law of Cosines

Goal A *to solve a triangle given two sides and the included angle, or three sides*

In the last section, we considered the Law of Sines and how it is used to solve triangles when we are given (1) one side and two angles, or (2) two sides and an angle opposite one of them. We now derive a formula, called the Law of Cosines, which will be useful if we are given (3) two sides and the included angle, or (4) three sides. Suppose we have $\triangle ABC$ positioned so that side b lies on the x-axis. Then

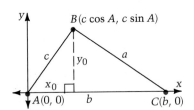

Figure 10.8 B has coordinates $x_0 = c \cos A$, $y_0 = c \sin A$.

$$x_0 = c \cos A, \quad \text{and} \quad y_0 = c \sin A. \qquad \blacksquare \cos A = \frac{x_0}{c}, \text{ and } \sin A = \frac{y_0}{c}.$$

By the distance formula, we have

$$\begin{aligned} a^2 &= (c \cos A - b)^2 + (c \sin A - 0)^2 \\ &= (c \cos A)^2 - 2(c \cos A)(b) + b^2 + (c \sin A)^2 \\ &= c^2 \cos^2 A - 2bc \cos A + b^2 + c^2 \sin^2 A \\ &= b^2 + c^2(\sin^2 A + \cos^2 A) - 2bc \cos A \\ &= b^2 + c^2 - 2bc \cos A \qquad \blacksquare \sin^2 A + \cos^2 A = 1. \end{aligned}$$

Note that the formula gives us a means of determining the square of one side of a triangle, given the other two sides and their included angle. There are three forms of this formula, as stated below.

Law of Cosines
In any triangle with angles A, B, and C, and opposite sides a, b, and c, respectively,

$$c^2 = a^2 + b^2 - 2ab \cos C,$$
$$a^2 = b^2 + c^2 - 2bc \cos A,$$
$$b^2 = a^2 + c^2 - 2ac \cos B.$$

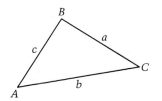

Figure 10.9

Example 1 Given $\triangle ABC$ with $a = 4$, $b = 6$, and $C = 120°$, find c.

Solution

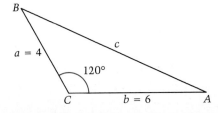

■ Begin by drawing a triangle that represents the given data.

$$c^2 = a^2 + b^2 - 2ab \cos C$$
$$c^2 = 4^2 + 6^2 - 2(4)(6) \cos 120°$$
$$c^2 = 16 + 36 - 48\left(-\frac{1}{2}\right) \qquad \blacksquare \ \cos 120° = -\cos 60° = -\tfrac{1}{2}.$$
$$c^2 = 52 + 24 = 76$$
$$c = \sqrt{76} \approx 8.7$$

Example 2 Given $\triangle ABC$ with $a = 6.00$, $b = 12.0$, and $c = 7.00$, find B.

Solution
$$b^2 = a^2 + c^2 - 2ac \cos B \qquad \blacksquare \text{ The form of the Law of Cosines that involves angle } B \text{ is used.}$$
$$12^2 = 6^2 + 7^2 - 2(6)(7) \cos B$$
$$\cos B = \frac{144 - 36 - 49}{-2(6)(7)} \approx -0.7024 \qquad \blacksquare \text{ Since } \cos B \text{ is negative, } B \text{ is the second quadrant angle whose reference angle has cosine function value } 0.7024.$$

Since $\cos 45°20' \approx 0.7024$, $B \approx 180° - 45°20' = 134°40'$.

Example 3 Highway 102 runs east-west and is intersected by Route 66, in a direction 20° north of due east. Car A is traveling along Highway 102 and is 4 miles east of the intersection. Car B is traveling eastbound on Route 66 and is 18 miles past the intersection. What is the distance between the two cars?

Solution

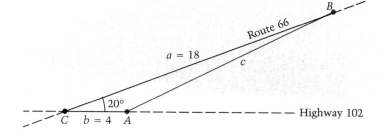

We wish to determine the distance c shown above.
$$c^2 = a^2 + b^2 - 2ab \cos C$$
$$c^2 = 18^2 + 4^2 - 2(4)(18) \cos 20°$$
$$c^2 = 324 + 16 - 144(0.9397) \approx 204.7$$
$$c \approx 14.31$$

The distance between the two cars is approximately 14 mi.

Example 4 Points B and C are on opposite sides of a reservoir. If the distance from point A to B is known to be 1.25 mi, and from point A to C is 1.15 mi, what is the distance between B and C if $\angle BAC = 55°10'$?

Solution

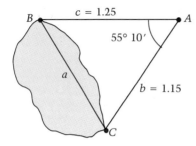

$a^2 = (1.15)^2 + (1.25)^2 - 2(1.15)(1.25)(\cos 55°10') \approx 1.24$

$a \approx 1.11$

- Draw a figure that represents the data.
- Apply Law of Cosines to $\triangle ABC$. The desired distance is a.
- The distance between B and C is approximately 1.11 mi.

Goal B *to solve problems involving the area of a triangle*

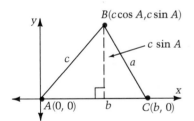

Figure 10.10

We now return to the figure we used to motivate the derivation of the Law of Cosines (see Figure 10.10). According to the formula for the area of a triangle, we can describe the area $\mathbb{A}$ of $\triangle ABC$ as

$$\mathbb{A} = \frac{1}{2}(\text{altitude})(\text{base}) = \frac{1}{2}(c \sin A)(b) = \frac{1}{2}bc \sin A.$$

In general, the area of a triangle is half the product of any two sides times the sine of the included angle. For $\triangle ABC$, $\mathbb{A}$ can be expressed three ways:

$$\mathbb{A} = \frac{1}{2}ab \sin C, \qquad \mathbb{A} = \frac{1}{2}ac \sin B, \qquad \mathbb{A} = \frac{1}{2}bc \sin A.$$

Example 5 Find the area of $\triangle ABC$ if $a = 13$ cm, $b = 10$ cm, and $C = 30°$.

Solution

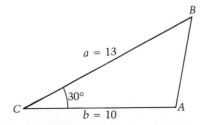

- Draw a triangle that represents the data.

$$\text{Area} = \frac{1}{2}(\text{product of two sides}) \cdot \sin(\text{included angle})$$

$$= \frac{1}{2}(ab)(\sin 30°)$$

$$= \frac{1}{2}(13 \cdot 10)\left(\frac{1}{2}\right) = 32.5$$

The area of the triangle is 32.5 cm².

Example 6 Find the area of $\triangle ABC$ if $a = 22$ in, $b = 16$ in, and $c = 18$ in.

Solution

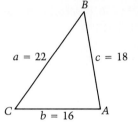

■ Draw a triangle that represents the data.

Since we are not given the lengths of two sides *and* the measure of an included angle, we first apply the Law of Cosines to determine one of the angles. We will find C.

$$c^2 = a^2 + b^2 - 2ab \cos C$$

$$\cos C = \frac{c^2 - a^2 - b^2}{-2ab} = \frac{a^2 + b^2 - c^2}{2ab} = \frac{22^2 + 16^2 - 18^2}{2(22)(16)} \approx 0.5909$$

$$C \approx 54°$$

Now that we have a and b and angle C, we can determine the area.

$$\text{Area} = \frac{1}{2}(ab) \sin 54° = \frac{1}{2}(22 \cdot 16)(0.8090) \approx 140 \text{ in}^2 \quad \text{(two significant digits)}$$

In the last example, it did not matter which of the three angles we chose in order to apply the formula for area; the result would have been the same. In addition, once we found that $\cos C \approx 0.5909$, we could have noted that since we need $\sin C$ in the formula for area, and since

$$\sin C = \sqrt{1 - \cos^2 C} \quad \blacksquare \text{ The Pythagorean Identity}$$

we could have determined $\sin C$ (necessary for computing the area) without ever determining the measure of C.

Exercise Set 10.3 = Calculator Exercises in the Appendix

Goal A

In exercises 1–8, solve $\triangle ABC$, given two sides and the included angle.

1. $a = 12$, $b = 10$, $C = 60°$
2. $a = 12$, $b = 10$, $C = 30°$
3. $b = 20$, $c = 16$, $A = 120°$
4. $b = 24$, $c = 30$, $A = 135°$
5. $a = 112$, $c = 96$, $B = 54°$
6. $a = 78$, $c = 125$, $B = 128°$
7. $A = 147°36'$, $b = 12.6$, $c = 16.3$
8. $B = 34°54'$, $a = 22.3$, $c = 18.2$

In exercises 9–16, solve $\triangle ABC$, given a, b, and c.

9. $a = 3.0$, $b = 4.0$, $c = 5.0$
10. $a = 5.0$, $b = 12.0$, $c = 13.0$
11. $a = 2.8$, $b = 2.7$, $c = 2.4$
12. $a = 3.6$, $b = 5.2$, $c = 4.8$
13. $a = 3.2$, $b = 4.8$, $c = 6.4$
14. $a = 9.0$, $b = 6.3$, $c = 5.4$
15. $a = 45.0$, $b = 30.0$, $c = 50.0$
16. $a = 30.0$, $b = 20.0$, $c = 35.0$

17. Points A and B are sighted from point C. If $C = 98°$, $AC = 128$ m, and $BC = 96$ m, how far apart are the points A and B?

18. Points A and B are sighted from point C. If $C = 36°$, $AC = 118$ ft, and $BC = 105$ ft, how far apart are the points A and B?

19. Two sides and the included angle of a parallelogram have measures 3.2, 4.8, and $54°24'$ respectively. Find the lengths of the diagonals.

20. The lengths of two sides of a parallelogram are 24.6 in and 38.2 in. The angle at one vertex has measure $108°42'$. Find the lengths of the diagonals.

21. A bridge is supported by triangular braces. If the sides of each brace have lengths 63 ft, 46 ft, and 40 ft, find the measure of the angle opposite the 46 ft side.

22. The measures of two sides of a parallelogram are 28 in and 42 in. If the longer diagonal has measure 58 in, find the measures of the angles at the vertices.

Goal B

In exercises 23–30, determine the area of $\triangle ABC$ using the given information.

23. $A = 60°$, $b = 12.6$, $c = 18.3$
24. $A = 45°$, $c = 23.7$, $b = 16.4$
25. $B = 37°12'$, $a = 10.9$, $c = 15.8$
26. $B = 24°54'$, $c = 10.5$, $a = 14.6$
27. $C = 112°$, $b = 44.6$, $a = 32.5$
28. $C = 118°$, $a = 18.7$, $b = 30.6$
29. $A = 13°30'$, $b = 254$, $c = 261$
30. $A = 66°24'$, $c = 0.231$, $b = 0.176$

Superset

31. The lengths of two sides of a triangle are 12 in and 16 in, and the area is 87.36 in². Solve the triangle.

32. Given $\triangle ABC$ with $a = 12$, $b = 16$, and $c = 10$, find the length of the altitude from vertex B.

33. Find the area of a triangle with vertices at the points with coordinates $(-5, 0)$, $(6, 3)$, and $(8, -5)$.

In exercises 34–36, assume you are given an isosceles triangle ABC with $a = c$ with $B = \theta$.

34. Find the length of the base b in terms of a and θ.

35. Show that $b = 2a \sin \frac{1}{2}\theta$.

36. Prove that $b^2 = 2a^2(1 - \cos \theta)$.

10.4 Vectors in the Plane

Goal A *to determine the magnitude and direction of a vector*

The statement "The distance between points A and B is 12 miles" tells us only the distance between A and B. The statement "An object undergoes a displacement of 12 miles due east from point A to point B" tells us both the distance between A and B and the direction in which the object travels. Quantities like distance, which indicate a magnitude (size) but no direction, are called **scalar quantities.** Quantities like displacement, which indicate both magnitude and direction, are called **vector quantities.**

Since a vector quantity contains information about magnitude and direction, it is convenient to represent it by an arrow (a directed line segment) in the xy-plane. The magnitude of the vector quantity is given by the length of the arrow. The direction of the vector quantity is given by the angle that the arrow makes with the horizontal in the direction of the positive x-axis.

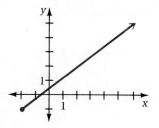
Figure 10.11

Example 1 Determine the magnitude and direction of the vector represented in Figure 10.11.

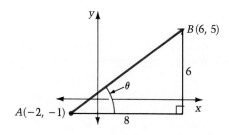

- Begin by drawing a right triangle with the arrow as its hypotenuse. Label the coordinates of the tail A and head B of the vector.

$$d = \sqrt{(x_2 - x_1)^2 + (y_2 - y_1)^2}$$
$$= \sqrt{(6 - (-2))^2 + (5 - (-1))^2}$$
$$= \sqrt{64 + 36} = 10$$

- The magnitude of the vector can be found by applying the distance formula to the coordinates of the tail $A(-2, -1)$ and head $B(6, 5)$ of the vector.

$$\tan \theta = \frac{\text{opposite}}{\text{adjacent}} = \frac{6}{8} = 0.7500$$
$$\theta \approx 37°$$

- The direction can be found by determining angle θ in the triangle.

The magnitude of the vector is 10, and it makes an angle of approximately 37° with the horizontal.

We usually use boldface letters to represent vector quantities.

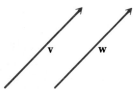
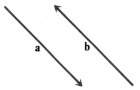
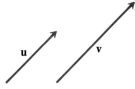

Vectors **v** and **w** are equal since they have the same magnitude and direction.

Vectors **x** and **y** have the same magnitude, but different directions.

Vectors **a** and **b** are said to have the same magnitude but *opposite* directions.

Vectors **u** and **v** have the same direction, but different magnitudes.

Figure 10.12

We will need to determine the sum of two vectors, and the product of a scalar (a real number) and a vector. The sum **a** + **b** of two vectors **a** and **b** is found by first moving **b** without changing its magnitude or direction, so that the tail of **b** is placed at the head of **a**. Then draw an arrow from the tail of **a** to the head of **b**.

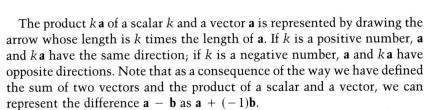

Figure 10.13

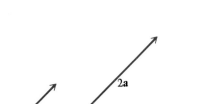

Figure 10.14

The product $k\mathbf{a}$ of a scalar k and a vector **a** is represented by drawing the arrow whose length is k times the length of **a**. If k is a positive number, **a** and $k\mathbf{a}$ have the same direction; if k is a negative number, **a** and $k\mathbf{a}$ have opposite directions. Note that as a consequence of the way we have defined the sum of two vectors and the product of a scalar and a vector, we can represent the difference **a** − **b** as **a** + (−1)**b**.

Example 2 Vectors **a** and **b** are represented at the right. Draw 2**a** + 3**b** and 2**b** − **a**.

Solution

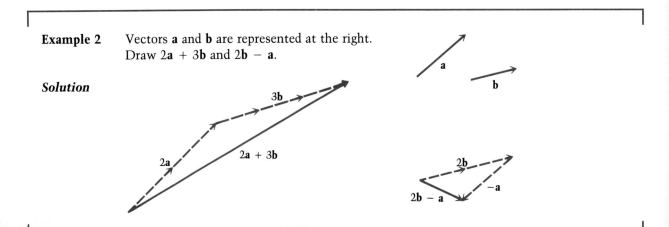

We have illustrated a method for adding (or subtracting) vectors by placing them in a tail-to-head arrangement. We can also add (or subtract) two vectors by placing them in a tail-to-tail arrangement. When positioned this way, the two vectors determine a parallelogram as shown in Figure 10.15. The sum **a** + **b** is found by drawing a vector from the point where the tails meet, along the diagonal of the parallelogram to the opposite vertex. The difference **a** − **b** is found by drawing a vector from the head of **b** to the head of **a**.

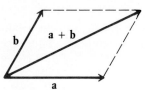 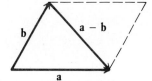

Figure 10.15

We have frequently referred to the tail and head of a vector. These points are alternately called the **initial point** and the **terminal point** of the vector, respectively. It is sometimes convenient to use these points when naming the vector. For example, if we know that vector **v** in Figure 10.16 has initial point A and terminal point B, we may refer to it as $\overrightarrow{AB}$.

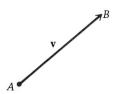

Figure 10.16 $\mathbf{v} = \overrightarrow{AB}$

Goal B *to describe vectors as ordered pairs*

One of the most common ways of naming a vector is with an ordered pair of numbers. The first number of the ordered pair represents the change in x from the initial point to the terminal point of the vector, and the second number represents the change in y from the initial point to the terminal point. To distinguish ordered pairs representing vectors from ordered pairs representing points, we will use angular brackets $\langle\,,\,\rangle$ when referring to vectors. For example, in Figure 10.17, we denote the vector $\overrightarrow{PQ}$ by $\langle 3, -2 \rangle$. This indicates that the change in x from P to Q is 3 units in the positive x-direction and the change in y is 2 units in the negative y-direction.

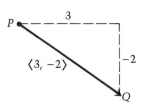

Figure 10.17

In a natural way, the x-coordinates and y-coordinates of two points P_1 and P_2 can be used to determine the vector $\overrightarrow{P_1 P_2}$.

Definition

Given two points, $P_1(x_1, y_1)$ and $P_2(x_2, y_2)$, the vector with initial point P_1 and terminal point P_2 is given by the ordered pair

$$\langle x_2 - x_1, y_2 - y_1 \rangle.$$

The numbers $x_2 - x_1$ and $y_2 - y_1$ are called the **scalar components** of the vector.

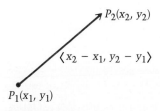

Figure 10.18

Example 3 (a) Describe the vector from $A(-3, 2)$ to $B(9, -7)$ as an ordered pair.
(b) If $\mathbf{v} = \langle 3, -5 \rangle$ is positioned in the plane so that its initial point is at $(1, 2)$, what are the coordinates of its terminal point?

Solution (a) $\overrightarrow{AB} = \langle 9 - (-3), -7 - 2 \rangle = \langle 12, -9 \rangle$.

(Note: $\overrightarrow{BA} = \langle -3 - 9, 2 - (-7) \rangle = \langle -12, 9 \rangle$, thus, the vector $\overrightarrow{BA}$ is the same as the vector $-\overrightarrow{AB}$.

(b) The terminal point is found by adding 3 to the x-coordinate of the initial point and -5 to the y-coordinate of the initial point. Thus, the terminal point has coordinates $(1 + 3, 2 - 5) = (4, -3)$.

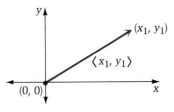

Figure 10.19

A vector that is positioned with its initial point at the origin $(0, 0)$ is called a **position vector**, or **radius vector**. In this case, if the vector is given by $\langle x_1, y_1 \rangle$, then the terminal point is (x_1, y_1). That is, the scalar components of a position vector are the same as the coordinates of the terminal point. The vector $\langle 0, 0 \rangle = \mathbf{0}$ is called the **zero vector**. We now formally define the notions of equality, addition, and scalar multiplication of vectors.

Definition
Given two vectors $\mathbf{a} = \langle a_1, a_2 \rangle$ and $\mathbf{b} = \langle b_1, b_2 \rangle$, and a scalar k, we have the following definitions.

Equality of vectors $\mathbf{a} = \mathbf{b}$ if and only if $a_1 = b_1$ and $a_2 = b_2$.
Addition of vectors $\mathbf{a} + \mathbf{b} = \langle a_1 + b_1, a_2 + b_2 \rangle$
Scalar Multiplication of a vector $k\mathbf{a} = \langle ka_1, ka_2 \rangle$

Example 4 Suppose $\mathbf{v} = \langle 2, 3 \rangle$, $\mathbf{w} = \langle 5, -4 \rangle$ and $\mathbf{j} = \langle 0, 1 \rangle$. Determine (a) $2\mathbf{v} + 3\mathbf{w}$ (b) $\mathbf{w} - 6\mathbf{j}$

Solution (a) $2\mathbf{v} + 3\mathbf{w} = 2\langle 2, 3 \rangle + 3\langle 5, -4 \rangle$ (b) $\mathbf{w} - 6\mathbf{j} = \langle 5, -4 \rangle + (-6)\langle 0, 1 \rangle$
$= \langle 4, 6 \rangle + \langle 15, -12 \rangle$ $= \langle 5, -4 \rangle + \langle 0, -6 \rangle$
$= \langle 4 + 15, 6 - 12 \rangle$ $= \langle 5 + 0, -4 - 6 \rangle$
$= \langle 19, -6 \rangle$ $= \langle 5, -10 \rangle$

The length of a vector $\mathbf{v}$ is called the **norm** of $\mathbf{v}$ and is denoted $\|\mathbf{v}\|$. Using the ordered pair definition of a vector and the Pythagorean Theorem, we have the following definition.

Definition

If $\mathbf{a} = \langle a_1, a_2 \rangle$, then $\|\mathbf{a}\|$ is the **norm** of $\mathbf{a}$, defined $\|\mathbf{a}\| = \sqrt{a_1^2 + a_2^2}$.

Note that the norm of a vector is a scalar quantity—it measures the length of the vector. A vector is called a **unit vector** if it has length equal to 1. Given a vector $\mathbf{a}$, we can determine a **unit vector in the direction of a**, denoted $\mathbf{u_a}$, by multiplying vector $\mathbf{a}$ by the scalar $\dfrac{1}{\|\mathbf{a}\|}$. That is,

$$\mathbf{u_a} = \frac{\mathbf{a}}{\|\mathbf{a}\|}$$

is the unit vector in the direction of $\mathbf{a}$.

Example 5 Suppose $\mathbf{v} = \langle 3, -4 \rangle$. Determine the following.

(a) $\|\mathbf{v}\|$ (b) $\mathbf{u_v}$ (c) a vector of length 2 in the direction of $\mathbf{v}$

Solution (a) $\|\mathbf{v}\| = \sqrt{3^2 + (-4)^2} = \sqrt{9 + 16} = 5$

(b) $\mathbf{u_v} = \dfrac{1}{\|\mathbf{v}\|}\mathbf{v} = \dfrac{1}{5}\langle 3, -4 \rangle = \left\langle \dfrac{3}{5}, -\dfrac{4}{5} \right\rangle$

Thus the unit vector in the direction of $\mathbf{v}$ is $\left\langle \dfrac{3}{5}, -\dfrac{4}{5} \right\rangle$.

(c) To determine a vector of length 2 in the direction of $\mathbf{v}$, multiply the unit vector (in that direction) by 2.

$$2\mathbf{u_v} = 2\left\langle \frac{3}{5}, -\frac{4}{5} \right\rangle = \left\langle \frac{6}{5}, -\frac{8}{5} \right\rangle$$

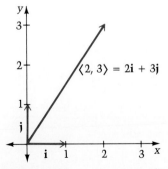

Figure 10.20

The unit vectors

$$\langle 1, 0 \rangle \quad \text{and} \quad \langle 0, 1 \rangle$$

occur so frequently that we give them the special names $\mathbf{i}$ and $\mathbf{j}$, respectively. Thus, $\mathbf{i}$ is a vector of length 1 in the positive x-direction, and $\mathbf{j}$ is a vector of length 1 in the positive y-direction. These two vectors are referred to as the **unit coordinate vectors**. For any real numbers a_1 and a_2,

$$a_1\mathbf{i} + a_2\mathbf{j} = a_1\langle 1, 0 \rangle + a_2\langle 0, 1 \rangle = \langle a_1, 0 \rangle + \langle 0, a_2 \rangle = \langle a_1, a_2 \rangle.$$

Thus, any vector $\mathbf{a} = \langle a_1, a_2 \rangle$ can be written as $\mathbf{a} = a_1\mathbf{i} + a_2\mathbf{j}$, which is referred to as a **linear combination** of $\mathbf{i}$ and $\mathbf{j}$.

Example 6 Suppose $\mathbf{v} = 2\mathbf{i} + 3\mathbf{j}$ and $\mathbf{w} = \mathbf{i} - 4\mathbf{j}$. Determine the following.
(a) $5\mathbf{v} - \mathbf{w}$ (b) $\|\mathbf{v}\|$ (c) a unit vector in the direction of $\mathbf{v}$

Solution (a) $5\mathbf{v} - \mathbf{w} = 5(2\mathbf{i} + 3\mathbf{j}) - (\mathbf{i} - 4\mathbf{j}) = 10\mathbf{i} + 15\mathbf{j} - \mathbf{i} + 4\mathbf{j} = 9\mathbf{i} + 19\mathbf{j}$

(b) $\|\mathbf{v}\| = \|2\mathbf{i} + 3\mathbf{j}\| = \sqrt{2^2 + 3^2} = \sqrt{4 + 9} = \sqrt{13}$

(c) $\mathbf{u}_\mathbf{v} = \dfrac{1}{\|\mathbf{v}\|}\mathbf{v} = \dfrac{1}{\sqrt{13}}(2\mathbf{i} + 3\mathbf{j}) = \dfrac{2}{\sqrt{13}}\mathbf{i} + \dfrac{3}{\sqrt{13}}\mathbf{j}$

Exercise Set 10.4 $\boxed{=}$ Calculator Exercises in the Appendix

Goal A

In exercises 1–8, determine the magnitude and direction of each vector.

1.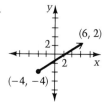

2.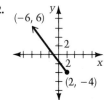

3.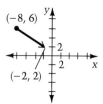

4.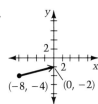

5.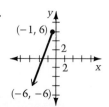

6.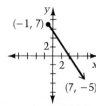

7.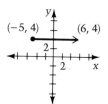

8.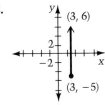

In exercises 9–20, using the given vectors, draw the vector sum or difference.

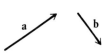

9. $\mathbf{a} + \mathbf{d}$ 10. $\mathbf{b} + \mathbf{c}$ 11. $\mathbf{a} - \mathbf{d}$

12. $\mathbf{b} - \mathbf{c}$ 13. $2\mathbf{a} - 3\mathbf{b}$ 14. $3\mathbf{d} - 4\mathbf{b}$

15. $\dfrac{1}{2}\mathbf{a} + \dfrac{1}{2}\mathbf{d}$ 16. $2\mathbf{b} + 2\mathbf{c}$ 17. $-2\mathbf{b}$

18. $-3\mathbf{c}$ 19. $\mathbf{a} + 2\mathbf{b} - \mathbf{c}$

20. $2\mathbf{a} - \mathbf{c} - 2\mathbf{d}$

Goal B

In exercises 21–26, points A and B are given. Describe the vector from A to B as an ordered pair.

21. $A(-1, -3)$, $B(4, 9)$ 22. $A(2, -2)$, $B(-4, 6)$
23. $A(-5, 9)$, $B(-2, 5)$ 24. $A(-10, -7)$, $B(-2, -1)$
25. $A(1, 4)$, $B(2, -4)$ 26. $A\left(-2, 3\frac{1}{2}\right)$, $B(-5, 4)$

In exercises 27–32, given the vector $\mathbf{v}$ and the coordinates of its initial point, find the coordinates of its terminal point.

27. $\mathbf{v} = \langle 3, -2 \rangle$, $(-1, 4)$
28. $\mathbf{v} = \langle -2, 3 \rangle$, $(-1, 4)$
29. $\mathbf{v} = \langle -4, -5 \rangle$, $\left(1\frac{1}{2}, 3\frac{1}{2}\right)$
30. $\mathbf{v} = \langle -5, 2 \rangle$, $\left(1\frac{1}{2}, 3\frac{1}{2}\right)$
31. $\mathbf{v} = \left\langle \frac{13}{2}, \frac{7}{2} \right\rangle$, $(-4, -1)$
32. $\mathbf{v} = \left\langle -\frac{3}{2}, -\frac{7}{2} \right\rangle$, $(1, -2)$

In exercises 33–40, for the vectors $\mathbf{u} = \langle 3, 6 \rangle$, $\mathbf{v} = \langle -8, 2 \rangle$, and $\mathbf{w} = \langle 2, -1 \rangle$, determine the following.

33. $3\mathbf{v} + 2\mathbf{w}$ 34. $2\mathbf{v} + \mathbf{u}$
35. $\frac{1}{2}\mathbf{v} + 3\mathbf{u}$ 36. $2\mathbf{w} + 3\mathbf{u}$
37. $-\frac{1}{2}\mathbf{v} + 3\mathbf{u}$ 38. $-4\mathbf{w} + \frac{1}{3}\mathbf{u}$
39. $3\mathbf{v} + 2\mathbf{w} - \frac{1}{3}\mathbf{u}$ 40. $\frac{1}{2}\mathbf{v} - 3\mathbf{w} + \frac{2}{3}\mathbf{u}$

In exercises 41–48, for each vector $\mathbf{v}$, determine $\|\mathbf{v}\|$, $\mathbf{u_v}$, and $-2\mathbf{v}$.

41. $\langle -6, 8 \rangle$ 42. $\langle -4, -3 \rangle$ 43. $\langle 0, -4 \rangle$
44. $\langle -6, 0 \rangle$ 45. $\langle 2\sqrt{2}, 2 \rangle$ 46. $\langle 3, \sqrt{7} \rangle$
47. $\left\langle \frac{8}{3}, -2 \right\rangle$ 48. $\left\langle -\frac{5}{2}, 6 \right\rangle$

In exercises 49–54, given vectors $\mathbf{v}$ and $\mathbf{w}$, determine $2\mathbf{v} - 3\mathbf{w}$ as a linear combination of the unit vectors $\mathbf{i}$ and $\mathbf{j}$. Determine $\|\mathbf{v}\|$ and $\mathbf{u_v}$.

49. $\mathbf{v} = \mathbf{i} + \mathbf{j}$, $\mathbf{w} = \mathbf{i} - 2\mathbf{j}$
50. $\mathbf{v} = 2\mathbf{i} - \mathbf{j}$, $\mathbf{w} = \mathbf{i} - 2\mathbf{j}$
51. $\mathbf{v} = 3\mathbf{i} - 2\mathbf{j}$, $\mathbf{w} = \mathbf{i} + 4\mathbf{j}$
52. $\mathbf{v} = 2\mathbf{i} - 3\mathbf{j}$, $\mathbf{w} = -\mathbf{i} + 4\mathbf{j}$
53. $\mathbf{v} = \frac{1}{2}\mathbf{i} + \frac{5}{2}\mathbf{j}$, $\mathbf{w} = \frac{7}{2}\mathbf{i} - \frac{1}{2}\mathbf{j}$
54. $\mathbf{v} = \frac{2}{3}\mathbf{i} + 4\mathbf{j}$, $\mathbf{w} = \frac{4}{3}\mathbf{i} + \frac{1}{2}\mathbf{j}$

Superset

In exercises 55–60, describe each radius vector as an ordered pair.

55. 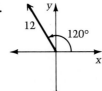 56.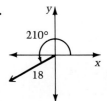

57. 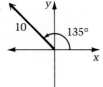 58.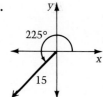

59. 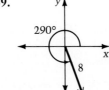 60.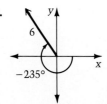

61. If $\mathbf{a} = 2\mathbf{i} + 3\mathbf{j}$, $\mathbf{b} = \mathbf{i} - 2\mathbf{j}$, and $\mathbf{c} = 4\mathbf{i} - \mathbf{j}$, find scalars r and s such that $\mathbf{c} = r\mathbf{a} - s\mathbf{b}$.

62. Find the angle between the radius vectors $\langle -2, 1 \rangle$ and $\langle 3, 4 \rangle$.

10.5 Vector Applications and the Dot Product

Goal A *to solve applied problems by using vector methods*

Often the quantities involved in applied problems are vectors, that is, quantities having both magnitude and direction. For example, the **velocity v** of an object is a vector quantity whose magnitude, $\|\mathbf{v}\|$, is the **speed** of the object and whose **direction** is the direction in which the object is moving.

In aviation, one frequently uses the concepts of *air speed* and *ground speed*. The **air speed** is the speed at which an airplane would fly in still air; the **ground speed** is the airplane's speed relative to the ground, after the effect of the wind has been accounted for. Thus, the (true) ground speed and true direction of an airplane are determined by forming the vector sum of the airplane's velocity vector and the wind's velocity vector.

Example 1 An airplane's air speed is set at 300 km/h and its bearing is set at N 90° E (i.e., due east). A 50 km/h wind is blowing with a bearing S 60° E (i.e., in a direction 60° east of due south). Determine the ground speed and true direction of the airplane.

Solution

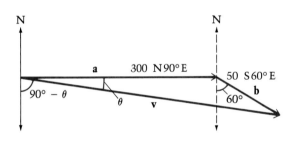

- Draw a diagram. Let **a** be the airplane's velocity, and let **b** be the wind velocity. The ground velocity **v** is the vector sum of vectors **a** and **b**.

$$\|\mathbf{v}\|^2 = 300^2 + 50^2 - 2(300)(50)\cos 150°$$

$$= 90{,}000 + 2500 - 30{,}000\left(-\frac{\sqrt{3}}{2}\right) \approx 118{,}481$$

$$\|\mathbf{v}\| \approx \sqrt{118{,}481} \approx 344 \text{ km/h}$$

- The Law of Cosines is used to find $\|\mathbf{v}\|$, the ground speed.

$$\frac{\sin \theta}{50} = \frac{\sin 150°}{344}$$

$$\sin \theta = \frac{50}{344}(\sin 150°) = \frac{50}{344}\left(\frac{1}{2}\right) \approx 0.0727$$

$$\theta \approx 4°10'$$

- The Law of Sines is used to find the true direction which is S (90° − θ) E. Note the angle formed by **a** and **b** is 90° + 60° = 150°.

The airplane's approximate ground speed is 344 km/h; its true bearing is S 85°50' E.

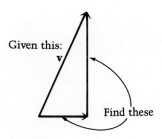

Given this:

Find these

Figure 10.21

Our next application is a common problem in physics and engineering: given a vector **v**, find two other vectors whose vector sum is **v**. The two "other" vectors are usually perpendicular to each other, and are called **vector components of v**. The process of determining these two component vectors is referred to as **resolving v** into perpendicular components.

Of special importance is the resolution of a vector **v** into horizontal and vertical components. (We will refer to these vector components as $\mathbf{v}_x$ and $\mathbf{v}_y$, respectively, as shown in Figure 10.22.) To do this, we position **v** with its tail at the origin of the xy-plane, and draw perpendiculars from the head of **v** to the x-axis and y-axis.

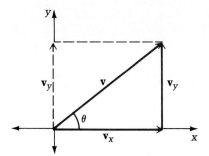

Figure 10.22

■ We move $\mathbf{v}_y$ to the position on the right in order to visualize it as one side in a right triangle with $\mathbf{v}_x$.

From the figure above, it is clear that

$$\cos \theta = \frac{\text{adjacent}}{\text{hypotenuse}} = \frac{\|\mathbf{v}_x\|}{\|\mathbf{v}\|}$$

and

$$\sin \theta = \frac{\text{opposite}}{\text{hypotenuse}} = \frac{\|\mathbf{v}_y\|}{\|\mathbf{v}\|}.$$

Thus,

$$\|\mathbf{v}_x\| = \|\mathbf{v}\| \cos \theta \quad \text{and} \quad \|\mathbf{v}_y\| = \|\mathbf{v}\| \sin \theta.$$

Since $\mathbf{v}_x$ is a vector with magnitude $\|\mathbf{v}\| \cos \theta$ in the direction of **i**, and $\mathbf{v}_y$ is a vector with magnitude $\|\mathbf{v}\| \sin \theta$ in the direction of **j**, we write

$$\mathbf{v}_x = (\|\mathbf{v}\| \cos \theta)\mathbf{i} \quad \text{and} \quad \mathbf{v}_y = (\|\mathbf{v}\| \sin \theta)\mathbf{j}$$

where θ is the angle formed by **v** and the positive x-axis. Note that

$$\mathbf{v} = \mathbf{v}_x + \mathbf{v}_y = (\|\mathbf{v}\| \cos \theta)\mathbf{i} + (\|\mathbf{v}\| \sin \theta)\mathbf{j},$$

and so $\mathbf{v}_x$ and $\mathbf{v}_y$ are truly vector components of **v**.

Example 2 Vector **v** has magnitude 8. Resolve **v** into horizontal and vertical components if
(a) it makes an angle of 60° with the positive x-axis;
(b) it makes an angle of 135° with the positive x-axis.

Solution (a)

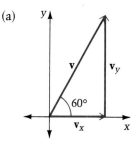

$$\mathbf{v_x} = (\|\mathbf{v}\| \cos \theta)\mathbf{i} = (8 \cdot \cos 60°)\mathbf{i} = (8)\left(\frac{1}{2}\right)\mathbf{i} = 4\mathbf{i}$$

$$\mathbf{v_y} = (\|\mathbf{v}\| \sin \theta)\mathbf{j} = (8 \cdot \sin 60°)\mathbf{j} = (8)\left(\frac{\sqrt{3}}{2}\right)\mathbf{j} = 4\sqrt{3}\mathbf{j}$$

$$\mathbf{v} = \mathbf{v_x} + \mathbf{v_y} = 4\mathbf{i} + 4\sqrt{3}\mathbf{j}.$$

(b)

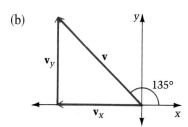

$$\mathbf{v_x} = (\|\mathbf{v}\| \cos \theta)\mathbf{i} = (8 \cdot \cos 135°)\mathbf{i} = \left[8 \cdot \left(-\frac{\sqrt{2}}{2}\right)\right]\mathbf{i} = -4\sqrt{2}\mathbf{i}$$

$$\mathbf{v_y} = (\|\mathbf{v}\| \sin \theta)\mathbf{j} = (8 \cdot \sin 135°)\mathbf{j} = \left(8 \cdot \frac{\sqrt{2}}{2}\right)\mathbf{j} = 4\sqrt{2}\mathbf{j}$$

$$\mathbf{v} = \mathbf{v_x} + \mathbf{v_y} = -4\sqrt{2}\mathbf{i} + 4\sqrt{2}\mathbf{j}.$$

To study an object under the influence of a force, it is useful to represent the force as a vector. We say that a force of one newton (N) is required to accelerate a mass of 1 kg at a rate of 1 m/s². One of the most common forces is **g**, the force that gravity exerts on an object (also called weight). If an object has a mass of M kg, then the magnitude of **g** is $9.8 \cdot M$ newtons.

When an object is not accelerating (either it is moving at a constant velocity or it is at rest), we say that **forces are in equilibrium.** This means that the vector sum of all forces is zero.

Example 3   A 400 kg piano is being rolled down a ramp. The ramp makes an angle of 30° with the level ground below. If we neglect friction, what is the force required to hold the piano stationary on the ramp?

Solution

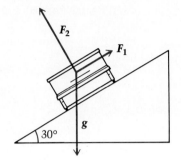

■ Begin by drawing a figure. Three forces act on the object: **g**, the force of gravity; $\mathbf{F}_1$, the force needed to hold the piano stationary ($\mathbf{F}_1$ is parallel to the ramp); $\mathbf{F}_2$, a force perpendicular to the surface of the ramp that keeps the piano from crashing through the ramp.

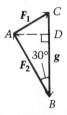

■ We draw a triangle representing the forces. Since there is no motion, the total vector sum is **0**. That is, $\mathbf{F}_1 + \mathbf{g} + \mathbf{F}_2 = \mathbf{0}$. Notice that $\mathbf{F}_1 + \mathbf{g} = -\mathbf{F}_2$.

The question asks us to determine $\mathbf{F}_1$. To determine $\angle B$ in the triangle above, we have drawn a perpendicular from A that meets side BC at point D. Given $\angle CAD = 30°$, and since $\angle CAB = 90°$, we conclude that $\angle DAB = 60°$. Thus, in right triangle ADB we have $\angle B = 30°$.

$$\sin 30° = \frac{\text{opposite}}{\text{hypotenuse}} = \frac{\|\mathbf{F}_1\|}{\|\mathbf{g}\|} = \frac{\|\mathbf{F}_1\|}{(9.8)(400)}$$

■ To determine $\mathbf{F}_1$, consider sin B in right triangle ABC.

$$\|\mathbf{F}_1\| = (\sin 30°)(9.8)(400) = \left(\frac{1}{2}\right)(3920) = 1960 \text{ N}$$

The force required to hold the piano on the ramp is 1960 N (roughly 441 lb).

Goal B   *to compute the dot product of two vectors*

Up to this point, the only type of vector multiplication that we have discussed is the product of a scalar k and a vector **v** (the result $k\mathbf{v}$ is a vector). We now define a product of two vectors in such a way that the product is a scalar, not a vector.

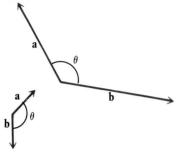

Figure 10.23

Definition

Let **a** and **b** be two nonzero vectors. The **dot product** of **a** and **b**, denoted **a** · **b**, is defined as

$$\mathbf{a} \cdot \mathbf{b} = \|\mathbf{a}\| \|\mathbf{b}\| \cos \theta$$

where θ is the angle between **a** and **b** such that $0° \leq \theta \leq 180°$.

There is an alternate definition of the dot product which involves the components of the two vectors **a** and **b**.

Alternate Definition

Let $\mathbf{a} = \langle a_1, a_2 \rangle$ and $\mathbf{b} = \langle b_1, b_2 \rangle$. Then

$$\mathbf{a} \cdot \mathbf{b} = a_1 b_1 + a_2 b_2.$$

Careful! The dot product of two vectors is a scalar, not a vector.

Taken together, the two definitions of dot product provide a way to determine the angle between two vectors. The technique is illustrated in Example 4. Note that when the angle between the two vectors is 90°, $\cos \theta = 0$, and so the dot product is 0. In addition, if the dot product of two nonzero vectors is 0, then $\theta = 90°$. Thus, if **a** and **b** are nonzero vectors,

$$\mathbf{a} \cdot \mathbf{b} = 0 \text{ if and only if } \mathbf{a} \text{ and } \mathbf{b} \text{ are perpendicular.}$$

Example 4 Find $\mathbf{v} \cdot \mathbf{w}$ if $\mathbf{v} = \langle 3, 2 \rangle$ and $\mathbf{w} = \langle 4, -5 \rangle$. Determine the angle between **v** and **w**.

Solution

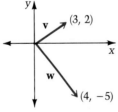

■ Although a figure is not necessary, it does help in visualizing the problem.

$\mathbf{v} \cdot \mathbf{w} = \langle 3, 2 \rangle \cdot \langle 4, -5 \rangle = (3)(4) + (2)(-5) = 12 - 10 = 2$

$\mathbf{v} \cdot \mathbf{w} = \|\mathbf{v}\| \|\mathbf{w}\| \cos \theta$

$2 = \|\langle 3, 2 \rangle\| \|\langle 4, -5 \rangle\| \cos \theta$

$2 = [\sqrt{3^2 + 2^2} \cdot \sqrt{4^2 + (-5)^2}] \cos \theta$

$\cos \theta = \dfrac{2}{\sqrt{13} \cdot \sqrt{41}} = \dfrac{2}{\sqrt{533}} \approx 0.0866$

$\theta \approx 85°$

■ Use $\mathbf{v} \cdot \mathbf{w}$ to approximate θ, the angle between **v** and **w**.

Exercise Set 10.5 □= Calculator Exercises in the Appendix

Goal A

In exercises 1–4, use that fact that eastbound airplanes "hitch a ride" on the jet stream.

1. A jet's air speed is 450 mph and its bearing is N 80° E. If the jet stream is blowing due east at 120 mph, what is the jet's ground speed and true bearing?

2. A jet's air speed is 450 mph and its bearing is S 80° W. If the jet stream is blowing due east at 120 mph, what is the jet's ground speed and true bearing?

3. A jet's ground speed is 410 mph and its true bearing is due north. If the jet stream is blowing due east at 120 mph, what is the jet's air speed and bearing?

4. A jet's ground speed is 510 mph and its true bearing is N 40° E. If the jet stream is blowing due east at 120 mph, what is the jet's air speed and bearing?

5. A 62 kg weight is on a ramp which is inclined at 40°. What is the force needed to hold the weight stationary on the ramp?

6. A crate containing a 300 kg wood stove is being unloaded from a delivery truck by sliding it down an inclined plank which is at an angle of 32° with the horizontal. What is the force needed to hold the crate at rest?

7. A 500 lb boat is being lowered into the water down an inclined ramp which is at an angle of 25° with the horizontal. What is the force needed to hold the boat at rest?

8. A force of 42 lb is required to hold a 250 lb block of granite from sliding down an inclined ramp. At what angle is the ramp inclined?

Goal B

In exercises 9–16, determine the dot product of the given vectors. Find the angle between the vectors to the nearest degree.

9. $\langle 3, 1 \rangle$, $\langle -2, 4 \rangle$
10. $\langle 1, 1 \rangle$, $\langle 3, -1 \rangle$
11. $\langle 2, -1 \rangle$, $\langle 2, -2 \rangle$
12. $\langle -1, 2 \rangle$, $\langle 4, 6 \rangle$
13. $\langle 1, 9 \rangle$, $\langle 9, -1 \rangle$
14. $\langle 4, 3 \rangle$, $\langle 3, -4 \rangle$
15. $\langle -5, -3 \rangle$, $\langle 4, -3 \rangle$
16. $\langle -10, 2 \rangle$, $\langle -3, -6 \rangle$

Superset

In exercises 17–22, two vectors are given. Determine whether they are parallel, perpendicular, or neither.

17. $\langle 2, -4 \rangle$, $\langle 2, 1 \rangle$
18. $\langle 1, -3 \rangle$, $\langle 3, -1 \rangle$
19. $\langle 1, 9 \rangle$, $\langle \frac{3}{2}, \frac{1}{6} \rangle$
20. $\langle 8, 6 \rangle$, $\langle \frac{1}{2}, -\frac{2}{3} \rangle$
21. $\langle -9, 6 \rangle$, $\langle 2, -\frac{4}{3} \rangle$
22. $\langle -\frac{1}{3}, \frac{2}{3} \rangle$, $\langle 4, -8 \rangle$

In physics, **work** is said to be done when a force applied to an object causes the object to move. In particular, if a constant force **F** is applied in the direction of motion, then the work W done by force **F** in moving the object a distance d is the scalar quantity $W = \|\mathbf{F}\| \cdot d$. However, if a constant force **F** is applied at an angle θ to the direction of the motion, then the work done by **F** in moving the object a distance d is

$$W = \begin{pmatrix} \text{Component of force in} \\ \text{direction of motion} \end{pmatrix} \cdot (\text{Distance})$$
$$= (\|\mathbf{F}\| \cos \theta) \cdot d$$

In exercises 23–26, solve each work problem.

23. A wagon loaded with 180 lb of patio bricks is pulled 100 yd over level ground by a handle which makes an angle of 43° with the horizontal. Find the work done if a force of 22 lb is exerted in pulling the wagon.

24. A large crate is pushed across a level floor. Find the work done in moving the crate 20 ft if a force of 35 lb is applied at an angle of 18° with the horizontal.

25. A wagon is used to haul three small children a distance of a half mile over level ground. The handle used to pull the wagon makes an angle of 35° with the horizontal. Find the work done if a force of 25 lb is exerted on the handle.

26. A box is pulled by exerting a force of 16 lb on a rope attached to the box that makes an angle of 40° with the horizontal. Determine the work done in pulling the box 46 ft.

10.6 Simple Harmonic Motion

Goal A *to solve problems involving simple harmonic motion*

We now use our knowledge of trigonometric functions to describe motion that repeats itself periodically, such as the up-and-down bobbing motion of a buoy in the ocean, or the back-and-forth motion of a simple pendulum.

As a model for this type of motion, we consider a mass m attached to a spring suspended from the ceiling (Figure 10.24). Initially the mass is at rest (Figure 10.24(a)). At that time, the mass is said to be in the **equilibrium position.** Next, the mass is pulled downward to a position A units below equilibrium, and then released (Figure 10.24(b)). The spring then causes the mass to move upward, through the equilibrium position until it reaches a point A units above equilibrium (Figures 10.24(c) and (d)). Then, the mass begins to move back downward towards the point from which it was released.

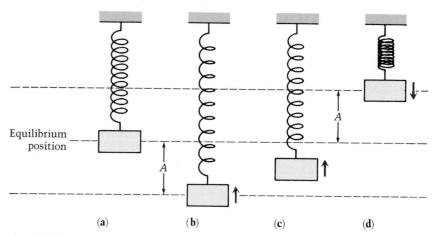

Figure 10.24

If the effects of friction in the spring are ignored, the up-and-down motion just described would continue forever. This "idealized" motion of the mass m is called *simple harmonic motion.* It is the basic model for a variety of physical phenomena, including the vibration of a guitar string, the oscillation of atoms in a molecule, the propagation in air of sound waves, as well as various types of electromagnetic waves, such as radio waves and television signals.

The mathematical model for simple harmonic motion describes an object's displacement x, measured from the equilibrium position, as a function of time t. (The equilibrium position corresponds to the point where $x = 0$.) The sine and cosine functions are the building blocks of this model.

Simple Harmonic Motion

If an object exhibits simple harmonic motion, its displacement x from the equilibrium position is a function of time t. This function can be described by an equation of the form

$$x = A \sin(\omega t - D),$$

or alternatively,

$$x = A \cos(\omega t - D).$$

Suppose that we are considering simple harmonic motion described by the equation $x = A \sin(\omega t - D)$ where ω and D are positive numbers. Note that the equation can be rewritten as

$$x = A \sin \omega \left(t - \frac{D}{\omega}\right).$$

Recall that this form was extremely useful when we graphed transformations of the trigonometric functions. In particular, it follows from our earlier definitions that

$$\text{the amplitude is } |A|, \text{ and the period is } \left|\frac{2\pi}{\omega}\right| = \frac{2\pi}{\omega}.$$

The **frequency** of the motion, denoted f, is the number of periods that are completed in one unit of time, and is described as the reciprocal of the period

$$f = \frac{1}{\text{period}} = \frac{1}{\frac{2\pi}{\omega}} = \frac{\omega}{2\pi}.$$

The graph of $x = A \sin(\omega t - D)$ is given below.

- Note that this graph is a translation $\frac{D}{\omega}$ units to the right of the graph of $x = A \sin \omega t$.

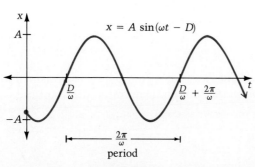

Figure 10.25

Example 1   An object suspended on a spring is pulled down 4 units from the equilibrium point. When released, the object is in simple harmonic motion, and makes one complete up-and-down oscillation every 2 sec.

(a) Determine the frequency of the motion;
(b) Write an equation for the object's displacement (x) as a function of time (t). Take displacements below the equilibrium point as positive, and those above as negative.

Solution   (a) $f = \dfrac{1}{\text{period}} = \dfrac{1}{2}$ ■ Frequency is the reciprocal of the period.

Note that this means the object makes $\frac{1}{2}$ of an oscillation each second.

(b) We need to determine values for A, ω, and D in the general formula

$$x = A \sin(\omega t - D).$$

$A = 4$ ■ Initial displacement is 4 units below equilibrium point, so A is (positive) 4. Remember that $|A|$ is the amplitude.

$f = \dfrac{\omega}{2\pi} = \dfrac{1}{2}$ ■ To find ω, we use the definition of f and the result from (a) that $f = \dfrac{1}{2}$.

$2\omega = 2\pi$

$\omega = \pi$

Now that we have found A and ω, we can determine D.

$x = 4 \sin(\pi t - D)$

$4 = 4 \sin(0 - D)$ ■ At release, $t = 0$ and $x = 4$. These values are substituted.

$1 = \sin(-D)$

$1 = -\sin D$ ■ Sine is odd: $\sin(-x) = -\sin x$.

$-1 = \sin D$

$D = \dfrac{3\pi}{2}$ ■ We have selected the smallest positive value of D for which $\sin D = -1$.

An equation for the object's displacement is $x = 4 \sin\left(\pi t - \dfrac{3\pi}{2}\right)$.

Example 2   Show that the equation $x = 3 \sin 2t - 4 \cos 2t$, where x is measured in inches and t in seconds, describes simple harmonic motion. Determine the amplitude, period, and frequency.

Solution We rewrite the equation in the form

$$x = A \sin(\omega t - D).$$

Recall: $\quad \sin(\omega t - D) = (\sin \omega t)(\cos D) - (\cos \omega t)(\sin D).$

Thus, $\quad A \sin(\omega t - D) = A[(\sin \omega t)(\cos D) - (\cos \omega t)(\sin D)]$
$$= (A \cos D)(\sin \omega t) - (A \sin D)(\cos \omega t)$$

Compare: $\quad x = (\quad 3 \quad)(\sin 2t) - (\quad 4 \quad)(\cos 2t)$

Therefore,

$$A \cos D = 3 \qquad A \sin D = 4 \qquad \text{and} \qquad \omega = 2$$
$$\cos D = \frac{3}{A} \qquad \sin D = \frac{4}{A}$$

Thus, since $\quad 1 = \cos^2 D + \sin^2 D = \dfrac{3^2}{A^2} + \dfrac{4^2}{A^2} = \dfrac{25}{A^2}$, we have $A = 5$.

Since $A = 5$, and $\omega = 2$, we have $x = 5 \sin(2t - D)$.

To determine D, notice that

$$\tan D = \frac{\sin D}{\cos D} = \frac{4A}{3A} = \frac{4}{3}, \text{ and so } D \approx 0.93 \text{ radians} \qquad \blacksquare \text{ Table 4}$$

Thus, $x = 5 \sin(2t - 0.93)$, which is an equation for simple harmonic motion. The amplitude is 5, the period $= \dfrac{2\pi}{2} = \pi$, and the frequency $= \dfrac{1}{\pi}$.

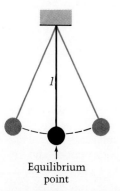

Figure 10.26

A pendulum in its simplest form consists of a point mass suspended by a "weightless" string of length l, such that the mass moves back and forth along a small arc. For small arcs the motion of the mass can be considered approximately straight-line motion. In this case the displacement x from the equilibrium point is given by the equation

$$x = A \cos \omega t,$$

where A is the initial displacement, and $\omega = \sqrt{\dfrac{g}{l}}$, with g the acceleration due to gravity: 32 ft/sec², or 9.8 m/sec².

Example 3 A pendulum in a grandfather clock is 2 ft long. It is released on an arc with initial displacement of 0.25 ft. (a) What is the period of the pendulum? (Assume that the acceleration due to gravity is 32 ft/sec².) (b) Write the equation for the motion of the pendulum and sketch the graph.

Solution (a) period $= \dfrac{2\pi}{\omega}$

period $= \dfrac{2\pi}{4} = \dfrac{\pi}{2}$ ▪ Since $\omega = \sqrt{\dfrac{g}{l}}$, we have $\omega = \sqrt{\dfrac{32}{2}} = \sqrt{16} = 4$.

This means that one back-and-forth oscillation takes approximately $\dfrac{\pi}{2} \approx 1.6$ sec.

(b) $x = A \cos \omega t$
$x = 0.25 \cos 4t$ ▪ 0.25 substituted for A; 4 for ω.

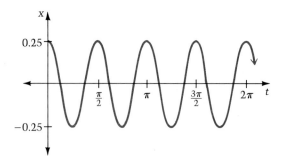

Exercise Set 10.6 = Calculator Exercises in the Appendix

Goal A

In exercises 1–6, a mass suspended from a spring is in simple harmonic motion, described by the given equation, where x is measured in inches and t in seconds. Use the graph of the motion to determine at each of the following times where the object is and whether it is moving up or down.
(a) $t = 0$, (b) $t = \tfrac{1}{2}$, (c) $t = 1$, (d) $t = 2$, (e) $t = 3$, (f) $t = 4$, (g) $t = 6$. (Note: take displacement below the equilibrium point as positive and above as negative.)

1. $x = \dfrac{1}{2} \sin\left(\dfrac{1}{2}\pi t\right)$

2. $x = \dfrac{3}{2} \cos\left(\dfrac{1}{2}\pi t\right)$

3. $x = 3 \cos 2\pi\left(t + \dfrac{1}{2}\right)$

4. $x = 3 \sin 2\pi\left(t - \dfrac{1}{2}\right)$

5. $x = 0.20 \sin 4\pi t$

6. $x = 0.33 \cos 6\pi t$

In exercises 7–10, each equation represents simple harmonic motion. Give the amplitude, period, and frequency. Sketch the graph.

7. $x = 6 \sin 2t$

8. $x = 4 \cos \dfrac{t}{2}$

9. $x = 3 \cos\left(\dfrac{t}{2} + 2\pi\right)$

10. $x = 5 \sin(2t - \pi)$

In exercises 11–16, show that each equation, where x is measured in centimeters and t in seconds, describes simple harmonic motion. Determine the amplitude, period, and frequency.

11. $x = \dfrac{\sqrt{3}}{2} \sin t + \dfrac{1}{2} \cos t$

12. $x = \sqrt{3} \sin t - \cos t$

13. $x = \sin t - \cos t$

14. $x = \sin t + \cos t$

15. $x = \dfrac{5}{2} \sin 2t - 6 \cos 2t$

16. $x = 4 \sin 2t + \dfrac{15}{2} \cos 2t$

In exercises 17–24, a situation describing simple harmonic motion is given. (a) Write an equation for the object's displacement (x) as a function of time (t). (b) Determine the frequency of the motion.

17. An object suspended on a spring is pulled down 6 in below the equilibrium point. When released, the object is in simple harmonic motion and makes one complete oscillation every 3 sec.

18. An object attached to a spring is released from a compressed position 4 cm above its position of equilibrium. When released, the object is in simple harmonic motion and makes one complete oscillation every second.

19. A weight of 50 g is attached to a spring. The weight is pulled down 6 cm below the equilibrium point and then released. The object is then in simple harmonic motion and has a period of 1.6 sec.

20. The weight in exercise 19 is pulled down 3 cm and released. It makes one complete oscillation every 0.8 sec.

21. A pendulum is 8 ft long and is released with an initial displacement of 0.60 ft.

22. A pendulum is 6 ft long and is released with an initial displacement of 0.62 ft.

23. A pendulum 1 m long is released with an initial displacement of 2 cm.

24. A pendulum 1.6 m long is released with an initial displacement of 6 cm.

Superset

In exercises 25–28, write an equation for simple harmonic motion, given the amplitude and period.

25. amplitude = 3, period = $\dfrac{4\pi}{3}$

26. amplitude = $\dfrac{3}{5}$, period = π

27. amplitude = $\dfrac{3}{2}$, period = 3π

28. amplitude = 4, period = $\dfrac{3\pi}{2}$

29. If the length of a pendulum is doubled, how does the period change?

30. How does the period of a pendulum compare with that of another pendulum which is one-fourth as long?

31. A guitar string is plucked so that a point on the string makes one complete oscillation every $\dfrac{1}{200}$ sec. If the string is plucked by lifting the point 0.01 cm and then releasing it, write an equation for the simple harmonic motion of the point.

32. Show that the motion of a particle which moves on a line according to the equation $x = 4 \sin 3t \cos 2t$ is the "sum" of two simple harmonic motions.

33. Is the motion represented by the equation $x = \sin^2 t$ simple harmonic motion?

34. A simple pendulum about 9.8 in long has a period of 1 sec at sea level. A pendulum 4 times as long has a period of 2 sec. One that is 9 times as long has a period of 3 sec, and so on. Verify each of these statements.

10.7 Polar Coordinates

Goal A *to plot points given their polar coordinates*

Up to this point, we have located points in a plane by specifying rectangular coordinates, such as $P(3, 4)$ or $Q(-1, 5)$. We now consider an alternate method for identifying points in a plane. The method depends on describing each point in terms of two numbers r and θ, known as polar coordinates. To simplify our work, we make the following definition:

Definition

A **θ-ray** is a ray which has its initial point at $(0, 0)$ and which makes an angle of θ with the positive x-axis. The ray in the direction opposite to that of a θ-ray is called the **opposite of the θ-ray**. Note that the opposite of the θ-ray is the $(\theta + \pi)$-ray.

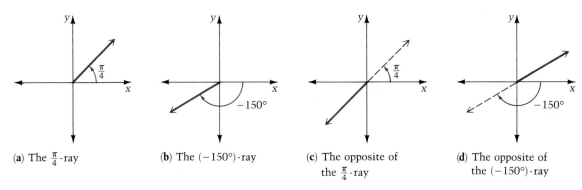

(a) The $\frac{\pi}{4}$-ray

(b) The $(-150°)$-ray

(c) The opposite of the $\frac{\pi}{4}$-ray

(d) The opposite of the $(-150°)$-ray

Figure 10.27

We construct what is called the *polar coordinate system* in the following way. We designate one point in the plane as the **origin** or **pole** (labeled O), and one ray emanating from O as the **polar axis**. Points in the plane can then be described in polar form as follows.

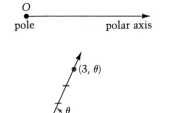

Definition

The **polar coordinates** (r, θ) specify a point that lies at a distance $|r|$ from O. If $r > 0$, the point lies on the θ-ray. If $r < 0$, the point lies on the opposite of the θ-ray. If $r = 0$, the point lies at O, regardless of the value of θ.

When plotting points given by polar coordinates, it is a common practice to use graph paper which displays concentric circles (centered at O) and rays emanating from O.

Figure 10.28

Example 1 Graph the following points in the polar plane.

(a) $A(3, 135°)$ (b) $B(-4, 60°)$ (c) $C(-3, 315°)$ (d) $D\left(2, -\frac{3\pi}{2}\right)$

(e) $E(-5, \pi)$ (f) $F\left(0, \frac{\pi}{12}\right)$

Solution

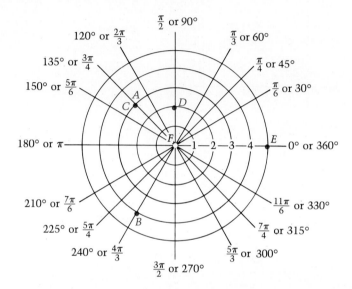

(a) Find the 135°-ray. Since 3 is positive, move 3 units away from the origin along that ray.
(b) Find the 60°-ray. Since -4 is negative, move 4 units away from the origin along the opposite of the 60°-ray.
(c) Find the 315°-ray. Since -3 is negative, move 3 units away from the origin along the opposite of the 315°-ray. Notice that this point coincides with $(3, 135°)$.
(d) Find the $\left(-\frac{3\pi}{2}\right)$-ray (it is the same as the $\frac{\pi}{2}$-ray). Since 2 is positive, move 2 units away from the origin on that ray.
(e) Find the π-ray. Since -5 is negative, move 5 units away from the origin on the opposite of the π-ray.
(f) Since $r = 0$, this point is at the origin, regardless of the value of θ.

The description of a point in polar coordinates is not unique. See, for example, points A and C in Example 1. The coordinates $(3, 135°)$ and $(-3, 315°)$ describe the same point. In addition, point $A(3, 135°)$ could alternately be described by adding any multiple of 360° to 135°, and keeping $r = 3$. In fact, for all integers n, any point (r, θ) can be described as

$$(r, \theta + n \cdot 360°), \quad \text{or} \quad (-r, (\theta + 180°) + n \cdot 360°).$$

Goal B *to translate ordered pairs and equations from rectangular form to polar form and vice versa*

When we superimpose the polar plane on the xy-plane, we discover some relationships that are useful in translating from one system to the other.

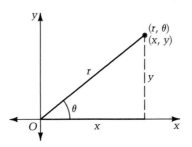

- Since $\sin\theta = \dfrac{y}{r}$, and $\cos\theta = \dfrac{x}{r}$, we have
$$y = r\sin\theta,$$
$$x = r\cos\theta.$$

Figure 10.28

The following four relationships are used to solve translation problems.

Rectangular ↔ Polar Relationships

If point P has rectangular coordinates (x, y) and polar coordinates (r, θ), then

$$x = r\cos\theta, \qquad \tan\theta = \frac{y}{x}, \quad \text{provided } x \neq 0,$$
$$y = r\sin\theta, \qquad x^2 + y^2 = r^2.$$

Example 2 Determine the xy-coordinates of the point having polar coordinates $\left(-1, \dfrac{\pi}{3}\right)$.

Solution $\quad x = r\cos\theta = (-1)\left(\cos\dfrac{\pi}{3}\right) = -\dfrac{1}{2} \qquad y = r\sin\theta = (-1)\left(\sin\dfrac{\pi}{3}\right) = -\dfrac{\sqrt{3}}{2}$

The rectangular coordinates of the polar point $\left(-1, \dfrac{\pi}{3}\right)$ are $\left(-\dfrac{1}{2}, -\dfrac{\sqrt{3}}{2}\right)$.

Example 3 Determine all polar coordinates of the point having rectangular coordinates $(-1, 2)$.

Solution $\quad r^2 = x^2 + y^2 = (-1)^2 + (2)^2 = 5 \qquad$ ■ First we find r.
$\quad\quad\quad\quad r = \pm\sqrt{5} \quad$ (two cases)

Case 1. Suppose $r = \sqrt{5}$: $\qquad x = r\cos\theta \qquad\qquad y = r\sin\theta$
$\quad\quad\quad\quad\quad\quad\quad\quad\quad\quad\quad\quad -1 = \sqrt{5}\cos\theta \qquad 2 = \sqrt{5}\sin\theta$

$$\cos\theta = -\frac{1}{\sqrt{5}} \approx -0.4472 \qquad \sin\theta = \frac{2}{\sqrt{5}} \approx 0.8944$$

Thus θ has 63°30′ as its reference angle (Table 4). Since $\cos\theta$ is negative and $\sin\theta$ is positive, θ is a second quadrant angle. Thus, $\theta = 180° - 63°30' = 116°30'$.

Case 2. If $r = -\sqrt{5}$, then $\theta = 116°30' + 180° = 296°30'$.

Thus, polar coordinates for the given point are of the form

$$(\sqrt{5}, 116°30' + n \cdot 360°) \quad \text{or} \quad (-\sqrt{5}, 296°30' + n \cdot 360), \quad \text{for all integers } n.$$

Certain graphs are represented by very simple equations in polar form. For example, the polar equation $r = c$ (where c is a constant) represents a circle of radius $|c|$, centered at the origin. Also, the polar equation $\theta = \alpha$ (where α is a constant) is the line formed by joining the α-ray and the opposite of the α-ray. The slope of such a line is $\tan\alpha$.

Example 4 Convert the following equations to rectangular form: (a) $r = 5$ (b) $r = 6 \sin\theta$

Solution (a)
$$r = 5$$
$$r^2 = 25 \qquad \blacksquare \text{ Both sides are squared.}$$
$$x^2 + y^2 = 25 \qquad \blacksquare \ r^2 \text{ is replaced by } x^2 + y^2.$$

Thus $r = 5$ is a circle with center at (0, 0) having radius 5.

(b)
$$r = 6\sin\theta$$
$$r^2 = 6(r\sin\theta) \qquad \blacksquare \text{ Both sides are multiplied by } r.$$
$$x^2 + y^2 = 6y \qquad \blacksquare \ r\sin\theta \text{ is replaced by } y; r^2 \text{ is replaced by } x^2 + y^2.$$
$$x^2 + (y^2 - 6y + \square) = \square \qquad \blacksquare \text{ We prepare to complete the square in } y.$$
$$x^2 + (y^2 - 6y + 9) = 9$$
$$x^2 + (y - 3)^2 = 9$$

Thus $r = 6\sin\theta$ is the polar equation of a circle of radius 3 centered at (0, 3).

Note that, in each problem in Example 4, we began by transforming the given equation so that one or more of the Rectangular ↔ Polar Relationships could be used. This is the key to handling such translation problems.

Example 5 Convert the following rectangular equations to polar form.

(a) $y = 5x$ (b) $x^2 - y^2 = 3$

Solution
(a) $y = 5x$
$\dfrac{y}{x} = 5$
$\tan \theta = 5$

(b) $x^2 - y^2 = 3$
$(r \cos \theta)^2 - (r \sin \theta)^2 = 3$
$r^2 \cos^2 \theta - r^2 \sin^2 \theta = 3$
$r^2 (\cos^2 \theta - \sin^2 \theta) = 3$
$r^2 \cos 2\theta = 3$ ■ $\cos^2 \theta - \sin^2 \theta = \cos 2\theta$.

Goal C *to graph a polar equation*

We end this section with the problem of graphing polar equations. In Example 6, we graph a polar equation by plotting a sufficient number of points to get a reasonable idea about the shape of the curve.

Example 6 Sketch the graph of the polar equation $r = 1 + 2 \cos \theta$.

Solution We will provide 4 charts showing values of r for selected values of θ, and will use these charts to sketch the graph in stages. Values of $\cos \theta$ and r are approximations.

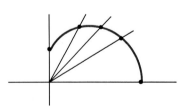

θ	0	$\dfrac{\pi}{6}$	$\dfrac{\pi}{4}$	$\dfrac{\pi}{3}$	$\dfrac{\pi}{2}$
r	3	2.7	2.4	2	1

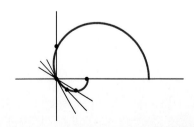

θ	$\dfrac{\pi}{2}$	$\dfrac{2\pi}{3}$	$\dfrac{3\pi}{4}$	$\dfrac{5\pi}{6}$	π
r	1	0	-0.4	-0.7	-1

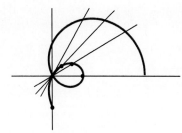

θ	π	$\dfrac{7\pi}{6}$	$\dfrac{5\pi}{4}$	$\dfrac{4\pi}{3}$	$\dfrac{3\pi}{2}$
r	-1	-0.7	-0.4	0	1

θ	$\dfrac{3\pi}{2}$	$\dfrac{5\pi}{3}$	$\dfrac{7\pi}{4}$	$\dfrac{11\pi}{6}$	2π
r	1	2	2.4	2.7	3

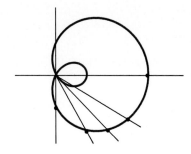

We now present a complete graph.

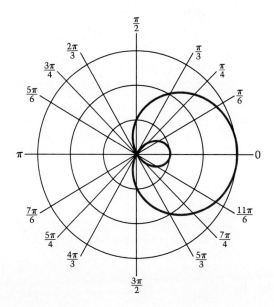

Exercise Set 10.7

Goal A

In exercises 1–16, graph the given point in the polar plane.

1. $(2, 120°)$
2. $(5, 30°)$
3. $(3, 135°)$
4. $(4, 210°)$
5. $(2, -120°)$
6. $(-3, 135°)$
7. $(-1, -270°)$
8. $(-5, -45°)$
9. $(-2, -420°)$
10. $(4, -570°)$
11. $(3, \pi)$
12. $\left(4, \dfrac{\pi}{4}\right)$
13. $\left(-2, \dfrac{5\pi}{6}\right)$
14. $\left(-5, -\dfrac{3\pi}{2}\right)$
15. $\left(-4, -\dfrac{\pi}{2}\right)$
16. $\left(-2, \dfrac{3\pi}{4}\right)$

Goal B

For exercises 17–32, determine the x- and y-coordinates of the polar points in exercises 1–16.

In exercises 33–44, the rectangular coordinates of the following points are given. Determine two polar representations (r, θ), with $0 \le \theta < 2\pi$, of each point.

33. $(0, 4)$
34. $(-2, 0)$
35. $(-1, -1)$
36. $(1, -1)$
37. $(\sqrt{3}, -1)$
38. $(-1, -\sqrt{3})$
39. $(-\sqrt{2}, \sqrt{2})$
40. $(\sqrt{2}, -\sqrt{2})$
41. $(-3, 4)$
42. $(-5, -12)$
43. $(4, 6)$
44. $(5, -10)$

In exercises 45–54, convert the given polar equation to rectangular form.

45. $r = 3$
46. $2r = 5$
47. $\theta = \dfrac{3\pi}{4}$
48. $\theta = \dfrac{\pi}{3}$
49. $r(\cos \theta - 2 \sin \theta) = 5$
50. $r(3 \sin \theta - 4 \cos \theta) = 5$
51. $r(1 + \cos \theta) = 1$
52. $r(1 + \sin \theta) = 2$
53. $r(1 - 2 \sin \theta) = 2$
54. $r(2 - \cos \theta) = 2$

In exercises 55–66, convert the given rectangular equation to polar form.

55. $x = 0$
56. $y = -5$
57. $x + y = 1$
58. $x + 4 = 4y$
59. $x^2 - 2x + y^2 = 0$
60. $x^2 - 6y - y^2 = 0$
61. $x^2 = 6y + 9$
62. $y^2 = 8x + 16$
63. $2xy = 1$
64. $xy = 1$
65. $x^2 + y^2 = 16y$
66. $x^2 + y^2 = 4x$

Goal C

In exercises 67–74, sketch the graph of the polar equation.

67. $r = -3$
68. $r^2 = 4$
69. $\theta^2 = \dfrac{\pi^2}{16}$
70. $\theta = \dfrac{5\pi}{4}$
71. $r = 3 \cos \theta$
72. $r = 4 \sin \theta$
73. $r = 2 + \sin \theta$
74. $r = 1 - \cos \theta$

Superset

75. Determine the polar equation of the line through the points $(-2, 0)$ and $(1, 2)$.
76. Determine the polar equation of the line through the points that have polar coordinates $(1, \dfrac{\pi}{2})$ and $(-2, \pi)$.
77. (a) Find the distance between the points $(-2, \dfrac{\pi}{3})$ and $(4, \dfrac{3\pi}{4})$. (Hint: first convert the points to rectangular coordinates.)
    (b) Sketch the points on the polar plane. Use the Law of Cosines to determine the distance between the two points.

In exercises 78–82, sketch the graph of each of the following polar equations for $0 \le \theta < \pi$.

78. $r = \theta$
79. $r = \dfrac{\theta}{2}$
80. $r(1 - \cos \theta) = 2$
81. $r(1 + \sin \theta) = 2$
82. $r = 4 \sin 2\theta$

10 / Applications in Trigonometry

Chapter Review & Test

Chapter Review

10.1 Applications Involving Right Triangles (pp. 404–409)

The process of determining the measures of all the sides and all the angles in a triangle is referred to as *solving the triangle*. A right triangle can be solved whenever one side and one of the acute angles are known, or whenever two sides are known. (p. 404)

10.2 Law of Sines (pp. 410–415)

Law of Sines: In any triangle with angles A, B, C, and opposite sides a, b, and c respectively,

$$\frac{\sin A}{a} = \frac{\sin B}{b} = \frac{\sin C}{c}.$$

The Law of Sines is used to solve oblique triangles when given (1) one side and two angles, or (2) two sides and the angle opposite one of them.

10.3 Law of Cosines (pp. 416–420)

Law of Cosines: In any triangle with angles A, B, and C, and opposite sides a, b, and c respectively,

$$a^2 = b^2 + c^2 - 2bc \cos A,$$
$$b^2 = a^2 + c^2 - 2ac \cos B,$$
$$c^2 = a^2 + b^2 - 2ab \cos C.$$

10.4 Vectors in the Plane (pp. 421–427)

A vector with initial point $P_1(x_1, y_1)$ and terminal point $P_2(x_2, y_2)$ is given by the ordered pair $\langle x_2 - x_1, y_2 - y_1 \rangle$. The numbers $x_2 - x_1$ and $y_2 - y_1$ are called the *scalar components* of the vector. (p. 423)

The length of a vector **v** is called the *norm* of **v** and is denoted $\|\mathbf{v}\|$. If $\mathbf{a} = \langle a_1, a_2 \rangle$, then $\|\mathbf{a}\| = \sqrt{a_1^2 + a_2^2}$. (p. 425)

10.5 Vector Applications and the Dot Product (pp. 428–433)

For a given vector **a**, the process of determining two component vectors **b** and **c** such that **b** is perpendicular to **c** and $\mathbf{b} + \mathbf{c} = \mathbf{a}$ is referred to as *resolving* **a** into perpendicular components. For a given vector **v**, the horizontal component vector is denoted $\mathbf{v}_x$ and the vertical component vector is denoted $\mathbf{v}_y$. (p. 429)

The *dot product* of two nonzero vectors $\mathbf{a} = \langle a_1, a_2 \rangle$ and $\mathbf{b} = \langle b_1, b_2 \rangle$ is denoted $\mathbf{a} \cdot \mathbf{b}$, and it can be defined in two equivalent ways: (1) $\mathbf{a} \cdot \mathbf{b} = \|\mathbf{a}\| \|\mathbf{b}\| \cos \theta$, where θ is the angle between $\mathbf{a}$ and $\mathbf{b}$ such that $0° \leq \theta \leq 180°$; (2) $\mathbf{a} \cdot \mathbf{b} = a_1 b_1 + a_2 b_2$. (p. 432)

10.6 Simple Harmonic Motion (pp. 434–439)

If an object exhibits *simple harmonic motion*, its displacement x from the equilibrium position is a function of t. This function can be described by either of the equations

$$x = A \sin(\omega t - D), \quad \text{or} \quad x = A \cos(\omega t - D).$$

10.7 Polar Coordinates (pp. 440–446)

The *polar coordinates* (r, θ) specify a point that lies at a distance $|r|$ from the origin. If $r > 0$, the point lies on the θ-ray. If $r < 0$, the point lies on the opposite of the θ-ray. If $r = 0$, the point lies at the origin, regardless of the value of θ. (p. 440) If point P has rectangular coordinates (x, y) and polar coordinates (r, θ), then $x = r \cos \theta$, $y = r \sin \theta$, $\tan \theta = \frac{y}{x}$ (provided $x \neq 0$) and $x^2 + y^2 = r^2$. Using these facts, equations in rectangular form may be converted to polar form, and vice versa. (p. 442)

Chapter Test

10.1A Solve the right triangle ABC using the given information about the triangle.

1. $B = 60°$, $b = 12$
2. $A = 45°$, $c = 18$
3. $A = 33°42'$, $b = 96.8$
4. $B = 55°12'$, $a = 78.4$
5. $a = 36.3$, $b = 46.9$
6. $a = 52.6$, $c = 134.1$

10.1B Solve the following applied problems.

7. What is the angle of elevation of the sun at the time that a 68 ft tall flagpole casts a shadow 25 ft long?

8. A 25 ft long ladder is leaning against a building. If the foot of the ladder is 6 ft from the base of the building (on level ground), what acute angle does the ladder make with the ground?

10.2A Solve $\triangle ABC$ using the given information about the triangle.

9. $A = 41°$, $B = 83°$, $c = 44.6$ 10. $A = 76°$, $C = 34°$, $b = 67.8$

10.2B Given the measure of one angle and two sides, state whether it is possible to have $\triangle ABC$. If it is, solve the triangle. (Determine both solutions if there is more than one.)

11. $A = 24°18'$, $c = 9.4$, $a = 6.2$
12. $B = 46°54'$, $a = 12.4$, $b = 10.6$

10.2C Solve the following applied problems.

13. A boat B is observed simultaneously by two Coast Guard stations S and T which are 550 yd apart. Angles STB and TSB are observed to be $132°48'$ and $21°42'$, respectively. Find the distance from the boat to station T.

14. Points A and B are 126 ft apart and on the same side of a river. A tree at point T, on the other side of the river, is observed, with angles ABT and BAT equal to $74°12'$ and $42°36'$, respectively. Find the distance from point T to point B.

10.3A Solve $\triangle ABC$, given the following.

15. $a = 1.52$, $b = 2.31$, $C = 119°$ 16. $b = 67.2$, $c = 34.9$, $A = 41°$
17. $a = 26.7$, $b = 34.2$, $c = 41.8$ 18. $a = 1.82$, $b = 2.63$, $c = 31.4$

10.3B Determine the area of $\triangle ABC$ using the given information.

19. $A = 30°$, $b = 18.2$, $c = 10.8$ 20. $a = 18$, $b = 24$, $c = 36$

10.4A The vector **a** has initial point $(0, 0)$ and terminal point $(-3, 3)$. The vector **b** has initial point $(0, 0)$ and terminal point $(1, 1)$.

21. (a) Determine the magnitude and direction of vector **a**.
(b) Draw $\mathbf{a} - 2\mathbf{b}$.

22. (a) Determine the magnitude and direction of vector **b**.
(b) Draw $-\mathbf{a} + 2\mathbf{b}$.

10.4B Points A and B are given. Describe the vector $\overrightarrow{AB}$ as an ordered pair.

23. $A(-3, -2)$, $B(2, 7)$ 24. $A(6, -3)$, $B(-1, 3)$

10.4C Use the vectors $\mathbf{a} = \left\langle -\frac{1}{2}, \frac{7}{2} \right\rangle$, $\mathbf{b} = \langle 6, -1 \rangle$ to determine the following.

25. $\|\mathbf{a}\|$ and $2\mathbf{a} - \frac{1}{2}\mathbf{b}$ 26. $\|\mathbf{b}\|$ and $\mathbf{b} - 4\mathbf{a}$

10.5A The magnitude $\|\mathbf{v}\|$ and the angle θ between the vector $\mathbf{v}$ and the positive x-axis are given. Find the horizontal and vertical components of $\mathbf{v}$.

27. $\|\mathbf{v}\| = 8.6,\ \theta = 148°$ 28. $\|\mathbf{v}\| = 10.8,\ \theta = 206°$

29. A plane's air speed is 500 mph and its bearing is N 75° E. An 8 mph wind is blowing in the direction N 70° W. Determine the ground speed and true bearing of the plane.

30. Find the force required to hold a 120 lb crate stationary on a ramp inclined at 25°.

10.5B Determine the dot product of the given vectors. Find the angle between the vectors to the nearest degree.

31. $\langle 3, -1 \rangle, \langle -2, -2 \rangle$ 32. $\langle 4, 6 \rangle, \langle -3, 2 \rangle$

10.6A The equation represents simple harmonic motion. Determine the amplitude, period, and frequency. Sketch the graph.

33. $x = 4 \sin\left(\dfrac{t}{2} - 2\pi\right)$ 34. $x = 6 \cos(2t + \pi)$

10.7A Graph the polar point and determine the x- and y-coordinates.

35. $(4, -135°)$ 36. $(-3, 210°)$

10.7B Convert the polar equation to rectangular form.

37. $r = 3 \sin \theta$ 38. $r = -4 \cos \theta$

10.7C Sketch the graph of the polar equation.

39. $r = 3 - 2 \cos \theta$ 40. $r = 5 - 4 \sin \theta$

Superset

Find the distance between the two polar points.

41. $(2, -120°), (4, -570°)$ 42. $\left(6, \dfrac{\pi}{3}\right), \left(-2, \dfrac{3\pi}{2}\right)$

43. Let $\mathbf{a} = \langle -3, 1 \rangle$, $\mathbf{b} = \langle 4, -3 \rangle$ and $\mathbf{c} = \langle -6, 7 \rangle$. Express $\mathbf{c}$ as a linear combination of $\mathbf{a}$ and $\mathbf{b}$.

44. Write the polar coordinates for the terminal point of the position vector $\mathbf{a}$ if $\mathbf{a} = -3\mathbf{i} + \sqrt{3}\mathbf{j}$.

45. Find the angle formed by an internal diagonal of a cube and one of its edges.

Cumulative Test: *Chapters 9 and 10*

In exercises 1–4, find the following trigonometric values without using Table 4 or a calculator.

1. sin 165° **2.** tan 285° **3.** $\tan\left(-\dfrac{\pi}{12}\right)$ **4.** $\cos\left(-\dfrac{5\pi}{12}\right)$

In exercises 5–8, given α, find sin 2α, cos 2α and tan 2α using the double-angle identities.

5. −315° **6.** 405° **7.** $\dfrac{5\pi}{4}$ **8.** $-\dfrac{11\pi}{6}$

In exercises 9–10, from the given information determine $\sin\dfrac{\alpha}{2}$ and $\cos\dfrac{\alpha}{2}$.

9. $\sin\alpha = \dfrac{8}{15}$, α in Quadrant II **10.** $\cos\alpha = -\dfrac{7}{24}$, α in Quadrant III.

In exercises 11–12, evaluate by using the conversion identities.

11. cos 52° + cos 16° **12.** sin 39° + sin 11°

In exercises 13–16, verify the identities.

13. $\dfrac{\cos 2x + \cos x}{\cos x + 1} = 2\cos x - 1$ **14.** $\dfrac{1 - \cos 2x}{2\cos x \sin x} = \tan x$

15. $\dfrac{(1 + \tan x)^2}{1 + \sin 2x} = 1 + \tan^2 x$ **16.** $\cot x + \tan x = \dfrac{2}{\sin 2x}$

In exercises 17–20, determine all the solutions of the equation in the interval $[0, \pi]$.

17. $\sin^2 x + \sin x = 2$ **18.** $2\cos^2 x - 3\cos x = 2$

19. $\sin 2x \cos x + \cos 2x \sin x = \dfrac{1}{2}$

20. $\cos 2x \cos x - \sin 2x \sin x = 0$.

In exercises 21–24, solve for x.

21. $\cos\left(\tan^{-1}\dfrac{5}{12}\right) = x$ **22.** $\tan\left(\arcsin\dfrac{8}{17}\right) = x$

23. $\cos^{-1}\dfrac{1}{2} + 2\sin^{-1}x = \pi$ **24.** $\tan^{-1}x + 2\tan^{-1}1 = \dfrac{3\pi}{4}$

In exercises 25–28, solve the triangle for the part indicated.

25. $B = 45°$, $C = 90°$, $b = 24.0$, find a.

26. $A = 60°$, $C = 90°$, $c = 18.0$, find b.

27. $A = 48°12'$, $B = 67°42'$, $c = 32.4$, find a.

28. $A = 56°48'$, $C = 25°18'$, $a = 76.8$, find b.

In exercises 29–30, solve each applied problem.

29. A surveyor measures a triangular lot. The three sides have lengths of 73 yd, 106 yd, and 145 yd. Find the area of the lot.

30. A painter is painting a large triangular sign with area of 18 ft². If one side is 6 ft long and another side 8 ft long, what is the length of the third side?

In exercises 31–32, $\mathbf{a} = \langle -1, 2 \rangle$, $\mathbf{b} = \langle 3, -1 \rangle$. Determine $\|\mathbf{c}\|$, $\mathbf{u_c}$, and $-\frac{1}{2}\mathbf{c}$.

31. $\mathbf{c} = 2\mathbf{a} + 3\mathbf{b}$ **32.** $\mathbf{c} = -\mathbf{a} + 4\mathbf{b}$

In exercises 33–34, solve each applied problem.

33. A ball is thrown due east at 15 yd/s from a train that is travelling due north at the rate of 20 yd/s. Find the speed of the ball and the direction of its path.

34. An airplane is headed due north at a speed of 420 mph. If the wind is blowing due east with a velocity of 32.0 mph, find the true bearing and ground speed of the plane.

In exercises 35–36, determine the dot product for the given vectors. Find the angle between the vectors to the nearest degree.

35. $\langle 2, -3 \rangle$, $\langle -9, -6 \rangle$ **36.** $\langle 1, -5 \rangle$, $\langle 4, -2 \rangle$

In exercises 37–38, the equation represents simple harmonic motion. Give the amplitude, period, and frequency, and state the position of the point at $t = \frac{\pi}{3}$.

37. $x = 4 \cos\left(\dfrac{t}{2} + 2\pi\right)$ **38.** $x = 6 \sin(2t - \pi)$

In exercises 39–40, the rectangular coordinates of a point are given. Determine a polar representation (r, θ), with $0 < \theta < \pi$.

39. $(-\sqrt{3}, -1)$ **40.** $(3\sqrt{3}, -3)$

Superset

41. If $\tan 2\theta = -\frac{24}{7}$, find $\sin \theta$ and $\cos \theta$, if θ is an acute angle.

42. Express $\sin x + \sin 3x + \sin 5x + \sin 7x$ as a product.

43. The diagonals of a parallelogram intersect at an angle of 40°24′. If the diagonals are 16.8 in and 22.4 in long, find the perimeter of the parallelogram.

44. A force of 6 N is applied in dragging an object. What is the work done in moving the object 10 m if the force is applied at an angle of 60° with the ground.

11

Complex Numbers

- Complex Numbers
  - Real Numbers
    - Rational Numbers
      - Integers
        - Negative Integers
        - 0
        - Positive Integers
      - Noninteger Fractions
    - Irrational Numbers
  - Imaginary Numbers
    - Pure Imaginary Numbers

11.1 Introduction to Complex Numbers

Goal A *to identify and simplify complex numbers*

When discussing sets of numbers in Chapter 1, we started with the set of positive integers and quickly enlarged that set to the set of all integers (positive, negative, and zero). The set of integers was then used to define the set of rational numbers, and the set of rational numbers was in turn joined to the set of irrational numbers to produce the set of all real numbers. At each step along the way, a certain set was enlarged to include more numbers. This process is called **extending the set.**

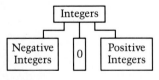

Figure 11.1

The need for extending sets becomes clear as we search for solutions to equations. For example, the equation

$$x + 5 = 1$$

cannot be solved over the set of positive integers. If the replacement set for x is extended to the set of all integers (thereby including zero and the negative integers, as shown in Figure 11.1), the equation then has a solution, -4.

Similarly, the equation $5x = 1$ cannot be solved over the set of integers. If the replacement set for x is extended to the set of rational numbers (thereby including noninteger fractions, as shown in Figure 11.2), the equation then has a solution, $\frac{1}{5}$.

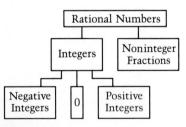

Figure 11.2

11 / Complex Numbers

Finally, the equation $x^2 = 2$ cannot be solved over the set of rational numbers. If the replacement set for x is extended to the set of real numbers (thereby including irrational numbers, as shown in Figure 11.3), the equation then has solutions, $\sqrt{2}$ and $-\sqrt{2}$.

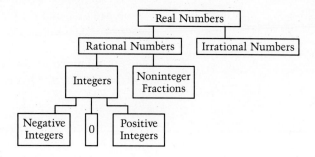

Figure 11.3

A quadratic equation such as

$$x^2 = -1,$$

however, has no real number solutions, for there is no real number whose square is -1. In order to solve such an equation, it will be necessary, once again, to extend the replacement set for x. To do this, we begin by defining a number that is a solution of the above equation.

Definition
The number i has the property that

$$i^2 = -1.$$

For this reason, i can be considered a square root of -1. Thus, we write

$$i = \sqrt{-1}.$$

Since $i^2 = -1$, we have invented a number i that is a solution of $x^2 = -1$. With a definition for $\sqrt{-1}$, we can define square roots of other negative numbers. Observe that

$$-2 = (-1)(2) = i^2(\sqrt{2})^2 = (i \cdot \sqrt{2})^2.$$

Thus, $i\sqrt{2}$ is a square root of -2. (Note that when i is multiplied by a radical, i is generally written first to avoid confusing a number like $\sqrt{2}i$ with $\sqrt{2i}$). In fact, we can use the number i to form square roots of any negative number.

Definition

If p is a positive real number, then

$$\sqrt{-p} = i\sqrt{p}.$$

As for any positive real number, $-p$ has two square roots. These roots are $i\sqrt{p}$ and $-i\sqrt{p}$. We refer to $i\sqrt{p}$ as the **principal square root of $-p$.**

Using this definition, we can rewrite expressions containing square roots of negative numbers. This is illustrated in the next example.

Example 1 Use the number i to rewrite each of the following.

(a) $\sqrt{-5}$ (b) $\sqrt{-16}$ (c) $\sqrt{-8}$

Solution (a) $\sqrt{-5} = i\sqrt{5}$ (b) $\sqrt{-16} = i\sqrt{16} = 4i$ (c) $\sqrt{-8} = i\sqrt{8} = 2i\sqrt{2}$

The equation $x^2 + 2x + 3 = 0$ has no real number solutions. Using the quadratic formula together with the definition of i, however, we can determine the solutions. Substituting $a = 1$, $b = 2$, and $c = 3$, we obtain the following (here we assume that the operations on real numbers also apply to complex numbers):

$$x = \frac{-b \pm \sqrt{b^2 - 4ac}}{2a}$$

$$= \frac{-2 \pm \sqrt{-8}}{2}$$

$$= \frac{-2 \pm i\sqrt{8}}{2}$$

$$= \frac{-2 \pm 2i\sqrt{2}}{2}$$

$$= -1 \pm i\sqrt{2}$$

Thus, the solutions are $-1 + i\sqrt{2}$ and $-1 - i\sqrt{2}$. The form of these two numbers motivates us to extend the set of real numbers to a larger set, called the *complex numbers*.

Definition

If a and b are real numbers, then any number of the form

$$a + bi$$

is called a **complex number**. The number a is called the **real part** of the complex number, and b is called the **imaginary part**.

We now consider three examples of complex numbers. The first is the real number 5. Since 5 can be written in the form

$$5 + 0i,$$

it is a complex number. Similarly, every real number a can be written in the form $a + 0i$. Thus, the set of real numbers is contained in the set of complex numbers.

The number $2 - 4i$ is also a complex number since it can be written as $2 + (-4)i$. Any complex number of the form $a + bi$, where the imaginary part b is not zero, is called an **imaginary number**. (Note that a complex number whose imaginary part is negative is usually written in the form $a - bi$.)

Finally, since $7i$ can be written

$$0 + 7i,$$

it is also a complex number. Since its imaginary part is not zero, it is also an imaginary number. Any number of the form bi, where $b \neq 0$, is called a **pure imaginary number**.

Example 2 Determine the real part and the imaginary part of each complex number.

(a) $4 + 7i$ (b) $\sqrt{3}$ (c) $-2 + i$ (d) $8 - 6i$ (e) $5i$ (f) 0

Solution
(a) The real part is 4; the imaginary part is 7.
(b) Since $\sqrt{3}$ can be written $\sqrt{3} + 0i$, the real part is $\sqrt{3}$ and the imaginary part is 0.
(c) Since $-2 + i$ can be written $-2 + 1i$, the real part is -2 and the imaginary part is 1.
(d) Since $8 - 6i$ can be written $8 + (-6)i$, the real part is 8 and the imaginary part is -6.
(e) Since $5i$ can be written $0 + 5i$, the real part is 0 and the imaginary part is 5.
(f) Since 0 can be written $0 + 0i$, the real part is 0 and the imaginary part is 0.

Thus, the set of complex numbers is formed by joining the set of real numbers and the set of imaginary numbers as shown in Figure 11.4.

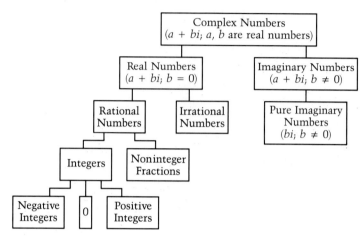

Figure 11.4

Goal B *to add, subtract, and multiply complex numbers*

The arithmetic of complex numbers follows patterns that we have already observed when working with binomials. In the case of complex numbers, however, we simplify results by using the fact that $i^2 = -1$.

Notice the following pattern:

$$i^1 = i \qquad\qquad i^5 = i^4(i) = (1)(i) = i$$
$$i^2 = -1 \qquad\qquad i^6 = i^4(i^2) = (1)(-1) = -1$$
$$i^3 = i^2(i) = -i \qquad\qquad i^7 = i^4(i^3) = (1)(-i) = -i$$
$$i^4 = (i^2)(i^2) = 1 \qquad\qquad i^8 = i^4(i^4) = (1)(1) = 1$$

It follows that

$$i^4, i^8, i^{12}, i^{16}, \ldots$$

are all equal to 1. In general, for any positive integer n, $i^{4n} = 1$. With this information, we can simplify a number like i^{31} very easily. Since 28 is the highest multiple of 4 less than 31, we can write the following:

$$i^{31} = i^{28}i^3 = (1)(-i) = -i.$$

Careful! When simplifying an expression containing the square root of a negative number, you should always begin by rewriting the expression in terms of i. For example,

$$\sqrt{-4} \cdot \sqrt{-9} = (2i)(3i) = i^2 \cdot 6 = -6.$$

Failure to do this may lead to serious errors. Note that the radical expression $\sqrt{-4} \cdot \sqrt{-9} \neq \sqrt{(-4)(-9)}$. If we assumed that this statement were true, then we would obtain the following incorrect statement:

$$\sqrt{-4} \cdot \sqrt{-9} = \sqrt{(-4)(-9)} = \sqrt{36} = 6.$$

Thus, $\sqrt{a} \cdot \sqrt{b} = \sqrt{ab}$ only if a and b are nonnegative.

Example 3   Simplify each expression.

(a) i^{64} (b) i^{10} (c) $\sqrt{-4} \cdot \sqrt{-8}$ (d) i^{37} (e) i^{127}

(f) $\sqrt{-9} \cdot \sqrt{-18} \cdot \sqrt{-2}$

Solution   (a) $i^{64} = 1$ ■ Since 64 is a multiple of 4, $i^{64} = 1$.

(b) $i^{10} = i^8 i^2 = (1)(-1) = -1$

(c) $\sqrt{-4} \cdot \sqrt{-8} = i\sqrt{4} \cdot i\sqrt{8}$
$= 2i \cdot 2i\sqrt{2}$
$= 4i^2\sqrt{2}$
$= -4\sqrt{2}$

(d) $i^{37} = i^{36}i = (1)(i) = i$

(e) $i^{127} = i^{124}i^3 = (1)(-i) = -i$

(f) $\sqrt{-9} \cdot \sqrt{-18} \cdot \sqrt{-2} = i\sqrt{9} \cdot i\sqrt{18} \cdot i\sqrt{2}$
$= 3i \cdot 3i\sqrt{2} \cdot i\sqrt{2}$
$= 9i^3(\sqrt{2})^2$
$= 18i^3$
$= -18i$

Definition

Two complex numbers $a + bi$ and $c + di$ are equal if and only if $a = c$ and $b = d$.

The previous definition states that two complex numbers are equal if and only if their real parts are equal and their imaginary parts are equal. Thus, if we are given that

$$a + 3i = -2 + (x - 5)i,$$

then equating the real parts, we have

$$a = -2,$$

and equating the imaginary parts, we have

$$3 = x - 5$$
$$x = 8.$$

Hence, the complex number represented by the expressions $a + 3i$ and $-2 + (x - 5)i$ is

$$-2 + 3i.$$

The sum of two complex numbers is a complex number whose real part is the sum of the real parts of the two numbers, and whose imaginary part is the sum of their imaginary parts. Their difference is found in a similar way.

Definition
If $a + bi$ and $c + di$ are complex numbers, then

$$(a + bi) + (c + di) = (a + c) + (b + d)i,$$

and

$$(a + bi) - (c + di) = (a - c) + (b - d)i.$$

We find the product of two complex numbers the same way we find the product of two binomials; we use the FOIL method.

$$(a + bi)(c + di) = ac + adi + bci + bdi^2$$
$$= ac + bdi^2 + (ad + bc)i$$
$$= ac - bd + (ad + bc)i \qquad \blacksquare \text{ Since } i^2 = -1.$$

Definition
If $a + bi$ and $c + di$ are complex numbers, then

$$(a + bi)(c + di) = (ac - bd) + (ad + bc)i$$

In the following examples, we shall practice the arithmetic of complex numbers.

Example 4 Simplify each expression.

(a) $(3 + 3i) + (7 - 5i)$ (b) $(6 - 2i) + 2i$ (c) $(10 + 7i) - (10 - 4i)$

(d) $(8 + i) - 17i$

Solution (a) $(3 + 3i) + (7 - 5i) = (3 + 7) + (3i - 5i)$ ■ Group like terms and combine.
$$= 10 + (-2i)$$
$$= 10 - 2i$$

(b) $(6 - 2i) + 2i = 6 + (-2i + 2i) = 6$

(c) $(10 + 7i) - (10 - 4i) = (10 - 10) + (7i + 4i)$
$$= 11i$$

(d) $(8 + i) - 17i = 8 + (i - 17i) = 8 - 16i$

Example 5 Simplify each expression.

(a) $5(2 - 7i)$ (b) $-6i(4 + 9i)$ (c) $(2 + i)(3 - 4i)$ (d) $(3 - 5i)^2$

Solution (a) $5(2 - 7i) = 5 \cdot 2 - 5 \cdot 7i$
$$= 10 - 35i$$

(b) $-6i(4 + 9i) = -24i - 54i^2$ ■ Replace i^2 with -1.
$$= -24i - 54(-1)$$
$$= 54 - 24i$$

(c) $(2 + i)(3 - 4i) = 2 \cdot 3 + 2(-4i) + 3i + i(-4i)$
$$= 6 - 8i + 3i - 4i^2$$ ■ Replace i^2 with -1.
$$= 6 + 4 + [(-8i) + 3i]$$
$$= 10 - 5i$$

(d) $(3 - 5i)^2 = (3 - 5i)(3 - 5i)$
$$= 9 - 15i - 15i + 25i^2$$
$$= 9 - 25 - (15i + 15i)$$
$$= -16 - 30i$$

Exercise Set 11.1

Goal A

In exercises 1–18, use the number i to rewrite each expression.

1. $\sqrt{-13}$
2. $\sqrt{-6}$
3. $\sqrt{-15}$
4. $\sqrt{-21}$
5. $\sqrt{-4}$
6. $\sqrt{-25}$
7. $\sqrt{-100}$
8. $\sqrt{-81}$
9. $\sqrt{-24}$
10. $\sqrt{-27}$
11. $\sqrt{-98}$
12. $\sqrt{-75}$
13. $-\sqrt{-9}$
14. $-\sqrt{-36}$
15. $-\sqrt{-20}$
16. $-\sqrt{-18}$
17. $-\sqrt{-63}$
18. $-\sqrt{-32}$

In exercises 19–32, determine the real part and the imaginary part of each complex number.

19. $2 + 5i$
20. $7 + i$
21. $-3 + 4i$
22. $5 - 8i$
23. $-1 - 2i$
24. $-5 - 10i$
25. $\frac{1}{2} + 6i$
26. $-7 - \frac{3}{5}i$
27. $6i$
28. 14
29. $-\sqrt{3}$
30. i
31. $\frac{1}{4} - i\sqrt{10}$
32. $-\sqrt{6} + \frac{2}{7}i$

Goal B

In exercises 33–94, simplify each expression.

33. i^{32}
34. i^{18}
35. i^{51}
36. i^{65}
37. $(-i)^{10}$
38. $(-i)^{43}$
39. $(-i)^{25}$
40. $(-i)^{52}$
41. $(2i)^3 \cdot i^{12}$
42. $(4i)^2 \cdot i^{50}$
43. $(3i)^4 \cdot i^{34}$
44. $(5i)^2 \cdot i^{15}$
45. $\sqrt{-7} \cdot \sqrt{-8}$
46. $\sqrt{-12} \cdot \sqrt{-3}$
47. $\sqrt{8} \cdot \sqrt{-2}$
48. $\sqrt{-5} \cdot \sqrt{10}$
49. $\sqrt{-2} \cdot \sqrt{-3} \cdot \sqrt{12}$
50. $\sqrt{-5} \cdot \sqrt{-15} \cdot \sqrt{-3}$
51. $(6 + 4i) + (2 + 5i)$
52. $(2 + 2i) + (7 + 3i)$
53. $(2 + 3i) + (-4 + 5i)$
54. $(5 - 3i) + (8 + 2i)$
55. $(7 - 9i) + (3 - 6i)$
56. $(4 - 4i) + (-5 + i)$
57. $(3 - 4i) + (-5 - 6i)$
58. $(-1 - i) + (10 - 9i)$
59. $(-10 - 5i) + (7 + 2i)$
60. $(1 + 6i) + (-1 + 6i)$
61. $(9 + 2i) - (4 + 3i)$
62. $(3 + i) - (8 + 4i)$
63. $(8 + 7i) - (6 - 7i)$
64. $(9 - 2i) - (3 + 5i)$
65. $(-11 + 7i) - (11 - 7i)$
66. $(-12 - 4i) - (6 - i)$
67. $7i + (2 + 8i)$
68. $9 + (1 - 5i)$
69. $2i - (12 - 2i)$
70. $(7 - 4i) - 10$
71. $4(-6 + 8i)$
72. $-2(5 - 9i)$
73. $3i(2 - i)$
74. $-4i(1 + 8i)$
75. $(5 + 2i)(3 + i)$
76. $(1 + 4i)(1 + 2i)$
77. $(-7 + 2i)(2 + 5i)$
78. $(9 - 3i)(-2 + i)$
79. $(-1 - i)(8 + 2i)$
80. $(2 + 6i)(-3 - 3i)$

81. $(5 + 4i)(5 - 4i)$
82. $(2 + 3i)(2 - 3i)$
83. $(1 + i)(1 - i)$
84. $(\sqrt{7} - i)(\sqrt{7} + i)$
85. $(2 + 3i)^2$
86. $(3 + 4i)^2$
87. $(-5 + i)^2$
88. $(6 - i)^2$
89. $(-1 - 2i)^2 \cdot i$
90. $(2i)(2 + 5i)^2$
91. $(3i)^3(-2 - 2i)$
92. $(-8 + 6i)(-3i^5)$
93. $(4 - 2i)^2(1 + i)$
94. $(-1 + 3i)^3$

Superset

In exercises 95–102, find real numbers x and y that satisfy each equation.

95. $2 + 3yi = -x + 9i$
96. $-4x + 7i = -8 - 7y$
97. $8x - 4i = 6 + 2yi$
98. $9 + 5yi = 6x - 7i$
99. $-7 + i = (x - y) + (2x + y)i$
100. $(2x + y) - 8i = 5 + (x - 3y)i$
101. $(x - y) + 4i = 8 - (x - 2y)i$
102. $3 - 2i = (x + y) - (3x - y)i$

In exercises 103–106, simplify each expression.

103. $i + i^2 + i^3 + \cdots + i^{48}$
104. $i + i^2 + i^3 + \cdots + i^{18}$
105. $i + i^2 + i^3 + \cdots + i^{29}$
106. $i + i^2 + i^3 + \cdots + i^{75}$

In exercises 107–114, determine whether each statement is true or false.

107. $3i$ is a solution of $x^2 + 9 = 0$.
108. $-3i$ is a solution of $x^2 + 9 = 0$.
109. $5 + i$ is a solution of $x^2 - 10x + 24 = 0$.
110. $5 - i$ is a solution of $x^2 - 10x + 24 = 0$.
111. $-3 + 2i$ is a solution of $x^2 + 6x + 13 = 0$.
112. $-3 - 2i$ is a solution of $x^2 + 6x + 13 = 0$.
113. Both $2 + 7i$ and $2 - 7i$ are solutions of the quadratic equation $x^2 - 4x + 53 = 0$.
114. Both $\frac{1}{2} + 3i$ and $\frac{1}{2} - 3i$ are solutions of the quadratic equation $4x^2 - 4x + 37 = 0$.

A complex number z is called an **nth root of one** or an **nth root of unity** if $z^n = 1$.

115. Show that $1, i, -1,$ and $-i$ are fourth roots of unity.
116. Show that $1, -\frac{1}{2} + \frac{\sqrt{3}}{2}i$, and $-\frac{1}{2} - \frac{\sqrt{3}}{2}i$ are cube roots of unity.
117. Show that $1, \frac{1}{2} + \frac{\sqrt{3}}{2}i, -\frac{1}{2} + \frac{\sqrt{3}}{2}i, -1, -\frac{1}{2} - \frac{\sqrt{3}}{2}i$, and $\frac{1}{2} - \frac{\sqrt{3}}{2}i$ are sixth roots of unity.
118. Show that the values $1, \frac{\sqrt{2}}{2} + \frac{\sqrt{2}}{2}i, i, -\frac{\sqrt{2}}{2} + \frac{\sqrt{2}}{2}i, -1, -\frac{\sqrt{2}}{2} - \frac{\sqrt{2}}{2}i, -i$, and $\frac{\sqrt{2}}{2} - \frac{\sqrt{2}}{2}i$ are eighth roots of unity.

In exercises 119–124, determine whether each statement is true or false. If false, illustrate this with an example.

119. If a is a positive real number and b is a negative real number, then $\sqrt{a} \cdot \sqrt{b} = \sqrt{ab}$.
120. If a and b are real numbers, then $\sqrt{a} \cdot \sqrt{b} = \sqrt{ab}$.
121. The product of two imaginary numbers is always imaginary.
122. The product of two imaginary numbers is always a real number.
123. If a and b are real numbers, then $(a + bi)(a - bi)$ is always a real number.
124. If a and b are real numbers, then $(a + bi)^2$ is always imaginary.

11.2 Properties of Complex Numbers

Goal A *to write the quotient of two complex numbers in the form $a + bi$*

By now you should have no difficulty solving linear equations like

$$3 + 4x = 2.$$

The last step in solving this equation involves multiplying both sides by the reciprocal of 4, namely $\frac{1}{4}$.

If we replaced the coefficient of x in the above equation with an irrational number, say $1 - \sqrt{5}$, the equation would then be

$$3 + (1 - \sqrt{5})x = 2,$$

and the last step in solving this equation would require multiplying both sides by the reciprocal of $1 - \sqrt{5}$. Recall that we can simplify by rationalizing the denominator, i.e., by multiplying the numerator and denominator by $1 + \sqrt{5}$.

$$\frac{1}{1 - \sqrt{5}} = \frac{1}{1 - \sqrt{5}} \cdot \frac{1 + \sqrt{5}}{1 + \sqrt{5}}$$

$$= \frac{1 + \sqrt{5}}{1 - 5}$$

$$= -\frac{1 + \sqrt{5}}{4}$$

Suppose instead, that the coefficient of x in the equation above were

$$3 - 2i.$$

Then the last step in solving the equation would involve multiplying both sides by the reciprocal of $3 - 2i$, $\frac{1}{3 - 2i}$. The reciprocal is a complex number and can be written in the form $a + bi$. To produce this form, we simply multiply the numerator and denominator by $3 + 2i$.

$$\frac{1}{3 - 2i} = \frac{1}{3 - 2i} \cdot \frac{3 + 2i}{3 + 2i} = \frac{3 + 2i}{9 - 4i^2} = \frac{3 + 2i}{9 + 4} = \frac{3}{13} + \frac{2i}{13}$$

The complex number $3 + 2i$ is called the **complex conjugate** of $3 - 2i$. In general, the complex conjugate of $a + bi$ is $a - bi$, and the complex conjugate of $a - bi$ is $a + bi$. Notice that since

$$(a + bi)(a - bi) = a^2 - b^2i^2 = a^2 + b^2,$$

the product of a complex number and its conjugate is a real number.

11 / Complex Numbers

Example 1 Write the reciprocal of each of the following in the form $a + bi$.

(a) $3 - 4i$ (b) $5 + 3i$

Solution (a) The reciprocal of $3 - 4i$ is $\dfrac{1}{3 - 4i}$.

$$\dfrac{1}{3 - 4i} = \dfrac{1}{3 - 4i} \cdot \dfrac{3 + 4i}{3 + 4i}$$ ■ Multiply the numerator and denominator by the conjugate of the denominator $3 - 4i$.

$$= \dfrac{3 + 4i}{9 - 16i^2}$$

$$= \dfrac{3 + 4i}{9 + 16}$$

$$= \dfrac{3}{25} + \dfrac{4}{25}i$$

(b) $\dfrac{1}{5 + 3i} = \dfrac{1}{5 + 3i} \cdot \dfrac{5 - 3i}{5 - 3i} = \dfrac{5 - 3i}{25 - 9i^2} = \dfrac{5 - 3i}{25 + 9} = \dfrac{5}{34} - \dfrac{3}{34}i$

We can use the complex conjugate in a similar way when we divide complex numbers.

Example 2 Simplify the quotient $\dfrac{2 - 3i}{-3 + i}$.

Solution

$$\dfrac{2 - 3i}{-3 + i} = \dfrac{2 - 3i}{-3 + i} \cdot \dfrac{-3 - i}{-3 - i}$$ ■ Multiply the numerator and denominator by the conjugate of the denominator $-3 + i$.

$$= \dfrac{-6 - 2i + 9i + 3i^2}{9 - i^2}$$ ■ We used the FOIL method to find the product in the numerator.

$$= \dfrac{-6 - 2i + 9i - 3}{9 - (-1)}$$ ■ We used the fact that $i^2 = -1$.

$$= -\dfrac{9}{10} + \dfrac{7}{10}i$$

11 / Complex Numbers

Goal B *to use the symbol $\bar{z}$ to represent the conjugate of a complex number*

If we let the variable z represent a complex number $a + bi$, it is customary to let $\bar{z}$ represent its complex conjugate $a - bi$; that is,

$$\overline{a + bi} = a - bi.$$

We can establish several facts about complex numbers and their conjugates by using the rules of arithmetic for complex numbers.

For example, the conjugate of the sum of two complex numbers is the sum of the conjugates of the two numbers:

$$\overline{z + w} = \bar{z} + \bar{w}.$$

To establish this fact, let $z = a + bi$ and $w = c + di$.

$\overline{z + w} = \overline{(a + bi) + (c + di)}$	■ Substitute $a + bi$ for z and $c + di$ for w. Then combine like terms.
$= \overline{(a + c) + (b + d)i}$	■ Now use the definition of a complex conjugate.
$= (a + c) - (b + d)i$	■ Rearrange the terms.
$= a - bi + c - di$	■ The expression is rewritten as the sum of two complex numbers.
$= \overline{a + bi} + \overline{c + di}$	■ The complex numbers are rewritten as conjugates.
$= \bar{z} + \bar{w}$	

Using the arithmetic of complex numbers, we can also show that the conjugate of the product of two complex numbers is the product of the conjugate of the two numbers, i.e., $\overline{z \cdot w} = \bar{z} \cdot \bar{w}$.

$\overline{z \cdot w} = \overline{(a + bi)(c + di)}$	■ Again, let $z = a + bi$ and $w = c + di$.
$= \overline{(ac - bd) + (ad + bc)i}$	
$= (ac - bd) - (ad + bc)i$	
$= ac - bd - adi - bci$	
$= ac - adi - bci + bdi^2$	■ $-bd$ is rewritten as bdi^2.
$= a(c - di) - bi(c - di)$	
$= (a - bi)(c - di)$	
$= \overline{(a + bi)} \, \overline{(c + di)}$	
$= \bar{z} \cdot \bar{w}$	

We now summarize some of the more useful properties of complex conjugates.

11 / Complex Numbers

Facts

If $\bar{z}$ and $\bar{w}$ are complex conjugates of the complex numbers z and w, respectively, then

1. $\overline{z + w} = \bar{z} + \bar{w}$
2. $\overline{z - w} = \bar{z} - \bar{w}$
3. $\overline{z \cdot w} = \bar{z} \cdot \bar{w}$
4. $\overline{\left(\dfrac{z}{w}\right)} = \dfrac{\bar{z}}{\bar{w}}$, for $w \neq 0$
5. $\overline{z^n} = (\bar{z})^n$, for every positive integer n
6. $\bar{z} = z$, if z is a real number

Example 3 If $z = 2 - 5i$ and $w = 3 + i$, verify that (a) $\overline{z + w} = \bar{z} + \bar{w}$ and (b) $\overline{z \cdot w} = \bar{z} \cdot \bar{w}$.

Solution (a) $\overline{z + w} = \overline{(2 - 5i) + (3 + i)}$
$= \overline{(2 + 3) + (1 - 5)i}$
$= \overline{5 - 4i}$
$= 5 + 4i$

■ Begin by substituting values for z and w. Then combine like terms.

$\bar{z} + \bar{w} = \overline{2 - 5i} + \overline{3 + i}$
$= 2 + 5i + 3 - i$
$= (2 + 3) + (5 - 1)i$
$= 5 + 4i$

■ Begin by substituting values for z and w. Then use the definition of conjugate.

Thus, since $\overline{z + w} = 5 + 4i$ and $\bar{z} + \bar{w} = 5 + 4i$, the two expressions $\overline{z + w}$ and $\bar{z} + \bar{w}$ are equal.

(b) $\overline{z \cdot w} = \overline{(2 - 5i)(3 + i)}$
$= \overline{6 + 2i - 15i - 5i^2}$
$= \overline{6 + 5 + 2i - 15i}$
$= \overline{11 - 13i}$
$= 11 + 13i$

■ Begin by substituting values for z and w.
■ Rearrange terms; let $i^2 = -1$.

$\bar{z} \cdot \bar{w} = \overline{2 - 5i} \cdot \overline{3 + i}$
$= (2 + 5i)(3 - i)$
$= 6 - 2i + 15i - 5i^2$
$= 6 + 5 + 13i$
$= 11 + 13i$

■ Begin by substituting values for z and w. Then use the definition of conjugate.
■ Rearrange terms; let $i^2 = -1$.

Thus, since $\overline{z \cdot w} = 11 + 13i$ and $\bar{z} \cdot \bar{w} = 11 + 13i$, we conclude $\overline{z \cdot w} = \bar{z} \cdot \bar{w}$.

11 / Complex Numbers

Goal C *to represent a complex number as a point in the complex plane*

Recall that in order to visualize the real numbers, we associated each number with a point on a line, called the real line. Complex numbers can be visualized as points in a plane called the **complex plane,** shown in Figure 11.5.

Every complex number $a + bi$ can be associated with an ordered pair (a, b) in the complex plane, where a is the real part of the complex number and b is the imaginary part. The horizontal axis in the complex plane is referred to as the real axis, since real numbers are graphed there. The vertical axis is called the imaginary axis. Pure imaginary numbers (numbers of the form bi) are graphed on the imaginary axis.

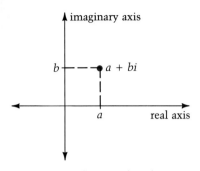

Figure 11.5 The Complex Plane.

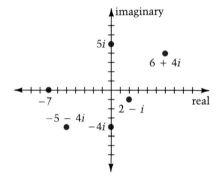

Figure 11.6

We will define the absolute value of a complex number $a + bi$ as the distance between the point corresponding to $a + bi$ and the origin. For example, the absolute value of $4 + 3i$ is the distance between the point $4 + 3i$ and the origin, and can be found by the Pythagorean Theorem:

$$\sqrt{4^2 + 3^2} = \sqrt{25} = 5.$$

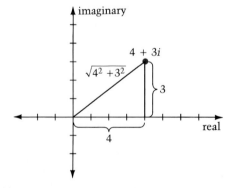

Figure 11.7

Example 4 Represent each of the following complex numbers as a point in the complex plane.
(a) $4 - 3i$ (b) $-3 + 4i$ (c) $-3 - 4i$ (d) $\overline{-2 - 3i}$

Solution (a)

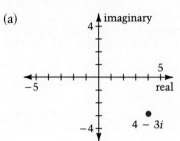

(b)

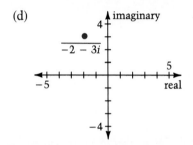

(c)

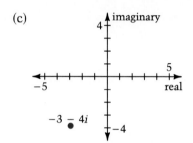

(d)

Definition
- The **absolute value** of a complex number $a + bi$, denoted $|a + bi|$, is the distance between the origin and the point associated with $a + bi$. That is, $|a + bi| = \sqrt{a^2 + b^2}$.

Example 5 Determine the absolute value of each of the following complex numbers.
(a) $2 - 3i$ (b) $-5i$ (c) $1 - i$

Solution (a) $|2 - 3i| = \sqrt{2^2 + (-3)^2} = \sqrt{4 + 9} = \sqrt{13}$

(b) $|-5i| = \sqrt{0^2 + (-5)^2} = \sqrt{25} = 5$

(c) $|1 - i| = \sqrt{1^2 + (-1)^2} = \sqrt{1 + 1} = \sqrt{2}$

Exercise Set 11.2

Goal A

In exercises 1–14, write the reciprocal of each complex number in the form $a + bi$.

1. $2 + 3i$
2. $1 + 4i$
3. $5 - 2i$
4. $3 - 7i$
5. $-1 + i$
6. $-1 - i$
7. $-2 - 4i$
8. $6 - 3i$
9. $5i$
10. $-8i$
11. $-\frac{1}{7}i$
12. $\frac{2}{3}i$
13. $\sqrt{3} - 2i$
14. $-\sqrt{11} + 5i$

In exercises 15–34, simplify each quotient.

15. $\dfrac{6 + 2i}{2 - i}$
16. $\dfrac{8 + i}{2 - 3i}$
17. $\dfrac{2 + 3i}{1 + 4i}$
18. $\dfrac{1 + 4i}{3 + 2i}$
19. $\dfrac{5 - i}{2 + 5i}$
20. $\dfrac{3 - 4i}{6 + i}$
21. $\dfrac{3 - 10i}{1 - i}$
22. $\dfrac{6 + 5i}{1 + i}$
23. $\dfrac{4 - 2i}{1 + 2i}$
24. $\dfrac{3 + i}{1 - 3i}$
25. $\dfrac{34}{5 - 3i}$
26. $\dfrac{-53}{7 - 2i}$
27. $\dfrac{-2i}{5 + 4i}$
28. $\dfrac{10i}{3 + i}$
29. $\dfrac{3 - 7i}{-7i}$
30. $\dfrac{-9 + 8i}{3i}$
31. $\dfrac{1}{-6 + 2i}$
32. $\dfrac{1}{-2 - 4i}$
33. $\dfrac{6}{i}$
34. $\dfrac{11}{-i}$

Goal B

In exercises 35–40, verify the following for the given values of the variables:
(a) $\overline{z + w} = \overline{z} + \overline{w}$
(b) $\overline{z - w} = \overline{z} - \overline{w}$
(c) $\overline{z \cdot w} = \overline{z} \cdot \overline{w}$
(d) $\overline{\left(\dfrac{z}{w}\right)} = \dfrac{\overline{z}}{\overline{w}}$

35. $z = 1 + 3i$ and $w = 2 - 4i$
36. $z = 5 + 2i$ and $w = 2 + 3i$
37. $z = 12 - 3i$ and $w = 5 + 9i$
38. $z = -11 + i$ and $w = 2 - 7i$
39. $z = -6 - 5i$ and $w = -1 - 4i$
40. $z = 9 - 9i$ and $w = -8 + 3i$

In exercises 41–46, verify the following for the given values of the variables: (a) $\overline{z^2} = (\overline{z})^2$; (b) $\overline{z^3} = (\overline{z})^3$.

41. $z = 2 - i$
42. $z = 3 + i$
43. $z = -2 + 6i$
44. $z = 3 + 4i$
45. $z = -5 - i$
46. $z = -2 - 8i$

Goal C

In exercises 47–72, represent each complex number as a point in the complex plane.

47. $2 + 5i$
48. $3 + 2i$
49. $2 - 5i$
50. $3 - 2i$
51. $-2 + 5i$
52. $-3 + 2i$
53. $-2 - 5i$
54. $3 - i$
55. $1 + 4i$
56. $1 + 5i$
57. $-3 + i$
58. $-4 + i$
59. 4
60. 5
61. -3
62. -6
63. $6i$
64. $-2i$
65. $-i$
66. i
67. $\overline{5 - 4i}$
68. $\overline{2 - 6i}$
69. $\overline{4 + 2i}$
70. $\overline{5 + 5i}$
71. $\overline{-3 - 6i}$
72. $\overline{-4 - 3i}$

In exercises 73–94, determine the absolute value of each complex number.

73. $7 + 2i$
74. $5 + 3i$
75. $7 - 2i$
76. $5 - 3i$
77. $-8 + 6i$
78. $-3 + 4i$
79. $-8 - 6i$
80. $-3 - 4i$
81. $3 + 3i$
82. $4 + 2i$
83. $7 - i$
84. $-8 + i$
85. $2i$
86. $-7i$
87. $-i$
88. i

89. $\sqrt{3} + i\sqrt{13}$
90. $\sqrt{6} + i\sqrt{3}$
91. $3 - i\sqrt{7}$
92. $\sqrt{5} - 2i$
93. $\sqrt{14} + i$
94. $2 + i\sqrt{10}$

Superset

In exercises 95–110, simplify each of the following expressions.

95. $\dfrac{1+i}{(1-2i)^2}$
96. $\dfrac{1-i}{(1+2i)^2}$
97. $\dfrac{5-i}{(1+i)(2-3i)}$
98. $\dfrac{-5-i}{(1-i)(2+3i)}$
99. $\dfrac{-1+8i}{(2-i)(3+2i)}$
100. $\dfrac{14+2i}{(2+i)(1+3i)}$
101. $\dfrac{4-3i}{(3+2i)(-1+i)}$
102. $\dfrac{5-i}{(3-2i)(1+i)}$
103. $\dfrac{1}{2+i} + \dfrac{-3}{1-3i}$
104. $\dfrac{-4}{1+3i} + \dfrac{2}{2-i}$
105. $\dfrac{2-3i}{1+i} - \dfrac{6+2i}{2-i}$
106. $\dfrac{5+2i}{3-i} - \dfrac{2-4i}{1+i}$
107. $\left|\dfrac{1-2i}{3+i}\right|$
108. $\left|\dfrac{1+2i}{3-i}\right|$
109. $\dfrac{|1-2i|}{|3+i|}$
110. $\dfrac{|1+2i|}{|3-i|}$

In exercises 111–118, find all complex numbers z that satisfy the given equation.

111. $(3 + 2i)z + i = 1 + 4i$
112. $(2 - 4i)z + 3i = -2 + 5i$
113. $(3 - 2i) + 5z = 3iz + (4 - 3i)$
114. $2iz - (8 - 5i) = z + (-10 + 7i)$
115. $2z + 5\overline{z} = 15 - 9i$
116. $7z - 3\overline{z} = -4 + 20i$
117. $z^2 = 8i$
118. $z^2 = -18i$

119. Let $z = a + bi$ and $w = c + di$. Show that $\overline{z - w} = \overline{z} - \overline{w}$.

120. Let $z = a + bi$ and $w = c + di$, where $w \neq 0$. Show that $\overline{\left(\dfrac{z}{w}\right)} = \dfrac{\overline{z}}{\overline{w}}$.

121. Let $z = a + bi$. Show that $\overline{\overline{z}} = z$.
122. Let $z = a + bi$. Show that $z \cdot \overline{z}$ is a real number.
123. Let $z = a + bi$. Show that $\tfrac{1}{2}(z + \overline{z}) = a$.
124. Let $z = a + bi$. Show that $\tfrac{1}{2}i(\overline{z} - z) = b$.
125. Let $z = a + bi$. Show that $|\overline{z}| = |z|$.
126. Let $z = a + bi$. Show that $|z|^2 = z \cdot \overline{z}$.

In exercises 127–130, determine whether each statement is true or false. If false, illustrate this with an example.

127. If z and w represent any complex numbers, then $|z - w| = |z| - |w|$.
128. If z and w represent any complex numbers, then $|z + w| = |z| + |w|$.
129. If z and w represent any complex numbers, then $|zw| = |z| \cdot |w|$.
130. If z and w represent any complex numbers, then $|z - w| = |w - z|$.

In exercises 131–138, represent z and $\overline{z}$ as points in the complex plane.

131. $z = 3 + 5i$
132. $z = 4 + 2i$
133. $z = -3 + 5i$
134. $z = -4 + 2i$
135. $z = 2 - 6i$
136. $z = 1 - 3i$
137. $z = -2 - 6i$
138. $z = -1 - 3i$

139. Let $z = 2 + 5i$ and $w = 4 - 7i$. Represent z, w, and $z + w$ as points in the complex plane.

140. Repeat exercise 139 with the complex numbers $z = -2 + 7i$ and $w = 8 - 4i$.

141. Let $z = 2 + i$ and $w = 3 - 2i$. Represent z, w, and zw as points in the complex plane.

142. Repeat exercise 141 with the complex numbers $z = -2 + 4i$ and $w = 1 + i$.

11.3 Zeros of Polynomial Functions

Goal A *to solve quadratic equations over the set of complex numbers*

We began this chapter by pointing out the need to extend the set of real numbers to the complex numbers so that any quadratic equation could be solved. The following examples show how to solve quadratic equations over the set of complex numbers.

Example 1 Solve each of the equations over the set of complex numbers.
(a) $x^2 = -9$ (b) $x^2 - 4x + 13 = 0$ (c) $x^2 = 3x - 4$

Solution
(a) $x^2 = -9$
$x = \pm i\sqrt{9}$ ■ -9 has two square roots, $i\sqrt{9}$ and $-i\sqrt{9}$.
$x = \pm 3i$ ■ There are two solutions, $3i$ and $-3i$.

(b) $x^2 - 4x + 13 = 0$
$x = \dfrac{4 \pm \sqrt{16 - 52}}{2}$ ■ We used the quadratic formula with $a = 1$, $b = -4$, and $c = 13$.
$x = \dfrac{4}{2} \pm \dfrac{\sqrt{-36}}{2}$
$x = \dfrac{4}{2} \pm \dfrac{6i}{2}$
$x = 2 \pm 3i$ ■ There are two solutions, $2 + 3i$ and $2 - 3i$.

(c) $x^2 = 3x - 4$ ■ Write the given equation in the form $ax^2 + bx + c = 0$.
$x^2 - 3x + 4 = 0$
$x = \dfrac{3 \pm \sqrt{9 - 16}}{2}$ ■ We used the quadratic formula with $a = 1$, $b = -3$, and $c = 4$.
$x = \dfrac{3}{2} \pm \dfrac{\sqrt{-7}}{2}$
$x = \dfrac{3}{2} \pm \dfrac{i\sqrt{7}}{2}$ ■ Note the two solutions are $\frac{3}{2} + \frac{i\sqrt{7}}{2}$ and $\frac{3}{2} - \frac{i\sqrt{7}}{2}$.

Notice that in the preceding examples whenever $a + bi$ was a solution, so was $a - bi$. For any quadratic equation with real coefficients, if $a + bi$ is a solution, so is $a - bi$; that is, the complex roots of a quadratic equation with real coefficients come in conjugate pairs.

11 / Complex Numbers

Goal B *to determine the complex zeros of a quadratic function*

One of the reasons for defining the complex numbers is to provide a set of numbers over which any quadratic equation can be solved. Consider the graph of the function

$$y = x^2 - 6x + 11$$

shown in Figure 11.8. (Note that the function is graphed, as always, in the real plane.) The function $y = x^2 - 6x + 11$ has no real zeros. Since the x-intercepts of the graph of a function correspond to the real zeros of the function, the parabola at the left never intersects the x-axis.

To determine the zeros of the function

$$y = x^2 - 6x + 11,$$

we solve the equation $x^2 - 6x + 11 = 0$ for x.

$$x = \frac{-(-6) \pm \sqrt{36 - 44}}{2} = \frac{6 \pm \sqrt{-8}}{2} = \frac{6 \pm 2i\sqrt{2}}{2} = 3 \pm i\sqrt{2}$$

The two zeros of the function are imaginary numbers. Thus, we refer to them as **complex zeros**.

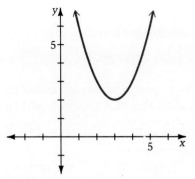

Figure 11.8 $y = x^2 - 6x + 11$

Example 2 Determine all the zeros of each function.

(a) $f(x) = x^2 + 3x + 6$ (b) $f(x) = (x - 2)(2x^2 - x + 4)$

Solution   (a) $x^2 + 3x + 6 = 0$ ■ Set $f(x)$ equal to zero and solve.

$$x = -\frac{3}{2} \pm \frac{i\sqrt{15}}{2}$$

There are two complex zeros, $-\frac{3}{2} + \frac{i\sqrt{15}}{2}$ and $-\frac{3}{2} - \frac{i\sqrt{15}}{2}$.

(b) $(x - 2)(2x^2 - x + 4) = 0$ ■ Set $f(x)$ equal to zero and solve.

$x - 2 = 0$ | $2x^2 - x + 4 = 0$
$x = 2$ | $x = \dfrac{1 \pm \sqrt{1 - 32}}{2(2)}$
         | $x = \dfrac{1 \pm i\sqrt{31}}{4}$

There are three zeros: one real zero, 2; and two complex zeros, $\dfrac{1 \pm i\sqrt{31}}{4}$.

Goal C *to solve higher degree polynomial equations*

Recall that an nth degree polynomial equation of the form

$$a_n x^n + a_{n-1} x^{n-1} + \cdots + a_1 x + a_0 = 0$$

can have at most n real number solutions. If we solve such an equation over the set of complex numbers, however, we shall find that we must modify this statement. Consider the equation

$$x^3 - 8 = 0.$$

After factoring, we have

$$(x - 2)(x^2 + 2x + 4) = 0.$$

If we solve this over the set of real numbers, we obtain only one solution, $x = 2$. (The equation $x^2 + 2x + 4 = 0$ has no real solutions.) Now let us solve $x^3 - 8 = 0$ over the set of complex numbers.

$$(x - 2)(x^2 + 2x + 4) = 0$$

$x - 2 = 0$	$x^2 + 2x + 4 = 0$
$x = 2$	$x = \dfrac{-2 \pm \sqrt{4 - 16}}{2}$
	$x = -1 \pm i\sqrt{3}$

Thus, we have found three solutions. It is no coincidence that we obtain three solutions when we solve a third degree polynomial equation over the set of complex numbers. We formalize this in the following rule.

Rule

Let

$$a_n x^n + a_{n-1} x^{n-1} + \cdots + a_1 x + a_0 = 0$$

be a polynomial equation of degree $n > 0$ with real coefficients. The equation has exactly n complex number solutions, provided each solution is counted according to its multiplicity.

Although we require that the polynomial equation in the above rule must have real coefficients, it can be shown that this rule is also true for polynomials with complex coefficients.

To understand the last sentence of the rule, let us look at the equation

$$(x - 7)^3 (x + 5)^2 (x^2 - x + 1) = 0.$$

Note that 7 is a solution of multiplicity three and -5 is a solution of multiplicity two. Thus, when counting solutions we should count 7 three times and -5 twice. Two more solutions of the equation are given by the factor $x^2 - x + 1$, namely $\frac{1}{2} \pm i\frac{\sqrt{3}}{2}$. Thus, the total number of solutions is seven. If you multiply the factors on the left side of the given equation, you will find that it is, as the rule predicts, a seventh degree polynomial.

Example 3   Verify the previous rule for the equation $x(x - 1)^2(x^2 + 2x + 2) = 0$.

Solution

$$x(x - 1)^2(x^2 + 2x + 2) = 0$$

$x = 0$	$(x - 1)^2 = 0$	$x^2 + 2x + 2 = 0$
	$x = 1$	$x = \dfrac{-2 \pm \sqrt{4 - 8}}{2}$
		$x = \dfrac{-2 \pm 2i}{2}$
		$x = -1 \pm i$

There are three roots of multiplicity one, 0, $-1 + i$, and $-1 - i$, and one root of multiplicity two, 1. Thus, there is a total of five roots. When the factors are multiplied, they yield $x^5 - x^3 - 2x^2 + 2x$, a fifth degree polynomial. The rule is therefore verified.

Recall that if you know that a number c is a root of a polynomial equation $p(x) = 0$, then you also know that $x - c$ is a factor of $p(x)$. Dividing $p(x)$ by $x - c$ might then produce a quotient polynomial that is more easily factored than the original. This method is demonstrated in the following example.

Example 4   The polynomial equation $x^4 - 4x^3 + 4x^2 + 3x - 6 = 0$ has 2 and -1 as solutions. Find all the solutions of the equation.

Solution   Since the equation is a fourth degree polynomial, we must find two roots (or one other root of multiplicity 2). Since 2 and -1 are solutions, $x - 2$ and $x + 1$ are factors of $x^4 - 4x^3 + 4x^2 + 3x - 6$. Thus, the product $(x - 2)(x + 1)$ or $x^2 - x - 2$ is also a factor. To determine the quotient polynomial we divide.

$$\begin{array}{r}x^2 - 3x + 3\\ x^2 - x - 2{\overline{\smash{\big)}\,x^4 - 4x^3 + 4x^2 + 3x - 6}}\\ \underline{-(x^4 - x^3 - 2x^2)}\\ -3x^3 + 6x^2 + 3x\\ \underline{-(-3x^3 + 3x^2 + 6x)}\\ 3x^2 - 3x - 6\\ \underline{-(3x^2 - 3x - 6)}\\ 0\end{array}$$

Thus, $x^4 - 4x^3 + 4x^2 + 3x - 6 = (x^2 - x - 2)(x^2 - 3x + 3)$.

$$x^4 - 4x^3 + 4x^2 + 3x - 6 = 0$$
$$(x^2 - x - 2)(x^2 - 3x + 3) = 0$$

$x^2 - x - 2 = 0$	$x^2 - 3x + 3 = 0$
$x = 2$ or $x = 1$	$x = \dfrac{3 \pm \sqrt{9 - 12}}{2}$
	$x = \dfrac{3}{2} \pm \dfrac{i\sqrt{3}}{2}$

- Each factor is set equal to zero.
- There are four solutions: 2, -1, $\frac{3}{2} + \frac{i\sqrt{3}}{2}$, and $\frac{3}{2} - \frac{i\sqrt{3}}{2}$.

For any polynomial equation with real coefficients, if a complex number $a + bi$ is a solution, then its conjugate $a - bi$ is also a solution. Thus, we say that the complex zeros of real polynomial equations come in conjugate pairs. The following example makes use of this fact.

Example 5   Determine a third degree polynomial equation with real coefficients that has 3 and $2 - i$ as two of its solutions.

Solution   Since the polynomial must have real coefficients, the complex roots must come in conjugate pairs. Since we know that $2 - i$ is a root, then $2 + i$ must also be a root. To build a polynomial, we multiply the three known factors:

$$(x - 3)[x - (2 - i)][x - (2 + i)] = (x - 3)(x^2 - (2 + i)x - (2 - i)x + 5)$$
$$= (x - 3)(x^2 - 4x + 5)$$
$$= x^3 - 7x^2 + 17x - 15$$

A third degree polynomial equation with real coefficients and 3 and $2 - i$ as two of its solutions is $x^3 - 7x^2 + 17x - 15 = 0$.

We now consider a streamlined technique, called **synthetic division,** which is useful in dividing a polynomial by a first degree polynomial of the form $x - c$, for any real number c. Below on the left, we show a problem that employs the technique used in Example 4. On the right, we show the same process, but with three simplifications: (1) the variables have been omitted; (2) terms usually "brought down" for convenience of computation have been omitted; and (3) the first number in each of the products has been deleted since it invariably combines with the number above it to produce 0. We note that omitting the variables is legitimate since it is the coefficients alone that are needed in completing the division process. Of course, we must agree to write the coefficients in order of descending powers of the variable.

$$
\begin{array}{r}
2x^2 + 5x - 1 \\
x - 3\overline{)\,2x^3 - x^2 - 16x + 7} \\
\underline{-(2x^3 - 6x^2)} \\
5x^2 - 16x \\
\underline{-(5x^2 - 15x)} \\
-x + 7 \\
\underline{-(-x + 3)} \\
4
\end{array}
\qquad
\begin{array}{r}
25-1 \\
1 - 3\overline{)\,2-1-167} \\
6 \\
\hline
5 \\
15 \\
\hline
-1 \\
-3 \\
\hline
4
\end{array}
$$

quotient → ; ← remainder →

We can further simplify the form on the right by moving all the numbers upward so as to occupy four lines. Since the divisor will always be of the form $x - c$, we need only write c (in this case 3). We show this further simplification below on the left. Notice that the bottom row contains all but the first coefficient of the quotient (in this case 2), and ends with the remainder. By placing this first coefficient in the bottom row, that row will contain all the coefficients of the quotient, followed by the remainder. Thus, the top row is unnecessary. This final simplification is shown on the right below.

$$
\begin{array}{r}
25-1 \\
3\overline{)\,2-1-167} \\
615-3 \\
\hline
5-14
\end{array}
\qquad
\begin{array}{r}
3\,|\ 2-1-167 \\
615-3 \\
\hline
25-14
\end{array}
$$

quotient with first coefficient missing

quotient   remainder

In the following example, we describe the steps involved in using synthetic division directly. Note that this technique relies only on the value of c and the coefficients of the polynomial to be divided by $x - c$.

Example 6 Use synthetic division to divide $2x^3 - x^2 - 16x + 7$ by $x - 3$.

Solution Begin by writing the top row: $\underline{c\,|}$, from the divisor $x - c$ (in this case $c = 3$), followed by the coefficients of the dividend.

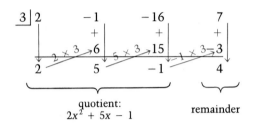

- Bring down the first coefficient, 2. Then successively multiply by 3 and add to the next coefficient.

quotient: $2x^2 + 5x - 1$

remainder

Example 7 Use synthetic division to divide $x^4 - 3x^3 + 2x - 1$ by $x + 2$.

Solution Since $x + 2 = x - (-2)$, in this case $c = -2$. For a missing power of x, record the coefficient 0.

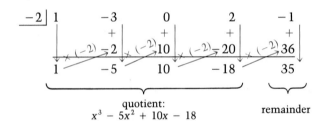

- Since there is no x^2-term, its coefficient must be recorded as 0.

quotient: $x^3 - 5x^2 + 10x - 18$

remainder

- The degree of the quotient is one less than that of the divisor.

We leave as a Superset exercise the proof of the following.

The Remainder Theorem

If a polynomial $f(x)$ is divided by $x - c$, then the remainder is equal to $f(c)$.

By the Remainder Theorem, we can use synthetic division to determine function values. For example, if $f(x) = 2x^3 - x^2 - 16x + 7$, then $f(x) = 4$ (by Example 6), and if $g(x) = x^4 - 3x^3 + 2x - 1$, then $g(-2) = 35$ (by Example 7).

Exercise Set 11.3

Goal A

In exercises 1–12, solve each equation over the set of complex numbers.

1. $x^2 = -16$
2. $x^2 = -49$
3. $x^2 - 6x + 25 = 0$
4. $x^2 - 4x + 29 = 0$
5. $x^2 - 4x + 5 = 0$
6. $x^2 - 2x + 5 = 0$
7. $x^2 + 3x + 4 = 0$
8. $x^2 + x + 1 = 0$
9. $4x^2 - 8x + 5 = 0$
10. $4x^2 - 4x + 5 = 0$
11. $4x^2 + 8x + 11 = 0$
12. $3x^2 + 4x + 2 = 0$

Goal B

In exercises 13–22, find all the zeros of each function.

13. $f(x) = 9x^2 + 64$
14. $f(x) = 4x^2 + 25$
15. $f(x) = 3x^2 - 4x + 2$
16. $f(x) = 3x^2 + x + 2$
17. $f(x) = -x^2 + 3x + 9$
18. $f(x) = -3x^2 - 3x - 1$
19. $f(x) = (x - 3)(x^2 - 6x + 11)$
20. $f(x) = (x + 8)(x^2 - 2x + 7)$
21. $f(x) = (3x - 1)(4x^2 + 5)$
22. $f(x) = (x - 11)(9x^2 + 10)$

Goal C

In exercises 23–28, a polynomial equation and one or more of its solutions are given. Find all the solutions.

23. $x^3 - 4x^2 + 5x - 6 = 0$; 3
24. $x^3 + 8x^2 + 20x + 25 = 0$; -5
25. $x^4 + 5x^3 + 11x^2 + 13x + 6 = 0$; $-2, -1$
26. $x^4 - x^2 + 4x - 4 = 0$; $1, -2$
27. $x^4 - 6x^3 + 13x^2 - 24x + 36 = 0$; $2i$
28. $x^4 - 4x^3 + 16x^2 - 24x + 20 = 0$; $1 + 3i$

29. Find a second degree polynomial equation with real coefficients that has $4 + 3i$ as one of its solutions.

30. Find a second degree polynomial equation with real coefficients that has $3 - 2i$ as one solution.

31. Find a third degree polynomial equation with real coefficients that has 4 and $-5i$ as solutions.

32. Find a third degree polynomial equation with real coefficients that has $2 - i\sqrt{5}$ and 4 as solutions.

In exercises 33–36, use synthetic division to verify the indicated function values.

33. $f(x) = 2x^3 - 3x^2 + 6x - 7$; $f(0) = -7$, $f(1) = -2$
34. $g(x) = x^4 - 3x^3 + 4x^2 - x - 1$; $g(-2) = 57$, $g(2) = 5$
35. $h(x) = x^5 - 3x^3 + x^2 - 4x + 10$; $h(-2) = 14$, $h(-1) = 17$
36. $G(x) = 1 - x + 8x^3 - 10x^5$; $G(-1) = 4$, $G(2) = -257$

Superset

The Rational Root Theorem says the following:
Let $a_n x^n + a_{n-1} x^{n-1} + \cdots + a_1 x + a_0 = 0$ be a polynomial equation with integral coefficients. If this equation has any rational solutions, then they must be of the form $\frac{c}{d}$, where c is a factor of a_0 and d is a factor of a_n. In exercises 37–40, solve each equation over the set of complex numbers. (Hint: Look for rational solutions.)

37. $x^4 + x^3 + 2x^2 + 4x - 8 = 0$
38. $x^4 - 5x^3 + 8x^2 - 10x + 12 = 0$
39. $2x^4 + x^3 - 2x^2 - 4x - 3 = 0$
40. $3x^4 - x^3 + 3x - 1 = 0$

In exercises 41–46, solve each equation over the set of complex numbers.

41. $x^4 + 1 = 0$
42. $x^4 + 16 = 0$
43. $x^4 - 25 = 0$
44. $x^4 - 49 = 0$
45. $x^4 + 5x^2 - 20 = 0$
46. $x^4 + 2x^2 - 8 = 0$

47. The Division Algorithm says the following:
When a polynomial $f(x)$ is divided by a nonzero polynomial $d(x)$, then there exist unique polynomials $q(x)$, the quotient, and $r(x)$, the remainder, such that $f(x) = d(x) \cdot q(x) + r(x)$, where $r(x) = 0$ or else the degree of $r(x)$ is one less than that of the divisor.

Use the Division Algorithm to prove the Remainder Theorem.

11 / Complex Numbers

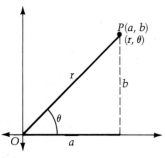

Figure 11.9

11.4 Trigonometric Form of Complex Numbers

Goal A *to express a complex number in trigonometric form*

In the last chapter we described how each point $P(a, b)$ can be represented by polar coordinates (r, θ), where r is the distance between P and the origin, and θ is the angle that ray OP makes with the positive x-axis. Any complex number $a + bi$ can be written in *polar* or *trigonometric form*, as illustrated below.

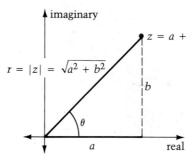

Figure 11.10

- $a = r \cos \theta$ since $\cos \theta = \dfrac{a}{r}$.
- $b = r \sin \theta$ since $\sin \theta = \dfrac{b}{r}$.
- Therefore, $z = (r \cos \theta) + (r \sin \theta)i$.

Definition

The complex number $z = a + bi$ has **trigonometric form**

$$z = r(\cos \theta + i \sin \theta)$$

where $r = |z| = \sqrt{a^2 + b^2}$, and $\tan \theta = \dfrac{b}{a}$. The number r is called the **modulus** of z, and θ is called the **argument** of z.

Example 1 Express each of the following complex numbers in trigonometric form. If rounding is necessary, express θ to the nearest multiple of $10'$.

(a) $2 + 2i$ (b) $-7 - 3i$

Solution (a)

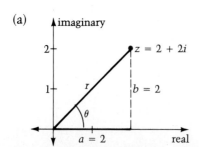

- Find r and θ, then apply the above definition.
- $r = |z| = \sqrt{2^2 + 2^2} = \sqrt{8} = 2\sqrt{2}$.
- Since $\tan \theta = \dfrac{2}{2} = 1$, and θ is a first quadrant angle,

$$\theta = \dfrac{\pi}{4}.$$

- Thus, $z = 2\sqrt{2}\left(\cos \dfrac{\pi}{4} + i \sin \dfrac{\pi}{4}\right)$.

(b)

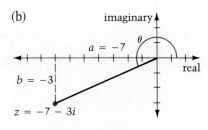

- $r = |z| = \sqrt{(-7)^2 + (-3)^2} = \sqrt{58}$.
- Since $\tan\theta = \dfrac{-3}{-7} \approx 0.4286$, θ is the third quadrant angle having 23°10′ (see Table 4) as its reference angle. Thus, $\theta \approx 180° + 23°10′ = 203°10′$.

$$z = \sqrt{58}\,(\cos 203°10′ + i \sin 203°10′)$$

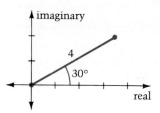

Figure 11.11

It is important to remember that the trigonometric form of a complex number is not unique. For example, all expressions of the form

$$z = 4[\cos(30° + k \cdot 360°) + i \sin(30° + k \cdot 360°)]$$

for every integer k, represent the same complex number, $2\sqrt{3} + 2i$.

Goal B *to determine the product and quotient of two complex numbers*

Suppose $z_1 = r_1(\cos\theta_1 + i \sin\theta_1)$ and $z_2 = r_2(\cos\theta_2 + i \sin\theta_2)$ are two complex numbers. Their product can be expressed as

$$\begin{aligned}
z_1 z_2 &= r_1(\cos\theta_1 + i \sin\theta_1) \cdot r_2(\cos\theta_2 + i \sin\theta_2) \\
&= r_1 r_2(\cos\theta_1 \cos\theta_2 + i \cos\theta_1 \sin\theta_2 \\
&\quad + i \sin\theta_1 \cos\theta_2 + i^2 \sin\theta_1 \sin\theta_2) \\
&= r_1 r_2[\cos\theta_1 \cos\theta_2 - \sin\theta_1 \sin\theta_2 \\
&\quad + i(\sin\theta_1 \cos\theta_2 + \cos\theta_1 \sin\theta_2)] \\
&= r_1 r_2[\cos(\theta_1 + \theta_2) + i \sin(\theta_1 + \theta_2)]
\end{aligned}$$

Rule

If $z_1 = r_1(\cos\theta_1 + i \sin\theta_1)$ and $z_2 = r_2(\cos\theta_2 + i \sin\theta_2)$ are two complex numbers, their product is given by

$$z_1 z_2 = r_1 r_2[\cos(\theta_1 + \theta_2) + i \sin(\theta_1 + \theta_2)].$$

The preceding rule tells us that the product of two complex numbers has modulus equal to the product of the two moduli, r_1 and r_2, and has argument equal to the sum of the two arguments θ_1 and θ_2.

Example 2 Find $z_1 z_2$ by using trigonometric forms. Express the answer in the form $a + bi$.
(a) $z_1 = 2(\cos 18° + i \sin 18°)$, $z_2 = 5(\cos 162° + i \sin 162°)$
(b) $z_1 = (-3 + 3i\sqrt{3})$, $z_2 = 1 - i\sqrt{3}$

Solution (a) $z_1 z_2 = 2 \cdot 5[\cos(18° + 162°) + i \sin(18° + 162°)]$
$= 10[\cos(180°) + i \sin(180°)] = 10[-1 + i(0)] = -10$

(b) First express each complex number in trigonometric form.

$$z_1 = 6\left(\cos \frac{2\pi}{3} + i \sin \frac{2\pi}{3}\right), \quad z_2 = 2\left(\cos \frac{5\pi}{3} + i \sin \frac{5\pi}{3}\right)$$

$$z_1 z_2 = 6 \cdot 2\left[\cos\left(\frac{2\pi}{3} + \frac{5\pi}{3}\right) + i \sin\left(\frac{2\pi}{3} + \frac{5\pi}{3}\right)\right] = 12\left[\cos\left(\frac{7\pi}{3}\right) + i \sin\left(\frac{7\pi}{3}\right)\right]$$

$$= 12\left(\cos \frac{\pi}{3} + i \sin \frac{\pi}{3}\right) = 12\left(\frac{1}{2} + i\frac{\sqrt{3}}{2}\right) = 6 + 6i\sqrt{3}$$

By using techniques of section 2, we can check the answer to part (b) of the above example.

$$(-3 + 3i\sqrt{3})(1 - i\sqrt{3}) = (-3)(1) + (-3)(-i\sqrt{3})$$
$$+ (3i\sqrt{3})(1) + (3i\sqrt{3})(-i\sqrt{3})$$
$$= -3 + 3i\sqrt{3} + 3i\sqrt{3} - 9i^2$$
$$= (-3 + 9) + i(3\sqrt{3} + 3\sqrt{3}) = 6 + 6i\sqrt{3}$$

In the exercises following this section, we suggest a way of proving the following rule for determining the quotient of two complex numbers.

Rule

If $z_1 = r_1(\cos \theta_1 + i \sin \theta_1)$ and $z_2 = r_2(\cos \theta_2 + i \sin \theta_2)$, then,

$$\frac{z_1}{z_2} = \frac{r_1}{r_2}[\cos(\theta_1 - \theta_2) + i \sin(\theta_1 - \theta_2)], \quad \text{for } z_2 \neq 0.$$

Example 3 Find $\frac{z_1}{z_2}$ where $z_1 = 3 - i\sqrt{3}$ and $z_2 = 4 + 4i$ by dividing trigonometric forms. Express the answer in trigonometric form.

Solution $z_1 = 2\sqrt{3}\left(\cos \frac{11\pi}{6} + i \sin \frac{11\pi}{6}\right), \quad z_2 = 4\sqrt{2}\left(\cos \frac{\pi}{4} + i \sin \frac{\pi}{4}\right)$ ■ Begin by expressing each complex number in trigonometric form.

$$\frac{z_1}{z_2} = \frac{2\sqrt{3}}{4\sqrt{2}}\left[\cos\left(\frac{11\pi}{6} - \frac{\pi}{4}\right) + i \sin\left(\frac{11\pi}{6} - \frac{\pi}{4}\right)\right]$$

$$= \frac{\sqrt{3}}{2\sqrt{2}} \cdot \frac{\sqrt{2}}{\sqrt{2}}\left[\cos\left(\frac{19\pi}{12}\right) + i \sin\left(\frac{19\pi}{12}\right)\right] = \frac{\sqrt{6}}{4}\left(\cos \frac{19\pi}{12} + i \sin \frac{19\pi}{12}\right)$$

11 / Complex Numbers

At this point you should verify that the answer in Example 3 is the same complex number you would find if you applied the techniques of section 2 to the quotient $\left(\dfrac{3 - i\sqrt{3}}{4 + 4i}\right)$. You can determine $\cos\dfrac{19\pi}{12}$ and $\sin\dfrac{19\pi}{12}$ exactly by using the appropriate sum and difference identities.

Goal C *to determine powers and roots of a complex number*

By successively multiplying a complex number $z = r(\cos\theta + i\sin\theta)$ by itself, we observe a simple pattern:

$$z^2 = r(\cos\theta + i\sin\theta) \cdot r(\cos\theta + i\sin\theta)$$
$$= r^2[\cos(\theta + \theta) + i\sin(\theta + \theta)]$$
$$= r^2(\cos 2\theta + i\sin 2\theta)$$

$$z^3 = z^2 \cdot z = r^2(\cos 2\theta + i\sin 2\theta) \cdot r(\cos\theta + i\sin\theta)$$
$$= r^3(\cos 3\theta + i\sin 3\theta)$$

This pattern holds for any positive integral power of z:

De Moivre's Theorem
For any positive integer n,

$$[r(\cos\theta + i\sin\theta)]^n = r^n(\cos n\theta + i\sin n\theta).$$

Example 4   Use De Moivre's Theorem to express in the form $a + bi$: (a) $(-1 + i\sqrt{3})^5$ (b) $(2 + 2i)^6$

Solution   (a) $-1 + i\sqrt{3} = 2\left(\cos\dfrac{2\pi}{3} + i\sin\dfrac{2\pi}{3}\right)$ ■ $-1 + i\sqrt{3}$ is expressed in trigonometric form.

$$(-1 + i\sqrt{3})^5 = \left[2\left(\cos\dfrac{2\pi}{3} + i\sin\dfrac{2\pi}{3}\right)\right]^5 = 2^5\left[\cos\left(5\cdot\dfrac{2\pi}{3}\right) + i\sin\left(5\cdot\dfrac{2\pi}{3}\right)\right]$$
$$= 32\left(\cos\dfrac{10\pi}{3} + i\sin\dfrac{10\pi}{3}\right) = 32\left[-\dfrac{1}{2} + i\left(-\dfrac{\sqrt{3}}{2}\right)\right] = -16 - 16i\sqrt{3}$$

(b) $2 + 2i = 2\sqrt{2}\left(\cos\dfrac{\pi}{4} + i\sin\dfrac{\pi}{4}\right)$ ■ $2 + 2i$ is expressed in trigonometric form.

$$(2 + 2i)^6 = \left[2\sqrt{2}\left(\cos\dfrac{\pi}{4} + i\sin\dfrac{\pi}{4}\right)\right]^6 = (2\sqrt{2})^6\left[\cos\left(6\cdot\dfrac{\pi}{4}\right) + i\sin\left(6\cdot\dfrac{\pi}{4}\right)\right]$$
$$= 512\left(\cos\dfrac{3\pi}{2} + i\sin\dfrac{3\pi}{2}\right) = 512[0 + i(-1)] = -512i$$

In Example 4, we found that $(2 + 2i)^6 = -512i$; that is, $2 + 2i$ is a sixth root of $-512i$. In general, if w and z are complex numbers such that $w^n = z$, then w is called an **nth root** of z.

We now consider how to determine such roots. For example, to find a cube root of $z = 1 - i\sqrt{3}$, we need $w = r(\cos\theta + i\sin\theta)$ such that

(1) $$[r(\cos\theta + i\sin\theta)]^3 = 1 - i\sqrt{3}.$$

By De Moivre's Theorem, $[r(\cos\theta + i\sin\theta)]^3 = r^3(\cos 3\theta + i\sin 3\theta)$. Also, we can rewrite $1 - i\sqrt{3}$ in general trigonometric form:

$$1 - i\sqrt{3} = 2[\cos(300° + k \cdot 360°) + i\sin(300° + k \cdot 360°)].$$

Thus, equation (1) can be rewritten as follows:

$$r^3(\cos 3\theta + i\sin 3\theta) = 2[\cos(300° + k \cdot 360°) + i\sin(300° + k \cdot 360°)].$$

We now determine r and θ so that we can write w in trigonometric form. It seems reasonable to set $r^3 = 2$ and $3\theta = 300° + k \cdot 360°$. Thus,

$$r = \sqrt[3]{2} \quad \text{and} \quad \theta = 100° + k \cdot 120°.$$

If $\quad k = 0, \quad w_0 = \sqrt[3]{2}(\cos 100° + i\sin 100°),$
$\quad\quad k = 1, \quad w_1 = \sqrt[3]{2}(\cos 220° + i\sin 220°),$
$\quad\quad k = 2, \quad w_2 = \sqrt[3]{2}(\cos 340° + i\sin 340°).$

For $k > 2$, the three complex numbers we found for $k = 0, 1, 2$, are repeated. Thus, there are exactly 3 distinct cube roots of $1 - i\sqrt{3}$ (see Figure 11.12). In general, any nonzero complex number has exactly n distinct nth roots.

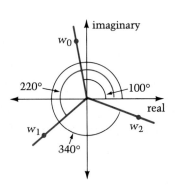

Figure 11.12 The three cube roots of $1 - i\sqrt{3}$. Note that they are equally spaced (in 120° increments) and lie at the same distance ($\sqrt[3]{2}$) from the origin.

The nth Root Theorem

The nth roots of the nonzero complex number $r(\cos\theta + i\sin\theta)$ are given by

$$\sqrt[n]{r}\left[\cos\left(\frac{\theta + k \cdot 360°}{n}\right) + i\sin\left(\frac{\theta + k \cdot 360°}{n}\right)\right]$$

where $k = 0, 1, 2, \ldots, n - 1$. We can replace 360° with 2π in the above statement to accommodate radian measure.

Example 5 Find and plot the 4 fourth roots of $16(\cos 120° + i\sin 120°)$.

Solution By the nth Root Theorem, there are exactly 4 fourth roots of this complex number:

$$w_0 = \sqrt[4]{16}\left(\cos\frac{120° + 0 \cdot 360°}{4} + i\sin\frac{120° + 0 \cdot 360°}{4}\right) = 2(\cos 30° + i\sin 30°)$$

$$w_1 = \sqrt[4]{16}\left(\cos\frac{120° + 1 \cdot 360°}{4} + i\sin\frac{120° + 1 \cdot 360°}{4}\right) = 2(\cos 120° + i\sin 120°)$$

$$w_2 = \sqrt[4]{16}\left(\cos\frac{120° + 2 \cdot 360°}{4} + i\sin\frac{120° + 2 \cdot 360°}{4}\right) = 2(\cos 210° + i\sin 210°)$$

$$w_3 = \sqrt[4]{16}\left(\cos\frac{120° + 3 \cdot 360°}{4} + i\sin\frac{120° + 3 \cdot 360°}{4}\right) = 2(\cos 300° + i\sin 300°)$$

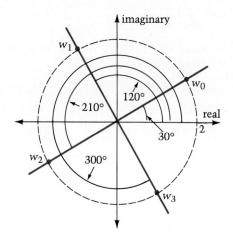

■ Note that the four fourth roots of $16(\cos 120° + i\sin 120°)$ are equally spaced $\left(\text{at } \frac{360°}{4} = 90° \text{ intervals}\right)$ on a circle centered at the origin, having radius 2.

If $z = 1$, then we refer to the n nth roots of z as the **nth roots of unity**.

Example 6 Determine the sixth roots of unity.

Solution First write $z = 1$ in trigonometric form. Since $z = 1 + 0i$, then $r = \sqrt{1^2 + 0^2} = 1$, and $\theta = 0°$. Thus $z = 1(\cos 0° + i\sin 0°)$. Each root has modulus $\sqrt[6]{1} = 1$, and argument determined by $\dfrac{0° + k \cdot 360°}{6}$ for $k = 0, 1, 2, 3, 4,$ and 5.

$w_0 = 1(\cos 0° + i\sin 0°) = 1$ $\qquad$ $w_1 = 1(\cos 60° + i\sin 60°) = \dfrac{1}{2} + \dfrac{\sqrt{3}}{2}i$

$w_2 = 1(\cos 120° + i\sin 120°) = -\dfrac{1}{2} + \dfrac{\sqrt{3}}{2}i$ $\qquad$ $w_3 = 1(\cos 180° + i\sin 180°) = -1$

$w_4 = 1(\cos 240° + i\sin 240°) = -\dfrac{1}{2} - \dfrac{\sqrt{3}}{2}i$

$w_5 = 1(\cos 300° + i\sin 300°) = \dfrac{1}{2} - \dfrac{\sqrt{3}}{2}i$

Exercise Set 11.4

Goal A

In exercises 1–20, express each of the following complex numbers in trigonometric form.

1. $3i$
2. $-2i$
3. -6
4. 8
5. $-1 + i\sqrt{3}$
6. $-\sqrt{3} + i$
7. $-2 - 2i$
8. $2 - 2i$
9. $-1 - i$
10. $1 + i$
11. $-3 + i\sqrt{3}$
12. $-3 - i\sqrt{3}$
13. $2\sqrt{3} - 2i$
14. $-2 + 2i\sqrt{3}$
15. $-3\sqrt{2} - 3i\sqrt{2}$
16. $3\sqrt{2} - 3i\sqrt{2}$
17. $5 + 12i$
18. $-8 + 15i$
19. $-24 - 7i$
20. $-7 - 4i$

Goal B

In exercises 21–28, determine the product by multiplying the trigonometric forms. Express the answer in the form $a + bi$.

21. $(-2 + 2i)(-2 - 2i)$
22. $(-\sqrt{3} + i)(\sqrt{3} - i)$
23. $(3 + i\sqrt{3})(1 - i\sqrt{3})$
24. $(1 - i\sqrt{3})(-3 - i\sqrt{3})$
25. $(\sqrt{3} + i\sqrt{3})(-2 + 2i)$
26. $(3\sqrt{3} + 3i)(3 + 3i\sqrt{3})$
27. $(3 - 4i)(3i)$
28. $(5 + 12i)(-2i)$

In exercises 29–36, find $\frac{z_1}{z_2}$ by using the rule for dividing trigonometric forms.

29. $z_1 = 6(\cos 300° + i \sin 300°)$
    $z_2 = 3(\cos 135° + i \sin 135°)$
30. $z_1 = 8\left(\cos \frac{7\pi}{6} + i \sin \frac{7\pi}{6}\right)$
    $z_2 = 4\left(\cos \frac{2\pi}{3} + i \sin \frac{2\pi}{3}\right)$
31. $z_1 = 1 + 3i$, $z_2 = 1 - i$
32. $z_1 = 1 - i\sqrt{3}$, $z_2 = 1 + i$
33. $z_1 = \sqrt{3} - 3i$, $z_2 = \sqrt{3} + i$
34. $z_1 = -3 - i\sqrt{3}$, $z_2 = 1 - i$
35. $z_1 = 2 + 2i$, $z_2 = 3 - i\sqrt{3}$
36. $z_1 = -2 - 2i$, $z_2 = -3 + 3i$

Goal C

In exercises 37–42, use De Moivre's Theorem to express in the form $a + bi$.

37. $(-2 + 2i)^4$
38. $(\sqrt{3} - i)^5$
39. $(\sqrt{3} + i)^3$
40. $(-1 - i)^6$
41. $(2\sqrt{3} - 2i)^{12}$
42. $(-3 - i\sqrt{3})^{10}$

In exercises 43–44, find the three cube roots of z.

43. $z = 3i$
44. $z = -2i$

In exercises 45–46, find the four fourth roots of z.

45. $z = -2 + 2i$
46. $z = -\sqrt{3} + i$

In exercises 47–48, find the five fifth roots of z.

47. $z = 1 + i$
48. $z = 1 + i\sqrt{3}$

49. Find (a) the fifth roots of unity, (b) the fourth roots of unity, (c) the eighth roots of unity.

50. Find (a) the eighth roots of -1, (b) the fifth roots of 32, (c) the fourth roots of 16.

Superset

In exercises 51–56, solve the equation over the set of complex numbers.

51. $x^2 + 2x + 3 = 0$
52. $x^2 + 4x + 5 = 0$
53. $x^3 - 2i = 0$
54. $x^3 + 1 = 0$
55. $x^5 + 32 = 0$
56. $x^6 - 64 = 0$

57. Prove that if $z = r(\cos \theta + i \sin \theta)$, $\bar{z} = r(\cos(-\theta) + i \sin(-\theta))$.

58. Prove that if $z = r(\cos \theta + i \sin \theta)$, $\frac{1}{z} = \frac{1}{r}(\cos \theta - i \sin \theta)$ for $z \neq 0 + 0i$. Use this result to prove the rule for dividing two complex numbers.

11 / Complex Numbers

Chapter Review & Test

Chapter Review

11.1 Introduction to Complex Numbers (pp. 454–463)

The number i has the property that $i^2 = -1$. For this reason, i can be considered a square root of -1. Thus, we write $i = \sqrt{-1}$. (p. 455)

If p is a positive real number, then $\sqrt{-p} = i\sqrt{p}$. Note that $-p$ has two square roots, $i\sqrt{p}$ and $-i\sqrt{p}$. We refer to $i\sqrt{p}$ as the *principal square root of* $-p$. (p. 456)

If a and b are real numbers, then any number of the form $a + bi$ is called a *complex number*. The number a is called the *real part* of the complex number, and b is called the *imaginary part*. (p. 457)

If $a + bi$ and $c + di$ are complex numbers, then

$$(a + bi) + (c + di) = (a + c) + (b + d)i,$$
$$(a + bi) - (c + di) = (a - c) + (b - d)i, \quad \text{and}$$
$$(a + bi)(c + di) = (ac - bd) + (ad + bc)i. \quad \text{(p. 460)}$$

11.2 Properties of Complex Numbers (p. 464–471)

In general, for real numbers a and b, the *complex conjugate* of $a + bi$ is $a - bi$. Since $(a + bi)(a - bi) = a^2 + b^2$, the product of a complex number and its conjugate is a real number. (p. 464)

Let z be a complex number $a - bi$. Then $\bar{z}$ represents the complex conjugate $a - bi$. (p. 466)

If $\bar{z}$ and $\bar{w}$ are complex conjugates of the complex numbers z and w, respectively, then:

1. $\overline{z + w} = \bar{z} + \bar{w}$
2. $\overline{z - w} = \bar{z} - \bar{w}$
3. $\overline{z \cdot w} = \bar{z} \cdot \bar{w}$
4. $\overline{\left(\dfrac{z}{w}\right)} = \dfrac{\bar{z}}{\bar{w}}$, for $w \neq 0$
5. $\overline{z^n} = (\bar{z})^n$, for every positive integer n
6. $\bar{z} = z$, if z is a real number

11.3 Zeros of Polynomial Functions (pp. 472–479)

Let $a_n x^n + a_{n-1} x^{n-1} + \cdots + a_1 x^1 + a_0 = 0$ be a polynomial equation of degree $n > 0$ with real coefficients. The equation has exactly n complex number solutions, provided each solution is counted according to its multiplicity. (p. 474)

The complex zeros of real polynomial equations come in conjugate pairs. (p. 476)

11.4 Trigonometric Form of Complex Numbers (pp. 480–486)

The complex number $z = a + bi$ has trigonometric form

$$z = r(\cos \theta + i \sin \theta),$$

where $r = \sqrt{a^2 + b^2}$ and $\tan \theta = \frac{b}{a}$. The number r is called the *modulus* of z and θ is called the *argument* of z. (p. 480)

If $z_1 = r_1(\cos \theta_1 + i \sin \theta_1)$ and $z_2 = r_2(\cos \theta_2 + i \sin \theta_2)$, then

$$z_1 z_2 = r_1 r_2 [\cos(\theta_1 + \theta_2) + i \sin(\theta_1 + \theta_2)],$$

$$\frac{z_1}{z_2} = \frac{r_1}{r_2}[\cos(\theta_1 - \theta_2) + i \sin(\theta_1 - \theta_2)] \quad \text{for } z_2 \neq 0.$$

De Moivre's Theorem: For any positive integer n,

$$[r(\cos \theta + i \sin \theta)]^n = r^n(\cos n\theta + i \sin n\theta). \qquad \text{(p. 483)}$$

The nth Root Theorem: The n nth roots of the nonzero complex number

$$r(\cos \theta + i \sin \theta)$$

are given by

$$\sqrt[n]{r}\left[\cos\left(\frac{\theta + k \cdot 360°}{n}\right) + i \sin\left(\frac{\theta + k \cdot 360°}{n}\right)\right]$$

where $k = 0, 1, 2, \ldots, n - 1$. We can replace 360° with 2π in the above statement to accommodate radian measure. (p. 484)

If $z = 1$, then the n nth roots of z are referred to as the nth *roots of unity*. (p. 485)

Chapter Test

11.1A Use the number i to rewrite each of the following.

1. $\sqrt{-11}$
2. $\sqrt{-49}$
3. $\sqrt{-48}$
4. $-\sqrt{-28}$

Determine the real part and the imaginary part of each complex number.

5. $3 - 9i$
6. $2 + \frac{1}{6}i$
7. $-3i$
8. $\sqrt{19}$

11.1B Simplify each expression.

9. $(-i)^{27}$
10. $\sqrt{-3} \cdot \sqrt{-21} \cdot \sqrt{-7}$
11. $(9 - 5i) + (-3 - 4i)$
12. $(6 - i)(7 + 3i)$
13. $(\sqrt{8} + i)(\sqrt{8} - i)$
14. $(4i)^2(2 - 2i)$

11.2A Write the reciprocal of each complex number in the form $a + bi$.

15. $-8 + i$
16. $4 - i\sqrt{3}$

Simplify each quotient.

17. $\dfrac{9 + i}{2 - 3i}$
18. $\dfrac{10 - 5i}{-4i}$

11.2B Verify the following for the given values of the variables:

(a) $\overline{z + w} = \overline{z} + \overline{w}$
(b) $\overline{z - w} = \overline{z} - \overline{w}$
(c) $\overline{z \cdot w} = \overline{z} \cdot \overline{w}$
(d) $\overline{\left(\dfrac{z}{w}\right)} = \dfrac{\overline{z}}{\overline{w}}$

19. $z = 3 + i$ and $w = 4 - 3i$
20. $z = -1 - i$ and $w = 10 + 8i$

Verify the following for the given values of the variables:

(a) $\overline{z^2} = (\overline{z})^2$
(b) $\overline{z^3} = (\overline{z})^3$

21. $z = 6 + i$
22. $z = 1 - 3i$

11.2C Represent each complex number as a point in the complex plane.

23. $5 - 12i$
24. -9
25. $-7i$
26. $-6 + 3i$

Determine the absolute value of each complex number.

27. $4 + 5i$
28. $7 - 8i$
29. $-9i$
30. $-3 - i\sqrt{4}$

11.3A Solve each equation over the set of complex numbers.

31. $x^2 + 29 = 10x$
32. $-6x - 4 = 9x^2$

11.3B Determine all the zeros of each of the following functions.

33. $f(x) = x^2 + x + 3$
34. $f(x) = 2x^2 - 5x + 4$
35. $f(x) = (3x^2 + 7x - 6)(2x^2 - x + 4)$
36. $f(x) = (x - 5)^2(-2x^2 + 4x - 3)$

11.3C Verify the rule in Goal C for each polynomial equation.

37. $(x - 4)(x + 3)(x^2 - 5x + 11) = 0$

38. $x^5(2x - 1)(x + 3) = 0$

39. $x(x - 2)^3(x^2 + 4x + 16) = 0$

40. $x^4(x - 1)(x^3 + 27)^2 = 0$

In each of the following, a polynomial equation and one of its solutions are given. Find all the solutions of each polynomial equation.

41. $x^3 - 4x^2 + 6x - 4 = 0$; 2

42. $x^4 - 6x^3 + 13x - 24x + 36 = 0$; $2i$

11.4A Express each of the following complex numbers in trigonometric form.

43. $3 - i\sqrt{3}$
44. $-4 - 4i$

11.4B Find $z_1 z_2$ and $\dfrac{z_1}{z_2}$ for each of the following complex numbers by using trigonometric forms. Express the answer in the form $a + bi$.

45. $z_1 = 6\sqrt{3} + 6i,\ z_2 = 1 - i\sqrt{3}$
46. $z_1 = -2 + 5i,\ z_2 = 7 - 2i$

11.4C Use De Moivre's Theorem to express in the form $a + bi$.

47. $\left(-\dfrac{\sqrt{3}}{2} + \dfrac{1}{2}i\right)^5$
48. $\left(-\dfrac{1}{2} + \dfrac{\sqrt{3}}{2}i\right)^4$

Superset

49. Find real numbers x and y that satisfy each equation.

   (a) $6 + 2xi = -3y - 10i$
   (b) $(3x - y) + 2i = 9 + (x + 2y)i$

50. Simplify each expression.

   (a) $\dfrac{1 + 5i}{(1 - 4i)^2}$
   (b) $\dfrac{4 + i}{1 - i} - \dfrac{1 - 3i}{2 + i}$

51. Find all complex numbers z that satisfy the equation $(2 - 7i)z + 4i = 5 - 3i$.

52. Let $z = 3 + i$ and $w = -2 - 5i$. Represent z, w, and $z + w$ as points in the complex plane.

53. Solve $x^4 + 81 = 0$ over the set of complex numbers.

12 Sequences, Series, and Limits

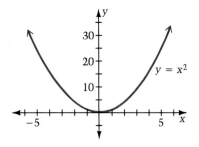

Figure 12.1

12.1 Arithmetic Sequences and Sums

Goal A *to determine the terms of a sequence*

Up to this point we have been concerned mainly with functions that have the set of real numbers (or intervals of real numbers) as their domains. One such function, $f(x) = x^2$, is graphed at the left. What would happen if we changed the domain from the set of all real numbers to the set of positive integers? The resulting graph would then consist of the set of discrete points (unconnected dots) shown in Figure 12.2. Such a function is an example of a *sequence*.

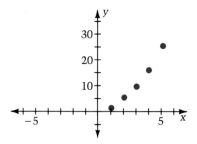

Figure 12.2

In general, a **sequence** is a function whose domain is the set of positive integers. To remind ourselves of this special domain, we usually use "n" instead of "x" to represent domain values.

Some values for the sequence a described by the equation

$$a(n) = n^2$$

n	$a(n)$
1	1
2	4
3	9
4	16
5	25
⋮	⋮
k	k^2
⋮	⋮

are given in the table at the left. We will adopt the common practice of referring to the listing 1, 4, 9, 16, 25, ... as the sequence, and to the individual values 1, 4, 9, etc., as the **terms of the sequence**. A note regarding notation is essential:

in place of the symbol $a(n)$, we shall use a_n.

Thus, the subscript n represents the domain value. Applying this new notation to the above sequence, we write:

$$a_1 = 1, \quad a_2 = 4, \quad a_3 = 9, \quad a_{10} = 100, \quad a_n = n^2.$$

Notice that a_1 is the first term of the sequence, a_2 the second term, a_3 the third term, and so on. We refer to a_n as the nth term, or the **general term of the sequence**. It can be used to state a rule for generating the specific terms of the sequence. Substituting the values 1, 2, 3, 4, 5, etc., for n in the rule produces the terms of the sequence. Several examples are displayed below.

Rule	Sequence
$a_n = 2n - 1$	1, 3, 5, 7, 9, ...
$a_n = 2^n$	2, 4, 8, 16, 32, ...
$a_n = 3 - 2^n$	1, −1, −5, −13, −29, ...
$a_n = (-1)^n$	−1, 1, −1, 1, −1, ...

Example 1 Find the first three terms and the twentieth term of the sequence whose general term is given by the rule $a_n = n^2 - n$.

Solution

$a_n = n^2 - n$ ■ We evaluate a_n for $n = 1, 2, 3,$ and 20.
$a_1 = 1^2 - 1 = 0$
$a_2 = 2^2 - 2 = 2$
$a_3 = 3^2 - 3 = 6$
$a_{20} = 20^2 - 20 = 380$

For the sequences discussed thus far, each term is found by evaluating an expression containing the term's subscript. A sequence can be defined another way: specify the first term of the sequence, and a rule for finding any other term by using the preceding term(s). A sequence described this way is said to be defined **recursively**.

Consider the sequence defined recursively as follows:

$$a_1 = 1, \quad a_n = 2(a_{n-1}) + 1 \quad \text{for } n > 1.$$

$a_1 = 1$
$a_2 = 2(1) + 1 = 3$
$a_3 = 2(3) + 1 = 7$
$a_4 = 2(7) + 1 = 15$
$a_5 = 2(15) + 1 = 31$

To find any term a_n after the first, multiply the preceding term, a_{n-1}, by 2 and then add 1. The first five terms of this sequence are 1, 3, 7, 15 and 31 (see the computations in the margin). Note that every term after the first is found by using the preceding term.

The following example specifies the first two terms of a sequence, and then the recursive rule.

$$a_1 = 1, \quad a_2 = 1, \quad a_n = a_{n-1} + a_{n-2} \quad \text{for } n > 2.$$

According to this rule, each term after the second is the sum of the two preceding terms. Thus, we obtain the sequence

1, 1, 2, 3, 5, 8, 13, 21, 34, 55, 89, 144, ...

Example 2 Find the first four terms of each sequence.

(a) $a_1 = 7, \quad a_n = a_{n-1} + 3 \quad \text{for } n > 1$ (b) $a_1 = 1, \quad a_n = a_{n-1} + 2n - 1 \quad \text{for } n > 1$

Solution

(a) $a_1 = 7$
$a_2 = 7 + 3 = 10$
$a_3 = 10 + 3 = 13$
$a_4 = 13 + 3 = 16$

(b) $a_1 = 1$
$a_2 = 1 + 2(2) - 1 = 4$
$a_3 = 4 + 2(3) - 1 = 9$
$a_4 = 9 + 2(4) - 1 = 16$

Goal B *to determine terms of an arithmetic sequence*

Notice that for the sequence 7, 10, 13, 16, . . . , the difference between successive terms is constant: it is always 3. Such a sequence is called an **arithmetic sequence**, and the constant difference, 3, is called the **common difference**. Every term of an arithmetic sequence after the first is the sum of the preceding term and the common difference.

Example 3 Find the first five terms of the arithmetic sequence
(a) whose first term is 20, and whose common difference is 4;
(b) whose first term is 10, and whose common difference is -3.

Solution   (a) $a_1 = 20$ (b) $a_1 = 10$
$a_2 = 20 + 4 = 24$ $a_2 = 10 + (-3) = 7$
$a_3 = 24 + 4 = 28$ $a_3 = 7 + (-3) = 4$
$a_4 = 28 + 4 = 32$ $a_4 = 4 + (-3) = 1$
$a_5 = 32 + 4 = 36$ $a_5 = 1 + (-3) = -2$

$a_1 = a_1 + 0d$
$a_2 = a_1 + 1d$
$a_3 = a_1 + 2d$
$a_4 = a_1 + 3d$

Suppose an arithmetic sequence has first term a_1 and common difference d. Then all the terms in the sequence can be expressed in terms of a_1 and d, as shown at the left. (We know that $a_3 = a_2 + d$; thus, $a_3 = (a_1 + d) + d = a_1 + 2d$. We find a_4 similarly.) Notice that the coefficient of d is one less than the subscript of the term. This suggests the following fact.

Fact

In an arithmetic sequence with first term a_1 and common difference d, the nth term can be described as follows:

$$a_n = a_1 + (n - 1)d \qquad (n \geq 1).$$

Example 4   Find the twelfth and thirtieth terms of the arithmetic sequence $-7, -2, 3, 8, \ldots$.

Solution $a_n = a_1 + (n - 1)d$ ■ Substitute -7 for a_1 and 5 for d. (d is the common difference.)
$a_{12} = -7 + (12 - 1)(5) = 48$
$a_{30} = -7 + (30 - 1)(5) = 138$

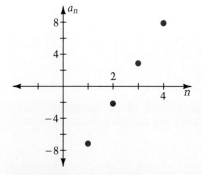

Figure 12.3

If we were to graph an arithmetic sequence, the discrete points would lie along a straight line. The graph of the sequence given in Example 4 is shown in Figure 12.3. Notice that for this sequence $a_1 = -7$ and $d = 5$. Thus, the general term is given by the rule

$$a_n = -7 + (n - 1)5, \quad \text{or simply} \quad a_n = -12 + 5n.$$

Therefore, in the xy-plane, the dots representing this sequence lie along the line $y = -12 + 5x$, a line with slope 5.

Goal C *to find the sum of the first n terms of an arithmetic sequence*

Suppose you are asked to determine the sum of the integers from 1 to 100. An easy way to find this sum is as follows. Begin by writing the sum; then write the sum in reverse order:

$$\begin{array}{c}1 + 2 + 3 + \cdots + 98 + 99 + 100 \\ 100 + 99 + 98 + \cdots + 3 + 2 + 1 \\ \hline 101 + 101 + 101 + \cdots + 101 + 101 + 101\end{array}$$

When we add the terms of the two sums, we obtain 100 addends, each equal to 101. Thus, twice the sum of the numbers from 1 to 100 is equal to 100(101). Therefore,

$$(1 + 2 + 3 + \cdots + 100) = \frac{1}{2}(100)(101) = 5050.$$

We can apply a similar method to find the sum of the first n terms of any arithmetic sequence.

$$\begin{array}{c}a_1 + (a_1 + d) + \cdots + [a_1 + (n-1)d] \\ [a_1 + (n-1)d] + [a_1 + (n-2)d] + \cdots + a_1 \\ \hline [2a_1 + (n-1)d] + [2a_1 + (n-1)d] + \cdots + [2a_1 + (n-1)d]\end{array}$$

The result is n addends, each equal to

$$[2a_1 + (n-1)d].$$

Thus, twice the sum of the first n terms is equal to

$$n[2a_1 + (n-1)d].$$

We conclude that the sum of the first n terms is

$$S_n = \frac{n}{2}[2a_1 + (n-1)d].$$

Since $2a_1 + (n-1)d = a_1 + [a_1 + (n-1)d] = a_1 + a_n$, we have

$$S_n = \frac{n}{2}(a_1 + a_n).$$

Fact

The sum S_n of the first n terms of an arithmetic sequence is

$$S_n = \frac{n}{2}[2a_1 + (n-1)d] \quad \text{or} \quad S_n = \frac{n}{2}(a_1 + a_n),$$

where a_1 is the first term, a_n is the nth term, and d is the common difference.

Example 5 Find the sum of the first 40 terms of the arithmetic sequence 11, 15, 19, 23, ...

Solution $S_n = \frac{n}{2}[2a_1 + (n-1)d]$ ■ Substitute 11 for a_1, 4 for d, and 40 for n.

$S_{40} = \frac{40}{2}[2(11) + (40-1)4]$

$S_{40} = 3560$

Alternative solution: to use the alternative formula for S_n, find the fortieth term, a_{40}.

$a_n = a_1 + (n-1)d$
$a_{40} = 11 + (40-1)4$
$a_{40} = 167$

$S_n = \frac{n}{2}(a_1 + a_n)$ ■ Now, substitute 11 for a_1, 167 for a_n, and 40 for n.

$S_{40} = \frac{40}{2}(11 + 167)$

$S_{40} = 3560$

Exercise Set 12.1   = Calculator Exercises in the Appendix

Goal A

In exercises 1–16, find the first four terms, the ninth term, and the twentieth term of the sequence whose general term is given by each rule.

1. $a_n = 3n - 5$
2. $a_n = 4 - n$
3. $a_n = \frac{n}{2n+1}$
4. $a_n = \frac{3n}{1+n}$
5. $a_n = 1 - \frac{1}{n}$
6. $a_n = (-1)^n \frac{1}{n}$
7. $a_n = n^2 + 3$
8. $a_n = 7 - n^2$
9. $a_n = (n-3)(n-4)$
10. $a_n = (n+2)(n+3)$
11. $a_n = n(n-1)(n-2)$
12. $a_n = n(n+6)(n+1)$
13. $a_n = \frac{1}{n} + \frac{1}{n+1}$
14. $a_n = \frac{1}{n} - \frac{1}{n+1}$
15. $a_n = 2^{-n}$
16. $a_n = (-2)^{-n}$

In exercises 17–26, find the first six terms of each sequence.

17. $a_1 = 3$, $a_n = a_{n-1} + 1$ for $n \geq 2$
18. $a_1 = 4$, $a_n = a_{n-1} - 1$ for $n \geq 2$
19. $a_1 = 1$, $a_n = 2a_{n-1} - 3$ for $n \geq 2$
20. $a_1 = 5$, $a_n = 2a_{n-1} - 3$ for $n \geq 2$
21. $a_1 = 3$, $a_n = (-1)^n a_{n-1} - 7$ for $n \geq 2$
22. $a_1 = 1$, $a_n = (-1)^n a_{n-1} + 2$ for $n \geq 2$
23. $a_1 = 1$, $a_2 = 3$, $a_n = a_{n-1} - a_{n-2}$ for $n \geq 3$

24. $a_1 = 3$, $a_2 = 1$, $a_n = a_{n-2} + a_{n-1}$ for $n \geq 3$

25. $a_1 = 1$, $a_2 = 3$, $a_n = a_{n-2}$ for $n \geq 3$

26. $a_1 = 1$, $a_2 = 4$, $a_n = a_{n-1} - a_{n-2}$ for $n \geq 3$

Goal B

In exercises 27–36, the first term a_1 and the common difference d of an arithmetic sequence are given. Find the first five terms of the sequence.

27. $a_1 = 4$, $d = 3$
28. $a_1 = 3$, $d = 6$
29. $a_1 = 5$, $d = -2$
30. $a_1 = -5$, $d = 2$
31. $a_1 = -2$, $d = 5$
32. $a_1 = 2$, $d = -5$
33. $a_1 = \frac{1}{2}$, $d = \frac{1}{4}$
34. $a_1 = \frac{1}{3}$, $d = \frac{1}{6}$
35. $a_1 = \frac{1}{4}$, $d = \frac{1}{8}$
36. $a_1 = \frac{1}{6}$, $d = \frac{1}{3}$

In exercises 37–46, find the fifteenth and thirtieth terms of each arithmetic sequence.

37. 3, 7, 11, 15, ...
38. 8, 6, 4, 2, ...
39. 9, 7, 5, 3, ...
40. 1, 6, 11, 16, ...
41. -7, -1, 5, 11, ...
42. -3, -1, 1, 3, ...
43. -2, -6, -10, -14, ...
44. -9, -6, -3, 0, ...
45. $\frac{1}{2}$, $\frac{5}{6}$, $\frac{7}{6}$, $\frac{3}{2}$, ...
46. $-\frac{1}{4}$, $\frac{1}{8}$, $\frac{1}{2}$, $\frac{7}{8}$, ...

Goal C

For exercises 47–66, find the sum of the first six terms and the sum of the first twenty terms of each of the arithmetic sequences specified in exercises 27–46.

Superset

In exercises 67–72, find a formula for the general term of the recursively defined sequence.

67. $a_1 = 1$, $a_n = a_{n-1} + 2n - 1$ for $n \geq 2$
68. $a_1 = 1$, $a_n = 2a_{n-1} + 1$ for $n \geq 2$
69. $a_1 = 1$, $a_n = 2a_{n-1}$ for $n \geq 2$
70. $a_1 = 1$, $a_n = -a_{n-1}$ for $n \geq 2$
71. $a_1 = 1$, $a_n = a_1 + a_2 + \cdots + a_{n-1}$ for $n \geq 2$
72. $a_1 = 1$, $a_n = \frac{1}{2}a_{n-1} + 1$ for $n \geq 2$

In exercises 73–76, the sum S_n of the first n terms of an arithmetic sequence is given. Find the first, third, and tenth terms of the sequence.

73. $S_n = n(n + 2)$
74. $S_n = n(n - 1)$
75. $S_n = n(7 - n)$
76. $S_n = 8n$

77. You are playing blackjack in Las Vegas. You bet $10 on the first hand and increase your bet by $5 every time you win. Fortunately, you win every hand you play. If you began with $50 and won ten hands, how much money do you have?

78. (Refer to exercise 77) How many hands in a row must you win before you have $810?

79. A new drive-in movie theatre has 20 spaces for cars in the first row, 22 spaces in the second row, 24 spaces in the third row, and so on, for a total of 18 rows. What is the capacity (number of cars) of the theatre?

80. If $b_1, b_2, b_3, \ldots, b_n$ are numbers such that $a, b_1, \ldots, b_n, c$ are successive terms of an arithmetic sequence, then $b_1, \ldots, b_n$ are the n **arithmetic means** between a and c. Finding such numbers when given a, c and n, is known as inserting arithmetic means. Find 3 arithmetic means between 4 and 11.

12.2 Geometric Sequences and Sums

Goal A *to determine the terms of a geometric sequence*

$a_{n+1} = a_n + d$
An arithmetic sequence

We have already seen that in an arithmetic sequence, each term is d units more than the preceding term, where d is some constant. There is another type of sequence that can be characterized just as easily. In this second type of sequence, each term is r *times* the preceding term, where r is some nonzero constant. Such a sequence is called a **geometric sequence** and r is called the **common ratio**. Thus, for a geometric sequence,

$a_{n+1} = ra_n$
A geometric sequence

$$a_{n+1} = ra_n.$$

1, 2, 4, 8, 16, ...
3, $\frac{9}{2}$, $\frac{27}{4}$, $\frac{81}{8}$, ...
5, -10, 20, -40, ...

Examples of geometric sequences are shown at the left. In the first sequence, each term (after the first) is 2 times the preceding term; thus, the common ratio is 2. In the second sequence, the common ratio is $\frac{3}{2}$, and in the third, it is -2.

Example 1 Determine whether each sequence is geometric.

(a) 2, -6, 18, -54, ...   (b) 2, 4, 8, 24, 40, ...

Solution (a) $\dfrac{-6}{2} = \dfrac{18}{-6} = \dfrac{-54}{18} = -3$

Since the ratio of each term to the preceding term is constant, the sequence is geometric.

(b) $\dfrac{4}{2} = 2$, $\dfrac{8}{4} = 2$, but $\dfrac{24}{8} = 3$

Since there is no constant ratio, the sequence is not geometric.

$a_1 = a_1 r^0$
$a_2 = a_1 r^1$
$a_3 = a_1 r^2$
$a_4 = a_1 r^3$

Suppose a geometric sequence has first term a_1 and common ratio r. Then all the terms of the sequence can be expressed in terms of a_1 and r, as shown at the left. (We know that $a_3 = a_2 r$; thus,

$$a_3 = (a_1 r)r = a_1 r^2.$$

We find a_4 similarly.) Notice that the exponent of r is always one less than the subscript of the term. This suggests the following fact.

Fact
In a geometric sequence with first term a_1 and common ratio r, the nth term can be described as follows:

$$a_n = a_1 r^{n-1} \quad (n \geq 1).$$

Example 2 Find the seventh term and the general term of each geometric sequence.

(a) 3, 6, 12, 24, . . .   (b) 2, -10, 50, -250, . . .

Solution (a) $r = \dfrac{6}{3} = 2$
   - Since we know the series is geometric, we find r by forming the ratio of any term to its preceding term.

$a_n = a_1 r^{n-1}$
   - To find the seventh term, substitute 3 for a_1, 2 for r, and 7 for n.

$a_7 = (3)(2)^{7-1}$

$a_7 = 192$

$a_n = 3(2)^{n-1}$
   - To find an expression for the nth term, we substituted 3 for a_1, and 2 for r, and let n remain a variable.

(b) $r = \dfrac{-10}{2} = -5$

$a_n = a_1 r^{n-1}$
   - To find the seventh term, substitute 2 for a_1, -5 for r, and 7 for n.

$a_7 = (2)(-5)^{7-1}$

$a_7 = 31{,}250$

$a_n = 2(-5)^{n-1}$
   - To find an expression for the nth term, we substituted 2 for a_1, -5 for r, and let n remain a variable.

Goal B *to find the sum of the first n terms of a geometric sequence*

Let S_n represent the sum of the first n terms of a geometric sequence. We represent this fact algebraically as follows:

$$S_n = a_1 + a_1 r + a_1 r^2 + \cdots + a_1 r^{n-1}.$$

Let us multiply both sides of this equation by r, and then subtract the result from the original equation.

$$\begin{array}{rl} S_n =& a_1 + a_1 r + a_1 r^2 + \cdots + a_1 r^{n-1} \\ -(S_n \cdot r) =& \quad\; -(a_1 r + a_1 r^2 + \cdots + a_1 r^{n-1} + a_1 r^n) \\ \hline S_n - S_n r =& a_1 + 0 + 0 + \cdots + 0 \;\; - a_1 r^n \end{array}$$

Notice that the terms on the right side of the second equation have been moved one place to the right to make the difference easier to compute. We conclude that

$$S_n - S_n r = a_1 - a_1 r^n$$

$$S_n (1 - r) = a_1 (1 - r^n)$$

$$S_n = \dfrac{a_1 (1 - r^n)}{1 - r} \qquad (r \neq 1).$$

12 / Sequences, Series, and Limits

Of course, if $r = 1$, then the denominator in the previous formula is zero, and this equation is undefined. In this case every term of the sequence is equal to a_1, and the sum of the first n terms is simply na_1.

Fact

The sum of the first n terms of a geometric sequence is

$$S_n = \frac{a_1(1 - r^n)}{1 - r}$$

where a_1 is the first term and $r \neq 1$ is the common ratio.

Example 3 Determine the sum of the first n terms and the sum of the first 10 terms of the geometric sequence 5, 15, 45, 135, ...

Solution The common ratio is 3. Thus, we can use the formula above with $a_1 = 5$ and $r = 3$.

$$S_n = \frac{5(1 - 3^n)}{1 - 3} = \frac{5(1 - 3^n)}{-2}$$

Now, we substitute $n = 10$.

$$S_{10} = \frac{5(1 - 3^{10})}{-2} = 147{,}620$$

Exercise Set 12.2

Goal A

In exercises 1–6, determine whether each sequence is geometric.

1. 2, 4, 6, 8, . . .
2. $-4, -2, -1, \frac{1}{2}, \ldots$
3. $12, 18, 27, \frac{81}{2}, \ldots$
4. $2, -3, 8, -12, \ldots$
5. $-1, -2, -0.04, 0.008, \ldots$
6. $-16, 24, -36, 54, \ldots$

In exercises 7–22, find the sixth, ninth, and general terms of each geometric sequence.

7. 1, 2, 4, 8, . . .
8. 1, 3, 9, 27, . . .
9. 2, 4, 8, 16, . . .
10. $8, -4, 2, -1, \ldots$

11. $\frac{1}{8}, \frac{1}{2}, 2, 8, \ldots$

12. $27, 18, 12, 8, \ldots$

13. $-81, 27, -9, 3, \ldots$

14. $\frac{1}{8}, -\frac{1}{4}, \frac{1}{2}, -1, \ldots$

15. $\frac{1}{3}, 1, 3, 9, \ldots$

16. $\sqrt{2}, 2, \sqrt{8}, 4, \ldots$

17. $-16, 24, -36, 54, \ldots$

18. $\sqrt{2}, \sqrt{6}, 3\sqrt{2}, 3\sqrt{6}, \ldots$

19. $0.1, -0.01, 0.001, -0.0001, \ldots$

20. $1, \frac{1}{3}, \frac{1}{9}, \frac{1}{27}, \ldots$

21. $\frac{25}{8}, \frac{5}{4}, \frac{1}{2}, \frac{1}{5}, \ldots$

22. $0.1, 0.02, 0.004, 0.0008, \ldots$

Goal B

For exercises 23–38, determine the sum of the first n terms and the sum of the first seven terms of each of the geometric sequences in exercises 7–22.

Superset

39. Suppose that $a_1, a_2, a_3, a_4, \ldots$ is a geometric sequence. Show that $a_1, a_3, a_5, \ldots$ and $a_2, a_4, a_6, \ldots$ are also geometric sequences.

40. Show that if $2^{a_1}, 2^{a_2}, 2^{a_3}, 2^{a_4}, \ldots$ is a geometric sequence, then $a_1, a_2, a_3, a_4, \ldots$ is an arithmetic sequence.

41. A rubber ball is dropped from a height of 27 feet and always bounces to a height that is $\frac{1}{3}$ the height of the previous fall. What distance does the ball fall the second time? the third time? the fifth time?

42. Every time a child jumps into a wading pool, one-tenth of the water splashes out of the pool. After the child has jumped in the pool 4 times, what percentage of the original amount of water remains in the pool?

43. (Refer to exercise 41) What is the total distance (up and down) the ball has traveled after the second fall? the third fall? the fifth fall?

44. The annual depreciation of an office computer facility is 20% of its value at the start of the year. Determine the value of the computer after 5 years, if its original cost was $60,000.

45. You have deposited $1000 in an account that earns 8% annually. You leave the interest in the account so that in subsequent years, you earn interest on your interest. How much interest do you earn the first year? the second year? the first five years?

46. Determine a value of r, not equal to 1, so that the third term of a geometric sequence is the arithmetic mean of the first and fifth terms of the sequence.

47. The **geometric mean** of two positive numbers, x and y, is defined as $\sqrt{xy}$. Suppose a and b are positive numbers with $a < b$. Show that the geometric mean of a and b is less than the arithmetic mean of a and b.

48. Let BD be the altitude from the vertex B to the hypotenuse AC of the right triangle ABC. Show that BD is the geometric mean of AD and DC. (Hint: Refer to exercise 47.)

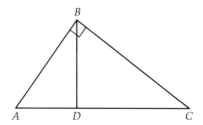

12.3 Series

Goal A *to use sigma notation to describe a sum*

We shall now consider an alternative way of writing the sum of the first n terms of a sequence, $a_1 + a_2 + \cdots + a_n$. The Greek letter *sigma*, written Σ, is used to express this sum in shorthand form:

$$\sum_{k=1}^{n} a_k$$

$$\sum_{k=1}^{3} a_k = a_1 + a_2 + a_3$$

$$\sum_{k=3}^{6} k^3 = 3^3 + 4^3 + 5^3 + 6^3$$

In the sigma notation above, the letter k is called an **index**. The statement "$k = 1$" below the sigma indicates that the sum starts with the first term of the sequence; the n above the sigma indicates that the sum ends with the nth term of the sequence. The index k increases by 1 from term to term and can be used as a subscript (as in the first sum at the left) or can be used to compute the terms of the sum (as in the second sum at the left).

Example 1 Rewrite each sum without using sigma notation; then compute each sum.

(a) $\sum_{k=1}^{4} 3k$ (b) $\sum_{k=3}^{5} (k^2 - 1)$

Solution (a) $\sum_{k=1}^{4} 3k = 3(1) + 3(2) + 3(3) + 3(4) = 3 + 6 + 9 + 12 = 30$

(b) $\sum_{k=3}^{5} (k^2 - 1) = (3^2 - 1) + (4^2 - 1) + (5^2 - 1) = 8 + 15 + 24 = 47$

Example 2 Use sigma notation to rewrite the following sums.

(a) $1 + 4 + 9 + 16 + 25$ (b) $7 + 9 + 11 + 13$

Solution (a) $1 + 4 + 9 + 16 + 25 = \sum_{k=1}^{5} k^2$

(b) $7 + 9 + 11 + 13 = \sum_{k=1}^{4} (2k + 5)$

Goal B *to compute partial sums of a series*

The expression
$$a_1 + a_2 + a_3 + \cdots + a_n + \cdots$$
is called an **infinite series.** In sigma notation this series is denoted
$$\sum_{k=1}^{\infty} a_k.$$

If a sequence is arithmetic, then its corresponding series is also called arithmetic; if a sequence is geometric, then its corresponding series is called geometric.

When does it make sense to talk about the "sum" of an infinite series? To help us answer this question, our strategy will be to look at the sum of the first term, then the sum of the first two terms, then the sum of the first three terms, etc., and determine whether there is a trend.

For the series
$$a_1 + a_2 + a_3 + \cdots + a_n + \cdots,$$
we find **partial sums** as follows:

first partial sum: $S_1 = a_1$
second partial sum: $S_2 = a_1 + a_2$
third partial sum: $S_3 = a_1 + a_2 + a_3$
.
.
.
nth partial sum: $S_n = a_1 + a_2 + a_3 + \cdots + a_n$

(Note that the subscript on S denotes the number of terms in the partial sum.)

For example, 1, 4, 9, 16, and 25 are the first five partial sums of the arithmetic series of odd integers, $1 + 3 + 5 + \cdots$:

$S_1 = 1$
$S_2 = 1 + 3 = 4$
$S_3 = 1 + 3 + 5 = 9$
$S_4 = 1 + 3 + 5 + 7 = 16$
$S_5 = 1 + 3 + 5 + 7 + 9 = 25$

As another example consider the geometric series
$$\frac{1}{2} + \frac{1}{4} + \frac{1}{8} + \frac{1}{16} + \frac{1}{32} + \frac{1}{64} + \cdots.$$

Its first five partial sums are $\frac{1}{2}, \frac{3}{4}, \frac{7}{8}, \frac{15}{16},$ and $\frac{31}{32}$.

Example 3   Compute the third and sixth partial sums of each series.

(a) $1 + 3 + 9 + 27 + \cdots + 3^{n-1} + \cdots$ (b) $\sum_{n=1}^{\infty} \left(-\frac{1}{2}\right)^{n-1}$

Solution   (a) $S_3 = 1 + 3 + 9 = 13$

$S_6 = \dfrac{1(1 - 3^6)}{1 - 3} = 364$ ■ Since the series is geometric, $S_n = \dfrac{a_1(1 - r^n)}{1 - r}$.

(b) $\sum_{n=1}^{\infty} \left(-\dfrac{1}{2}\right)^{n-1} = 1 - \dfrac{1}{2} + \dfrac{1}{4} - \dfrac{1}{8} + \cdots + \left(-\dfrac{1}{2}\right)^{n-1} + \cdots$

$S_3 = 1 - \dfrac{1}{2} + \dfrac{1}{4} = \dfrac{3}{4}$

$S_6 = \dfrac{1\left(1 - \left(-\frac{1}{2}\right)^6\right)}{1 + \frac{1}{2}} = \dfrac{21}{32}$ ■ Since the series is geometric, $S_n = \dfrac{a_1(1 - r^n)}{1 - r}$.

Goal C   *to compute the sum of a series, if the sum exists*

For the arithmetic series

$$1 + 3 + 5 + 7 + 9 + \cdots,$$

1, 4, 9, 16, 25, ...

$\dfrac{1}{2}, \dfrac{3}{4}, \dfrac{7}{8}, \dfrac{15}{16}, \dfrac{31}{32}, \ldots$

we know that the first five partial sums are 1, 4, 9, 16, and 25. We can consider these values to be the first five terms of a **sequence of partial sums**, as shown at the left. For the geometric series $\frac{1}{2} + \frac{1}{4} + \frac{1}{8} + \frac{1}{16} + \frac{1}{32} + \frac{1}{64} + \cdots$, we found the first five partial sums to be $\frac{1}{2}, \frac{3}{4}, \frac{7}{8}, \frac{15}{16}$, and $\frac{31}{32}$. The sequence of partial sums for this series is also shown at the left.

Notice that for the arithmetic series, the terms in the sequence of partial sums, 1, 4, 9, 16, 25, ..., are getting arbitrarily large (eventually larger than any number you can imagine). Thus, we say that this series "goes to infinity or has no sum." For the geometric series, the terms in the sequence of partial sums get arbitrarily close to 1 (eventually as close to 1 as you wish). In this case, we say that the sum of this series is 1, and write:

$$\dfrac{1}{2} + \dfrac{1}{4} + \dfrac{1}{8} + \dfrac{1}{16} + \cdots = 1 \quad \text{or} \quad \sum_{k=1}^{\infty} \dfrac{1}{2^k} = 1.$$

The sum of a series is determined by the behavior of the sequence of its partial sums.

Definition

The infinite series

$$a_1 + a_2 + a_3 + \cdots + a_n + \cdots$$

is said to have **sum** S, if the terms in the sequence of partial sums

$$S_1, S_2, S_3, \ldots, S_n, \ldots,$$

get arbitrarily close to S as the value of n increases.

We now consider geometric series. Whether or not a geometric series has a sum depends on the value of the common ratio r.

Fact

The sum of the geometric series $a_1 + a_1 r + a_1 r^2 + \cdots$ is

$$S = \frac{a_1}{1 - r}$$

provided $|r| < 1$. If $|r| \geq 1$, the series has no sum.

Example 4 Find the sum, if it exists, of each geometric series.

(a) $\dfrac{16}{125} + \dfrac{4}{25} + \dfrac{1}{5} + \dfrac{1}{4} + \cdots$ (b) $\displaystyle\sum_{k=2}^{\infty} 3(2)^{-k}$

Solution (a) The series has no sum since $r = \tfrac{5}{4}$, and $\tfrac{5}{4} \geq 1$.

(b) $\displaystyle\sum_{k=2}^{\infty} 3(2)^{-k} = \dfrac{3}{4} + \dfrac{3}{8} + \dfrac{3}{16} + \dfrac{3}{32} + \cdots$ ■ Begin by writing out the first few terms. Notice that $a_1 = \tfrac{3}{4}$ and $r = \tfrac{1}{2}$.

$$S = \frac{a_1}{1 - r} = \frac{\tfrac{3}{4}}{1 - \tfrac{1}{2}} = \frac{3}{2}$$

A repeating decimal represents a rational number. We can determine the fractional form of this rational number by viewing the decimal expansion as a geometric series. For example, the repeating decimal $0.333\ldots$ can be viewed as a geometric series with $a_1 = \tfrac{3}{10}$ and $r = \tfrac{1}{10}$, as shown at the left.

$$0.333\ldots = \frac{3}{10} + \frac{3}{10^2} + \frac{3}{10^3} + \cdots$$

Computing the sum S of this infinite series, we obtain

$$S = \frac{a_1}{1-r} = \frac{\frac{3}{10}}{1-\frac{1}{10}} = \frac{\frac{3}{10}}{\frac{9}{10}} = \frac{1}{3}.$$

Thus, $\frac{1}{3}$ is the fractional equivalent of $0.333\ldots$.

Example 5 Find the fraction represented by the decimal expansion for $x = 0.7\overline{21}$.

Solution

$x = 0.7\overline{21} = \dfrac{7}{10} + 0.0\overline{21}$ ■ Write x as the sum of a nonrepeating part and a repeating part.

$0.0\overline{21} = \dfrac{21}{10^3} + \dfrac{21}{10^5} + \dfrac{21}{10^7} + \cdots$ ■ Write the repeating part as a geometric series.

$a_1 = \dfrac{21}{10^3}, \quad r = \dfrac{1}{10^2}$

$x = \dfrac{7}{10} + \dfrac{a_1}{1-r} = \dfrac{7}{10} + \dfrac{\frac{21}{10^3}}{1-\frac{1}{10^2}} = \dfrac{7}{10} + \dfrac{7}{330} = \dfrac{119}{165}$

Another type of series whose sum we can compute is called the **telescoping series**. Consider the following:

$$\sum_{n=1}^{\infty}\left(\frac{1}{n} - \frac{1}{n+1}\right) = \left(1 - \frac{1}{2}\right) + \left(\frac{1}{2} - \frac{1}{3}\right) + \cdots + \left(\frac{1}{n} - \frac{1}{n+1}\right) + \cdots.$$

The fourth partial sum, S_4, for this series is computed at the left. Notice that almost every term in the sum cancels with some other term. Viewed this way, the partial sum S_n collapses, like a telescope, from

$$S_n = \left(1 - \frac{1}{2}\right) + \left(\frac{1}{2} - \frac{1}{3}\right) + \cdots + \left(\frac{1}{n-1} - \frac{1}{n}\right) + \left(\frac{1}{n} - \frac{1}{n+1}\right)$$

to

$$S_n = 1 - \frac{1}{n+1}.$$

As n increases, the value of $\frac{1}{n+1}$ gets closer and closer to 0, and thus S_n gets closer and closer to 1. We conclude that the sum of the original series is equal to 1.

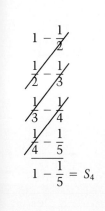

Example 6 Find the sum of the telescoping series

$$\left(\frac{1}{3} - \frac{1}{5}\right) + \left(\frac{1}{4} - \frac{1}{6}\right) + \cdots + \left(\frac{1}{n+2} - \frac{1}{n+4}\right) + \cdots.$$

Solution $S_n = \left(\frac{1}{3} - \frac{1}{5}\right) + \left(\frac{1}{4} - \frac{1}{6}\right) + \left(\frac{1}{5} - \frac{1}{7}\right) + \left(\frac{1}{6} - \frac{1}{8}\right) + \left(\frac{1}{7} - \frac{1}{9}\right) + \cdots$

$\qquad + \left(\frac{1}{n-1} - \frac{1}{n+1}\right) + \left(\frac{1}{n} - \frac{1}{n+2}\right) + \left(\frac{1}{n+1} - \frac{1}{n+3}\right) + \left(\frac{1}{n+2} - \frac{1}{n+4}\right)$

$S_n = \frac{1}{3} + \frac{1}{4} - \frac{1}{n+3} - \frac{1}{n+4}$ ■ Note $\frac{1}{n+3}$ and $\frac{1}{n+4}$ approach zero as n increases.

Thus, $S = \frac{1}{3} + \frac{1}{4} = \frac{7}{12}$.

$1 - \frac{1}{3} + \frac{1}{5} - \frac{1}{7} + \frac{1}{9} - \cdots = \frac{\pi}{4}$

$1 + \frac{1}{2} + \frac{1}{3} + \frac{1}{4} + \frac{1}{5} + \frac{1}{6} + \cdots$

There are many other types of series, but their sums (if they exist) are not always easily found. Some of the results are rather surprising. For example, the sum of the first series at the left is equal to $\frac{\pi}{4}$, whereas the second series (the *harmonic* series) has no sum.

Exercise Set 12.3   = Calculator Exercises in the Appendix

Goal A

In exercises 1–8, rewrite each sum without using sigma notation; then compute each sum.

1. $\sum_{n=1}^{5} (3n - 2)$
2. $\sum_{n=1}^{6} (9 - 2n)$
3. $\sum_{n=2}^{4} (17 - n^2)$
4. $\sum_{n=3}^{5} (7n - n^2)$
5. $\sum_{n=1}^{4} 3$
6. $\sum_{n=2}^{5} (6 - n)$
7. $\sum_{n=1}^{5} (-1)^n (n - 1)$
8. $\sum_{n=3}^{6} 2^n(n)$

In exercises 9–16, use sigma notation to rewrite each sum.

9. $3 + 4 + 5 + 6$
10. $7 + 8 + 9 + 10$
11. $1 + 4 + 9 + 16 + 25$
12. $1 + 8 + 27 + 64$
13. $2 + 4 + 6 + 8 + 10 + 12$
14. $4 + 8 + 12 + 16 + 20$
15. $-1 + 1 + (-1) + 1$
16. $5 + 5 + 5 + 5$

Goal B

In exercises 17–26, compute the third and sixth partial sums of each series.

17. $1 + 4 + 9 + 16 + \cdots$
18. $1 - 8 + 27 - 64 + \cdots$

19. $1 + 4 + 16 + 64 + \cdots$

20. $1 + 6 + 36 + 216 + \cdots$

21. $1 - \frac{1}{3} + \frac{1}{9} - \frac{1}{27} + \cdots$

22. $1 + \frac{2}{5} + \frac{4}{25} + \frac{8}{125} + \cdots$

23. $\sum_{n=1}^{\infty} 5\left(\frac{2}{3}\right)^{n-1}$

24. $\sum_{n=1}^{\infty} 7\left(\frac{3}{5}\right)^{n-1}$

25. $\sum_{n=1}^{\infty} (2^n - n)$

26. $\sum_{n=1}^{\infty} (n^2 - 8)$

Goal C

In exercises 27–38, find the sum, if it exists, of each series.

27. $25 - 10 + 4 - \frac{8}{5} + \cdots$

28. $64 + 48 + 36 + 27 + \cdots$

29. $27 + 36 + 48 + 64 + \cdots$

30. $1 - \frac{3}{2} + \frac{9}{4} - \frac{27}{8} + \cdots$

31. $\sum_{n=1}^{\infty} 4\left(-\frac{2}{5}\right)^{n-1}$

32. $\sum_{n=1}^{\infty} 5\left(-\frac{6}{5}\right)^{n-1}$

33. $\sum_{n=1}^{\infty} 6\left(\frac{4}{3}\right)^{n-1}$

34. $\sum_{n=1}^{\infty} 7\left(\frac{1}{3}\right)^{n-1}$

35. $\left(\frac{1}{2} - \frac{1}{5}\right) + \left(\frac{1}{3} - \frac{1}{6}\right) + \left(\frac{1}{4} - \frac{1}{7}\right) + \left(\frac{1}{5} - \frac{1}{8}\right) + \cdots$

36. $\left(\frac{1}{4} - \frac{1}{8}\right) + \left(\frac{1}{5} - \frac{1}{9}\right) + \left(\frac{1}{6} - \frac{1}{10}\right) + \left(\frac{1}{7} - \frac{1}{11}\right) + \cdots$

37. $\sum_{n=1}^{\infty} \left(\frac{1}{n+2} - \frac{1}{n}\right)$

38. $\sum_{n=1}^{\infty} \left(\frac{1}{n+3} - \frac{1}{n}\right)$

In exercises 39–46, find the fraction represented by the decimal expansion of each of the following.

39. $0.215\overline{1}$

40. $0.568\overline{8}$

41. $0.437\overline{171}$

42. $0.236\overline{262}$

43. $0.132\overline{132}$

44. $0.395\overline{395}$

45. $0.0032\overline{32}$

46. $0.09\overline{09}$

Superset

47. Find a geometric series whose first term is 4 and whose sum is 3.

48. Find a geometric series whose second term is -2 and whose sum is 1.

In exercises 49–54 find the first four terms and the general term of the series whose nth partial sum is given.

49. $S_n = 2^n - 1$

50. $S_n = n^2 + 2n$

51. $S_n = 2n^2 + n$

52. $S_n = \frac{1}{2}(3^n - 1)$

53. $S_n = n^2 - 4n$

54. $S_n = 2n^2 - n$

In exercises 55–60, use sigma notation to rewrite each series.

55. $4 + 7 + 10 + 13 + 16 + \cdots$

56. $3 + 5 + 7 + 9 + 11 + \cdots$

57. $2 + 5 + 10 + 17 + 26 + \cdots$

58. $3 + 5 + 9 + 17 + 33 + \cdots$

59. $2 + 8 + 26 + 80 + \cdots$

60. $3 + 10 + 29 + 66 + 127 + \cdots$

In exercises 61–62, find the sum of each series. (Hint: write each term of the series as the difference of two fractions.)

61. $\frac{1}{2 \cdot 3} + \frac{1}{3 \cdot 4} + \frac{1}{4 \cdot 5} + \frac{1}{5 \cdot 6} + \frac{1}{6 \cdot 7} + \cdots$

62. $\frac{1}{3 \cdot 5} + \frac{1}{4 \cdot 6} + \frac{1}{5 \cdot 7} + \frac{1}{6 \cdot 8} + \frac{1}{7 \cdot 9} + \cdots$

12.4 Limits

Goal A *to evaluate the limit of a function described by a graph*

When we plot a point, we focus on the value of *y* for a particular value of *x*. When we find the *limit* of a function, however, we look at the larger picture; i.e., we determine whether the *y*-values exhibit any trend as the *x*-coordinates increase or decrease through an entire set of values.

For example, in the four statements below we characterize the behavior of the *y*-coordinates of the graph of

$$y = \frac{x-1}{x-2}$$

(shown at the left) as the *x*-coordinates get arbitrarily large, arbitrarily small, and arbitrarily close to some fixed value.

A. As *x* takes on larger and larger values, the corresponding *y*-values get closer and closer to 1.
B. As *x* takes on smaller and smaller values, the corresponding *y*-values get closer and closer to 1.
C. As the *x*-values get closer and closer to 2 from the right, the corresponding *y*-values increase without limit.
D. As the *x*-values get closer and closer to 2 from the left, the corresponding *y*-values decrease without limit.

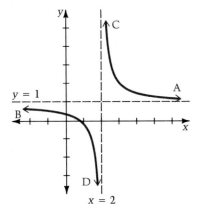

Figure 12.4

Now, using limit notation, we restate these four statements. Remember that since the function being discussed is

$$y = \frac{x-1}{x-2},$$

the symbols *y* and $\frac{x-1}{x-2}$ are interchangeable.

A. $\lim\limits_{x \to +\infty} \dfrac{x-1}{x-2} = 1$ ■ The limit of *x* − 1 divided by *x* − 2, as *x* "goes to" positive infinity.

B. $\lim\limits_{x \to -\infty} \dfrac{x-1}{x-2} = 1$ ■ The limit of *x* − 1 divided by *x* − 2, as *x* "goes to" negative infinity.

C. $\lim\limits_{x \to 2^+} \dfrac{x-1}{x-2} = +\infty$ ■ The limit of *x* − 1 divided by *x* − 2, as *x* approaches 2 from the right.

D. $\lim\limits_{x \to 2^-} \dfrac{x-1}{x-2} = -\infty$ ■ The limit of *x* − 1 divided by *x* − 2, as *x* approaches 2 from the left.

The "+" in $x \to 2^+$ and the "−" in $x \to 2^-$ indicate whether the *x*-values approach from the right or the left. We say that the first two limits are equal to 1 and the last two do not exist.

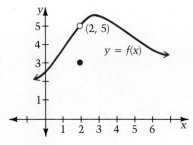

Figure 12.5

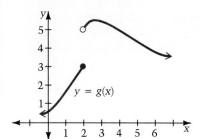

Figure 12.6

In Figure 12.5, $f(2) = 3$. But as the x-values approach 2 from both the right and left, the corresponding y-values of the graph approach 5. When determining limits, we are concerned with what happens to $f(x)$ as x approaches a certain value, and are not concerned with what happens at the x-value. We write:

$$\lim_{x \to 2^+} f(x) = 5 \quad \text{and} \quad \lim_{x \to 2^-} f(x) = 5.$$

Since the limit from the right and limit from the left are equal, we can drop the "$+$" and "$-$" and simply write

$$\lim_{x \to 2} f(x) = 5.$$

In Figure 12.6, the graph of $y = g(x)$ jumps at $x = 2$. Here, the limit from the right and limit from the left are not equal. We write:

$$\lim_{x \to 2^+} g(x) = 5 \quad \text{and} \quad \lim_{x \to 2^-} g(x) = 3.$$

Since these two limits are not equal, we say that the limit of $g(x)$, as x approaches 2, does not exist.

Example 1 For the function sketched below, evaluate each limit.

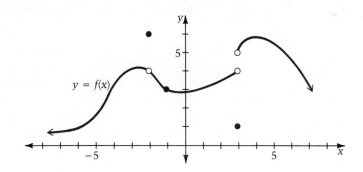

(a) $\lim_{x \to -2} f(x)$ (b) $\lim_{x \to -1^+} f(x)$ (c) $\lim_{x \to 3^+} f(x)$ (d) $\lim_{x \to 3^-} f(x)$ (e) $\lim_{x \to 3} f(x)$

Solution (a) Since $\lim_{x \to -2^+} f(x) = \lim_{x \to -2^-} f(x) = 4$, we write $\lim_{x \to -2} f(x) = 4$.
(b) 3
(c) 5
(d) 4
(e) Since $\lim_{x \to 3^+} f(x) \neq \lim_{x \to 3^-} f(x)$, we say $\lim_{x \to 3} f(x)$ does not exist.

Example 2 Evaluate $\lim_{x \to +\infty} f(x)$ and $\lim_{x \to -\infty} f(x)$ for the functions graphed below.

(a) 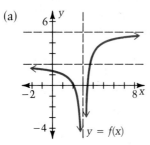 (b) 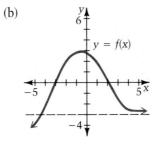 (c)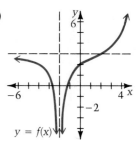

Solution
(a) $\lim_{x \to +\infty} f(x) = 5$; $\lim_{x \to -\infty} f(x) = 2$.
(b) $\lim_{x \to +\infty} f(x) = -3$; $\lim_{x \to -\infty} f(x) = -\infty$; thus, the limit does not exist.
(c) $\lim_{x \to +\infty} f(x) = +\infty$; thus, the limit does not exist; $\lim_{x \to -\infty} f(x) = 3$.

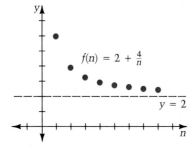

Figure 12.7

Careful! Remember that $+\infty$ and $-\infty$ are not numbers. When we say that a limit equals $+\infty$ or $-\infty$, we are simply specifying the reason the limit does not exist.

Goal B *to determine the limit of a sequence*

Since a sequence is also a function, we can discuss its limit. Figures 12.7 and 12.8 suggest that the limit of a sequence can be found by determining the limit of the corresponding function of x as $x \to +\infty$. The set of discrete dots in Figure 12.7 follows the same pattern as the curve in Figure 12.8. Since

$$\lim_{x \to +\infty} \left(2 + \frac{4}{x}\right) = 2,$$

we conclude that the limit of the sequence is 2, and we write

$$\lim_{n \to +\infty} \left(2 + \frac{4}{n}\right) = 2.$$

Thus, the sequence whose general term is $2 + \frac{4}{n}$, i.e.,

$$6, 4, 3\frac{1}{3}, 3, \ldots, 2 + \frac{4}{n}, \ldots,$$

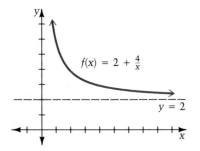

Figure 12.8

has a limit of 2.

Suppose we want to find the limit of a sequence whose general term is a rational expression. In such a case, we need not sketch the graph of the associated function. Instead, we can apply the Second Rule of Rational Functions (Section 4.3) to find the limit.

$a_n = \dfrac{3n^2 + 5}{2n^2 + 5n + 2}$

Suppose we are given the general term at the left. By the Second Rule of Rational Functions, we know that since the degrees of the numerator and denominator are equal, the line $y = \frac{3}{2}$ is a horizontal asymptote. Thus,

$$\lim_{x \to +\infty} \frac{3x^2 + 5}{2x^2 + 5x + 2} = \frac{3}{2}, \quad \text{and so} \quad \lim_{n \to +\infty} \frac{3n^2 + 5}{2n^2 + 5n + 2} = \frac{3}{2}.$$

Example 3   Evaluate the following limits

(a) $\lim\limits_{n \to +\infty} \left(\dfrac{2n - 1}{3n + 4}\right)^2$ (b) $\lim\limits_{n \to +\infty} \dfrac{2n^2}{4 - n^3}$ (c) $\lim\limits_{n \to +\infty} \dfrac{7n - n^2 + 2}{4n + 3}$

Solution   (a) $\lim\limits_{n \to +\infty} \left(\dfrac{2n - 1}{3n + 4}\right)^2 = \lim\limits_{n \to +\infty} \dfrac{4n^2 - 4n + 1}{9n^2 + 24n + 16} = \dfrac{4}{9}$ ■ Second Rule of Rational Functions (ii) is used.

(b) $\lim\limits_{n \to +\infty} \dfrac{2n^2}{4 - n^3} = 0$ ■ Second Rule of Rational Functions (i) is used.

(c) $\lim\limits_{n \to +\infty} \dfrac{7n - n^2 + 2}{4n + 3}$ does not exist   ■ Second Rule of Rational Functions (iii) is used.

Exercise Set 12.4   Calculator Exercises in the Appendix

Goal A

In exercises 1–4, evaluate the following limits.

(a) $\lim\limits_{x \to -1} f(x)$ (b) $\lim\limits_{x \to 2} f(x)$ (c) $\lim\limits_{x \to 3} f(x)$

1.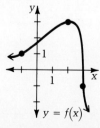

2.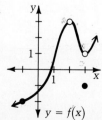

3.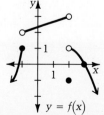

4.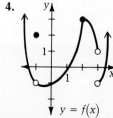

In exercises 5–8, evaluate the following:
(a) $\lim_{x \to -\infty} f(x)$ (b) $\lim_{x \to +\infty} f(x)$

5.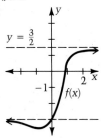

6.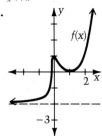

7.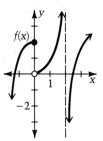

8.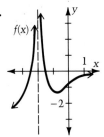

Goal B

In exercises 9–16, evaluate each limit.

9. $\lim_{n \to +\infty} \dfrac{3n}{2n - 1}$

10. $\lim_{n \to +\infty} 3^{-n}$

11. $\lim_{n \to +\infty} 2^{-n}$

12. $\lim_{n \to +\infty} \dfrac{n^3}{7n + 3}$

13. $\lim_{n \to +\infty} \dfrac{7n^2}{4 + 5n}$

14. $\lim_{n \to +\infty} \dfrac{7n^2}{n^2 + 4}$

15. $\lim_{n \to +\infty} \dfrac{8n^2}{n^4 + 1}$

16. $\lim_{n \to +\infty} \dfrac{5n^2}{2n^5 + 3}$

Superset

In exercises 17–24, express the sum of each series in the form

$$\lim_{n \to +\infty} S_n,$$

where S_n is the nth partial sum of the series.

17. $1 - \dfrac{1}{2} + \dfrac{1}{4} - \dfrac{1}{8} + \cdots$

18. $1 + \dfrac{1}{3} + \dfrac{1}{9} + \dfrac{1}{27} + \cdots$

19. $\sum_{n=1}^{\infty} 6\left(\dfrac{3}{4}\right)^{n-1}$

20. $\sum_{n=1}^{\infty} 7\left(-\dfrac{2}{5}\right)^{n-1}$

21. $\left(\dfrac{1}{2} - \dfrac{1}{3}\right) + \left(\dfrac{1}{3} - \dfrac{1}{4}\right) + \left(\dfrac{1}{4} - \dfrac{1}{5}\right) + \cdots$

22. $\left(1 - \dfrac{1}{3}\right) + \left(\dfrac{1}{2} - \dfrac{1}{4}\right) + \left(\dfrac{1}{3} - \dfrac{1}{5}\right) + \cdots$

23. $\sum_{n=1}^{\infty} \left(\dfrac{1}{n+4} - \dfrac{1}{n+1}\right)$

24. $\sum_{n=1}^{\infty} \left(\dfrac{1}{n+6} - \dfrac{1}{n+5}\right)$

25. Let $M(x)$ be the slope of the line that passes through the points $A(0, 0)$ and $B(x, x^2)$ on the curve $y = x^2$.

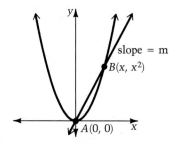

Notice that as $x \to 0$, point B slides along the curve towards point A. Compute $\lim_{x \to 0} M(x)$.

26. Repeat exercise 25 with the point $A(1, 1)$.

27. Repeat exercise 25 with the point $A(3, 9)$.

28. Repeat exercise 25 with the point $A(a, a^2)$, for any number a.

29. Repeat exercise 25 with the curve $y = x^3$.

30. Repeat exercise 29 with the point $A(a, a^3)$, for any number a.

12.5 Mathematical Induction

Goal A *to use mathematical induction to prove that a statement is true for all positive integers*

Imagine that Figure 12.9 represents an infinite row of dominoes. The dominoes are set up so that if any one of them falls to the right, it will necessarily strike the one next to it. Thus the next domino falls, which in turn causes yet another one to fall, and so forth, forever. Pick any domino, and eventually it will fall.

Of course, in order to start the row of dominoes falling, we must specify that the first domino falls, since it has no predecessor to strike it.

Now consider the following two statements:

1. We know that the first domino falls.
2. If we know that any domino falls, then we are sure that the one following it also falls.

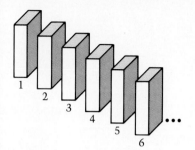

Figure 12.9

Taken together, these two statements assure us that *all* the dominoes will fall; i.e., any domino you pick, no matter how far out in the row, will eventually fall.

These two statements illustrate a method known as *mathematical induction*. The method is very useful for proving statements that we claim are true for every positive integer. For example, using the method of mathematical induction, we can prove that the following statement is true for all positive integers n:

$$1 + 2 + 3 + \cdots + n = \frac{n(n + 1)}{2}.$$

Principle of Mathematical Induction

A statement is true for all positive integers n if the following two requirements are met:

1. the statement is true for

$$n = 1$$

(the first domino falls);
2. if you assume the statement is true for $n = k$, then you can prove that the statement is true for

$$n = k + 1$$

(if any domino falls, the one next to it also falls).

Example 1 Prove the following by mathematical induction: for all positive integers n,

$$1 + 2 + 3 + \cdots + n = \frac{n(n+1)}{2}.$$

Solution

Requirement 1
When $n = 1$, the given statement becomes

$$1 = \frac{1(1+1)}{2},$$

which is true. Thus, Requirement 1 is met.

Requirement 2
Assumption: the given statement is true for $n = k$; that is, we assume

$$1 + 2 + 3 + \cdots + k = \frac{k(k+1)}{2}.$$

To be proved: the given statement with $n = k + 1$; that is, we must prove that

$$1 + 2 + 3 + \cdots + k + (k+1) = \frac{(k+1)(k+2)}{2}.$$

Proof:

$$1 + 2 + 3 + \cdots + k = \frac{k(k+1)}{2}$$ ■ Begin by stating the assumption.

$$1 + 2 + 3 + \cdots + k + (k+1) = \frac{k(k+1)}{2} + (k+1)$$ ■ Add the $(k+1)$st term of the sum to both sides of the equation and simplify.

$$= \frac{k^2 + k}{2} + \frac{2(k+1)}{2}$$

$$= \frac{k^2 + 3k + 2}{2}$$

$$1 + 2 + 3 + \cdots + k + (k+1) = \frac{(k+1)(k+2)}{2}$$

Thus, Requirement 2 is met. Therefore, by the Principle of Mathematical Induction, the given statement is true for all positive integers.

Note that Requirement 2 does not say that the statement is true for $n = k$. Rather, it says that *if* the statement is true for $n = k$, then we can conclude that it is also true for $n = k + 1$. Thus, Requirement 2 does not say that the kth domino has fallen. It does say that if we know that the kth domino has fallen, then we can conclude that the $(k + 1)$st domino has also fallen.

Example 2 Prove the following statement: for all positive integers n,

$$\frac{1}{1 \cdot 2} + \frac{1}{2 \cdot 3} + \cdots + \frac{1}{n(n + 1)} = \frac{n}{n + 1}.$$

Solution

Requirement 1
If $n = 1$, the given statement becomes

$$\frac{1}{1 \cdot 2} = \frac{1}{1 + 1},$$

which is true. Thus, Requirement 1 is met.

Requirement 2
Assumption ($n = k$): $\quad \dfrac{1}{1 \cdot 2} + \dfrac{1}{2 \cdot 3} + \cdots + \dfrac{1}{k(k + 1)} = \dfrac{k}{k + 1}.$

To be proved ($n = k + 1$): $\quad \dfrac{1}{1 \cdot 2} + \dfrac{1}{2 \cdot 3} + \cdots + \dfrac{1}{(k + 1)(k + 2)} = \dfrac{k + 1}{k + 2}$

Proof:

$$\frac{1}{1 \cdot 2} + \frac{1}{2 \cdot 3} + \cdots + \frac{1}{k(k + 1)} = \frac{k}{k + 1} \quad \blacksquare \text{ Begin with assumption. Then add the } (k + 1)\text{st term to both sides.}$$

$$\frac{1}{1 \cdot 2} + \cdots + \frac{1}{k(k + 1)} + \frac{1}{(k + 1)(k + 2)} = \frac{k}{k + 1} + \frac{1}{(k + 1)(k + 2)}$$

$$= \frac{k}{k + 1} \cdot \frac{k + 2}{k + 2} + \frac{1}{(k + 1)(k + 2)}$$

$$= \frac{k^2 + 2k + 1}{(k + 1)(k + 2)}$$

$$= \frac{\cancel{(k + 1)}(k + 1)}{\cancel{(k + 1)}(k + 2)}$$

$$= \frac{k + 1}{k + 2}$$

Thus, Requirement 2 is met. Therefore, by the Principle of Mathematical Induction, the given statement is true for all positive integers.

Exercise Set 12.5

Goal A

In exercises 1–16, prove each statement by mathematical induction. (Assume $n \geq 1$, unless stated otherwise.)

1. $1 + 3 + 5 + 7 + \cdots + (2n - 1) = n^2$
2. $2 + 4 + 6 + 8 + \cdots + 2n = n(n + 1)$
3. $2 + 5 + 8 + 11 + \cdots + (3n - 1) = \dfrac{n(3n + 1)}{2}$
4. $1 + 4 + 7 + 10 + \cdots + (3n - 2) = \dfrac{n(3n - 1)}{2}$
5. $1 + 4 + 9 + \cdots + n^2 = \dfrac{n(n + 1)(2n + 1)}{6}$
6. $1 + 8 + 27 + 64 + \cdots + n^3 = \dfrac{n^2(n + 1)^2}{4}$
7. $1 + 5 + 9 + \cdots + (4n - 3) = n(2n - 1)$
8. $7 + 10 + 13 + \cdots + (4 + 3n) = \dfrac{n}{2}(11 + 3n)$
9. $\dfrac{1}{2} + \dfrac{1}{4} + \dfrac{1}{8} + \cdots + \dfrac{1}{2^n} = 1 - \dfrac{1}{2^n}$
10. $\dfrac{1}{2} - \dfrac{1}{4} - \dfrac{1}{8} - \cdots - \dfrac{1}{2^n} = \dfrac{1}{2^n}$
11. $\dfrac{1}{3} + \dfrac{1}{9} + \dfrac{1}{27} + \cdots + \dfrac{1}{3^n} = \dfrac{1}{2}\left(1 - \dfrac{1}{3^n}\right)$
12. $\dfrac{1}{2} - \dfrac{1}{4} + \dfrac{1}{8} - \dfrac{1}{16} + \cdots + (-1)^{n-1}\dfrac{1}{2^n} = \dfrac{1}{3}\left(1 + (-1)^{n-1}\dfrac{1}{2^n}\right)$
13. $\dfrac{1}{1 \cdot 3} + \dfrac{1}{3 \cdot 5} + \dfrac{1}{5 \cdot 7} + \cdots + \dfrac{1}{(2n - 1)(2n + 1)} = \dfrac{n}{2n + 1}$
14. $\dfrac{1}{1 \cdot 4} + \dfrac{1}{4 \cdot 7} + \cdots + \dfrac{1}{(3n - 2)(3n + 1)} = \dfrac{n}{3n + 1}$
15. $\dfrac{1}{3} + \dfrac{1}{15} + \dfrac{1}{35} + \cdots + \dfrac{1}{4n^2 - 1} = \dfrac{n}{2n + 1}$
16. $\dfrac{1}{1 \cdot 2 \cdot 3} + \dfrac{1}{2 \cdot 3 \cdot 4} + \cdots + \dfrac{1}{n(n + 1)(n + 2)} = \dfrac{n(n + 3)}{4(n + 1)(n + 2)}$

Superset

In exercises 17–30, prove each statement by mathematical induction. (Assume $m, n \geq 1$ unless stated otherwise.)

17. $x^m x^n = x^{m+n}$
18. $(ab)^n = a^n b^n$
19. $2n \leq 2^n$
20. $1 + 2n \leq 3^n$
21. $\dfrac{1}{\sqrt{1}} + \dfrac{1}{\sqrt{2}} + \dfrac{1}{\sqrt{3}} + \cdots + \dfrac{1}{\sqrt{n}} > \sqrt{n}$
22. $a(b_1 + \cdots + b_n) = ab_1 + \cdots + ab_n, \; n \geq 2$
23. $a + (a + d) + (a + 2d) + \cdots + [a + (n - 1)d] = \dfrac{n}{2}[2a + (n - 1)d]$
24. $a + ar + ar^2 + \cdots + ar^{n-1} = \dfrac{a(1 - r^n)}{1 - r}$
25. If $x > 1$, then $x^n > 1$.
26. If $0 < x < 1$, then $0 < x^n < 1$.
27. $n^2 + n$ is a multiple of 2.
28. $n^3 + 2n$ is a multiple of 3.
29. 4 is a factor of $5^n - 1$.
30. 5 is a factor of $6^n - 1$.

12.6 The Binomial Theorem

Goal A   *to use the Binomial Theorem to expand expressions of the form* $(a + b)^n$

Suppose you wished to expand

$$(9x^2 + 6y)^{17},$$

that is, to write it as a sum of terms in x and y. It is the purpose of this section to provide you with a streamlined "formula" for determining such expansions. This formula is contained in a theorem, called the Binomial Theorem, which we will prove at the end of this section.

To motivate the Binomial Theorem, let us write the expansions of the expression $(a + b)^n$ for values of $n = 1, 2, 3, 4,$ and 5.

$n = 1$ $(a + b)^1 = a + b$
$n = 2$ $(a + b)^2 = a^2 + 2ab + b^2$
$n = 3$ $(a + b)^3 = a^3 + 3a^2b + 3ab^2 + b^3$
$n = 4$ $(a + b)^4 = a^4 + 4a^3b + 6a^2b^2 + 4ab^3 + b^4$
$n = 5$ $(a + b)^5 = a^5 + 5a^4b + 10a^3b^2 + 10a^2b^3 + 5ab^4 + b^5$

For these five values of n, we observe some patterns in the expansions:

- There are a total of $(n + 1)$ terms in the expansion of $(a + b)^n$.
- The first term is always a^n and the last term is always b^n.
- As you read each expansion from left to right, the exponent of a starts with n and decreases by one in each succeeding term; the exponent of b starts with 0 and increases by one in each succeeding term.
- In each term, the sum of the exponents of a and b is always n.
- In each expansion, the coefficient of the first term, a^n, is 1. Thereafter, the coefficient of any term is: the coefficient of the preceding term times the exponent of a in the preceding term, divided by the number of the preceding term.

Let us illustrate this last observation. In the expansion of $(a + b)^4$, the coefficient of the third term $(6a^2b^2)$ is 6. It is computed as follows:

$$\frac{\text{(coefficient of the second term)} \times \text{(exponent of } a \text{ in the second term)}}{\text{number of the preceding term}}$$

$$= \frac{4 \times 3}{2} = 6$$

Example 1 Use the observations on the previous page to guess the expansion for $(a + b)^7$.

Solution

Term:	first	second	third	fourth	fifth	sixth	seventh	eight
	$1a^7$	$7a^6b$	$21a^5b^2$	$35a^4b^3$	$35a^3b^4$	$21a^2b^5$	$7ab^6$	$1b^7$
		$\frac{1 \cdot 7}{1}$	$\frac{7 \cdot 6}{2}$	$\frac{21 \cdot 5}{3}$	$\frac{35 \cdot 4}{4}$	$\frac{35 \cdot 3}{5}$	$\frac{21 \cdot 2}{6}$	$\frac{7 \cdot 1}{7}$

Note that in this case $n = 7$, thus the first term is a^7 and the last term is b^7. For terms after the first, the power on a drops by one from term to term, and the power on b increases by 1; the sum of the two powers is always 7. The coefficient of each term after the first is the product of the coefficient and the power of a in the preceding term, divided by the number of the coefficient and the power of a in the preceding term, divided by the number of the preceding term. Thus we guess that

$$(a + b)^7 = a^7 + 7a^6b + 21a^5b^2 + 35a^4b^3 + 35a^3b^4 + 21a^2b^5 + 7ab^6 + b^7.$$

You should verify by (tedious) multiplication that this guess is, in fact, true.

There is a more common way of describing coefficients in expansions of $(a + b)^n$. This description involves *factorials*. For any positive integer n, we define **n factorial**, writen **n!**, as the product of the first n positive integers:

$$n! = n \cdot (n - 1) \cdot (n - 2) \cdot \cdots \cdot 3 \cdot 2 \cdot 1.$$

For example, $6! = 6 \cdot 5 \cdot 4 \cdot 3 \cdot 2 \cdot 1 = 720$, $(7 - 4)! = 3! = 3 \cdot 2 \cdot 1 = 6$, and $1! = 1$. In addition, we define **zero factorial** to be equal to 1: $0! = 1$.

Definition

If n is any positive integer, and if k is any positive integer such that $0 \leq k \leq n$, the **binomial coefficient** $\binom{n}{k}$ is $\binom{n}{k} = \frac{n!}{(n-k)!k!}$.

Example 2 Evaluate the following: $\binom{3}{0}$, $\binom{3}{1}$, $\binom{3}{2}$, and $\binom{3}{3}$.

Solution

$$\binom{3}{0} = \frac{3!}{(3-0)!0!} = \frac{3!}{3!0!} = \frac{1}{1} = 1 \qquad \binom{3}{1} = \frac{3!}{(3-1)!1!} = \frac{3!}{2!1!} = \frac{3 \cdot 2 \cdot 1}{(2 \cdot 1)1} = 3$$

$$\binom{3}{2} = \frac{3!}{(3-2)!2!} = \frac{3!}{1!2!} = \frac{3 \cdot 2 \cdot 1}{(1) \cdot 2 \cdot 1} = 3 \qquad \binom{3}{3} = \frac{3!}{(3-3)!3!} = \frac{3!}{0!3!} = \frac{1}{1} = 1$$

Notice that in Example 2, what we have determined are the binomial coefficients in the expansion of $(a + b)^3$: $1a^3 + 3a^2b + 3ab^2 + 1b^3$. You should verify that the coefficients for the expansion of $(a + b)^7$, which we determined in Example 1, can be expressed in terms of our definition of binomial coefficients as follows:

$$(a + b)^7 = \binom{7}{0}a^7 + \binom{7}{1}ab^6 + \binom{7}{2}a^5b^2 + \binom{7}{3}a^4b^3$$
$$+ \binom{7}{4}a^3b^4 + \binom{7}{5}a^2b^5 + \binom{7}{6}ab^6 + \binom{7}{7}b^7$$

Our earlier guesswork, together with the definition of binomial coefficients, leads us to the following theorem. A proof of this theorem, using mathematical induction, appears at the end of this section.

Binomial Theorem

For all complex numbers a and b, and for all positive integers n, then

$$(a + b)^n = \binom{n}{0}a^nb^0 + \binom{n}{1}a^{n-1}b^1 + \binom{n}{2}a^{n-2}b^2$$
$$+ \cdots + \binom{n}{n-1}a^1b^{n-1} + \binom{n}{n}a^0b^n.$$

Using sigma notation, we write this as $\sum_{k=0}^{n} \binom{n}{k}a^{n-k}b^k$.

Example 3 Use the Binomial Theorem to expand (a) $(2 + t)^5$ (b) $(2x - 4y^2)^3$.

Solution (a) In the Binomial Theorem take $a = 2$, $b = t$, and $n = 5$.

$$(2 + t)^5 = \binom{5}{0}2^5t^0 + \binom{5}{1}2^4t^1 + \binom{5}{2}2^3t^2 + \binom{5}{3}2^2t^3 + \binom{5}{4}2^1t^4 + \binom{5}{5}2^0t^5$$
$$= (1)(32) \cdot 1 + (5)(16)t + (10)(8)t^2 + (10)(4)t^3 + (5)(2)t^4 + (1)(1)t^5$$
$$= 32 + 80t + 80t^2 + 40t^3 + 10t^4 + t^5$$

(b) In the Binomial Theorem take $a = 2x$, $b = -4y^2$, and $n = 3$.

$$(2x - 4y^2)^3 = \binom{3}{0}(2x)^3(-4y^2)^0 + \binom{3}{1}(2x)^2(-4y^2)^1 + \binom{3}{2}(2x)^1(-4y^2)^2 + \binom{3}{3}(2x)^0(-4y^2)^3$$
$$= (1)(8x^3) \cdot 1 + (3)(4x^2)(-4y^2) + (3)(2x)(16y^4) + (1)(1)(-64y^6)$$
$$= 8x^3 - 48x^2y^2 + 96xy^4 - 64y^6$$

Note that mth term in the expansion of $(a + b)^n$ is given by

$$\binom{n}{m-1} a^{n-(m-1)} b^{m-1}.$$

For example, the fifth term in the expansion of $(a + b)^7$ is $\binom{7}{4}a^3b^4$. As for any term in a binomial expansion, the bottom number in the binomial coefficient (here 4), is the power on b.

Example 4   Determine the fourth term in the expansion of $\left(\sqrt{x} - \dfrac{y}{2}\right)^8$. Assume $x > 0$.

Solution   The fourth term in the expansion of $(a + b)^8$ is $\binom{8}{3}a^5b^3$. Here $a = \sqrt{x}$ and $b = -\dfrac{y}{2}$. Since

$$\binom{8}{3}(\sqrt{x})^5\left(-\dfrac{y}{2}\right)^3 = \dfrac{8!}{5!3!}(x^{5/2})\left(-\dfrac{y}{2}\right) = 56(x^{5/2})\left(-\dfrac{1}{8}y^3\right) = -7x^{5/2}y^3,$$

the desired term is $-7x^{5/2}y^3$.

Sometimes the binomial coefficients are displayed in a triangular array, known as **Pascal's Triangle**.

```
           1    1                              (1 0)  (1 1)
        1    1    1                        (2 0)  (2 1)  (2 2)
     1    3    3    1                  (3 0)  (3 1)  (3 2)  (3 3)
  1    4    6    4    1            (4 0)  (4 1)  (4 2)  (4 3)  (4 4)
1    5   10   10    5    1     (5 0)  (5 1)  (5 2)  (5 3)  (5 4)  (5 5)
```

Notice that each entry other than the 1's is the sum of the two entries to the right and left of it in the row above. For instance $\binom{4}{1} + \binom{4}{2} = \binom{5}{2}$. In general, $\binom{n}{k-1} + \binom{n}{k} = \binom{n+1}{k}$, a result that we prove next.

Example 5 For $1 \leq k \leq n$, verify that $\binom{n}{k-1} + \binom{n}{k} = \binom{n+1}{k}$.

Solution
$$\binom{n}{k-1} + \binom{n}{k} = \frac{n!}{(n-k+1)!(k-1)!} + \frac{n!}{(n-k)!k!}$$
$$= \frac{k \cdot n!}{(n-k+1)!k!} + \frac{(n-k+1)n!}{(n-k+1)!k!}$$
$$= \frac{n!(k+n-k+1)}{(n-k+1)!k!} = \frac{n!(n+1)}{(n+1-k)!k!}$$
$$= \frac{(n+1)!}{(n+1-k)!k!} = \binom{n+1}{k}$$

We are now in a position to prove the Binomial Theorem:

$$(a+b)^n = \binom{n}{0}a^n b^0 + \binom{n}{1}a^{n-1}b^1 + \cdots + \binom{n}{n-1}a^1 b^{n-1} + \binom{n}{n}a^0 b^n.$$

Proof: We use mathematical induction, and thus must show that two requirements are met.

Requirement 1: Show that the statement in the theorem is true for $n = 1$. If $n = 1$, the statement becomes $(a+b)^1 = \binom{1}{0}a^1 b^0 + \binom{1}{1}a^0 b^1$, which is true since the expression on the left simplifies to yield $a + b$.

Requirement 2: Show that if we assume the statement to be true when $n = N$, then we can conclude it is also true when $n = N + 1$. We use our assumption to write

$$(a+b)^{N+1} = (a+b)(a+b)^N$$
$$= (a+b)\binom{N}{0}a^N b^0 + \binom{N}{1}a^{N-1}b^1$$
$$+ \cdots + \binom{N}{k-1}a^{N-k+1}b^{k-1} + \binom{N}{k}a^{N-k}b^k + \binom{N}{N}a^0 b^N.$$

Next we carry out the multiplication and collect like terms to write

$$(a+b)^{N+1} = \binom{N}{0}a^{N+1}b^0 + \binom{N}{0}a^N b^1 + \binom{N}{1}a^N b^1 + \binom{N}{1}a^{N-1}b^2$$
$$+ \cdots + \binom{N}{k-1}a^{N-k+1}b^k + \binom{N}{k}a^{N-k+1}b^k + \cdots + \binom{N}{N}a^0 b^{N+1}$$
$$= \binom{N}{0}a^{N+1}b^0 + \cdots + \left[\binom{N}{k-1} + \binom{N}{k}\right]a^{N-k+1}b^k + \cdots + \binom{N}{N}a^0 b^{N+1}.$$

By virtue of our work in Example 5, we can rewrite this last statement as

$$(a + b)^{N+1} = \binom{N}{0}a^{N+1}b^0 + \cdots + \binom{N+1}{k}a^{N+1-k}b^k + \cdots + \binom{N}{N}a^0 b^{N+1}.$$

We can rewrite the first and last coefficients since

$$\binom{N}{0} = \binom{N+1}{0} \quad \text{and} \quad \binom{N}{N} = \binom{N+1}{N+1}.$$

Thus,

$$(a + b)^{N+1} = \binom{N+1}{0}a^{N+1}b^0 + \cdots$$
$$+ \binom{N+1}{k}a^{N+1-k}b^k + \cdots + \binom{N+1}{N+1}a^0 b^{N+1},$$

and Requirement 2 is met, thus completing the proof.

Exercise Set 12.6 ▭ Calculator Exercises in the Appendix

Goal A

In exercises 1–16, use the Binomial Theorem to expand the given expression.

1. $(x + y)^6$
2. $(x + y)^8$
3. $(a - b)^7$
4. $(a - b)^6$
5. $(3 + 2p)^4$
6. $(2 + 5s)^3$
7. $(2 + 3p)^4$
8. $(5 + 2s)^3$
9. $\left(\dfrac{t}{3} - 1\right)^3$
10. $\left(\dfrac{a}{2} - 1\right)^4$
11. $\left(z + \dfrac{1}{z}\right)^5$
12. $\left(p - \dfrac{1}{p}\right)^4$
13. $(1 - \sqrt{x})^6$
14. $(1 + \sqrt{t})^5$
15. $(s^{1/3} + s^{2/3})^3$
16. $(p^{1/4} + p^{3/4})^4$

In exercises 17–22, use the Binomial Theorem to write the first four terms in the expansion of the given expression.

17. $(1 + x)^{12}$
18. $(1 - y)^{10}$
19. $(2 - s)^8$
20. $(3 + p)^9$
21. $(1 + 2p)^{11}$
22. $(1 + 3x)^{12}$

In exercises 23–30, find the indicated term in the expansion of the given expression.

23. fourth term of $(1 + t)^{20}$
24. fifth term of $(1 - s)^{18}$
25. fifth term of $\left(x - \dfrac{1}{x}\right)^8$
26. third term of $(2 + y)^{10}$
27. tenth term of $(y + \sqrt{y})^{11}$
28. seventh term of $\left(2t - \dfrac{1}{2t}\right)^{12}$
29. ninth term of $\left(\dfrac{p}{3} + 3p\right)^{16}$
30. fourteenth term of $(2y + 1)^{14}$

Superset

31. If P dollars are invested at an interest rate i (expressed as a decimal), and the interest is compounded n times per year, then the amount, A, of money in the account after t years is

$$A = P\left(1 + \dfrac{i}{n}\right)^{nt}.$$

The *effective rate of interest* (expressed as a decimal) is

$$\left(1 + \dfrac{i}{n}\right)^n - 1$$

Use the first four terms in the binomial expansion of $(1 + \dfrac{i}{n})^n$ to obtain an approximate expression for the effective rate of interest.

32. Use the first five terms of a binomial expansion to estimate the value after one year of $1000, invested in a savings account that earns interest at the rate of 6% compounded daily. (Hint: in exercise 31, take $n = 360$.)

33. Show that $\binom{n}{0} + \binom{n}{1} + \binom{n}{2} + \cdots + \binom{n}{n+1} + \binom{n}{n}$ $= 2^n$ for $n \geq 1$. (Hint: use the Binomial Theorem to expand $(1 + x)^n$ and evaluate the two expressions when $x = 1$.)

34. Show that $\binom{n}{0} + \binom{n}{1} + \binom{n}{2} + \binom{n}{3} + \cdots + (-1)^n \binom{n}{n} = 0$ for $n \geq 1$. (Hint: see exercise 33.)

35. Evaluate the quantity $\binom{n}{0} + 2\binom{n}{1} + 4\binom{n}{2} + \cdots + 2^n \binom{n}{n}$ for $n \geq 1$. (Hint: see exercise 33.)

36. Express the sum $\binom{n}{n} + \binom{n+1}{n} + \binom{n+2}{n} + \cdots + \binom{n+k}{n}$ as a single binomial coefficient. (Hint: observe the position of the addends in Pascal's Triangle.)

37. Suppose p is a prime. Show that p divides each of the binomial coefficients $\binom{p}{1}, \binom{p}{2}, \binom{p}{3}, \binom{p}{4}, \ldots, \binom{p}{p-1}$. Show that a comparable statement is not true if p is not prime.

Chapter Review & Test

Chapter Review

12.1 Arithmetic Sequences and Sums (pp. 490–496)

A *sequence* is a function whose domain is the set of positive integers. (p. 490)   The function values are called the *terms of the sequence.* We refer to a_n as the *n*th term, or the general term of the sequence. (p. 491)

A sequence is defined *recursively* if we specify its first term, along with a rule for finding any other term by using the preceding term(s). (p. 491)

Successive terms of an *arithmetic sequence* differ by a constant called the *common difference.* (p. 492)   In an arithmetic sequence with first term a_1 and common difference d, the general term is given by the equation $a_n = a_1 + (n-1)d$. (p. 493)   The sum of the first n terms of such a sequence is given by the equation

$$S_n = \frac{n}{2}[2a_1 + (n-1)d] = \frac{n}{2}(a_1 + a_n).$$

12.2 Geometric Sequences and Sums (pp. 497–500)

In a *geometric sequence,* each term is r times the preceding term, where r is some nonzero constant called the *common ratio.* In a geometric sequence with first term a_1 and common ratio r, the general term is given by the equation $a_n = a_1 r^{n-1}$. (p. 497)   The sum of the first n terms of such a sequence is given by the equation

$$S_n = \frac{a_1(1-r^n)}{1-r}.$$

12.3 Series (pp. 501–507)

The expression $a_1 + a_2 + \cdots + a_n + \cdots$ is called an *infinite series.* The *n*th *partial sum* of a series is the sum of the first n terms of the series. (p. 502)   The series $a_1 + a_2 + \cdots + a_n + \cdots$ has sum S, if the terms in the sequence of partial sums, $S_1, S_2, \ldots, S_n, \ldots$, get arbitrarily close to S as the value of n increases. (p. 504)

The sum of the geometric series $a_1 + a_1 r + a_1 r^2 + \ldots$ is

$$S = \frac{a_1}{1-r},$$

provided $|r| < 1$. If $|r| \geq 1$, the series has no sum. (p. 504)

12.4 Limits (pp. 508–512)

We read the statement

$$\lim_{x \to c} f(x) = L$$

as, "The limit of $f(x)$, as x approaches c, is L." This means that, as the x-values get closer and closer to the value c, the values of $f(x)$ get closer and closer to the value L. If the values of $f(x)$ do not approach a fixed value, then the limit does not exist. (p. 508)

To find the limit of a sequence whose general term is a rational expression, we use the Second Rule of Rational Functions to find the horizontal asymptote of the associated function. (p. 511)

12.5 Mathematical Induction (pp. 513–516)

Principle of Mathematical Induction:
A statement is true for all positive integers n if the following two requirements are met:

1. the statement is true for $n = 1$;
2. if you assume the statement is true for $n = k$, then you can prove that the statement is true for $n = k + 1$.

12.6 The Binomial Theorem (pp. 517–523)

For any positive integer n, we define n *factorial*, written $n!$, as the product of the first n positive integers:

$$n! = n \cdot (n - 1) \cdot (n - 2) \cdot \ldots \cdot 3 \cdot 2 \cdot 1.$$

The *zero factorial* is equal to 1: $0! = 1$. (p. 518)

If n is any positive integer, and if k is any positive integer such that $0 \leq k \leq n$, the *binomial coefficient* $\binom{n}{k}$ is

$$\binom{n}{k} = \frac{n!}{(n - k)!k!}.$$

Binomial Theorem: For all complex numbers a and b, and for all positive integers n, then

$$(a + b)^n = \binom{n}{0}a^n b^0 + \binom{n}{1}a^{n-1}b^1 + \binom{n}{2}a^{n-2}b^2$$
$$+ \cdots + \binom{n}{n-1}a^1 b^{n-1} + \binom{n}{n}a^0 b^n.$$

Chapter Test

12.1A Find the first five terms and the ninth term of the sequence whose general term is given by each rule.

1. $a_n = (n - 2)(n - 4)$ **2.** $a_n = 2^n - n$

Find the first six terms of each sequence.

3. $a_1 = 2, a_n = 5 + (-1)^n a_{n-1}$ for $n \geq 2$

4. $a_1 = 1, a_2 = 3, a_3 = 2, a_n = a_{n-1} + a_{n-2} + a_{n-3}$ for $n \geq 4$

12.1B Find the first five terms of the arithmetic sequence whose first term a_1 and common difference d are given.

5. $a_1 = 7, d = -5$ **6.** $a_1 = \dfrac{1}{6}, d = \dfrac{1}{2}$

Find the eleventh and twenty-first terms of each arithmetic sequence.

7. $3, 9, 15, 21, \ldots$ **8.** $4, \dfrac{9}{4}, \dfrac{1}{2}, -\dfrac{5}{4}, \ldots$

12.1C Find the sum of the first twelve terms of each arithmetic sequence.

9. $2, 5, 8, 11, 14, \ldots$ **10.** $\dfrac{2}{5}, 1, \dfrac{8}{5}, \dfrac{11}{5}, \ldots$

12.2A Determine whether each sequence is geometric.

11. $16, 12, 9, 6, \dfrac{9}{2}, \ldots$ **12.** $3, -6, 12, -24, 48, \ldots$

Find the sixth, ninth, and general terms of each geometric sequence.

13. $\dfrac{27}{10}, -\dfrac{9}{5}, \dfrac{6}{5}, -\dfrac{4}{5}, \ldots$ **14.** $\sqrt{2}, 4\sqrt{128}, 32, \ldots$

12.2B Find the sum of the first n terms and the sum of the first five terms of each geometric sequence.

15. $2, \dfrac{4}{3}, \dfrac{8}{9}, \dfrac{16}{27}, \ldots$ **16.** $4, -12, 36, -108, \ldots$

12.3A Rewrite each sum without using sigma notation; then compute each sum.

17. $\displaystyle\sum_{n=1}^{5} (5n - 1)$ **18.** $\displaystyle\sum_{n=2}^{6} (n^2 - 3n)$

Use sigma notation to rewrite each sum.

19. $4 + 9 + 16 + 25 + 36$

20. $5 + 9 + 13 + 17$

12.3B Compute the third and sixth partial sums of each series.

21. $1 + 2 + 4 + 8 + 16 + \cdots$

22. $\sum_{n=1}^{\infty} 3\left(\frac{1}{2}\right)^{n-1}$

12.3C Compute the sum, if it exists, of each series.

23. $4 - 2 + 1 - \frac{1}{2} + \frac{1}{4} - \cdots$

24. $\sum_{n=1}^{\infty} 7\left(\frac{3}{4}\right)^{n-1}$

25. $\left(\frac{1}{3} - \frac{1}{4}\right) + \left(\frac{1}{4} - \frac{1}{5}\right) + \cdots$

26. $\sum_{n=1}^{\infty} \left(\frac{1}{n+3} - \frac{1}{n}\right)$

Find the fraction represented by the decimal expansion of each of the following.

27. $0.37\overline{272}$

28. $0.415\overline{415}$

12.4A Evaluate the following limits for the function whose graph is shown on the right.

29. $\lim_{x \to -2} f(x)$ **30.** $\lim_{x \to 1} f(x)$

31. $\lim_{x \to 3} f(x)$ **32.** $\lim_{x \to 6} f(x)$

33. $\lim_{x \to -\infty} f(x)$ **34.** $\lim_{x \to +\infty} f(x)$

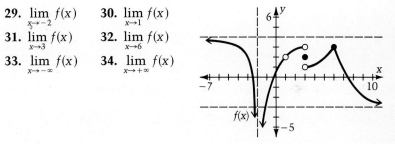

12.4B Evaluate each limit.

35. $\lim_{n \to +\infty} \frac{7n - n^2}{5n^2 + 1}$

36. $\lim_{n \to +\infty} \frac{8n}{n^3 + 1}$

37. $\lim_{n \to +\infty} \frac{3n^2}{7n + 2}$

12.5A Prove each statement by mathematical induction. (Assume $n \geq 1$.)

38. $3 + 6 + 9 + 12 + \ldots + 3n = \frac{3n(n+1)}{2}$

39. $0 + 3 + 8 + 15 + \ldots + (n^2 - 1) = \frac{n(2n+5)(n-1)}{6}$

12.6A Use the Binomial Theorem to expand the given expression.

40. $(a - b)^5$

41. $(3 + 5t)^4$

Superset

40. The sum of the first n terms of a sequence is $S_n = n(3 - n)$. Find the first, fourth, and ninth terms of the sequence.

41. Suppose that $a_1, a_2, a_3, \ldots$ is a geometric sequence of positive numbers. Show that $\frac{1}{a_1}, \frac{1}{a_2}, \frac{1}{a_3}, \ldots$ is also a geometric sequence.

42. Find the first four terms and the general term of the series whose nth partial sum is given.

(a) $S_n = n^2 + 4n$ (b) $S_n = \frac{1}{2}(3n^2 + 5n)$

43. Prove each statement by mathematical induction.

(a) $(x^m)^n = x^{mn}, \quad m, n \geq 1$ (b) $\dfrac{x^m}{x^n} = x^{m-n}, \quad m > n > 1$

Cumulative Test: *Chapters 11 and 12*

In exercises 1–2, simplify each expression.

1. $(5 - 3i)(-2 + 4i)$

2. $(i - 7)(2 - i)i$

In exercises 3–4, simplify each quotient.

3. $\dfrac{6 + 5i}{-3i}$

4. $\dfrac{5i}{3 - 2i}$

In exercises 5–6, represent the following complex numbers as a point in the complex plane.

5. $5 - 3i$

6. $8 + i$

In exercises 7–8, determine the absolute value of each complex number.

7. $6i$

8. $-5 + 9i$

In exercises 9–10, solve the following equations over the set of complex numbers.

9. $2x^2 - 6x + 5 = 0$

10. $5x^2 = -2x - 3$

In exercises 11–12, a polynomial equation and one of its solutions is given. Find all the solutions to the polynomial equation.

11. $x^3 + x^2 - 16x - 16 = 0,\ -1$

12. $x^4 + 4x^3 - 8x^2 + 16x - 48 = 0,\ 2i$

In exercises 13–14, express the following complex numbers in trigonometric form.

13. $2 - 2\sqrt{3}i$

14. $-\sqrt{2} + \sqrt{2}i$

In exercises 15–16, using trigonometric forms determine $z_1 z_2$ and $\dfrac{z_1}{z_2}$.

15. $z_1 = 2 - 2i\sqrt{3},\ z_2 = -1 + i$

16. $z_1 = -\sqrt{2} + i\sqrt{2},\ z_2 = \sqrt{3} + i$

In exercises 17–18, use DeMoivre's Theorem to express in the form $a + bi$.

17. $z = (1 + i)^4$

18. $z = (\sqrt{3} + i)^5$

In exercises 19–20, find the ninth and twentieth terms of the following arithmetic sequences.

19. $20, 13, 6, -1, \ldots$

20. $\dfrac{3}{2}, \dfrac{9}{4}, 3, \dfrac{15}{4}, \ldots$

In exercises 21–22, find the sum of the first eleven terms of the following arithmetic sequences.

21. $\dfrac{4}{5}, \dfrac{7}{5}, 2, \dfrac{13}{5}, \ldots$

22. $4, -1, -6, -11, \ldots$

In exercises 23–24, find the eighth, eleventh and general terms of the following geometric sequences.

23. $-\dfrac{1}{3}, 1, -3, 9, -27, \ldots$

24. $\dfrac{1}{2}, -\dfrac{5}{4}, \dfrac{25}{8}, -\dfrac{125}{16}, \ldots$

In exercises 25–26, find the sum of the first n terms and the sum of the first six terms of the following geometric sequences.

25. $4, 1, \dfrac{1}{4}, \dfrac{1}{16}, \ldots$

26. $2, 0.2, 0.02, 0.002, \ldots$

In exercises 27–28, compute the third and sixth partial sums of the following series.

27. $\displaystyle\sum_{n=1}^{\infty} (2^n - 1)$

28. $\displaystyle\sum_{n=1}^{\infty} (n^2 + n)$

In exercises 29–30, compute the sum, if it exists, of each series.

29. $4 - 10 + 25 - \dfrac{125}{2} + \cdots$

30. $\displaystyle\sum_{n=1}^{\infty} \left(\dfrac{1}{2n-1} - \dfrac{1}{2n+1}\right)$

In exercises 31–32, prove the following statements by mathematical induction. (Assume $n \geq 1$.)

31. $4 + 7 + 10 + 13 + \cdots + (3n + 1) = \dfrac{n(3n+5)}{2}$

32. $5 + 9 + 13 + 17 + \cdots + (4n + 1) = n(2n + 3)$

Superset

33. Solve the following logarithmic equations.
    (a) $\log_2(\log_2 4^x) = 1$
    (b) $\log_3 x - \log_9 2x = 0$

34. Use the fact that $\log_b x = \dfrac{\log x}{\log b}$ to estimate $\log_2 12$.

35. Simplify each expression. Write your answer in the form $a + bi$, a, or bi.
    (a) $\dfrac{5 + i}{3 - 2i} - \dfrac{7 + 2i}{3i}$
    (b) $\dfrac{|1 - i|}{|-2 - i|}$

36. Solve $5x^3 - 10x^2 + 25x - 50 = 0$ over the set of complex numbers.

Appendix

Calculator Exercises

Exercise Set 1.2

In exercises 1–8, evaluate each expression; then round to the nearest tenth. Use the following "order of operations" convention: evaluate expressions inside parentheses or brackets first; perform multiplications and divisions before additions and subtractions; when an expression involves a fraction (as in exercises 5–8), evaluate the numerator and denominator separately.

1. $(-3.5) - (7.1)(-4.6)$
2. $5(-6.2) - 3(-5.4)$
3. $8.7 \div (2 - 9.3) - 5$
4. $(6.1 - 4.7) \div [5(8.1 - 9.7)]$
5. $\dfrac{3.7 - 6.25}{(0.89)(-10.3)}$
6. $\dfrac{4(16.021 - 21.06)}{9.7 - 3.85}$
7. $\dfrac{6(5.01 - 8) - 10.2}{0.914 - 3(2.8 \div 7)}$
8. $\dfrac{0.003[85.6 \div (-0.01)]}{(0.4 - 7)(6 - 3.01) - 0.2}$

In exercises 9–12, evaluate each expression to the nearest whole number.

9. $[3697(618 - 9721)] \div 3178$
10. $80{,}000 \div (6291 + 899) - 3520$

11. $\dfrac{15{,}928 + 16(615 \div 12)}{1 - 3(3007 \div 48) - 5608}$

12. $\dfrac{87{,}600 - 3(729 \div 0.01)}{6281(1 - 0.007)}$

Exercise Set 1.3

In exercises 1–4, use the given symbol to write an inequality relating the two expressions.

1. $<$: $|8.01 - 9.26|$, $\dfrac{6.8 + 5.7}{10.02}$

2. $>$: $\dfrac{6.7 - |3 - 7.5|}{16}$, $|0.28 - 5| \div 40$

3. $>$: $|8.14 - 9.011|$, $\dfrac{(6.07)(0.889)}{1 + |1 - 6.19|}$

4. $<$: $\left|5.908 - \dfrac{(16.1)(8.2)}{17.04}\right|$, $\dfrac{5 - 3(9.07)}{7.9 - 20}$

In exercises 5–10, assume that points A, B, C, D, and E correspond to the real numbers 0.1, 0.01, -0.1, -0.01, and 0.005 respectively, on the real number line.

5. Determine the number corresponding to the point halfway between C and E.

6. Determine the number corresponding to the point halfway between D and E.

7. Determine the number corresponding to the point that is one-fourth of the distance from D to A.

8. Determine the number corresponding to the point that is one-eighth of the distance from B to E.

9. Which point (A, B, C, D, or E) is between the points corresponding to $\frac{1}{60} + \frac{1}{100}$ and $\frac{1}{60} - \frac{1}{100}$?

10. Which point (A, B, C, D, or E) is closest to the point corresponding to $\frac{111}{800} - \frac{57}{407}$?

Exercise Set 1.4

In exercises 1–4, evaluate each expression to the nearest hundredth. (**Careful:** Your calculator may not raise a negative number to a power. You may have to keep track of the sign of your calculations yourself.)

1. $(5.82)^2 - 6(3 - 4.71)^3$

2. $(3.75 - 6.8)^4 - (9.08)^2$

Calculator Exercises

3. $\left[\dfrac{6.28 - 1.59}{(1.08)^2}\right]^3$

4. $(1.3)^{13} - (6.91 \div 5.75)^{13}$

In exercises 5–8, determine which expression in the given pair represents the greater value.

5. $(1 - 0.004)^2$, $1 - (0.004)^2$

6. $(1 + 0.02)^{-3}$, $1 + (0.02)^{-3}$

7. $1 + 0.02$, $(1 + 0.02)^2$

8. $1 - 0.02$, $(1 - 0.02)^2$

In exercises 9–12, insert the correct inequality sign ($>$ or $<$) between the given numbers.

9. $\dfrac{6.2 \times 10^{-8}}{9.1 \times 10^{-9}}$, 6.9

10. $\dfrac{3.2 \times 10^{-40}}{7.5 \times 10^{-38}}$, 0.004

11. $\dfrac{1.8 \times 10^{56}}{1.1 \times 10^{60}}$, 0.001

12. $\dfrac{5.4 \div 10^{-10}}{1.9 \div 10^{-12}}$, $\dfrac{1}{5}$

▣ Exercise Set 1.7

In exercises 1–6, determine which expression in the given pair represents the greater value.

1. $1 + 0.02$, $\sqrt{1 + 0.02}$

2. $1 - 0.02$, $\sqrt{1 - 0.02}$

3. $\sqrt[3]{16.82}$, $\sqrt{7}$

4. $\sqrt[8]{900}$, $\sqrt[9]{800}$

5. $\sqrt[8]{8}$, $\sqrt[4]{4}$

6. $\sqrt{0.8}$, $(0.8)^2$

7. Using a calculator, determine $\sqrt{2}$. Take the result and press the square root key ten more times. What pattern do you notice in the displayed values?

8. Perform the same procedure described in exercise 7 with the value $\sqrt{3}$. Do you notice the same pattern?

9. Using a calculator, determine 2^2. Take the result and press the x^2 key ten more times. Are the displayed values getting closer and closer to any specific value?

10. Take successive square roots of $\tfrac{1}{2}$ and successive squares of $\tfrac{1}{2}$. Do you notice any patterns?

11. How would you determine the value of 5^9 if the only operations on your calculator were $+$, $-$, $\times$, $\div$, and x^2?

12. How would you determine the value of $\sqrt[1024]{2}$ if the only operations on your calculator were $+$, $-$, $\times$, $\div$, and $\sqrt{x}$?

Exercise Set 1.8

In exercises 1–6, evaluate each expression to the nearest hundredth.

1. $\left(\dfrac{5.15 - 7.28}{3.01}\right) \div (6.37 - 2.89)$

2. $1 \div \left(\dfrac{4.95 - 6.82}{9.15}\right)$

3. $\dfrac{\sqrt{20.5} - 20.5}{\sqrt{10.5} - 10.5}$

4. $\dfrac{\sqrt{1.05} - 1.05}{\sqrt{0.95} - 0.95}$

5. $\dfrac{1 - \dfrac{1}{1 + 0.2}}{1 + \dfrac{1}{1 - 0.2}}$

6. $1 + \dfrac{1}{1 + \dfrac{1}{1 + 0.03}}$

7. Add up the first ten terms in the infinite sum $\frac{1}{2} + \frac{1}{4} + \frac{1}{8} + \frac{1}{16} + \frac{1}{32} + \cdots$ (the denominators are successive powers of 2). What pattern do you notice as you add more and more terms?

8. Add up the fifteen terms $1 + \frac{1}{2} + \frac{1}{3} + \frac{1}{4} + \cdots + \frac{1}{14} + \frac{1}{15}$. Do you notice any limiting pattern here, such as the one observed in exercise 7?

Exercise Set 2.1

In exercises 1–8, solve the given equations. Round solutions to the nearest tenth.

1. $8.2x - 6.1 = 3.8(x - 5)$

2. $5(2.7 - 9x) = 16.3 - 20.5x$

3. $3.95x + 16.33(x - 7) = 52.18 - 0.05x$

4. $2.25x - 5(16.8 - x) = x(83.04 - 16.91)$

5. $\dfrac{1}{x - 3.1} = \dfrac{6.03}{9.28}$

6. $\dfrac{8.75}{x + 2.05} = \dfrac{10.25}{11 - x}$

In exercises 7–8, solve the given equations.

7. $|x - 8.25| = 0.005$

8. $|x - 9.01| = 0.0001$

In exercises 9–12, determine the value of s for the given values of s_0, v_0, a, and t, if

$$s = s_0 + v_0 t + \tfrac{1}{2} a t^2$$

Round answers to the nearest tenth.

9. $s_0 = 50$, $v_0 = 22.5$, $a = 32$, $t = 2.1$

10. $s_0 = 100$, $v_0 = 88.8$, $a = 32$, $t = 2.5$

11. $s_0 = 10.5$, $v_0 = 65.7$, $a = 9.8$, $t = 60$

12. $s_0 = 12.5$, $v_0 = 82.5$, $a = 9.8$, $t = 60$

Exercise Set 2.2

In exercises 1–6, use the discriminant to determine the nature of the roots without solving.

1. $3.7x^2 - 4.5x + 1.4 = 0$
2. $8.7x^2 + 0.08x - 0.3 = 0$
3. $0.2x^2 + 1.6x + 3.2 = 0$
4. $18.7 + 5.5x^2 = 19.2x$
5. $1897x^2 + 1650x + 374 = 0$
6. $508x^2 + 482x + 110 = 0$

In exercises 7–12, use the quadratic formula to approximate the solutions of the given equation to the nearest tenth.

7. $4x^2 + 7x + 2 = 0$
8. $5x^2 + 4x - 4 = 0$
9. $2x^2 + 2x - 5 = 0$
10. $x^2 - 5x + 1 = 0$
11. $6.2x^2 - 8.1x + 2.5 = 0$
12. $16.85x^2 - 1.5x - 20 = 0$

Exercise Set 2.4

In exercises 1–8, use interval notation to describe the solution set of the given inequality. Unless otherwise specified, round endpoint values to the nearest tenth.

1. $15.8x - 3(x + 2.7) < 46.3$
2. $5.6(3.5 - x) \geq 1 - 2.8x$
3. $|3x - 7| < 0.001$ (round to nearest thousandth)
4. $|8x - 6.2| < 0.001$ (round to nearest ten-thousandth)
5. $|1 - 5.4x| \geq 3.7$
6. $|9.8 - 0.7x| \leq 19.6$
7. $9x^2 - 125 < 1$
8. $2.5x^2 - x \geq 1.8$

Exercise Set 2.5

1. A man 6′3″ tall is walking toward a street light that is $24\frac{1}{2}$ ft tall. How far away from the light is the man when his shadow is the same length as his height?

2. Three thousand feet of fencing are to be used to enclose a rectangular (or square) lot. If the lot is to have an area of five hundred thousand square feet, what should the dimensions be?

3. Two people leave their respective apartments at the same time to drive to a concert. Each must drive 42.5 mi to the concert. If, on the trip, one person averages 8 mph faster than the other, and arrives at the concert 8.5 min before the other, what were the average speeds of the two drivers?

4. A group of students rowed 6 mi downstream and 6 mi back in 1 hr 47 min. If the current flowed at a rate of 4.9 mph, what was the rate of the students in still water?

Exercise Set 3.2

In exercises 1–6, determine $f(x)$ for the x-values 1, 1.2, 1.4, 1.6, 1.8, and 2.

1. $f(x) = 6x - 5$
2. $f(x) = 0.02x - 3.76$
3. $f(x) = x^2 - 5x + 3.66$
4. $f(x) = 3.5x^2 - 0.05x + 1$
5. $f(x) = \sqrt{\dfrac{x - 0.08}{2}}$ (round to nearest thousandth)
6. $f(x) = \sqrt{\dfrac{16 - 5.2x}{0.05}}$ (round to nearest thousandth)

Exercise Set 3.4

In exercises 1–8, determine the slope-intercept form ($y = mx + b$) of the line that is described. Round the values of m and b to the nearest hundredth after all computations are performed.

1. line passes through (3.41, 8.22) and (−1.73, 4.56)
2. line passes through (−7.80, −5.49) and (5, 2.11)
3. line passes through (−1, 5.37) and (2.12, 1.48)
4. line passes through (0, 0.82) and (1.03, −0.08)
5. line passes through (5.71, 3.08) and is parallel to the graph of $y = -0.89x + 1.7$
6. line passes through (6.88, −0.21) and is parallel to the graph of $4.80x - 7.68y + 8 = 0$
7. line passes through (0.53, −0.72) and is perpendicular to the graph of $1.20x + 6.30y - 8.95 = 0$
8. line passes through (−1.78, −6.81) and is perpendicular to the graph of $6.30x - 2.8y - 5 = 0$

In exercises 9–16, (a) carefully graph the given linear function over the interval $0 \le x \le 5$; (b) use your graph to estimate $f(2.5)$; (c) use your calculator to determine $f(2.5)$ exactly.

9. $f(x) = 2x - 7$
10. $f(x) = 5 - 3x$
11. $f(x) = -2.2x + 7.4$
12. $f(x) = 9.6x - 13.8$

13. $f(x) = 0.2x + 5.1$

14. $f(x) = 0.03x - 1.2$

15. $f(x) = \pi - \sqrt{2}x$

16. $f(x) = \sqrt{2}x - \pi$

Exercise Set 3.5

In exercises 1–8, (a) carefully graph the given quadratic function over the interval $0 \le x \le 5$; (b) use your graph to estimate $f(2.5)$; (c) use your calculator to determine $f(2.5)$ exactly.

1. $f(x) = 4 - 0.2x^2$

2. $f(x) = 3 + 0.3x^2$

3. $f(x) = x^2 - 3x + 4$

4. $f(x) = x^2 - x - 3$

5. $f(x) = 0.4x^2 - 1.1x$

6. $f(x) = 10.3x - 2.1x^2$

7. $f(x) = 0.3x^2 - x - 0.43$

8. $f(x) = 0.2x^2 - 0.08x - 3.92$

The height $h(t)$ of an object shot upwards (from ground level) with a velocity of v_0 ft/sec can be described as a function of time (t): $h(t) = v_0 t - 16t^2$. In exercises 9–12, determine the time at which the object reaches its maximum height, and the maximum height achieved. (After computations, round answers to the nearest tenth.)

9. $v_0 = 80.5$ ft/sec

10. $v_0 = 22.6$ ft/sec

11. $v_0 = 50.2$ ft/sec

12. $v_0 = 197.4$ ft/sec

Exercise Set 4.2

Recall that if f is a polynomial function, and if $f(c_1)$ and $f(c_2)$ have different signs, then f has a zero somewhere between c_1 and c_2. In exercises 1–8, the function f has a zero between the given values of c_1 and c_2. Determine whether the zero lies in the interval $[c_1, c_1 + 0.25]$, $[c_1 + 0.25, c_1 + 0.5]$, $[c_1 + 0.5, c_1 + 0.75]$ or $[c_1 + 0.75, c_2]$.

1. $f(x) = x^3 - x^2 + x + 7$; $c_1 = -2$, $c_2 = -1$

2. $f(x) = 3x^3 + x - 21$; $c_1 = 1$, $c_2 = 2$

3. $f(x) = 16 - x^3 - x^4$; $c_1 = 1$, $c_2 = 2$

4. $f(x) = x^5 - 3x^4 - x + 10$; $c_1 = 1$, $c_2 = 2$

5. $f(x) = x^4 - 2x^2 - 5x - 10$; $c_1 = 2$, $c_2 = 3$

6. $f(x) = x^6 - x^3 + 2x - 18$; $c_1 = -2$, $c_2 = -1$

7. $f(x) = 20 + x - x^2 + x^3 - x^4$; $c_1 = 2$, $c_2 = 3$

8. $f(x) = 4x^4 - 3x^3 + 2x^2 - x$; $c_1 = 0$, $c_2 = 1$

Exercise Set 4.4

The distance d between the point $P(x_0, y_0)$ and the line $Ax + By + C = 0$ is given by the formula

$$d = \frac{|Ax_0 + By_0 + C|}{\sqrt{A^2 + B^2}}.$$

In exercises 1–4, determine the distance between the given point and the given line. Round answers to the nearest hundredth.

1. $3x - 4y + 12 = 0$; $P(3.5, -1.5)$
2. $x + 7y - 15 = 0$; $P(-1.8, -4)$
3. $0.3x - 2.1y + 31.8 = 0$; $P(-5, 100)$
4. $605x - 30y - 515 = 0$; $P(10.5, -12.8)$

The unit circle is a circle centered at $(0, 0)$ with radius equal to 1. Therefore, the unit circle is the set of all points in the plane that lie at a distance of exactly 1 unit from the origin. In exercises 5–12, use the distance formula to determine whether the given point lies inside or outside the unit circle.

5. $(0.65, 0.85)$
6. $(-0.25, -0.75)$
7. $(-0.6, -0.7)$
8. $(-0.9, -0.1)$
9. $\left(\frac{\sqrt{3}}{3}, -\frac{\sqrt{3}}{3}\right)$
10. $\left(-\frac{\sqrt{6}}{3}, -\frac{\sqrt{6}}{4}\right)$
11. $\left(\frac{\sqrt{2}}{2}, \frac{\sqrt{3}}{2}\right)$
12. $\left(-\frac{1}{\pi}, \frac{\pi}{4}\right)$

Exercise Set 5.1

In exercises 1–4, solve the given system of equations. (Depending on the method of solution, answers may vary slightly.)

1. $\begin{cases} 5.7x - 8y = 12.2 \\ 6x + 2.7y = 20 \end{cases}$
   Round answers to the nearest hundredth.

2. $\begin{cases} 10x + 22.5y = -18.6 \\ 14.3x - 0.25y = 40 \end{cases}$
   Round answers to the nearest tenth.

3. $\begin{cases} 1607x - 5925y = -856 \\ 47x + 219y = 360 \end{cases}$
   Round answers to the nearest thousandth.

4. $\begin{cases} 0.003x - 1.02y = 0.05 \\ -0.008x + 1.83y = 0.008 \end{cases}$
Round answers to the nearest hundredth.

In exercises 5–10, determine the coordinates of the points of intersection of the given line and the unit circle $x^2 + y^2 = 1$. Round answers to the nearest hundredth.

5. $y = x$

6. $y = -\dfrac{\sqrt{3}}{3}x$

7. $y = \sqrt{3}x$

8. $y = 0.27x$

9. $y = \sqrt{2} - x$

10. $2y = \sqrt{3} + 1 - 2x$

Exercise Set 5.3

The distance d between the point $P(x_0, y_0, z_0)$ and the plane described by the equation $Ax + By + Cz + D = 0$ is given by the formula

$$d = \frac{|Ax_0 + By_0 + Cz_0 + D|}{\sqrt{A^2 + B^2 + C^2}}.$$

In exercises 1–4, determine the distance between the given point and the given plane. Round answers to the nearest hundredth.

1. $2x + y + z + 5 = 0$; $P(3, 5, 8)$

2. $x - 3y + 7z - 15 = 0$; $P(13, 15, -27)$

3. $2.5x - 18.6z = 7$; $P(-6.1, 22.7, 31.5)$

4. $12.2y + 15.8z = -4.8$; $P(14.7, 19.5, -3.4)$

In exercises 5–8, determine the points of intersection of the given plane with (a) the x-axis, (b) the y-axis, and (c) the z-axis. Round the coordinates to the nearest tenth.

5. $6x + 8y - z + 17 = 0$

6. $3.2x - 0.2y + 8z - 15.7 = 0$

7. $5x + 3.7y - 8.2z = 1$

8. $\sqrt{2}x + \sqrt{3}y - \sqrt{5}z + 1 = 0$

Exercise Set 6.1

1. Use the y^x-key on your calculator to determine 2^x for the x-values 3.10, 3.11, 3.12, 3.13, 3.14, 3.15. Which of these six values most closely approximates 2^π?

2. Which value is greater, 3^π or π^3?

In exercises 3–12, use a calculator to estimate each of the following to the nearest hundredth.

3. $2^{\sqrt{2}}$
4. $2^{\sqrt{3}}$
5. $3^{\sqrt{2}+\sqrt{3}}$
6. $3^{1-\sqrt{2}}$
7. $\pi^{2-\sqrt{2}}$
8. $\pi^{3+\sqrt{3}}$
9. $(2.72)^{-0.25}$
10. $(2.72)^{-3.14}$
11. $\sqrt{2}^{\sqrt{2}}$
12. π^{π}

In exercises 13–16, an equation and three values are given. Determine which of the three values best approximates a solution of the equation.

13. $3^x = 1 + x$; $-0.41, -0.17, 1.15$
14. $2^{x-1} = 1 - x$; $-1.15, 0.35, 1.5$
15. $2^{x-2} = \sqrt{x - 1}$; $0.5, 1.0, 1.5$
16. $2^{x+1} = x^2 - x + 1$; $-0.8, -0.4, 0.4$

Exercise Set 6.4

In exercises 1–12, solve the given equation by taking the logarithm of each side and then solving for x. Round answers to the nearest thousandth. Check answers by using the y^x key on your calculator.

1. $6^x = 3$
2. $8^x = 50$
3. $(2.1)^x = 10$
4. $2^x = 100$
5. $3^x = 4^{x-1}$
6. $(3.5)^x = (2.2)^{x+1}$
7. $x^5 = 7x$, $x > 0$
8. $x^8 = 3x$, $x \neq 0$
9. $(x - 1)^5 = 0.0015$, $x > 1$
10. $(x + 1)^8 = 37.92$, $x > -1$
11. $15x^6 = x$, $x > 0$
12. $0.01x^7 = x$, $x > 0$

Exercise Set 6.5

1. Suppose $18,500 is placed in an account that advertises a 9.13% annual interest rate compounded continuously.

   (a) How much is the account worth after $4\frac{1}{2}$ years? (b) How long will it take for the account to be worth three times the initial investment?

2. If 86.2 gm of Carbon-14 were present in an organism when it was alive, and 38.7 gm are present in the remains today, how long ago did the organism live?

3. A new water treatment installation continuously reduces the percentage of pollutant in the water by $\frac{1}{50}$ per day ($k = -0.02$). How many days does it take to cut the pollutant in the water to less than 1% of what it was originally?

4. A certain species of bird is in danger of extinction. It is estimated that only 900 of these creatures are still alive. Estimates five years ago placed the population at 1200. Experts claim that once the population drops below 200, this bird's situation will be irreversible. In how many years will this happen? (Assume that the population changes at a rate proportional to its size.)

Exercise Set 7.1

In exercises 1–10, $\triangle ABC$ is a right triangle, c is the length of the hypotenuse and a and b are the lengths of the legs. Given the lengths of two sides, use the Pythagorean Theorem to find the length of the third side.

1. $a = 42.72 \quad b = 31.68$ **2.** $a = 34.35 \quad b = 63.45$

3. $a = 293.4 \quad c = 568.3$ **4.** $b = 349.6 \quad c = 638.2$

5. $a = 0.7184 \quad b = 0.6319$ **6.** $a = 1.231 \quad b = 1.893$

7. $b = 2462 \quad c = 4091$ **8.** $a = 5000 \quad c = 5663$

9. $a = 12.427 \quad c = 22.645$ **10.** $b = 79.351 \quad c = 83.381$

Exercise Set 7.2

In exercises 1–10, use the given information about the right triangle ABC from exercises 1–10 in Exercise Set 2.1 to find the values of the six trigonometric ratios of $\angle A$.

In exercises 11–14, given an equilateral triangle ABC with side of length s and altitude of length h, find the indicated part.

11. If $s = 6.348$, find h. **12.** If $h = 15.259$, find s.

13. If $h = 25.9808$, find the perimeter of $\triangle ABC$.

14. If $h = 9.4873$, find the area of $\triangle ABC$.

In exercises 15–20, α and β are complementary. From the given information, find the indicated trigonometric ratios.

15. $\sin \alpha = 0.3846$, find $\cos \alpha$ and $\cos \beta$.

16. $\cos \alpha = 0.4286$, find $\sin \alpha$ and $\tan \beta$.

17. $\csc \alpha = 1.1333$, find $\tan \alpha$ and $\cos \beta$.

18. $\sec \alpha = 1.3810$, find $\tan \beta$ and $\tan \alpha$.

19. $\cos \alpha = 0.9459$, find $\tan \alpha$ and $\sin \beta$.

20. $\sin \alpha = 0.2800$, find $\cos \alpha$ and $\tan \alpha$.

Exercise Set 7.3

In exercises 1–6, the given point is on the terminal side of angle α in standard position. Determine the values of the six trigonometric functions of α.

1. $(-5\sqrt{7}, 3\sqrt{3})$
2. $(-4\sqrt{5}, -7\sqrt{5})$
3. $(25.34, -72.67)$
4. $(2.872, -6.449)$
5. $(-13.856, -5.657)$
6. $(-11.180, 6.471)$

Exercise Set 7.4

In exercises 1–4, two approximate trigonometric function values of an angle are given. Find the other four trigonometric function values rounded to four decimal places.

1. $\sin 32° = 0.5299$, $\cos 32° = 0.8480$
2. $\tan 66° = 2.2460$, $\sec 66° = 2.4586$
3. $\tan 55° = 1.4281$, $\sin 55° = 0.8192$
4. $\tan 23° = 0.4245$, $\csc 23° = 2.5593$

In exercises 5–8, use the Reciprocal and Pythagorean Identities to determine the values of the other five trigonometric functions of α.

5. $\sin \alpha = 0.5124$
6. $\cos \alpha = 0.3051$
7. $\cos \alpha = 0.8236$
8. $\sin \alpha = 0.7645$

A calculator will compute the trigonometric function values of angles measured either in degrees or radians. You can select the units for the angular measure with a degree/radian key (DRG on some calculators, DEG on others). Be sure the calculator is in the degree mode for the next exercise set.

In exercises 9–14, use a calculator to determine the following.

9. $\sin 31.2°$
10. $\cos 43.8°$
11. $\tan 21.66°$
12. $\tan 49.33°$
13. $\cos 52.25°$
14. $\sin 80.46°$

To use a calculator to evaluate the cotangent, secant and cosecant functions, it is necessary to use the Reciprocal Identities. For example:

$$\csc 36° = \frac{1}{\sin 36°}.$$

To take a reciprocal of a number x, use the reciprocal key, $1/x$. (Be sure that the calculator is in the degree mode.)

Calculator Exercises

In exercises 15–20, use a calculator to determine the following.

15. csc 56.3° **16.** csc 10.5° **17.** cot 48.2°

18. sec 15.7° **19.** sec 45.5° **20.** cot 78.9°

To find an angle α to two decimal places such that $\sin \alpha = 0.5402$, we use the inverse key, INV.

ENTER .5402 PRESS INV PRESS sin
DISPLAY 32.69725475

Thus, $\alpha = 32.70°$. (If you did not get this answer, check to be sure you have your calculator in the degree mode.)

The inverse key followed by the sine key "undoes" what the sine key "does". The INV key, when followed by the sin, cos, or tan key calculates the smallest angle in Quadrant I having the positive trigonometric function value entered in the display (negative trigonometric function values entered in the display produce an angle in Quadrant IV).

In exercises 21–26, use a calculator to determine the acute angle α in degrees to two decimal places.

21. $\sin \alpha = 0.4695$ **22.** $\cos \alpha = 0.8480$

23. $\cos \alpha = 0.7698$ **24.** $\sin \alpha = 0.4669$

25. $\cot \alpha = 1.072$ **26.** $\sec \alpha = 3.078$

In exercises 27–36, evaluate each of the following to four decimal places.

27. cos 408° **28.** tan 597° **29.** tan 212° **30.** sin 320°

31. sin(−1134°) **32.** cos(−666°) **33.** cos 196.2° **34.** sin 196.2°

35. csc(−287°) **36.** sec(−136°)

In exercises 37–44, find α in degrees to two decimal places.

37. $\cos \alpha = -0.7145$, α in Quadrant II

38. $\sin \alpha = -0.7593$, α in Quadrant III

39. $\tan \alpha = -10.988$, α in Quadrant IV

40. $\tan \alpha = -4.915$, α in Quadrant II

41. $\sin \alpha = -0.6211$, α in Quadrant IV

42. $\cos \alpha = 0.9898$, α in Quadrant IV

43. $\cot \alpha = 3.420$, α in Quadrant III

44. $\csc \alpha = -1.070$, α in Quadrant III

Exercise Set 8.1

If P is the terminal point of a standard arc of length s, then

P is in Quadrant I when $0 < s < \dfrac{\pi}{2}$,

P is in Quadrant II when $\dfrac{\pi}{2} < s < \pi$,

P is in Quadrant III when $\pi < s < \dfrac{3\pi}{2}$,

P is in Quadrant IV when $\dfrac{3\pi}{2} < s < 2\pi$,

P is in Quadrant I when $2\pi < s < \dfrac{5\pi}{2}$, etc.

In exercises 1–10, given that the real number is represented as a standard arc, determine the quadrant in which the standard arc terminates.

1. 25 **2.** 32 **3.** $7\sqrt{11}$ **4.** $6\sqrt{3}$ **5.** -416
6. -128 **7.** $3\sqrt{5} + 21$ **8.** $2\sqrt{7} - 34$ **9.** $\sqrt{\pi} + 24$ **10.** $\pi^2 + 10$

In exercises 11–14, the point (x, y) is on the unit circle. You are given one of its coordinates and the quadrant in which the point lies. Determine an approximate value of the other coordinate to four decimal places.

11. $x = 0.3672$, Quadrant IV **12.** $y = 0.7218$, Quadrant II
13. $y = -0.4516$, Quadrant III **14.** $x = -0.6711$, Quadrant II

In exercises 15–22, convert the measure of the angle from degrees to radians or vice versa. Approximate all answers to three decimal places.

15. 35.8° **16.** 169.2° **17.** 248°36′ **18.** $-312°48′$
19. 12.5664 **20.** 15.7080 **21.** -5.4264 **22.** -18.1681

Exercise Set 8.2

In exercises 1–14, determine the function values to four decimal places. (Be sure the calculator is in the radian mode.)

1. sin 1.0417 **2.** cos 0.4189 **3.** cos 2.0944 **4.** sin 5.2360
5. sin -7.3304 **6.** cos -8.6394 **7.** tan 26.1799 **8.** tan -13.1947
9. cos 10.5000 **10.** sin 5.0000 **11.** sec 8.500 **12.** csc 3.7000
13. cot 0.1768 **14.** cot 27.3200

Calculator Exercises

Exercise Set 8.3

In exercises 1–6, evaluate the given function for the x-values 1, −3, 21, 7.1416, −15.7080, and 235.

1. $f(x) = \sin(x + \pi)$
2. $f(x) = \cos(x - \pi)$
3. $f(x) = -\sin\left(x - \dfrac{\pi}{4}\right)$
4. $f(x) = -\cos\left(x + \dfrac{\pi}{4}\right)$
5. $f(x) = \tan\left(x - \dfrac{\pi}{2}\right)$
6. $f(x) = \cot\left(x - \dfrac{\pi}{6}\right)$

Exercise Set 8.4

In exercises 1–4, evaluate the given function for the x-values 1, −2.5, 4.512, 24.67, −31.56, and 49.223.

1. $f(x) = 2\sin\left(\dfrac{1}{2}x + \dfrac{\pi}{4}\right)$
2. $f(x) = 2\cos(2x + \pi)$
3. $f(x) = \sqrt{3}\cos\left(-\dfrac{1}{2}x - \dfrac{\pi}{2}\right)$
4. $f(x) = \dfrac{3}{2}\sin\left(-x + \dfrac{\pi}{2}\right)$

Exercise Set 8.5

In exercises 1–8, evaluate the expression. Express answers to four decimal places, with angles expressed in radians.

1. $\sin^{-1}\left(\dfrac{\sqrt{5}}{12}\right)$
2. $\arcsin\left(\dfrac{\sqrt{7}}{3}\right)$
3. $\arctan(-\sqrt{11})$
4. $\tan^{-1}\left(\dfrac{\sqrt{17}}{\sqrt{3}}\right)$
5. $\sin(\arctan(3.1562))$
6. $\cos(\arcsin(-0.7246))$
7. $\tan^{-1}(\tan(36.7°))$
8. $\tan(\tan^{-1}(-24))$

Exercise Set 9.2

We saw that the trigonometric values for angles of 15° and 75° can be calculated exactly using the special angles and the Sum and Difference Identities. In exercises 1–6, compare the trigonometric function values found on the calculator, with the value found by using the special angles and identities.

1. sin 15°   2. cos 15°   3. cos 75°   4. sin 75°   5. tan 15°   6. tan 75°

In exercises 7–12, suppose $\sin \alpha = -0.3472$, α a third quadrant angle, and $\cos \beta = -0.6561$, β a second quadrant angle. Compute the given expression in two ways: first determine α and β, then evaluate the trigonometric function value. Second, use the appropriate sum or difference identity.

7. $\sin(\alpha + \beta)$ **8.** $\cos(\alpha + \beta)$ **9.** $\cos(\alpha - \beta)$

10. $\sin(\alpha - \beta)$ **11.** $\tan(\alpha + \beta)$ **12.** $\tan(\alpha - \beta)$

Exercise Set 9.3

In exercises 1–4, suppose that θ is acute and $\sin \theta = 0.3846$. Compute the given trigonometric function value.

1. $\sin 2\theta$ **2.** $\cos 2\theta$ **3.** $\cot 2\theta$ **4.** $\tan 2\theta$

In exercises 5–8, suppose that θ is a second quadrant angle and $\cos \theta = -0.6154$. Compute the given trigonometric function value.

5. $\cos \dfrac{\theta}{2}$ **6.** $\sin \dfrac{\theta}{2}$ **7.** $\tan \dfrac{\theta}{2}$ **8.** $\cot \dfrac{\theta}{2}$

Exercise Set 9.5

In exercises 1–10, solve the equation for x in the interval $[0°, 360°)$.

1. $12 \sin^2 x - \sin x - 1 = 0$ **2.** $15 \cos^2 x + \cos x - 6 = 0$

3. $\sin x + 3 \cos x = 0$ **4.** $4 \cos^2 x + 2 \sin x = 3$

5. $2 \cos 2x - 2 \sin^2 x = 1$ **6.** $\tan^2 2x + 2 \tan 2x = 1$

7. $\sec^2 x + 2 \tan x - 2 = 0$ **8.** $3 \tan^2 x + 2 \sec x = 5$

9. $\cos^2 x - 0.97 \cos x + 0.18 = 0$ **10.** $\sin^2 x + 0.24 \sin x - 0.1081 = 0$

Exercise Set 10.1

In exercises 1–10, solve the right triangle ABC with right angle at C.

1. $a = 23.47$, $A = 32.8°$ **2.** $a = 1.831$, $B = 58.2°$

3. $b = 1293.0$, $A = 47.73°$ **4.** $b = 2.828$, $A = 12.62°$

5. $c = 19.800$, $A = 28.19°$ **6.** $c = 505.07$, $B = 41.38°$

7. $a = 2.3261$, $b = 2.4495$ **8.** $a = 84.853$, $c = 305.123$

9. $b = 0.9847$, $a = 0.2231$ **10.** $b = 398.24$, $c = 1763.66$

11. A 52 ft long guy wire runs from level ground to the top of an antenna. The guy wire makes an angle of $68°36'$ with the antenna. How far out from the base of the antenna is the guy wire anchored?

12. A railroad track passes below a long, level highway bridge and at a right angle to it. At the same instant that a train traveling 60 mph passes under the bridge, a car traveling 45 mph on the bridge is directly above the train. If the highway bridge is 100 feet above the track, find the distance between the train and car 10 seconds later. (Note that 60 mph = 88 ft/s and 45 mph = 66 ft/s.)

Exercise Set 10.2

In exercises 1–10, solve triangle ABC.

1. $A = 101.7°$, $B = 19.3°$, $c = 118.5$
2. $A = 42.8°$, $C = 51.1°$, $b = 92.75$
3. $B = 75.6°$, $C = 33.2°$, $a = 4384.0$
4. $A = 48.6°$, $B = 94.1°$, $c = 282.84$
5. $a = 14.95$, $b = 17.43$, $A = 58.6°$
6. $b = 0.4731$, $a = 0.4347$, $B = 78.4°$
7. $a = 6241$, $c = 8233$, $C = 31.7°$
8. $a = 3.3569$, $c = 2.7973$, $A = 62.5°$
9. $b = 756.2$, $c = 541.6$, $B = 112.3°$
10. $b = 464.5$, $c = 637.3$, $C = 73.8°$

11. A wind storm caused a 64 ft tall telephone pole to lean due east at an angle of 63°48′ to the ground. A 106 ft long guy wire is then attached to the top of the pole and anchored in the ground due west of the foot of the pole. How far from the foot of the pole is the guy wire anchored?

12. Two lighthouses are located 49 mi apart on the coast of Massachusetts, one near Salem and the other in Provincetown on Cape Cod. The Provincetown lighthouse is on a bearing of S46°E from the Salem lighthouse. A fishing boat in distress radios an SOS signal. The bearing of the signal from the Salem station is S55°12′E and from Provincetown N47°36′E. How far is the ship from Provincetown?

Exercise Set 10.3

In exercises 1–10, solve the triangle ABC.

1. $a = 5808$, $b = 7920$, $C = 109.3°$
2. $a = 39.12$, $b = 36.37$, $C = 80.7°$
3. $b = 9.591$, $c = 6.856$, $A = 51.5°$
4. $b = 207.36$, $c = 214.48$, $A = 98.4°$

5. $a = 0.9674$, $c = 0.8834$, $B = 61.6°$

6. $a = 8347.0$, $c = 6723.0$, $B = 68.71°$

7. $a = 2.2365$, $b = 3.4641$, $c = 3.1623$

8. $a = 46.90$, $b = 38.73$, $c = 33.17$

9. $a = 0.2500$, $b = 0.1887$, $c = 0.3448$

10. $a = 3218.8$, $b = 3555.3$, $c = 3806.7$

11. Determine the perimeter of a regular octagon inscribed in a circle of radius 24.6 in.

12. Determine the perimeter of a regular 12-gon inscribed in a circle of radius 35 cm.

In exercises 13–16, find the area of triangle ABC.

13. $A = 41.8°$, $b = 35.49$, $c = 42.78$

14. $B = 69.6°$, $a = 479.6$, $c = 404.9$

15. $C = 105.1°$, $a = 3.784$, $b = 3.465$

16. $a = 8.485$, $b = 5.568$, $c = 7.616$

17. Find the area of the octagon in exercise 11.

18. Two sides and the included angle of a parallelogram have measures 17.88, 21.91 and 54.4°, respectively. Find the area of the parallelogram.

Exercise Set 10.4

In exercises 1–4, points A and B are given. Describe the vector from A to B as an ordered pair and determine its magnitude and direction.

1. $A(2.37, -1.89)$, $B(-2.75, 3.45)$
2. $A(-2.64, -3.31)$, $B(3.73, 1.80)$
3. $A(-5.57, -2.48)$, $B(1.96, -3.73)$
4. $A(7.57, 3.09)$, $B(-3.78, -1.07)$

In exercises 5–8, vector $\mathbf{a}$ has magnitude $\|\mathbf{a}\|$ and direction θ, vector $\mathbf{b}$ has magnitude $\|\mathbf{b}\|$ and direction φ. Express vector $\mathbf{c} = \mathbf{a} + \mathbf{b}$ as an ordered pair.

5. $\|\mathbf{a}\| = 9.89$, $\theta = 87°$
   $\|\mathbf{b}\| = 2.79$, $\varphi = 73°$

6. $\|\mathbf{a}\| = 65.4$, $\theta = 42°$
   $\|\mathbf{b}\| = 78.8$, $\varphi = 64°$

7. $\|\mathbf{a}\| = 979.8$, $\theta = 205.6°$
   $\|\mathbf{b}\| = 346.4$, $\varphi = 81.4°$

8. $\|\mathbf{a}\| = 286.2$, $\theta = 101.2°$
   $\|\mathbf{b}\| = 303.0$, $\varphi = 293.6°$

Exercise Set 10.5

In exercises 1–4, the magnitude $\|\mathbf{a}\|$ and direction θ of the vector $\mathbf{a}$ are given. Find $\mathbf{a}_x$ and $\mathbf{a}_y$.

1. $\|\mathbf{a}\| = 9.89, \theta = 87°$
2. $\|\mathbf{a}\| = 65.4, \theta = 42°$
3. $\|\mathbf{a}\| = 979.8, \theta = 205.6°$
4. $\|\mathbf{a}\| = 286.2, \theta = 101.2$

5. Along a stretch of the Mississippi River, the water flows directly south with a current of 5 mph. A person who can row a boat at 4 mph in still water attempts to row due west across the river. What is the actual direction and speed of the boat?

6. Two forces of 26 newtons and 40 newtons act on a point in the plane. If the angle between the forces is 73.6°, find the magnitude of the resultant force and its angle with the smaller of the two given forces.

7. An object weighing 100 lb is suspended between two trees by two ropes attached at the top of the object. The rope off to the left makes an angle of 58° with the vertical and the rope off to the right an angle of 27° with the vertical. Find the tension (magnitude of the force vector) along each rope.

Exercise Set 10.6

In exercises 1–4, an object is in simple harmonic motion described by the given equation, where x is measured in cm and t in seconds. Determine where the object is and whether it is moving up or down at $t = 0$, $t = 1.35$, $t = 2.57$, and $t = 12.39$. (Note: Take displacement below the equilibrium point as positive and above as negative.)

1. $x = 0.56 \sin 4.2t$
2. $x = 9.32 \cos(0.37t)$
3. $x = 0.24 \cos\left(\dfrac{t}{0.25} + 1.05\right)$
4. $x = 2.36 \sin(1.25t - 0.79)$

In exercises 5–8, the point is given in polar form. Determine the rectangular coordinates of the point.

5. $(-5.38, 231.7°)$
6. $(4.32, 325.9°)$
7. $(2.19, 2.59)$
8. $(-3.07, 0.93)$

In exercises 9–12, the point is written with rectangular coordinates. Determine the polar coordinates of the given point for θ in the interval $[0, \pi)$.

9. $(-3.83, 5.25)$
10. $(4.14, -3.86)$
11. $(-5.20, -3.12)$
12. $(7.46, 2.39)$

In exercises 13–16, for each of the polar equations determine the values of r for the following values of θ: 0, 1, 1.5, 2.25, 3.14, 4, 4.71, 5.12, 6.28.

13. $r = 3 + \sin\theta$ **14.** $r = 1 - 2\cos\theta$

15. $r = 1.4\theta$ **16.** $r = 1.5\sin 5\theta$

Exercise Set 12.1

In exercises 1–4, determine the fiftieth and one hundredth terms in the given arithmetic sequence; then compute the sum of the first 100 terms.

1. $-8.12, -4.04, 0.04, \ldots$ **2.** 9.76, 9.09, 8.42, ...

3. 6595, 6153, 5711, ... **4.** 18, 1800, 3582, ...

5. Determine x so that the following is an arithmetic sequence: 8.65, x, -4.11, ...

6. Determine x so that the following is an arithmetic sequence: 10^{-2}, x, 10^{-4}, ...

7. Determine the sum of all the odd numbers between 10,000 and 11,500.

8. Determine the sum of all the even numbers between 10,001 and 11,499.

Exercise Set 12.3

In exercises 1–4, determine the sum of the given geometric series.

1. $3.2 + 1.6 + 0.8 + 0.4 + \cdots$

2. $2 - 0.5 + 0.125 - 0.03125 + \cdots$

3. $\sum_{k=3}^{\infty} (1.63)(0.3)^k$ (approximate the sum to the nearest thousandth)

4. $\sum_{k=2}^{\infty} (2.8)(0.25)^k$ (approximate the sum to the nearest thousandth)

5. Suppose a wealthy friend agrees to pay you a penny on January 1, two cents on January 2, four cents on January 3, and so forth, doubling on any given day the amount given on the previous day. How much money will you receive during the month of January?

6. (Refer to exercise 5) How much will you receive on January 21?

7. (Refer to exercise 5) How much will you receive during the third week of January (the 15th thru the 21st)?

8. (Refer to exercise 5) Determine the first day on which the daily gift exceeds $1000 (use logarithms).

Calculator Exercises

Exercise Set 12.4

In exercises 1–4, for the given function f, evaluate the expression

$$\frac{f(2+h) - f(2)}{h}$$

for the h-values 0.01, 0.001, and 0.0001. As the h-values get smaller, what limiting value does the expression seem to be approaching?

1. $f(x) = x^2$ **2.** $f(x) = x^3$ **3.** $f(x) = x^4$ **4.** $f(x) = x^5$

5. For values of x such that $|x| < 1$, the infinite series $1 + x + x^2 + x^3 + \cdots$ has sum equal to $\dfrac{1}{1-x}$.

(a) Use the first five terms of the series with $x = 0.2$ to approximate $\dfrac{1}{1-0.2}$. What is the difference between the approximation and the exact value of $\dfrac{1}{1-0.2}$?

(b) Use the first five terms of the series with $x = 0.5$ to approximate $\dfrac{1}{1-0.5}$. What is the difference between the approximation and the exact value of $\dfrac{1}{1-0.5}$?

(c) Repeat the process outlined in parts (a) and (b) for the x-values 0.6, 0.75, and 0.8.

(d) Look at the differences between the approximate values and the actual values for the x-values 0.2, 0.5, 0.6, 0.75, and 0.8. What do you conclude?

6. For values of x such that $-1 < x \leq 1$, the infinite series

$$x - \frac{x^2}{2} + \frac{x^3}{3} - \frac{x^4}{4} + \frac{x^5}{5} - \cdots$$

has sum equal to $\ln(1+x)$. Repeat the process outlined in exercise 5. That is, for each of the x-values 0.2, 0.5, 0.6, 0.75, and 0.8, use the first five terms of the series above to approximate $\ln(1+x)$. Then use a calculator to evaluate $\ln(1+x)$. (Recall that the calculator value is also an approximation.)

7. Evaluate the first ten partial sums of the series

$$\sum_{n=1}^{\infty} \left(\frac{1}{n+1} - \frac{1}{n+2} \right).$$

(The sum of the series is 0.5.)

8. Evaluate the first ten partial sums of the series

$$\sum_{n=2}^{\infty} \frac{1}{n^2 - 1}.$$

(The sum of the series is 0.75.)

Exercise Set 12.6

For specific values of x and y, if the value of y is relatively small compared to the value of x, the quantity $(x + y)^n$ can be approximated by adding the first few terms in the expansion. In exercises 1–6, use the first four terms of the expansion to approximate the given quantity. Then, approximate the quantity by using a calculator, and compare the two estimates.

1. Expand $(1 + 0.01)^8$ to approximate $(1.01)^8$.

2. Expand $(1 + 0.02)^{10}$ to approximate $(1.02)^{10}$.

3. Expand $(1 - 0.02)^{10}$ to approximate $(0.98)^{10}$.

4. Expand $(1 - 0.01)^8$ to approximate $(0.99)^8$.

5. Expand $(2 + 0.001)^7$ to approximate $(2.001)^7$.

6. Expand $(3 + 0.001)^9$ to approximate $(3.001)^9$.

7. There is a rule stating that as long as nx is close to zero, $(1 + x)^n \approx 1 + nx$. Use a calculator to approximate $(1.002)^{15}$, and then use the rule above to approximate the same expression.

8. Repeat exercise 7 with the expression $(1.007)^{10}$.

9. Factorials are important in certain series approximations of function values. For example, the infinite series

$$\sum_{n=0}^{\infty} \frac{x^n}{n!} = 1 + x + \frac{x^2}{2!} + \frac{x^3}{3!} + \frac{x^4}{4!} + \cdots$$

has e^x as its sum for all real numbers x. Use the first five terms of the series to approximate $e^{0.2}$, $e^{0.5}$, $e^{0.6}$, $e^{0.75}$, and $e^{0.8}$; then use a calculator to approximate these same expressions.

10. In probability theory the binomial theorem is very important in establishing facts such as $1 = 1^5 = (0.6 + 0.4)^5$. Expand the rightmost expression and verify that it equals 1.

Tables

Table 1 Common Logarithms

x	0	1	2	3	4	5	6	7	8	9
1.0	.0000	.0043	.0086	.0128	.0170	.0212	.0253	.0294	.0334	.0374
1.1	.0414	.0453	.0492	.0531	.0569	.0607	.0645	.0682	.0719	.0755
1.2	.0792	.0828	.0864	.0899	.0934	.0969	.1004	.1038	.1072	.1106
1.3	.1139	.1173	.1206	.1239	.1271	.1303	.1335	.1367	.1399	.1430
1.4	.1461	.1492	.1523	.1553	.1584	.1614	.1644	.1673	.1703	.1732
1.5	.1761	.1790	.1818	.1847	.1875	.1903	.1931	.1959	.1987	.2014
1.6	.2041	.2068	.2095	.2122	.2148	.2175	.2201	.2227	.2253	.2279
1.7	.2304	.2330	.2355	.2380	.2405	.2430	.2455	.2480	.2504	.2529
1.8	.2553	.2577	.2601	.2625	.2648	.2672	.2695	.2718	.2742	.2765
1.9	.2788	.2810	.2833	.2856	.2878	.2900	.2923	.2945	.2967	.2989
2.0	.3010	.3032	.3054	.3075	.3096	.3118	.3139	.3160	.3181	.3201
2.1	.3222	.3243	.3263	.3284	.3304	.3324	.3345	.3365	.3385	.3404
2.2	.3424	.3444	.3464	.3483	.3502	.3522	.3541	.3560	.3579	.3598
2.3	.3617	.3636	.3655	.3674	.3692	.3711	.3729	.3747	.3766	.3784
2.4	.3802	.3820	.3838	.3856	.3874	.3892	.3909	.3927	.3945	.3962
2.5	.3979	.3997	.4014	.4031	.4048	.4065	.4082	.4099	.4116	.4133
2.6	.4150	.4166	.4183	.4200	.4216	.4232	.4249	.4265	.4281	.4298
2.7	.4314	.4330	.4346	.4362	.4378	.4393	.4409	.4425	.4440	.4456
2.8	.4472	.4487	.4502	.4518	.4533	.4548	.4564	.4579	.4594	.4609
2.9	.4624	.4639	.4654	.4669	.4683	.4698	.4713	.4728	.4742	.4757
3.0	.4771	.4786	.4800	.4814	.4829	.4843	.4857	.4871	.4886	.4900
3.1	.4914	.4928	.4942	.4955	.4969	.4983	.4997	.5011	.5024	.5038
3.2	.5051	.5065	.5079	.5092	.5105	.5119	.5132	.5145	.5159	.5172
3.3	.5185	.5198	.5211	.5224	.5237	.5250	.5263	.5276	.5289	.5302
3.4	.5315	.5328	.5340	.5353	.5366	.5378	.5391	.5403	.5416	.5428
3.5	.5441	.5453	.5465	.5478	.5490	.5502	.5514	.5527	.5539	.5551
3.6	.5563	.5575	.5587	.5599	.5611	.5623	.5635	.5647	.5658	.5670
3.7	.5682	.5694	.5705	.5717	.5729	.5740	.5752	.5763	.5775	.5786
3.8	.5798	.5809	.5821	.5832	.5843	.5855	.5866	.5877	.5888	.5899
3.9	.5911	.5922	.5933	.5944	.5955	.5966	.5977	.5988	.5999	.6010
4.0	.6021	.6031	.6042	.6053	.6064	.6075	.6085	.6096	.6107	.6117
4.1	.6128	.6138	.6149	.6160	.6170	.6180	.6191	.6201	.6212	.6222
4.2	.6232	.6243	.6253	.6263	.6274	.6284	.6294	.6304	.6314	.6325
4.3	.6335	.6345	.6355	.6365	.6375	.6385	.6395	.6405	.6415	.6425
4.4	.6435	.6444	.6454	.6464	.6474	.6484	.6493	.6503	.6513	.6522
4.5	.6532	.6542	.6551	.6561	.6571	.6580	.6590	.6599	.6609	.6618
4.6	.6628	.6637	.6646	.6656	.6665	.6675	.6684	.6693	.6702	.6712
4.7	.6721	.6730	.6739	.6749	.6758	.6767	.6776	.6785	.6794	.6803
4.8	.6812	.6821	.6830	.6839	.6848	.6857	.6866	.6875	.6884	.6893
4.9	.6902	.6911	.6920	.6928	.6937	.6946	.6955	.6964	.6972	.6981
5.0	.6990	.6998	.7007	.7016	.7024	.7033	.7042	.7050	.7059	.7067
5.1	.7076	.7084	.7093	.7101	.7110	.7118	.7126	.7135	.7143	.7152
5.2	.7160	.7168	.7177	.7185	.7193	.7202	.7210	.7218	.7226	.7235
5.3	.7243	.7251	.7259	.7267	.7275	.7284	.7292	.7300	.7308	.7316
5.4	.7324	.7332	.7340	.7348	.7356	.7364	.7372	.7380	.7388	.7396

Table 1 Common Logarithms (*continued*)

x	0	1	2	3	4	5	6	7	8	9
5.5	.7404	.7412	.7419	.7427	.7435	.7443	.7451	.7459	.7466	.7474
5.6	.7482	.7490	.7497	.7505	.7513	.7520	.7528	.7536	.7543	.7551
5.7	.7559	.7566	.7574	.7582	.7589	.7597	.7604	.7612	.7619	.7627
5.8	.7634	.7642	.7649	.7657	.7664	.7672	.7679	.7686	.7694	.7701
5.9	.7709	.7716	.7723	.7731	.7738	.7745	.7752	.7760	.7767	.7774
6.0	.7782	.7789	.7796	.7803	.7810	.7818	.7825	.7832	.7839	.7846
6.1	.7853	.7860	.7868	.7875	.7882	.7889	.7896	.7903	.7910	.7917
6.2	.7924	.7931	.7938	.7945	.7952	.7959	.7966	.7973	.7980	.7987
6.3	.7993	.8000	.8007	.8014	.8021	.8028	.8035	.8041	.8048	.8055
6.4	.8062	.8069	.8075	.8082	.8089	.8096	.8102	.8109	.8116	.8122
6.5	.8129	.8136	.8142	.8149	.8156	.8162	.8169	.8176	.8182	.8189
6.6	.8195	.8202	.8209	.8215	.8222	.8228	.8235	.8241	.8248	.8254
6.7	.8261	.8267	.8274	.8280	.8287	.8293	.8299	.8306	.8312	.8319
6.8	.8325	.8331	.8338	.8344	.8351	.8357	.8363	.8370	.8376	.8382
6.9	.8388	.8395	.8401	.8407	.8414	.8420	.8426	.8432	.8439	.8445
7.0	.8451	.8457	.8463	.8470	.8476	.8482	.8488	.8494	.8500	.8506
7.1	.8513	.8519	.8525	.8531	.8537	.8543	.8549	.8555	.8561	.8567
7.2	.8573	.8579	.8585	.8591	.8597	.8603	.8609	.8615	.8621	.8627
7.3	.8633	.8639	.8645	.8651	.8657	.8663	.8669	.8675	.8681	.8686
7.4	.8692	.8698	.8704	.8710	.8716	.8722	.8727	.8733	.8739	.8745
7.5	.8751	.8756	.8762	.8768	.8774	.8779	.8785	.8791	.8797	.8802
7.6	.8808	.8814	.8820	.8825	.8831	.8837	.8842	.8848	.8854	.8859
7.7	.8865	.8871	.8876	.8882	.8887	.8893	.8899	.8904	.8910	.8915
7.8	.8921	.8927	.8932	.8938	.8943	.8949	.8954	.8960	.8965	.8971
7.9	.8976	.8982	.8987	.8993	.8998	.9004	.9009	.9015	.9020	.9025
8.0	.9031	.9036	.9042	.9047	.9053	.9058	.9063	.9069	.9074	.9079
8.1	.9085	.9090	.9096	.9101	.9106	.9112	.9117	.9122	.9128	.9133
8.2	.9138	.9143	.9149	.9154	.9159	.9165	.9170	.9175	.9180	.9186
8.3	.9191	.9196	.9201	.9206	.9212	.9217	.9222	.9227	.9232	.9238
8.4	.9243	.9248	.9253	.9258	.9263	.9269	.9274	.9279	.9284	.9289
8.5	.9294	.9299	.9304	.9309	.9315	.9320	.9325	.9330	.9335	.9340
8.6	.9345	.9350	.9355	.9360	.9365	.9370	.9375	.9380	.9385	.9390
8.7	.9395	.9400	.9405	.9410	.9415	.9420	.9425	.9430	.9435	.9440
8.8	.9445	.9450	.9455	.9460	.9465	.9469	.9474	.9470	.9484	.9489
8.9	.9494	.9499	.9504	.9509	.9513	.9518	.9523	.9528	.9533	.9538
9.0	.9542	.9547	.9552	.9557	.9562	.9566	.9571	.9576	.9581	.9586
9.1	.9590	.9595	.9600	.9605	.9609	.9614	.9619	.9624	.9628	.9633
9.2	.9638	.9643	.9647	.9652	.9657	.9661	.9666	.9671	.9675	.9680
9.3	.9685	.9689	.9694	.9699	.9703	.9708	.9713	.9717	.9722	.9727
9.4	.9731	.9736	.9741	.9745	.9750	.9754	.9759	.9763	.9768	.9773
9.5	.9777	.9782	.9786	.9791	.9795	.9800	.9805	.9809	.9814	.9818
9.6	.9823	.9827	.9832	.9836	.9841	.9845	.9850	.9854	.9859	.9863
9.7	.9868	.9872	.9877	.9881	.9886	.9890	.9894	.9899	.9903	.9908
9.8	.9912	.9917	.9921	.9926	.9930	.9934	.9939	.9943	.9948	.9952
9.9	.9956	.9961	.9965	.9969	.9974	.9978	.9983	.9987	.9991	.9996

Table 2 Values of e^x and e^{-x}

x	e^x	e^{-x}	x	e^x	e^{-x}	x	e^x	e^{-x}
.00	1.00000	1.00000	.40	1.49182	.67032	.80	2.22554	.44032
.01	1.01005	.99005	.41	1.50682	.66365	.85	2.33965	.42741
.02	1.02020	.98020	.42	1.52196	.65705	.90	2.45960	.40657
.03	1.03045	.97045	.43	1.53726	.65051	.95	2.58571	.38674
.04	1.04081	.96079	.44	1.55271	.64404	1.00	2.71828	.36788
.05	1.05127	.95123	.45	1.56831	.63763	1.10	3.00416	.33287
.06	1.06184	.94176	.46	1.58407	.63128	1.20	3.32011	.30119
.07	1.07251	.93239	.47	1.59999	.62500	1.30	3.66929	.27253
.08	1.08329	.92312	.48	1.61607	.61878	1.40	4.05519	.24659
.09	1.09417	.91393	.49	1.63232	.61263	1.50	4.48168	.22313
.10	1.10517	.90484	.50	1.64872	.60653	1.60	4.95302	.20189
.11	1.11628	.89583	.51	1.66529	.60050	1.70	5.47394	.18268
.12	1.12750	.88692	.52	1.68203	.59452	1.80	6.04964	.16529
.13	1.13883	.87810	.53	1.69893	.58860	1.90	6.68589	.14956
.14	1.15027	.86936	.54	1.71601	.58275	2.00	7.38905	.13533
.15	1.16183	.86071	.55	1.73325	.57695	2.10	8.16616	.12245
.16	1.17351	.85214	.56	1.75067	.57121	2.20	9.02500	.11080
.17	1.18530	.84366	.57	1.76827	.56553	2.30	9.97417	.10025
.18	1.19722	.83527	.58	1.78604	.55990	2.40	11.02316	.09071
.19	1.20925	.82696	.59	1.80399	.55433	2.50	12.18248	.08208
.20	1.22140	.81873	.60	1.82212	.54881	3.00	20.08551	.04978
.21	1.23368	.81058	.61	1.84043	.54335	3.50	33.11545	.03020
.22	1.24608	.80252	.62	1.85893	.53794	4.00	54.59815	.01832
.23	1.25860	.79453	.63	1.87761	.53259	4.50	90.01713	.01111
.24	1.27125	.78663	.64	1.89648	.52729	5.00	148.41316	.00674
.25	1.28403	.77880	.65	1.91554	.52205	5.50	224.69193	.00409
.26	1.29693	.77105	.66	1.93479	.51685	6.00	403.42879	.00248
.27	1.30996	.76338	.67	1.95424	.51171	6.50	665.14163	.00150
.28	1.32313	.75578	.68	1.97388	.50662	7.00	1096.63316	.00091
.29	1.33643	.74826	.69	1.99372	.50158	7.50	1808.04241	.00055
.30	1.34986	.74082	.70	2.01375	.49659	8.00	2980.95799	.00034
.31	1.36343	.73345	.71	2.03399	.49164	8.50	4914.76884	.00020
.32	1.37713	.72615	.72	2.05443	.48675	9.00	8130.08392	.00012
.33	1.39097	.71892	.73	2.07508	.48191	9.50	13359.72683	.00007
.34	1.40495	.71177	.74	2.09594	.47711	10.00	22026.46579	.00005
.35	1.41907	.70469	.75	2.11700	.47237			
.36	1.43333	.69768	.76	2.13828	.46767			
.37	1.44773	.69073	.77	2.15977	.46301			
.38	1.46228	.68386	.78	2.18147	.45841			
.39	1.47698	.67706	.79	2.20340	.45384			

Table 3 Natural Logarithms

x	ln x	x	ln x	x	ln x
		4.5	1.5041	9.0	2.1972
0.1	−2.3026	4.6	1.5261	9.1	2.2083
0.2	−1.6094	4.7	1.5476	9.2	2.2192
0.3	−1.2040	4.8	1.5686	9.3	2.2300
0.4	−0.9163	4.9	1.5892	9.4	2.2407
0.5	−0.6931	5.0	1.6094	9.5	2.2513
0.6	−0.5108	5.1	1.6292	9.6	2.2618
0.7	−0.3567	5.2	1.6487	9.7	2.2721
0.8	−0.2231	5.3	1.6677	9.8	2.2824
0.9	−0.1054	5.4	1.6864	9.9	2.2925
1.0	0.0000	5.5	1.7047	10	2.3026
1.1	0.0953	5.6	1.7228	11	2.3979
1.2	0.1823	5.7	1.7405	12	2.4849
1.3	0.2624	5.8	1.7579	13	2.5649
1.4	0.3365	5.9	1.7750	14	2.6391
1.5	0.4055	6.0	1.7918	15	2.7081
1.6	0.4700	6.1	1.8083	16	2.7726
1.7	0.5306	6.2	1.8245	17	2.8332
1.8	0.5878	6.3	1.8405	18	2.8904
1.9	0.6419	6.4	1.8563	19	2.9444
2.0	0.6931	6.5	1.8718	20	2.9957
2.1	0.7419	6.6	1.8871	25	3.2189
2.2	0.7885	6.7	1.9021	30	3.4012
2.3	0.8329	6.8	1.9169	35	3.5553
2.4	0.8755	6.9	1.9315	40	3.6889
2.5	0.9163	7.0	1.9459	45	3.8067
2.6	0.9555	7.1	1.9601	50	3.9120
2.7	0.9933	7.2	1.9741	55	4.0073
2.8	1.0296	7.3	1.9879	60	4.0943
2.9	1.0647	7.4	2.0015	65	4.1744
3.0	1.0986	7.5	2.0149	70	4.2485
3.1	1.1314	7.6	2.0281	75	4.3175
3.2	1.1632	7.7	2.0412	80	4.3820
3.3	1.1939	7.8	2.0541	85	4.4427
3.4	1.2238	7.9	2.0669	90	4.4998
3.5	1.2528	8.0	2.0794	100	4.6052
3.6	1.2809	8.1	2.0919	110	4.7005
3.7	1.3083	8.2	2.1041	120	4.7875
3.8	1.3350	8.3	2.1163	130	4.8676
3.9	1.3610	8.4	2.1282	140	4.9416
4.0	1.3863	8.5	2.1401	150	5.0106
4.1	1.4110	8.6	2.1518	160	5.0752
4.2	1.4351	8.7	2.1633	170	5.1358
4.3	1.4586	8.8	2.1748	180	5.1930
4.4	1.4816	8.9	2.1861	190	5.2470

Table 4 Values of Trigonometric Functions

α (degrees)	α (radians)	sin α	cos α	tan α	cot α	sec α	csc α		
0°00′	.0000	.0000	1.0000	.0000	—	1.000	—	1.5708	**90°00′**
10	.0029	.0029	1.0000	.0029	343.8	1.000	343.8	1.5679	50
20	.0058	.0058	1.0000	.0058	171.9	1.000	171.9	1.5650	40
30	.0087	.0087	1.0000	.0087	114.6	1.000	114.6	1.5621	30
40	.0116	.0116	.9999	.0116	85.94	1.000	85.95	1.5592	20
50	.0145	.0145	.9999	.0145	68.75	1.000	68.76	1.5563	10
1°00′	.0175	.0175	.9998	.0175	57.29	1.000	57.30	1.5533	**89°00′**
10	.0204	.0204	.9998	.0204	49.10	1.000	49.11	1.5504	50
20	.0233	.0233	.9997	.0233	42.96	1.000	42.98	1.5475	40
30	.0262	.0262	.9997	.0262	38.19	1.000	38.20	1.5446	30
40	.0291	.0291	.9996	.0291	34.37	1.000	34.38	1.5417	20
50	.0320	.0320	.9995	.0320	31.24	1.001	31.26	1.5388	10
2°00′	.0349	.0349	.9994	.0349	28.64	1.001	28.65	1.5359	**88°00′**
10	.0378	.0378	.9993	.0378	26.43	1.001	26.45	1.5330	50
20	.0407	.0407	.9992	.0407	24.54	1.001	24.56	1.5301	40
30	.0436	.0436	.9990	.0437	22.90	1.001	22.93	1.5272	30
40	.0465	.0465	.9989	.0466	21.47	1.001	21.49	1.5243	20
50	.0495	.0494	.9988	.0495	20.21	1.001	20.23	1.5213	10
3°00′	.0524	.0523	.9986	.0524	19.08	1.001	19.11	1.5184	**87°00′**
10	.0553	.0552	.9985	.0553	18.07	1.002	18.10	1.5155	50
20	.0582	.0581	.9983	.0582	17.17	1.002	17.20	1.5126	40
30	.0611	.0610	.9981	.0612	16.35	1.002	16.38	1.5097	30
40	.0640	.0640	.9980	.0641	15.60	1.002	15.64	1.5068	20
50	.0669	.0669	.9978	.0670	14.92	1.002	14.96	1.5039	10
4°00′	.0698	.0698	.9976	.0699	14.30	1.002	14.34	1.5010	**86°00′**
10	.0727	.0727	.9974	.0729	13.73	1.003	13.76	1.4981	50
20	.0756	.0756	.9971	.0758	13.20	1.003	13.23	1.4952	40
30	.0785	.0785	.9969	.0787	12.71	1.003	12.75	1.4923	30
40	.0814	.0814	.9967	.0816	12.25	1.003	12.29	1.4893	20
50	.0844	.0843	.9964	.0846	11.83	1.004	11.87	1.4864	10
5°00′	.0873	.0872	.9962	.0875	11.43	1.004	11.47	1.4835	**85°00′**
10	.0902	.0901	.9959	.0904	11.06	1.004	11.10	1.4806	50
20	.0931	.0929	.9957	.0934	10.71	1.004	10.76	1.4777	40
30	.0960	.0958	.9954	.0963	10.39	1.005	10.43	1.4748	30
40	.0989	.0987	.9951	.0992	10.08	1.005	10.13	1.4719	20
50	.1018	.1016	.9948	.1022	9.788	1.005	9.839	1.4690	10
6°00′	.1047	.1045	.9945	.1051	9.514	1.006	9.567	1.4661	**84°00′**
10	.1076	.1074	.9942	.1080	9.255	1.006	9.309	1.4632	50
20	.1105	.1103	.9939	.1110	9.010	1.006	9.065	1.4603	40
30	.1134	.1132	.9936	.1139	8.777	1.006	8.834	1.4573	30
40	.1164	.1161	.9932	.1169	8.556	1.007	8.614	1.4544	20
50	.1193	.1190	.9929	.1198	8.345	1.007	8.405	1.4515	10
		cos α	sin α	cot α	tan α	csc α	sec α	α (radians)	α (degrees)

Table 4 Values of Trigonometric Functions (*continued*)

α (degrees)	α (radians)	sin α	cos α	tan α	cot α	sec α	csc α		
7°00′	.1222	.1219	.9925	.1228	8.144	1.008	8.206	1.4486	**83°00′**
10	.1251	.1248	.9922	.1257	7.953	1.008	8.016	1.4457	50
20	.1280	.1276	.9918	.1287	7.770	1.008	7.834	1.4428	40
30	.1309	.1305	.9914	.1317	7.596	1.009	7.661	1.4399	30
40	.1338	.1334	.9911	.1346	7.429	1.009	7.496	1.4370	20
50	.1376	.1363	.9907	.1376	7.269	1.009	7.337	1.4341	10
8°00′	.1396	.1392	.9903	.1405	7.115	1.010	7.185	1.4312	**82°00′**
10	.1425	.1421	.9899	.1435	6.968	1.010	7.040	1.4283	50
20	.1454	.1449	.9894	.1465	6.827	1.011	6.900	1.4254	40
30	.1484	.1478	.9890	.1495	6.691	1.011	6.765	1.4224	30
40	.1513	.1507	.9886	.1524	6.561	1.012	6.636	1.4195	20
50	.1542	.1536	.9881	.1554	6.435	1.012	6.512	1.4166	10
9°00′	.1571	.1564	.9877	.1584	6.314	1.012	6.392	1.4137	**81°00′**
10	.1600	.1593	.9872	.1614	6.197	1.013	6.277	1.4108	50
20	.1629	.1622	.9868	.1644	6.084	1.013	6.166	1.4079	40
30	.1658	.1650	.9863	.1673	5.976	1.014	6.059	1.4050	30
40	.1687	.1679	.9858	.1703	5.871	1.014	5.955	1.4021	20
50	.1716	.1708	.9853	.1733	5.769	1.015	5.855	1.3992	10
10°00′	.1745	.1736	.9848	.1763	5.671	1.015	5.759	1.3963	**80°00′**
10	.1774	.1765	.9843	.1793	5.576	1.016	5.665	1.3934	50
20	.1804	.1794	.9838	.1823	5.485	1.016	5.575	1.3904	40
30	.1833	.1822	.9833	.1853	5.396	1.017	5.487	1.3875	30
40	.1862	.1851	.9827	.1883	5.309	1.018	5.403	1.3846	20
50	.1891	.1880	.9822	.1914	5.226	1.018	5.320	1.3817	10
11°00′	.1920	.1908	.9816	.1944	5.145	1.019	5.241	1.3788	**79°00′**
10	.1949	.1937	.9811	.1974	5.066	1.019	5.164	1.3759	50
20	.1978	.1965	.9805	.2004	4.989	1.020	5.089	1.3730	40
30	.2007	.1994	.9799	.2035	4.915	1.020	5.016	1.3701	30
40	.2036	.2022	.9793	.2065	4.843	1.021	4.945	1.3672	20
50	.2065	.2051	.9787	.2095	4.773	1.022	4.876	1.3643	10
12°00′	.2094	.2079	.9781	.2126	4.705	1.022	4.810	1.3614	**78°00′**
10	.2123	.2108	.9775	.2156	4.638	1.023	4.745	1.3584	50
20	.2153	.2136	.9769	.2186	4.574	1.024	4.682	1.3555	40
30	.2182	.2164	.9763	.2217	4.511	1.024	4.620	1.3526	30
40	.2211	.2193	.9757	.2247	4.449	1.025	4.560	1.3497	20
50	.2240	.2221	.9750	.2278	4.390	1.026	4.502	1.3468	10
13°00′	.2269	.2250	.9744	.2309	4.331	1.026	4.445	1.3439	**77°00′**
10	.2298	.2278	.9737	.2339	4.275	1.027	4.390	1.3410	50
20	.2327	.2306	.9730	.2370	4.219	1.028	4.336	1.3381	40
30	.2356	.2334	.9724	.2401	4.165	1.028	4.284	1.3352	30
40	.2385	.2363	.9717	.2432	4.113	1.029	4.232	1.3323	20
50	.2414	.2391	.9710	.2462	4.061	1.030	4.182	1.3294	10
		cos α	sin α	cot α	tan α	csc α	sec α	α (radians)	α (degrees)

Table 4 Values of Trigonometric Functions (*continued*)

α (degrees)	α (radians)	sin α	cos α	tan α	cot α	sec α	csc α		
14°00′	.2443	.2419	.9703	.2493	4.011	1.031	4.134	1.3265	**76°00′**
10	.2473	.2447	.9696	.2524	3.962	1.031	4.086	1.3235	50
20	.2502	.2476	.9689	.2555	3.914	1.032	4.039	1.3206	40
30	.2531	.2504	.9681	.2586	3.867	1.033	3.994	1.3177	30
40	.2560	.2532	.9674	.2617	3.821	1.034	3.950	1.3148	20
50	.2589	.2560	.9667	.2648	3.776	1.034	3.906	1.3119	10
15°00′	.2618	.2588	.9659	.2679	3.732	1.035	3.864	1.3090	**75°00′**
10	.2647	.2616	.9652	.2711	3.689	1.036	3.822	1.3061	50
20	.2676	.2644	.9644	.2742	3.647	1.037	3.782	1.3032	40
30	.2705	.2672	.9636	.2773	3.606	1.038	3.742	1.3003	30
40	.2734	.2700	.9628	.2805	3.566	1.039	3.703	1.2974	20
50	.2763	.2728	.9621	.2836	3.526	1.039	3.665	1.2945	10
16°00′	.2793	.2756	.9613	.2867	3.487	1.040	3.628	1.2915	**74°00′**
10	.2822	.2784	.9605	.2899	3.450	1.041	3.592	1.2886	50
20	.2851	.2812	.9596	.2931	3.412	1.042	3.556	1.2857	40
30	.2880	.2840	.9588	.2962	3.376	1.043	3.521	1.2828	30
40	.2909	.2868	.9580	.2994	3.340	1.044	3.487	1.2799	20
50	.2938	.2896	.9572	.3026	3.305	1.045	3.453	1.2770	10
17°00′	.2967	.2924	.9563	.3057	3.271	1.046	3.420	1.2741	**73°00′**
10	.2996	.2952	.9555	.3089	3.237	1.047	3.388	1.2712	50
20	.3025	.2979	.9546	.3121	3.204	1.048	3.356	1.2683	40
30	.3054	.3007	.9537	.3153	3.172	1.049	3.326	1.2654	30
40	.3083	.3035	.9528	.3185	3.140	1.049	3.295	1.2625	20
50	.3113	.3062	.9520	.3217	3.108	1.050	3.265	1.2595	10
18°00′	.3142	.3090	.9511	.3249	3.078	1.051	3.236	1.2566	**72°00′**
10	.3171	.3118	.9502	.3281	3.047	1.052	3.207	1.2537	50
20	.3200	.3145	.9492	.3314	3.018	1.053	3.179	1.2508	40
30	.3229	.3173	.9483	.3346	2.989	1.054	3.152	1.2479	30
40	.3258	.3201	.9474	.3378	2.960	1.056	3.124	1.2450	20
50	.3287	.3228	.9465	.3411	2.932	1.057	3.098	1.2421	10
19°00′	.3316	.3256	.9455	.3443	2.904	1.058	3.072	1.2392	**71°00′**
10	.3345	.3283	.9446	.3476	2.877	1.059	3.046	1.2363	50
20	.3374	.3311	.9436	.3508	2.850	1.060	3.021	1.2334	40
30	.3403	.3338	.9426	.3541	2.824	1.061	2.996	1.2305	30
40	.3432	.3365	.9417	.3574	2.798	1.062	2.971	1.2275	20
50	.3462	.3393	.9407	.3607	2.773	1.063	2.947	1.2246	10
20°00′	.3491	.3420	.9397	.3640	2.747	1.064	2.924	1.2217	**70°00′**
10	.3520	.3448	.9387	.3673	2.723	1.065	2.901	1.2188	50
20	.3549	.3475	.9377	.3706	2.699	1.066	2.878	1.2159	40
30	.3578	.3502	.9367	.3739	2.675	1.068	2.855	1.2130	30
40	.3607	.3529	.9356	.3772	2.651	1.069	2.833	1.2101	20
50	.3636	.3557	.9346	.3805	2.628	1.070	2.812	1.2072	10
		cos α	sin α	cot α	tan α	csc α	sec α	α (radians)	α (degrees)

Table 4 Values of Trigonometric Functions (*continued*)

α (degrees)	α (radians)	sin α	cos α	tan α	cot α	sec α	csc α		
21°00′	.3665	.3584	.9336	.3839	2.605	1.071	2.790	1.2043	**69°00′**
10	.3694	.3611	.9325	.3872	2.583	1.072	2.769	1.2014	50
20	.3723	.3638	.9315	.3906	2.560	1.074	2.749	1.1985	40
30	.3752	.3665	.9304	.3939	2.539	1.075	2.729	1.1956	30
40	.3782	.3692	.9293	.3973	2.517	1.076	2.709	1.1926	20
50	.3811	.3719	.9283	.4006	2.496	1.077	2.689	1.1897	10
22°00′	.3840	.3746	.9272	.4040	2.475	1.079	2.669	1.1868	**68°00′**
10	.3869	.3773	.9261	.4074	2.455	1.080	2.650	1.1839	50
20	.3898	.3800	.9250	.4108	2.434	1.081	2.632	1.1810	40
30	.3927	.3827	.9239	.4142	2.414	1.082	2.613	1.1781	30
40	.3956	.3854	.9228	.4176	2.394	1.084	2.595	1.1752	20
50	.3985	.3881	.9216	.4210	2.375	1.085	2.577	1.1723	10
23°00′	.4014	.3907	.9205	.4245	2.356	1.086	2.559	1.1694	**67°00′**
10	.4043	.3934	.9194	.4279	2.337	1.088	2.542	1.1665	50
20	.4072	.3961	.9182	.4314	2.318	1.089	2.525	1.1636	40
30	.4102	.3987	.9171	.4348	2.300	1.090	2.508	1.1606	30
40	.4131	.4014	.9159	.4383	2.282	1.092	2.491	1.1577	20
50	.4160	.4041	.9147	.4417	2.264	1.093	2.475	1.1548	10
24°00′	.4189	.4067	.9135	.4452	2.246	1.095	2.459	1.1519	**66°00′**
10	.4218	.4094	.9124	.4487	2.229	1.096	2.443	1.1490	50
20	.4247	.4120	.9112	.4522	2.211	1.097	2.427	1.1461	40
30	.4276	.4147	.9100	.4557	2.194	1.099	2.411	1.1432	30
40	.4305	.4173	.9088	.4592	2.177	1.100	2.396	1.1403	20
50	.4334	.4200	.9075	.4628	2.161	1.102	2.381	1.1374	10
25°00′	.4363	.4226	.9063	.4663	2.145	1.103	2.366	1.1345	**65°00′**
10	.4392	.4253	.9051	.4699	2.128	1.105	2.352	1.1316	50
20	.4422	.4279	.9038	.4734	2.112	1.106	2.337	1.1286	40
30	.4451	.4305	.9026	.4770	2.097	1.108	2.323	1.1257	30
40	.4480	.4331	.9013	.4806	2.081	1.109	2.309	1.1228	20
50	.4509	.4358	.9001	.4841	2.066	1.111	2.295	1.1199	10
26°00′	.4538	.4384	.8988	.4877	2.050	1.113	2.281	1.1170	**64°00′**
10	.4567	.4410	.8975	.4913	2.035	1.114	2.268	1.1141	50
20	.4596	.4436	.8962	.4950	2.020	1.116	2.254	1.1112	40
30	.4625	.4462	.8949	.4986	2.006	1.117	2.241	1.1083	30
40	.4654	.4488	.8936	.5022	1.991	1.119	2.228	1.1054	20
50	.4683	.4514	.8923	.5059	1.977	1.121	2.215	1.1025	10
27°00′	.4712	.4540	.8910	.5095	1.963	1.122	2.203	1.0996	**63°00′**
10	.4741	.4566	.8897	.5132	1.949	1.124	2.190	1.0966	50
20	.4771	.4592	.8884	.5169	1.935	1.126	2.178	1.0937	40
30	.4800	.4617	.8870	.5206	1.921	1.127	2.166	1.0908	30
40	.4829	.4643	.8857	.5243	1.907	1.129	2.154	1.0879	20
50	.4858	.4669	.8843	.5280	1.894	1.131	2.142	1.0850	10
		cos α	sin α	cot α	tan α	csc α	sec α	α (radians)	α (degrees)

Table 4 Values of Trigonometric Functions (*continued*)

α (degrees)	α (radians)	sin α	cos α	tan α	cot α	sec α	csc α		
28°00′	.4887	.4695	.8829	.5317	1.881	1.133	2.130	1.0821	**62°00′**
10	.4916	.4720	.8816	.5354	1.868	1.134	2.118	1.0792	50
20	.4945	.4746	.8802	.5392	1.855	1.136	2.107	1.0763	40
30	.4974	.4772	.8788	.5430	1.842	1.138	2.096	1.0734	30
40	.5003	.4797	.8774	.5467	1.829	1.140	2.085	1.0705	20
50	.5032	.4823	.8760	.5505	1.816	1.142	2.074	1.0676	10
29°00′	.5061	.4848	.8746	.5543	1.804	1.143	2.063	1.0647	**61°00′**
10	.5091	.4874	.8732	.5581	1.792	1.145	2.052	1.0617	50
20	.5120	.4899	.8718	.5619	1.780	1.147	2.041	1.0588	40
30	.5149	.4924	.8704	.5658	1.767	1.149	2.031	1.0559	30
40	.5178	.4950	.8689	.5696	1.756	1.151	2.020	1.0530	20
50	.5207	.4975	.8675	.5735	1.744	1.153	2.010	1.0501	10
30°00′	.5236	.5000	.8660	.5774	1.732	1.155	2.000	1.0472	**60°00′**
10	.5265	.5025	.8646	.5812	1.720	1.157	1.990	1.0443	50
20	.5294	.5050	.8631	.5851	1.709	1.159	1.980	1.0414	40
30	.5323	.5075	.8616	.5890	1.698	1.161	1.970	1.0385	30
40	.5352	.5100	.8601	.5930	1.686	1.163	1.961	1.0356	20
50	.5381	.5125	.8587	.5969	1.675	1.165	1.951	1.0327	10
31°00′	.5411	.5150	.8572	.6009	1.664	1.167	1.942	1.0297	**59°00′**
10	.5440	.5175	.8557	.6048	1.653	1.169	1.932	1.0268	50
20	.5469	.5200	.8542	.6088	1.643	1.171	1.923	1.0239	40
30	.5498	.5225	.8526	.6128	1.632	1.173	1.914	1.0210	30
40	.5527	.5250	.8511	.6168	1.621	1.175	1.905	1.0181	20
50	.5556	.5275	.8496	.6208	1.611	1.177	1.896	1.0152	10
32°00′	.5585	.5299	.8480	.6249	1.600	1.179	1.887	1.0123	**58°00′**
10	.5614	.5324	.8465	.6289	1.590	1.181	1.878	1.0094	50
20	.5643	.5348	.8450	.6330	1.580	1.184	1.870	1.0065	40
30	.5672	.5373	.8434	.6371	1.570	1.186	1.861	1.0036	30
40	.5701	.5398	.8418	.6412	1.560	1.188	1.853	1.0007	20
50	.5730	.5422	.8403	.6453	1.550	1.190	1.844	.9977	10
33°00′	.5760	.5446	.8387	.6494	1.540	1.192	1.836	.9948	**57°00′**
10	.5789	.5471	.8371	.6536	1.530	1.195	1.828	.9919	50
20	.5818	.5495	.8355	.6577	1.520	1.197	1.820	.9890	40
30	.5847	.5519	.8339	.6619	1.511	1.199	1.812	.9861	30
40	.5876	.5544	.8323	.6661	1.501	1.202	1.804	.9832	20
50	.5905	.5568	.8307	.6703	1.492	1.204	1.796	.9803	10
34°00′	.5934	.5592	.8290	.6745	1.483	1.206	1.788	.9774	**56°00′**
10	.5963	.5616	.8274	.6787	1.473	1.209	1.781	.9745	50
20	.5992	.5640	.8258	.6830	1.464	1.211	1.773	.9716	40
30	.6021	.5664	.8241	.6873	1.455	1.213	1.766	.9687	30
40	.6050	.5688	.8225	.6916	1.446	1.216	1.758	.9657	20
50	.6080	.5712	.8208	.6959	1.437	1.218	1.751	.9628	10
		cos α	sin α	cot α	tan α	csc α	sec α	α (radians)	α (degrees)

Table 4 Values of Trigonometric Functions (*continued*)

α (degrees)	α (radians)	sin α	cos α	tan α	cot α	sec α	csc α		
35°00′	.6109	.5736	.8192	.7002	1.428	1.221	1.743	.9599	**55°00′**
10	.6138	.5760	.8175	.7046	1.419	1.223	1.736	.9570	50
20	.6167	.5783	.8158	.7089	1.411	1.226	1.729	.9541	40
30	.6196	.5807	.8141	.7133	1.402	1.228	1.722	.9512	30
40	.6225	.5831	.8124	.7177	1.393	1.231	1.715	.9483	20
50	.6254	.5854	.8107	.7221	1.385	1.233	1.708	.9454	10
36°00′	.6283	.5878	.8090	.7265	1.376	1.236	1.701	.9425	**54°00′**
10	.6312	.5901	.8073	.7310	1.368	1.239	1.695	.9396	50
20	.6341	.5925	.8056	.7355	1.360	1.241	1.688	.9367	40
30	.6370	.5948	.8039	.7400	1.351	1.244	1.681	.9338	30
40	.6400	.5972	.8021	.7445	1.343	1.247	1.675	.9308	20
50	.6429	.5995	.8004	.7490	1.335	1.249	1.668	.9279	10
37°00′	.6458	.6018	.7986	.7536	1.327	1.252	1.662	.9250	**53°00′**
10	.6487	.6041	.7969	.7581	1.319	1.255	1.655	.9221	50
20	.6516	.6065	.7951	.7627	1.311	1.258	1.649	.9192	40
30	.6545	.6088	.7934	.7673	1.303	1.260	1.643	.9163	30
40	.6574	.6111	.7916	.7720	1.295	1.263	1.636	.9134	20
50	.6603	.6134	.7898	.7766	1.288	1.266	1.630	.9105	10
38°00′	.6632	.6157	.7880	.7813	1.280	1.269	1.624	.9076	**52°00′**
10	.6661	.6180	.7862	.7860	1.272	1.272	1.618	.9047	50
20	.6690	.6202	.7844	.7907	1.265	1.275	1.612	.9018	40
30	.6720	.6225	.7826	.7954	1.257	1.278	1.606	.8988	30
40	.6749	.6248	.7808	.8002	1.250	1.281	1.601	.8959	20
50	.6778	.6271	.7790	.8050	1.242	1.284	1.595	.8930	10
39°00′	.6807	.6293	.7771	.8098	1.235	1.287	1.589	.8901	**51°00′**
10	.6836	.6316	.7753	.8146	1.228	1.290	1.583	.8872	50
20	.6865	.6338	.7735	.8195	1.220	1.293	1.578	.8843	40
30	.6894	.6361	.7716	.8243	1.213	1.296	1.572	.8814	30
40	.6923	.6383	.7698	.8292	1.206	1.299	1.567	.8785	20
50	.6952	.6406	.7679	.8342	1.199	1.302	1.561	.8756	10
40°00′	.6981	.6428	.7660	.8391	1.192	1.305	1.556	.8727	**50°00′**
10	.7010	.6450	.7642	.8441	1.185	1.309	1.550	.8698	50
20	.7039	.6472	.7623	.8491	1.178	1.312	1.545	.8668	40
30	.7069	.6494	.7604	.8541	1.171	1.315	1.540	.8639	30
40	.7098	.6517	.7585	.8591	1.164	1.318	1.535	.8610	20
50	.7127	.6539	.7566	.8642	1.157	1.322	1.529	.8581	10
41°00′	.7156	.6561	.7547	.8693	1.150	1.325	1.524	.8552	**49°00′**
10	.7185	.6583	.7528	.8744	1.144	1.328	1.519	.8523	50
20	.7214	.6604	.7509	.8796	1.137	1.332	1.514	.8494	40
30	.7243	.6626	.7490	.8847	1.130	1.335	1.509	.8465	30
40	.7272	.6648	.7470	.8899	1.124	1.339	1.504	.8436	20
50	.7301	.6670	.7451	.8952	1.117	1.342	1.499	.8407	10
		cos α	sin α	cot α	tan α	csc α	sec α	α (radians)	α (degrees)

Table 4 Values of Trigonometric Functions (*continued*)

α (degrees)	α (radians)	sin α	cos α	tan α	cot α	sec α	csc α		
42°00′	.7330	.6691	.7431	.9004	1.111	1.346	1.494	.8378	**48°00′**
10	.7359	.6713	.7412	.9057	1.104	1.349	1.490	.8348	50
20	.7389	.6734	.7392	.9110	1.098	1.353	1.485	.8319	40
30	.7418	.6756	.7373	.9163	1.091	1.356	1.480	.8290	30
40	.7447	.6777	.7353	.9217	1.085	1.360	1.476	.8261	20
50	.7476	.6799	.7333	.9271	1.079	1.364	1.471	.8232	10
43°00′	.7505	.6820	.7314	.9325	1.072	1.367	1.466	.8203	**47°00′**
10	.7534	.6841	.7294	.9380	1.066	1.371	1.462	.8174	50
20	.7563	.6862	.7274	.9435	1.060	1.375	1.457	.8145	40
30	.7592	.6884	.7254	.9490	1.054	1.379	1.453	.8116	30
40	.7621	.6905	.7234	.9545	1.048	1.382	1.448	.8087	20
50	.7560	.6926	.7214	.9601	1.042	1.386	1.444	.8058	10
44°00′	.7679	.6947	.7193	.9657	1.036	1.390	1.440	.8029	**46°00′**
10	.7709	.6967	.7173	.9713	1.030	1.394	1.435	.7999	50
20	.7738	.6988	.7153	.9770	1.024	1.398	1.431	.7970	40
30	.7767	.7009	.7133	.9827	1.018	1.402	1.427	.7941	30
40	.7796	.7030	.7112	.9884	1.012	1.406	1.423	.7912	20
50	.7825	.7050	.7092	.9942	1.006	1.410	1.418	.7883	10
45°00′	.7854	.7071	.7071	1.000	1.000	1.414	1.414	.7854	**45°00′**
		cos α	sin α	cot α	tan α	csc α	sec α	α (radians)	α (degrees)

Answers to Exercises

Chapter 1

Exercise Set 1.1, page 5

1. $8x = 47$ **3.** $x + 5 = 2x$ **5.** $\frac{1}{3} + \frac{3}{5}x = 2x$ **7.** $n + (n + 1) = 3n$ **9.** $x + y = 11$

11. $2x = y - 15$ **13.** $x + 6 = 2y - 8$ **15.** $\frac{1}{2}(x + y) = xy + 19$ **17.** $10n$ **19.** $\frac{d}{12}$ **21.** $3n + 7m$

23. $x + 22 = 36$ **25.** $99 + 0.20m = 198$

Exercise Set 1.2, page 10

1. rational, real   **3.** integer, rational, real   **5.** rational, real   **7.** rational, real   **9.** True

11. False   **13.** True   **15.** False   **17.** False   **19.** $-5, \frac{1}{5}$ **21.** $\frac{2}{3}, -\frac{3}{2}$ **23.** 0, none   **25.** $-\pi, \frac{1}{\pi}$

27. $-x, \frac{1}{x}$ **29.** $-\frac{1}{x}, x$ **31.** -2 **33.** -2 **35.** -3 **37.** 14   **39.** $\frac{1}{4}x = 5\left(\frac{1}{x}\right)$

41. $3\left(x + \dfrac{1}{x}\right) = x + 4$ **43.** $-(x + y) = xy - 5$ **45.** False, $(2)(3)(2)(4) \neq (2)(3)(4)$

47. True   **49.** False, $\dfrac{1}{2(2)} \neq \left(\dfrac{1}{2}\right)(2)$ **51.** $3(7 + 2 - 5)$ **53.** $[(12 - 3) \cdot 8] \div 6$

Exercise Set 1.3, page 15

1. 4   **3.** 0   **5.** $-\dfrac{1}{5}$ **7.** 13   **9.** 30   **11.** $0 > -11$ **13.** $\dfrac{3}{5} > \dfrac{4}{7}$ **15.** $-0.624 > -\dfrac{5}{8}$

17. $2x - 7 > 20$ **19.** $x + 6 > 0$ **21.** [number line]   **23.** [number line]   **25.** [number line]

27. 12   **29.** 18   **31.** 18   **33.** $a = 2, b = -5,$ and $c = 5$.   **35.** False, $x = -1$

37. True   **39.** True   **41.** False, $|-1| \neq -1$

Exercise Set 1.4, page 20

1. 625   **3.** -625 **5.** $\dfrac{1}{8}$ **7.** 0   **9.** Not defined   **11.** $\dfrac{1}{9}$ **13.** $\dfrac{3}{2}$ **15.** 8   **17.** 9   **19.** 151, 263

21. $\dfrac{1}{6^2}$ **23.** $\left(\dfrac{7}{3}\right)^5$ **25.** $\dfrac{1}{a^2}$ **27.** x^3 **29.** -3^4 **31.** $\dfrac{1}{x}$ **33.** 8^{-3} **35.** $\dfrac{1}{4^{-5}}$ **37.** x^{-5}

39. -3^{-8} **41.** $\left(\dfrac{3}{2}\right)^{-5}$ **43.** y^{-1} **45.** 34   **47.** -31 **49.** -125 **51.** $-\dfrac{4}{45}$ **53.** 3^7 **55.** 3^0

57. 3^{30} **59.** 3^{18} **61.** $-4x^3y^3$ **63.** $14x^4y$ **65.** $\dfrac{15y}{x^2}$ **67.** $\dfrac{12xy}{z^5}$ **69.** $8x^3$ **71.** $36x^2y^4z^2$

73. $-125z^{12}$ **75.** $\dfrac{2x}{3yz}$ **77.** $\dfrac{7x^3z^2}{10y^5}$ **79.** $\dfrac{x^7}{yz^2}$ **81.** $\dfrac{27z^3}{8x^3y^3}$ **83.** $\dfrac{243z^{15}}{x^{10}}$ **85.** False   **87.** False

89. True   **91.** False   **93.** False   **95.** $\dfrac{4x^3}{yz^2}$ **97.** $\dfrac{16x^8y}{z^5}$ **99.** $\dfrac{1}{x^9y^6z^{26}}$ **101.** $\dfrac{1}{(x + y)^3}$

103. True   **105.** True   **107.** True   **109.** True   **111.** $4\pi x^6 \text{ cm}^2$ **113.** $25p^{-2} \text{ in}^2$ **115.** $100\pi n^4 \text{ cm}^3$

117. $V_{\text{cylinder}} > V_{\text{sphere}}$

Exercise Set 1.5, page 26

1. 1 term; degree 8 3. 3 terms; degree 2 5. 4 terms; degree 3 7. 4 terms; degree 5
9. 1 term; degree 0 11. $3x^3 - 5x^2 + 10x - 4$ 13. $4u^4 - 3u^2 + u$ 15. $6x^4 + x^2 - 8x + 11$
17. $-2v^2 - 6v + 20$ 19. $10x^3 - 14x^2 + 2x$ 21. $b^2 - a^2$ 23. $21x^2 - 5x - 6$
25. $20m^2 - 17m + \frac{3}{2}$ 27. $90x^3 + 135x^2y + 50xy^2$ 29. $3x^3 + \frac{47}{5}x^2 - \frac{54}{5}x - \frac{8}{5}$
31. $8m^4 - 6m^3 + 11m^2 - m - 12$ 33. $3x^4 - 2x^3 - 6x^2 + 7x - 2$ 35. $x^4 - y^4$
37. $9a^2 + 42a + 49$ 39. $25x^2 + 30xy + 9y^2$ 41. $x^3 + 6x^2 + 12x + 8$ 43. $-8x$ 45. $x^2 + 4x + 4$
47. $x^2 - 2x$ 49. $36\pi x^3 + 72\pi x^2 + 48\pi x + \frac{32\pi}{3}$ 51. $2x^3 - x^2 + 8x - 4$

Exercise Set 1.6, page 34

1. $5ab$ 3. $2xy^2$ 5. 15 7. $x^2 + 3x + 5$ 9. $-6a + 7b$ 11. $-m + n - 2$
13. $-5m + 10n + 25mn$ 15. $m(5 - m)$ 17. $v(4v + 3)(v - 2)$ 19. $(x + 10)(x - 10)$
21. $(2m + 0.5)(2m - 0.5)$ 23. $(x + y)(x - y)$ 25. Cannot be factored 27. Cannot be factored
29. $(x - 2)(x^2 + 2x + 4)$ 31. $(1 - 4m)(1 + 4m + 16m^2)$ 33. $(2x + 3y)(4x^2 - 6xy + 9y^2)$
35. $b(0.1a + b)(0.01a^2 - 0.1ab + b^2)$ 37. $(x + 7)^2$ 39. $(x + 5)^2$ 41. $(2v + 3)^2$ 43. $3(a - 3)^2$
45. $(2x + y)(x - y)$ 47. $(m - 2)(m^2 + 1)$ 49. $(x + 4)(x + 3)$ 51. Cannot be factored
53. $(z + 7)(z - 2)$ 55. $(p + 3)(p - 2)$ 57. $(2x + 1)(x - 3)$ 59. $(d + 12)(d + 3)$
61. $(5t + 1)(3t - 8)$ 63. $(3x + 4y)(2x - y)$ 65. $(x^2 + 2)(x + 2)(x - 2)$
67. $y(3x + 4y)(3x - 4y)$ 69. $x^3(x + 7)(x - 2)$ 71. $(x^2 + 5y)(x^2 - 5y)$
73. Cannot be factored 75. $(x - 2 + y)(x - 2 - y)$ 77. $x^n(1 - 8x)$ 79. $(x^n + y^n)^2$
81. $(a^x - b^x)(a^x - b^x)$ 83. $x(x^n + y^n)(x^n - y^n)$ 85. $(x^n - y^n)(x^{2n} + x^ny^n + y^{2n})$
87. $(a + b + c)^2$ 89. $x + 4$ 91. $2c + 5$ 93. $4a^2 + 6a + 9$ 95. $3x + 7$
97. $(10 - 9)(10 + 9) = 19$ 99. $(1000 - 999)(1000 + 999) = 1999$ 101. $(100 - 90)(100 + 90) = 1900$

Exercise Set 1.7, page 43

1. 10　　**3.** $\frac{6}{7}$　　**5.** 0.04　　**7.** Not a real number　　**9.** 5　　**11.** Not a real number　　**13.** $2\sqrt[3]{5}$

15. $-3\sqrt{3}$　　**17.** $\frac{3\sqrt{10}}{10}$　　**19.** $\frac{\sqrt[3]{-75}}{3}$　　**21.** $\frac{2\sqrt{3}}{3}$　　**23.** $2x\sqrt[3]{x}$　　**25.** $\frac{\sqrt{xy}}{|y|}$　　**27.** $2xy\sqrt{2y}$

29. $2b\sqrt[4]{2a^2bc}$　　**31.** $2m^2n\sqrt[5]{2m^2n^3}$　　**33.** $\frac{x\sqrt{x}}{y^2}$　　**35.** $\frac{\sqrt[3]{5x^2y}}{y}$　　**37.** $5xyz$　　**39.** $\sqrt{3}$

41. $-13\sqrt{5}$　　**43.** $4\sqrt{5}$　　**45.** $\sqrt[3]{150}$　　**47.** $\frac{\sqrt{6}}{3}$　　**49.** $3\sqrt{2} - 2\sqrt{3}$　　**51.** -1

53. $8 - 2\sqrt{7}$　　**55.** $2pq^2\sqrt[3]{9p^2}$　　**57.** $\frac{a\sqrt{6a}}{3b^2}$　　**59.** $(3 - 2x)\sqrt{x} + x$　　**61.** $x + \sqrt{xy}$

63. $x - y$　　**65.** 1000　　**67.** 4　　**69.** $\frac{6\sqrt{30}}{25}$　　**71.** 9　　**73.** $2\sqrt[5]{2}$　　**75.** $20x^{4/3}$　　**77.** $6a^{5/6}$

79. $-20x^{4/3}y^2$　　**81.** $512x^{-3/2}y^{3/4}z^3$　　**83.** $8x^{3/2}y^{3/4}$　　**85.** $\frac{1}{2}x^{1/2}y^2$　　**87.** $\frac{\sqrt{3}}{3}x^{-1/2}$　　**89.** $\sqrt{x}$

91. $\sqrt[12]{m^4v^3}$　　**93.** $\sqrt[8]{z}$　　**95.** $\sqrt[12]{x^2y}$　　**97.** $\sqrt[30]{196a^{-15}}$　　**99.** False, $\sqrt{(-1)^2} \neq -1$

101. False, $\sqrt{1+1} \neq \sqrt{1} + \sqrt{1}$　　**103.** False, $\frac{1}{\sqrt{1}+\sqrt{1}} \neq \frac{1}{\sqrt{1}} + \frac{1}{\sqrt{1}}$　　**105.** False, $2^{1/1} \neq \frac{1}{2^1}$

107. False, $\sqrt[3]{2} \cdot \sqrt{2} \neq \sqrt[6]{2^2}$　　**109.** False, $\sqrt{2^2 - 1} \neq 2 - 1$　　**111.** The square root is multiplied by $\sqrt{2}$.

113. Multiply it by 8.　　**115.** yes; yes; no　　**117.** No, because $\sqrt{5 - 2\sqrt{6}}$ is positive, but $\sqrt{2} - \sqrt{3}$ is negative.

119. Since both sides of the original equation are positive, and after we square both sides we obtain an identity, $\sqrt{\pi} \cdot \sqrt{3 + 2\sqrt{2}} = \sqrt{\pi} + \sqrt{2\pi}$.

Exercise Set 1.8, page 49

1. $x - 3$　　**3.** $3m - 9$　　**5.** $\frac{v(v-2)}{v-3}$　　**7.** $\frac{x^2 + 3x + 9}{x + 3}$　　**9.** $\frac{2(a+2)}{3a - 2}$　　**11.** $\frac{2 - x^2}{3(x - 2)}$　　**13.** $\frac{15 - v}{2(2v + 1)}$

15. $m^2(n-4)$ **17.** $-\dfrac{15}{a^2b}$ **19.** $\dfrac{x^2-1}{x^2}$ **21.** $\dfrac{x}{2y}$ **23.** $\dfrac{3}{4mn}$ **25.** $\dfrac{6(v-2)}{5}$ **27.** $-\dfrac{a}{b}$

29. $\dfrac{2(v+10)(v+3)}{v(v-3)}$ **31.** $\dfrac{-v-14}{(v+2)(v-2)}$ **33.** $\dfrac{2x^3+2x^2-35x-70}{(2x+7)(x-5)(x+3)}$ **35.** $\dfrac{\sqrt{v}}{4v}$

37. $\dfrac{5+4\sqrt{t}-t}{1-t}$ **39.** $\dfrac{x(\sqrt{x}+2\sqrt{y})}{x-4y}$ **41.** $\dfrac{x+2-\sqrt{y}}{x^2+4x+4-y}$ **43.** $u+2$ **45.** $\dfrac{v-1}{v+1}$

47. $\dfrac{-m^2-2m-2}{m^2+m+2}$ **49.** False, $\dfrac{1}{1}+\dfrac{1}{1}\neq\dfrac{1}{1+1}$ **51.** False, $(1+1)^{-2}\neq 1^{-2}+1^{-2}$

53. True   **55.** $\dfrac{4x-1}{2x^2}$ **57.** $\dfrac{x+1}{x}$ **59.** 3.0   **61.** \$28,142

Chapter Test, page 52

1. $2x+13=56$ **3.** integer, rational, real   **5.** irrational, real   **7.** False, $6-(8+2)=6-8-2=-4$

9. 2   **11.** (number line: -50, 0, 10, 35)   **13.** 15   **15.** 9   **17.** x^7 **19.** $\dfrac{15y^2}{x^3}$ **21.** $\dfrac{125y^6}{8x^3}$ **23.** -7

25. $3n^3-6n^2+17n+3$ **27.** $24m^2+34m-10$ **29.** 14   **31.** $xy(6x-2y+7xy)$

33. $(2z-3)(4z^2+6z+9)$ **35.** 0.02   **37.** $2\sqrt{6}xyz$ **39.** $-\sqrt{3}$ **41.** x^2-y **43.** $4x^{2/3}y^{4/5}$

45. $\dfrac{m(m-3)}{3m-1}$ **47.** $\dfrac{9a^4}{35b}$ **49.** -1 **51.** (a) False, $3+(-1)=2$, (b) False, $|(-3)+2|\neq|-3|+|2|$, (c) False, $\sqrt[4]{(-2)^4}\neq -2$ **53.** $(4-\pi)x^2-(8-2\pi)x+(4-\pi)$

Chapter 2

Exercise Set 2.1, page 61

1. 2 is not a solution, -2 is a solution.   **3.** 2 is not a solution, -2 is not a solution.

5. 2 is a solution, -2 is not a solution.   **7.** 2 is not a solution, -2 is not a solution.   **9.** -4

11. $\dfrac{2}{7}$ **13.** $\dfrac{19}{11}$ **15.** 14   **17.** No solutions   **19.** 12   **21.** $-\dfrac{2}{7}$ **23.** $\dfrac{21}{25}$ **25.** $\dfrac{7+b}{3}$

27. $\dfrac{2-c}{ab}$ **29.** $\dfrac{a}{m-5}$ **31.** The set of all real numbers   **33.** $\dfrac{2}{2b-2-a}$ **35.** $\dfrac{a}{a-b-2}$ **37.** -1

39. 10   **41.** $-\dfrac{7}{9}$ **43.** $-\dfrac{3}{2}$ **45.** $\dfrac{5}{8}$ **47.** The set of all real numbers except 0.   **49.** $\dfrac{5}{3}$ **51.** -1

53. -4 **55.** No solutions   **57.** 1   **59.** $\dfrac{24}{11}$ **61.** $\dfrac{5}{2}, -\dfrac{11}{2}$ **63.** $\dfrac{1}{3}, 5$ **65.** $-1, \dfrac{17}{5}$

67. $\dfrac{1}{2}$ **69.** 5   **71.** $F = \dfrac{9}{5}C + 32$ **73.** $W = \dfrac{P}{2} - L$ **75.** $a = \dfrac{2s}{t^2}$ **77.** $h = \dfrac{P - 2lw}{2(w + l)}$

79. $v_0 = \dfrac{1}{t}\left(s - s_0 - \dfrac{1}{2}at^2\right)$ **81.** $n = \dfrac{t - a}{d} + 1$ **83.** -6 **85.** 5   **87.** $xy - \dfrac{2x}{5} = 10$

89. $x = 8, x + 2 = 10$

Exercise Set 2.2, page 68

1. $-8, -1$ **3.** $5, -2$ **5.** 4   **7.** $8, -8$ **9.** 3   **11.** $\dfrac{2}{5}, -4$ **13.** $\dfrac{7}{4}, -\dfrac{7}{4}$ **15.** $0, -10$

17. $-\dfrac{3}{2}, \dfrac{1}{2}$ **19.** $-\dfrac{1}{5}, \dfrac{8}{3}$ **21.** $2, -2$ **23.** $-12, 2$ **25.** $4, -1$ **27.** $9, -6$ **29.** $-\dfrac{1}{12}, 3$ **31.** $\dfrac{2}{3}$

33. $3, -7$ **35.** -7 **37.** $2, 3$ **39.** $-\dfrac{3}{2}, -\dfrac{5}{2}$ **41.** $\dfrac{2 + \sqrt{19}}{5}, \dfrac{2 - \sqrt{19}}{5}$ **43.** $-4, -1$

45. 4   **47.** No real number solutions   **49.** $8, -8$ **51.** $-\dfrac{1}{2}$

53. $3, -\dfrac{2}{3}$ **55.** $\sqrt{11}, -\sqrt{11}$ **57.** $\dfrac{-3 + \sqrt{29}}{2}, \dfrac{-3 - \sqrt{29}}{2}$ **59.** $-\dfrac{1}{6}$

61. No real number solutions   **63.** $-4 + \sqrt{2}, -4 - \sqrt{2}$ **65.** $\dfrac{9}{2}, 0$ **67.** $\dfrac{4 + \sqrt{22}}{3}, \dfrac{4 - \sqrt{22}}{3}$

69. No real number solutions   **71.** 2 different real number roots   **73.** No real number roots

75. 2 different real number roots   **77.** 2 different real number roots   **79.** No such value exists

81. 2, −2   **83.** −3, 1   **85.** $x^2 + 8x - 10 = 0$ **87.** After $2\frac{1}{2}$ seconds the ball will be 620 feet above the ground and traveling up. After 3 seconds the ball will be 620 feet above the ground and traveling down.

Exercise Set 2.3, page 74

1. $0, \frac{2}{3}, 4$ **3.** 0   **5.** $0, \frac{7 \pm \sqrt{17}}{4}$ **7.** $\pm 3, \frac{1}{2}$ **9.** $\frac{2}{3}$ **11.** $-\frac{1}{3}, 1$ **13.** $0, \pm\frac{2\sqrt{5}}{5}, 4$ **15.** $-\frac{1}{2}, 3$

17. $\frac{1}{2}, 1$ **19.** 0, 7   **21.** 1   **23.** $\frac{9}{2}$ **25.** No real number solutions   **27.** $\pm\sqrt{7}$ **29.** $\sqrt[3]{5}, \sqrt[3]{-2}$

31. $\pm\left(\frac{-1 \pm \sqrt{5}}{2}\right)^{1/2}$ **33.** $\pm\sqrt[4]{2}$ **35.** $\frac{3}{4}$ **37.** −2, 0   **39.** $\frac{1}{125}, 8$ **41.** $\frac{2}{5}$

43. 9   **45.** 4   **47.** 7, 3   **49.** 19   **51.** 0, 1   **53.** 3   **55.** 25   **57.** 7.8 ft   **59.** 4, 1   **61.** $\frac{1}{64}$

Exercise Set 2.4, page 84

1. (a) $\{x | 3 < x < 8\}$, (b) [number line]   **3.** (a) $\{x | -2.5 < x \leq 6\}$, (b) [number line]

5. (a) $\{x | x < -2\}$, (b) [number line]   **7.** (a) $\{x | -\sqrt{3} \leq x \leq \sqrt{3}\}$, (b) [number line]

9. (a) $\{x | x > 4\}$, (b) $(4, +\infty)$ (c) [number line]   **11.** (a) $\{v | v \leq -1\}$, (b) $(-\infty, -1]$, (c) [number line]

13. (a) $\left\{x | x > \frac{6}{5}\right\}$, (b) $\left(\frac{6}{5}, +\infty\right)$, (c) [number line]   **15.** (a) $\left\{x | x \leq \frac{13}{23}\right\}$, (b) $\left(-\infty, \frac{13}{23}\right]$, (c) [number line]

17. No solutions   **19.** (a) $\{y | y \text{ is a real number}\}$, (b) $(-\infty, +\infty)$, (c) [number line]

21. [number line]   **23.** [number line]   **25.** [number line]

27. (a) $\{x | x > 1 \text{ or } x < -1\}$, (b) [number line]   **29.** (a) $\left\{v | -\frac{7}{2} \leq v \leq \frac{7}{2}\right\}$, (b) [number line]

31. (a) $\{y | y \text{ is a real number}\}$, (b) [number line]   **33.** (a) $\left\{x | x > 5 \text{ or } x < -\frac{5}{3}\right\}$, (b) [number line]

35. (a) $\{u \mid -\frac{5}{3} \leq u \leq 7\}$, (b) [number line from $-\frac{5}{3}$ to 7] **37.** No solutions

39. (a) $\{s \mid s = 7\}$, (b) [number line at 0 and 7]   **41.** $(-\infty, -3)$ and $(0, +\infty)$ **43.** $(-\infty, 0]$ and $[6, +\infty)$

45. $(-\infty, -1]$ and $[8, +\infty)$ **47.** $\left[\dfrac{2 - \sqrt{2}}{2}, \dfrac{2 + \sqrt{2}}{2}\right]$ **49.** $(-\infty, -3)$ and $(0, 3)$ **51.** $(-\infty, 1)$ and $(3, +\infty)$

53. $(-\infty, 3)$ and $(5, +\infty)$ **55.** $\left(-\infty, \dfrac{1}{3}\right]$ and $[1, +\infty)$ **57.** $(-\infty, 0)$ and $\left(\dfrac{9}{2}, +\infty\right)$

59. False; $0 < 1$, but $(0)(-1) > (1)(-1)$ **61.** False; $-2 < -1$, but $\dfrac{-2}{-1} > 1$ **63.** $(1.725, 1.775)$

65. $(4.6\overline{3}, 4.7)$ **67.** [number line]   **69.** There is no graph because the two intervals have no common points.   **71.** [number line]

Exercise Set 2.5, page 91

1. 5   **3.** 11, 12, 13   **5.** The sports car costs $14,904, the station wagon costs $11,904, and the subcompact costs $5,952.   **7.** 30 ft × 30 ft   **9.** $7000 in the first account and $3000 in the second account.

11. 160 mph   **13.** 0.375 mph   **15.** $23\frac{1}{3}$ ft   **17.** 27 in^2

Chapter Test

1. -5 is not a solution   **3.** $-\dfrac{7}{5}$ **5.** $\dfrac{b-a}{3}$ **7.** -3 **9.** $-\dfrac{4}{3}, 4$ **11.** ± 9 **13.** $\dfrac{7}{2}, -4$

15. No real number solutions   **17.** $\dfrac{-4 \pm \sqrt{11}}{5}$ **19.** $\dfrac{5}{2}$ **21.** $0, \dfrac{-3 \pm \sqrt{13}}{2}$ **23.** $0, 7$ **25.** $-\dfrac{\sqrt[3]{9}}{3}, \sqrt[3]{2}$

27. $\pm\left(2 \pm \dfrac{\sqrt{3}}{2}\right)^{1/2}$ **29.** $\dfrac{1}{27}$ **31.** $\pm\dfrac{3\sqrt{5}}{5}$ **33.** $\{x \mid -3.5 \leq x < 2.5\}$ **35.** $\{x \mid x \geq -8\}$ **37.** [number line]

39. [number line]   **41.** $\left\{x \mid x < -\dfrac{5}{2} \text{ or } x > \dfrac{11}{2}\right\}$, [number line]   **43.** $(-2, 5)$ **45.** 11 ft, 4 ft

Answers to Exercises

47. (a) Since the step where we square both sides of the equation produces an equation that has no solutions, by the Power Property, the original equation has no solutions. (b) There are no solutions because in order for $\sqrt{-x}$ to be a real number, $x \leq 0$. But if $x \leq 0$, then $\sqrt{x-1}$ is not a real number.

Cumulative Test: Chapters 1 and 2, page 97

1. $2[n + (n + 1)] = 3(n + 1) - 8$ 3. -3 5. $\dfrac{5y^7}{6x^3z^6}$ 7. $2x^3 + 11x^2 - 25x + 6$ 9. $(5x - 4)(3x - 2)$

11. $30x^2y^3\sqrt{x}$ 13. $-32x^3y^{12/5}$ 15. $\dfrac{4}{x(x+2)(x-2)}$ 17. $\dfrac{x + 4\sqrt{x} + 3}{x - 1}$ 19. $\dfrac{3 + 4b}{11}$ 21. 2

23. $\dfrac{2}{3}, 4$ 25. $\dfrac{-1 + \sqrt{22}}{7}, \dfrac{-1 - \sqrt{22}}{7}$ 27. $2, -2, -\dfrac{2}{7}$ 29. $\dfrac{3}{5}$ 31. $\left\{x \mid x > \dfrac{2}{5}\right\}$ 33. $\left\{x \mid -\dfrac{1}{8} \leq x \leq 0\right\}$

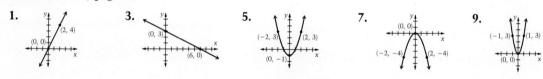

35. 6, 8   37. (a) False, $-[-(-5)] \neq -(-5)$ (b) False, $-1(|0 - 1|) < -1(0) - (-1)(1)$ (c) True

Chapter 3

Exercise Set 3.1, page 107

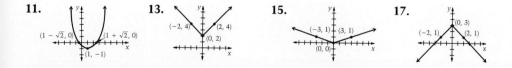

19. (a) (3, 2), (b) (−3, −2), (c) (−3, 2) **21.** (a) (4, −3), (b) (−4, 3), (c) (−4, −3)

23. (a) (0, −5), (b) (0, 5), (c) (0, −5)   **25.**   **27.**   **29.**

31. no, yes, no   **33.** yes, yes, yes   **35.** no, yes, no   **37.** no, no, no   **39.** yes, yes, yes   **41.** no, no, yes

43. **45.** **47.** **49.** **51.** (5, 2)

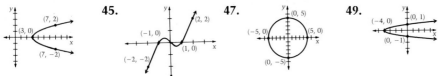

53. (−6, −3)   **55.** (3, 8)   **57.**   **59.**

Exercise Set 3.2, page 113

1. function: {(1, A), (2, B), (3, C), (4, A)}; domain: {1, 2, 3, 4}; range: {A, B, C}

3. function: {(81, 10), (82, 20), (83, 20), (84, 35), (85, 30)}; domain: {81, 82, 83, 84, 85}; range: {10, 20, 30, 35}

5.

h	t	(h, t)
-1	-3	$(-1, -3)$
0	-1	$(0, -1)$
3	5	$(3, 5)$
4	7	$(4, 7)$

7.

x	y	(x, y)
-5	41	$(-5, 41)$
-2	11	$(-2, 11)$
1	-1	$(1, -1)$
4	5	$(4, 5)$

9. $14, 8, -7, 8 - 3c^2, 2 - 3h$
11. $8, 0, 50, 2c^4, 8 + 8h + 2h^2$
13. $20, 6, 6, c^4 - 5c^2 + 6, -h + h^2$
15. $10, 10, 10, 10, 10$

17. Domain: $[-3, +\infty)$ **19.** Domain: $(-\infty, 4]$ **21.** Domain: $\left[\dfrac{9}{2}, +\infty\right)$

23. Domain: all real numbers except -4 **25.** Domain: all real numbers except 7.

27. (a) 26, (b) 12, (c) $4t^2 - 8t + 5$ **29.**   **31.** odd **33.** neither even nor odd **35.** even

37. (a) symmetry with respect to origin, (b) symmetry with respect to y-axis

Exercise Set 3.3, page 122

1.   **3.**   **5.**   **7.**   **9.**

11.   **13.** 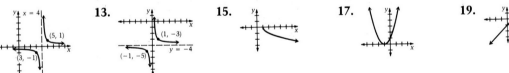 **15.**   **17.**   **19.**

Answers to Exercises

21.
23.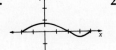
25.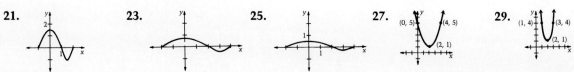
27.
29.

31.
33.
35.
37.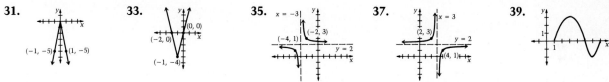
39.

41.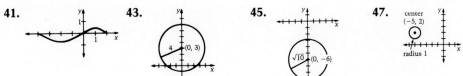
43.
45.
47.

Exercise Set 3.4, page 128

1. $m = 5$ **3.** $m = -\dfrac{1}{3}$ **5.** $m = -\dfrac{1}{4}$ **7.** $m = 0$ **9.** $m = 0$

11. $m = \dfrac{1}{8}$ **13.** $m = 1.1$ **15.** (a) $y - 8 = 5(x - 3)$, (b) $y = 5x - 7$, (c) $5x - y - 7 = 0$

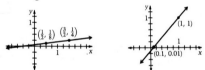
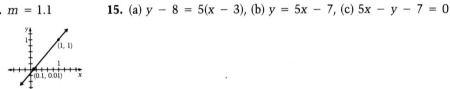

17. (a) $y - (-1) = -\dfrac{1}{3}(x - 8)$, (b) $y = -\dfrac{1}{3}x + \dfrac{5}{3}$, (c) $\dfrac{1}{3}x + y - \dfrac{5}{3} = 0$

19. (a) $y - (-6) = -\dfrac{1}{4}(x - (-5))$, (b) $y = -\dfrac{1}{4}x - \dfrac{29}{4}$, (c) $\dfrac{1}{4}x + y + \dfrac{29}{4} = 0$

21. (a) $y - 5 = 0(x - 2)$, (b) $y = 5$, (c) $y - 5 = 0$ **23.** (a) $y - 0 = 0(x - 3)$, (b) $y = 0$, (c) $y = 0$

25. (a) $y - \dfrac{1}{8} = \dfrac{1}{8}\left(x - \dfrac{1}{2}\right)$, (b) $y = \dfrac{1}{8}x + \dfrac{1}{16}$, (c) $\dfrac{1}{8}x - y + \dfrac{1}{16} = 0$

27. (a) $y - 1 = 1.1(x - 1)$, (b) $y = 1.1x - 0.1$, (c) $1.1x - y - 0.1 = 0$

29. (a) $y - 5 = -2(x - 3)$, (b) $y = -2x + 11$, (c) $2x + y - 11 = 0$

31. (a) $y - 1 = -\frac{2}{3}(x - 1)$, (b) $y = -\frac{2}{3}x + \frac{5}{3}$, (c) $\frac{2}{3}x + y - \frac{5}{3} = 0$

33. (a) $y - (-2) = 0(x - 2)$, (b) $y = -2$, (c) $y + 2 = 0$

35. (a) $y - 3 = \frac{1}{2}(x - 0)$, (b) $y = \frac{1}{2}x + 3$, (c) $\frac{1}{2}x - y + 3 = 0$

37. (a) $y - (0) = -2(x - 0)$, (b) $y = -2x$, (c) $2x + y = 0$

39. (a) $y - (-2) = \frac{3}{7}(x - 1)$, (b) $y = \frac{3}{7}x - \frac{17}{7}$, (c) $\frac{3}{7}x - y - \frac{17}{7} = 0$

41. (a) $y - (-8) = 0(x - (-2))$, (b) $y = -8$, (c) $y + 8 = 0$

43. (a) $y - (-2) = -\frac{1}{6}(x - 5)$, (b) $y = -\frac{1}{6}x - \frac{7}{6}$, (c) $\frac{1}{6}x + y + \frac{7}{6} = 0$

45. (a) $y - (-10) = 0(x - 5)$, (b) $y = -10$, (c) $y + 10 = 0$

47. The slope of the segment with endpoints (6, 3) and (5, 1) is 2. The slope of the segment with endpoints (1, 3) and (5, 1) is $-\frac{1}{2}$. Since the slopes are negative reciprocals, the segments are perpendicular and the triangle is a right triangle.

49. The ant did not pass over (1, 5) since this point is not on the line. The ant did pass over (1, 7).

51. $y = \frac{7}{3} - \frac{x}{3}$, Domain: $[1, +\infty)$ 53. $y = \frac{8}{7}x + \frac{100}{7}$

Exercise Set 3.5, page 134

1. $y = 3(x - 1)^2 + 4$. The vertex is (1, 4), the axis of symmetry is $x = 1$, and the parabola opens upward.

3. $y = -2(x - (-6))^2 + 0$. The vertex is (−6, 0), the axis of symmetry is $x = -6$, and the parabola opens downward.

5. $y = 2(x - 3)^2 + 0$. The vertex is (3, 0), the axis of symmetry is $x = 3$, and the parabola opens upward.

7. $y = -1(x - (-5))^2 + (-1)$. The vertex is (−5, −1), the axis of symmetry is $x = -5$, and the parabola opens downward.

A50 Answers to Exercises

9. $y = 1(x - 3)^2 + 10$. The vertex is $(3, 10)$, the axis of symmetry is $x = 3$, and the parabola opens upward.

11. $y = -\frac{1}{3}(x - (-1))^2 + \left(-\frac{5}{3}\right)$. The vertex is $(-1, -\frac{5}{3})$, the axis of symmetry is $x = -1$, and the parabola opens downward.

13. $y = \frac{1}{3}(x - 9)^2 + (-1)$. The vertex is $(9, -1)$, the axis of symmetry is $x = 9$, and the parabola opens upward.

15. $y = -\frac{2}{7}(x - (-6))^2 + 0$. The vertex is $(-6, 0)$, the axis of symmetry is $x = -6$, and the parabola opens downward.

17. $y = 1(x - 0)^2 + 4$. The vertex is $(0, 4)$, the axis of symmetry is $x = 0$, and the parabola opens upward.

19. None 21. -6 23. 3 25. None 27. None 29. None 31. $9 \pm \sqrt{3}$ 33. -6 35. None

37. 39. 41. 43. 45.

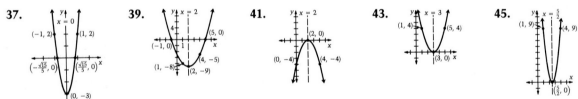

47.   49.   51. 0, maximum 53. 9, maximum 55. $\pm 2\sqrt{6}$ 57. $\dfrac{125}{9}$

59. (a) x, $12 - x$, (b) $P(x) = 12x - x^2$, (c) 6, 6

Exercise Set 3.6, page 140

1. 3   3. 1   5. $-\dfrac{1}{3}$ 7. 23   9. 23   11. 3   13. $-\dfrac{199}{25}$ 15. $-\dfrac{31}{4}$ 17. $-\dfrac{127}{16}$ 19. $-\dfrac{4}{17}$

21. $-\dfrac{2}{9}$ 23. 8   25. -8 27. $x^2 - 6x + 10$, domain: all real numbers

29. $\left|\dfrac{1}{x-3}\right|$, domain: all real numbers except 3   31. $1 - \sqrt{x-5}$, domain: $[5, +\infty)$

33. $\dfrac{1}{x^2+9}$, domain: all real numbers   35. $\dfrac{1}{\sqrt{x}}$, domain: $(0, +\infty)$ 37. $(f \circ j)(x)$

39. $(h \circ k)(x)$ 41. $(k \circ f \circ j)(x)$ 43. $(k \circ f \circ f)(x)$ 45. $y = -5 - 15t$ 47. 50 seconds

49. $2x + 3, [-1, 1]$ 51. $\sqrt{x+3}, [-3, 0]$ 53. Suppose f and g are odd. Then $(f \circ g)(-x) = f(g(-x)) = -(f \circ g)(x)$.

Chapter Test, page 144

1.   3. 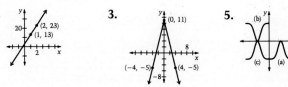   5.   7. yes, yes, yes   9. no, no, yes

11. function: {(A, 10), (B, 20), (C, 30), (D, 40), (E, 40), (F, 40), (G, 30), (H, 50)}; domain: {A, B, C, D, E, F, G, H}; range: {10, 20, 30, 40, 50}   13. $12, 4, 4 - h^6, -b^3 + 3b^2 - 3b + 5$

15.  **17.** **19.** 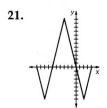 **21.** **23.**

25. $m = -5$ **27.** (a) $y - 4 = -1(x - (-2))$, (b) $y = -x + 2$, (c) $x + y - 2 = 0$

29. $y = 6(x - (-4))^2 + 9$. The vertex is $(-4, 9)$, the axis of symmetry is $x = -4$, and the parabola opens up.

31. **33.** 3 **35.** $\dfrac{10}{3}$ **37.** $\dfrac{1}{2 - \sqrt{x + 6}}$, domain: all real numbers greater than or equal to -6, except -2.

39. $(j \circ f \circ h)(x)$ **41.** $3h + 14$ **43.** $-22, \dfrac{11}{3}$ **45.** $y = \dfrac{1}{8}x + \dfrac{41}{8}$

Chapter 4

Exercise Set 4.1, page 155

1. yes; $f(x) = -x^3 + 7x + 2$; $-x^3$ 3. no 5. yes; $H(x) = -2x^5 + 3x^3 + 9x^2 + 1$; $-2x^5$
7. no 9. yes; $f(x) = 9x^2 + 13x + 11$; $9x^2$ 11. yes; $k(x) = -\sqrt{5}x^4 + 3x^3 + 7x - 3$; $-\sqrt{5}x^4$
13. Type III 15. Type II 17. Type I 19. Type II 21. Type III 23. Type IV 25. Type IV
27. Type I 29. No. 2 31. No. 1 33. No. 1 35. No. 2

37.   39.   41. 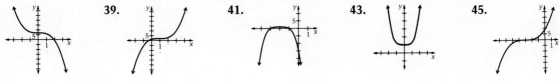 43.   45.

47. No; graph has a sharp corner 49. Yes 51. Yes 53. No. 2 55. No. 1

Exercise Set 4.2, page 165

1. No. 5 3. No. 6 5. No. 6 7. No. 3 9. $-3, 5, -7$ 11. $0, \frac{2}{3}, -2, -1$ 13. $5, -\frac{5}{2}$
15. $1, 1 \pm \sqrt{2}$ 17. $-3, 0, -4, -1$ 19. 5

21.   23.   25.   27. 1, 3   29. 0, 1, 2

31. 0, −2, 2   **33.** $\frac{\sqrt{2}}{2}, -\frac{\sqrt{2}}{2}, \sqrt{3}, -\sqrt{3}$  **35.** −1, 1, −2   **37.** 1, 0, −2, 2

39. (−3, −2), (0, 1), (3, 4)   **41.** zeros at −2, −1, and 1

43. Possible rational roots: ±1, ±2, ±3, ±6; Actual rational roots: −1, 2, −3

45. Possible rational roots: ±1, ±$\frac{1}{2}$, ±5, ±$\frac{5}{2}$; Actual rational roots: $\frac{5}{2}$

47. Possible rational roots: ±1, ±2, ±3, ±4, ±6, ±12; Actual rational roots: 3

Exercise Set 4.3, page 174

1. Vertical: none
  Horizontal: $y = 0$
3. Vertical: $x = 0, x = 3$
  Horizontal: $y = 0$
5. Vertical: $x = -9$
  Horizontal: $y = 0$
7. Vertical: $x = -2, x = -1$
  Horizontal: $y = 0$

9. Vertical: $x = -3, x = 3$
  Horizontal: $y = 1$
11. Vertical: $x = -\frac{1}{5}, x = -1$
  Horizontal: $y = 2$
13. Vertical: $x = \frac{1}{2}, x = -1$
  Horizontal: none

15.   **17.** 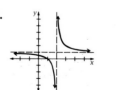 **19.**   **21.**

Answers to Exercises A55

23.  25.   27.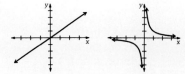

29. 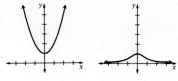 31.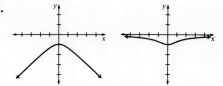

33. 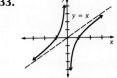 35.   37.

Exercise Set 4.4, page 187

1. vertex: (0, 0)
axis of symmetry: $y = 0$
x-intercepts: 0
y-intercepts: 0

3. vertex: (0, 0)
axis of symmetry: $y = 0$
x-intercepts: 0
y-intercepts: 0

5. vertex: (5, 3)
axis of symmetry: $y = 3$
x-intercepts: -13
y-intercepts: $3 \pm \frac{\sqrt{10}}{2}$

7. vertex: $(\frac{3}{2}, -1)$
axis of symmetry: $y = -1$
x-intercepts: $\frac{7}{2}$
y-intercepts: none

9. vertex: $(-4, 3)$
axis of symmetry: $y = 3$
x-intercepts: -13
y-intercepts: none

11. vertex: $(3, -1)$
axis of symmetry: $y = -1$
x-intercepts: $\frac{8}{3}$
y-intercepts: $2, -4$

13. center: (0, 0)
radius: 4

15. center: $(-1, 3)$
radius: 3

17. center: (2, 0)
radius: 2

19. center: $(-2, 1)$
radius: 4

21. center: $(\frac{1}{2}, -\frac{1}{2})$
radius: $\sqrt{3}$

23. center: $(\frac{1}{5}, 3)$
radius: 3

25. center: (0, 0)
vertices: (3, 0), $(-3, 0)$, (o, 4), $(0, -4)$

27. center: $(2, -3)$
vertices: $(4, -3)$, $(0, -3)$, (2, 7), $(2, -13)$

29. center: (5, 0)
vertices: $(5 + \sqrt{10}, 0), (5 - \sqrt{10}, 0), (5, 4), (5, -4)$

31. center: (0, 0)
vertices: $(\frac{11}{5}, 0), (-\frac{11}{5}, 0), (0, 4), (0, -4))$

33. center: (1, 2)
vertices: (4, 2), (−2, 2), (1, 6), (1, −2)

35. center: (−3, 0)
vertices: $(-3 + \sqrt{2}, 0), (-3, -\sqrt{2}, 0), (-3, 4), (-3, -4)$

37. center: (−1, 1)
vertices: (2, 1), (−4, 1), (−1, 2), (−1, 0)

39. center: (0, 0)
vertices: (4, 0), (−4, 0)
asymptotes: $y = \frac{3}{4}x, y = -\frac{3}{4}x$
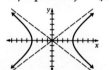

41. center: (0, 0)
vertices: $(\sqrt{2}, 0), (-\sqrt{2}, 0)$
asymptotes: $y = \sqrt{5}x, y = -\sqrt{5}x$

43. center: (1, 2)
vertices: (1, 7), (1, −3)
asymptotes: $y = \frac{5}{2}x - \frac{1}{2}, y = -\frac{5}{2}x + \frac{9}{2}$

45. center: $(2, 3)$
vertices: $(\frac{13}{4}, 3), (\frac{3}{4}, 3)$
asymptotes: $y = \frac{7}{10}x + \frac{8}{5}, y = -\frac{7}{10}x + \frac{22}{5}$

47. center: $(0, 0)$
vertices: $(5, 0), (-5, 0)$
asymptotes: $y = \frac{2}{5}x, y = -\frac{2}{5}x$

49. center: $(0, \frac{3}{2})$
vertices: $(0, \frac{7}{2}), (0, -\frac{1}{2})$
asymptotes: $y = \frac{1}{4}x + \frac{3}{2}, y = -\frac{1}{4}x + \frac{3}{2}$

51. center: $(-2, -1)$
vertices: $(-2, 1), (-2, -3)$
asymptotes: $y = \frac{2}{3}x + \frac{1}{3}, y = -\frac{2}{3}x - \frac{7}{3}$

53. circle

55. parabola

57. circle

59. $(x + 1)^2 + (y - 2)^2 = 18$

61. $(x + 3)^2 + (y + 3)^2 = 9$ **63.** yes   **65.** no   **67.** yes   **69.** no

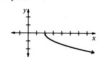

Chapter Test, page 190

1. yes; $f(x) = 3x^2 - 2x + 5$; $3x^2$ 3. no   5. Type III   7. Type I   9. No. 4   11. No. 3

13. No. 2   15. $0, 5, \frac{3}{2}$ 17. $2, -3, -1$ 19. Vertical: $x = \frac{1}{3}, x = \frac{1}{2}$; Horizontal: $y = \frac{1}{2}$ 21.

23. vertex: $(0, 1)$
    axis of symmetry: $y = 1$
    x-intercepts: 2
    y-intercepts: 1

25. center: $(2, -3)$
    radius: 5

27. center: $(5, 0)$
    vertices: $(10, 0), (0, 0), (5, 2), (5, -2)$

29. center: $(1, -2)$
    vertices: $(1, -2 + 2\sqrt{6}), (1, -2 - 2\sqrt{6})$
    asymptotes: $y = \sqrt{6}x - 2 - \sqrt{6}, y = -\sqrt{6}x - 2 + \sqrt{6}$

31. (a) (b) (c) (d)

33. (a)   (b) Parabola: $y = 2 - (x - 1)^2$ (c) $(1, 2)$

Chapter 5

Exercise Set 5.1, page 203

1. (a) no (b) no   **3.** (a) yes (b) yes   **5.** $(2, -1)$ **7.** $\left(\frac{1}{2}, \frac{1}{4}\right)$ **9.** All points on the line $4x + 3y = 2$

11. No solutions   **13.** $\left(\frac{2}{5}, \frac{7}{15}\right)$ **15.** All points on the line $y = 4 - 3x$

17. No solutions   **19.** $(\sqrt{3}, -2)$ **21.** $(3, 0), (1, -2)$ **23.** $(1, 1)$

25. No real number solutions   **27.** No real number solutions   **29.** $(3, 3)$

31. $(1, -2), (3, 2)$ **33.** 17 quarters, 8 dimes   **35.** $500

Exercise Set 5.2, page 209

1. The region on or to the left of the line $x = 0$ (the y-axis).   **3.** The region above the line $y = -x + 1$.

5. The region on or below the line $y = 3x - 1$.   **7.** The region inside and on the circle $x^2 + y^2 = 9$.

9. The region outside or on the circle $(x - 1)^2 + y^2 = 1$.   **11.** The region inside the circle $(x + 4)^2 + (y - 7)^2 = 9$.

13. The region inside or on the parabola $y = x^2$.   **15.** The region outside or on the parabola $x = (y + 3)^2 + 2$.

17. The region inside or on the V-shaped graph $y = |x|$.   **19.** The region outside the V-shaped graph $x = |6 - y|$.

21. The regions outside or on the branches of the hyperbola $xy = 1$.

Answers to Exercises

23. The region outside the branches of the hyperbola $y(x + 2) = 4$.

25. The region below the line $y = 1$ and on or to the left of the line $x = -2$.

27. The region below the line $y = 1$ and to the right of the line $x + y = 1$.

29. The region above or on the line $5y + 2x = 10$ and above the line $3y - x = 3$.

31. The interior and boundary of the triangular region defined by the corner points $(0, 0)$, $(3, 6)$, and $(6, 6)$.

33. The interior of the triangular region defined by the corner points $(1, -1)$, $(3, 3)$, and $(4, 2)$.

35. The interior of the triangular region defined by the corner points $(-4, 2)$, $(2, 4)$, and $(4, -2)$, and the line segment between $(-4, 2)$ and $(4, -2)$, and the line segment between $(4, -2)$ and $(2, 4)$.

37. The interior and boundary of the quadrilateral defined by the corner points $(1, -1)$, $(1, 1)$, $(3, 5)$ and $(9, -1)$.

39. The interior of the triangular region defined by the corner points $(0, -3)$, $(4, 1)$, and $(6, 0)$.

41. **43.** **45.** **47.** **49.** $\begin{cases} y \leq \dfrac{4}{3}x + 2 \\ y > \dfrac{2}{3}x \\ y < 2 \end{cases}$

Exercise Set 5.3, page 215

1. **3.** **5.** **7.** **9.**

11. 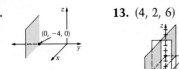   **13.** (4, 2, 6)   **15.** (−3, 7, −1)   **17.** (−2, −10, 0)

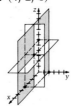 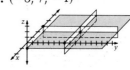 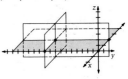

19. (3, 1, −2)   **21.** $\left(\frac{36}{17}, -\frac{9}{17}, \frac{10}{17}\right)$ **23.** $\left(\frac{15}{7}, \frac{31}{21}, -\frac{23}{21}\right)$ **25.** (1, 3, −2)   **27.** No solutions

29. $\left(c, c + \frac{12}{7}, c - \frac{11}{7}\right)$, where c is any real number   **31.** $\left(\frac{1}{2}, -\frac{2}{3}, \frac{5}{6}\right)$ **33.** $\left(\frac{5}{12}, -\frac{7}{12}, -\frac{15}{4}\right)$

35. No solutions   **37.** 24   **39.** 223   **41.** 7

Exercise Set 5.4, page 224

1. −1   **3.** 0   **5.** $\begin{bmatrix} 3 & 7 & | & 8 \\ 2 & 9 & | & 1 \end{bmatrix}$ **7.** $\begin{bmatrix} 2 & -1 & 1 & | & 2 \\ 1 & 3 & 0 & | & 5 \\ 1 & 0 & -3 & | & 1 \end{bmatrix}$ **9.** $\begin{cases} 12x + 9y = 6 \\ 8x + 6y = 4 \end{cases}$

11. $\begin{cases} -x + y - 2z = 1 \\ x + y - z = 7 \\ 2x + z = 4 \end{cases}$ **13.** $\begin{bmatrix} -2 & 7 & | & 8 \\ 3 & -1 & | & 7 \end{bmatrix}, \begin{bmatrix} -2 & 7 \\ 3 & -1 \end{bmatrix}, \begin{bmatrix} 8 \\ 7 \end{bmatrix}$ **15.** $\begin{bmatrix} 8 & 4 & | & 5 \\ 6 & 3 & | & 7 \end{bmatrix}, \begin{bmatrix} 8 & 4 \\ 6 & 3 \end{bmatrix}, \begin{bmatrix} 5 \\ 7 \end{bmatrix}$

17. $\begin{bmatrix} 3 & 5 & | & 1 \\ 4 & 3 & | & 5 \end{bmatrix}, \begin{bmatrix} 3 & 5 \\ 4 & 3 \end{bmatrix}, \begin{bmatrix} 1 \\ 5 \end{bmatrix}$ **19.** $\begin{bmatrix} 5 & -1 & 1 & | & -1 \\ 1 & 3 & 0 & | & 7 \\ 0 & 1 & 2 & | & -6 \end{bmatrix}, \begin{bmatrix} 5 & -1 & 1 \\ 1 & 3 & 0 \\ 0 & 1 & 2 \end{bmatrix}, \begin{bmatrix} -1 \\ 7 \\ -6 \end{bmatrix}$

21. $\begin{bmatrix} 3 & 2 & -1 & | & 6 \\ -1 & -1 & 1 & | & 1 \\ 1 & 2 & 2 & | & 5 \end{bmatrix}, \begin{bmatrix} 3 & 2 & -1 \\ -1 & -1 & 1 \\ 1 & 2 & 2 \end{bmatrix}, \begin{bmatrix} 6 \\ 1 \\ 5 \end{bmatrix}$ **23.** $\begin{bmatrix} 1 & -10 & 14 & | & 2 \\ -2 & -1 & 2 & | & 5 \\ -1 & 3 & -4 & | & 1 \end{bmatrix}, \begin{bmatrix} 1 & -10 & 14 \\ -2 & -1 & 2 \\ -1 & 3 & -4 \end{bmatrix}, \begin{bmatrix} 2 \\ 5 \\ 1 \end{bmatrix}$

25. $\begin{bmatrix} 3 & -6 & 2 & | & -1 \\ 1 & 4 & 1 & | & 0 \\ 3 & 2 & -2 & | & 5 \end{bmatrix}, \begin{bmatrix} 3 & -6 & 2 \\ 1 & 4 & 1 \\ 3 & 2 & -2 \end{bmatrix}, \begin{bmatrix} -1 \\ 0 \\ 5 \end{bmatrix}$ **27.** $(5, -1)$ **29.** $\left(\frac{3}{2}, \frac{7}{6}, \frac{1}{6}\right)$

31. All points on the line $x + \frac{3}{4}y = \frac{1}{2}$ **33.** No solutions **35.** $(3, 2)$ **37.** No solutions

39. $(2, -1)$ **41.** $(1, 2, -4)$ **43.** $(5, -3, 3)$ **45.** $\left(c, 11 + 5c, 8 + \frac{7}{2}c\right)$, where c is any real number

47. $\left(\frac{3}{4}, \frac{1}{8}, -\frac{5}{4}\right)$ **49.** All real numbers except 1 and 3 **51.** No real number values

53. No real number values

Exercise Set 5.5, page 233

1. 3 **3.** 6 **5.** 0 **7.** 2 **9.** $\left(\frac{2}{3}, \frac{1}{3}\right)$ **11.** No unique solution **13.** $\left(\frac{4}{9}, \frac{1}{9}\right)$

15. No unique solution **17.** $\left(-\frac{7}{5}, \frac{1}{5}\right)$ **19.** $(1, -2)$ **21.** $\frac{9}{2}$ **23.** No real number values **25.** $5, -3$

27. (a) 0 (b) 0 **29.** (a) 13 (b) 13 **31.** (a) -18 (b) -18 **33.** 0 **35.** 13 **37.** -18 **39.** $(3, 1, -2)$

41. $\left(\frac{37}{17}, -\frac{9}{17}, \frac{10}{17}\right)$ **43.** $\left(\frac{15}{7}, \frac{31}{21}, -\frac{23}{21}\right)$ **45.** $(1, 3, -2)$ **47.** No unique solution **49.** No unique solution

51. $\left(\frac{1}{2}, -\frac{2}{3}, \frac{5}{6}\right)$ **53.** $\left(\frac{5}{12}, -\frac{7}{16}, -\frac{15}{4}\right)$ **55.** No unique solution **57.** 0 **59.** 4

61. Let $A = \begin{bmatrix} 0 & 0 & 0 \\ 1 & 2 & 3 \\ 4 & 5 & 6 \end{bmatrix}$; $\det A = 0$

63. $\det A = \begin{vmatrix} -1 & 0 & 4 \\ 3 & 1 & -2 \\ 0 & 1 & 1 \end{vmatrix} = 9$; $\det B \begin{vmatrix} 3 & 1 & -2 \\ -1 & 0 & 4 \\ 0 & 1 & 1 \end{vmatrix} = -9$

65. We know from exercise 63 that $\det A = 9$. $\det B = \begin{vmatrix} -1 & 0 & 4 \\ 3 & 1 & -2 \\ -2 & 1 & 9 \end{vmatrix} = 9$

67. Let $A = \begin{bmatrix} 0 & 1 & 2 \\ 0 & 3 & 4 \\ 0 & 5 & 6 \end{bmatrix}$; $\det A = 0$

69. We know from exercise 63 that $\det A = 9$. $\det B = \begin{vmatrix} 0 & -1 & 4 \\ 1 & 3 & -2 \\ 1 & 0 & 1 \end{vmatrix} = -9$

71. We know from exercise 63 that $\det A = 9$. $\det B = \begin{vmatrix} -1 & 0 & 2 \\ 3 & 1 & 4 \\ 0 & 1 & 1 \end{vmatrix} = 9$

73. Calculating the determinant and setting it equal to zero produces $y - y_0 = \frac{y_1 - y_0}{x_1 - x_0}(x - x_0)$.

Exercise Set 5.6, page 240

1. 300 donuts, 300 brownies, $81   **3.** 8000 45's, 3000 LP's, $6320   **5.** 20 multiple choice, 10 true/false

7. To minimize: 15 amateurs, 20 professionals; to maximize: 10 amateurs, 50 professionals

9. 20.8 pounds of brand A, none of brand B   **11.** 6 of Package I, 10 of Package II

13. 128 acres of alfalfa, no beans, $11,520 from F_2 to S_1, none from F_2 to S_2 **15.** 20 gallons from F_1 to S_1, 20 gallons from F_1 to S_2, 5 gallons from F_2 to S_1, none from F_2 to S_2

17. All the points in the feasible region that lie on the line $800 = 0.08b + 0.12s$ yield earnings of $800. As E increases, there are fewer and fewer points in the feasible region that lie on the line. When $E > 1000$ or $E < 480$, the line does not intersect the feasible region.

19. Since $x^2 + y^2$ is the square of the distance from the origin to any point (x, y), the maximum value of $E = x^2 + y^2$ will occur at the point in the region that is farthest from the origin and the minimum value of $E = x^2 + y^2$ will occur at the point in the region that is closest to the origin.

21. The minimum value would occur at the boundary point that is closest to the line $y = x$.

23. 4 tables, 3 chairs

Chapter Test, page 244

1. no   **3.** no   **5.** All points on the line $5x - 2y = -1$ **7.** $\left(\dfrac{29}{26}, -\dfrac{24}{13}\right)$ **9.** No real number solutions

11. The region outside the circle $x^2 + y^2 = 4$.   **13.** The region on or above the line $y = 5 - 3x$ and below the line $y = x - 2$.   **15.** The quadrilateral region defined by the corner points $(1, -2)$, $(2, -2)$, $(1, -9)$, and $(5, -5)$, including the line segments from $(2, -2)$ to $(5, -5)$ and from $(5, -5)$ to $(1, -9)$, and not including the line segments from $(1, -9)$ to $(1, -2)$ and from $(1, -2)$ to $(2, -2)$.

17. 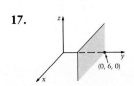   **19.** $\left(0, -\dfrac{3}{19}, \dfrac{10}{19}\right)$ **21.** $\left[\begin{array}{cc|c} 4 & -3 & 7 \\ 1 & 2 & -3 \end{array}\right], \left[\begin{array}{cc} 4 & -3 \\ 1 & 2 \end{array}\right], \left[\begin{array}{c} 7 \\ -3 \end{array}\right]$ **23.** $\left(\dfrac{5}{7}, \dfrac{8}{7}\right)$

25. -6 **27.** (a) -30 (b) -30 **29.** $\left(0, -\dfrac{3}{19}, \dfrac{10}{19}\right)$ **31.** 7000 hamburgers, 9000 hotdogs

33. (a) All real numbers except -2, (b) -2, (c) No real number values

35. (a) $\begin{cases} y \geq -\dfrac{9}{2}x - 13 \\ y < -\dfrac{9}{2}x + \dfrac{37}{2} \\ y < 5 \\ y \geq -4 \end{cases}$ (b) $\begin{cases} y < \dfrac{13}{5}x + \dfrac{56}{5} \\ y \leq -\dfrac{4}{7}x + \dfrac{34}{7} \\ y \leq -3x + 17 \\ y > \dfrac{3}{14}x - \dfrac{11}{2} \end{cases}$

Cumulative Test: Chapters 3–5, page 247

1. Symmetry with respect to the y-axis **3.** $6, 11, h^4 - 2h^2 + 3, h^2 - 10h + 27$ **5.**

7.   **9.** (a) $y - (-2) = -\dfrac{3}{2}(x - 0)$ (b) $y = -\dfrac{3}{2}x - 2$ (c) $\dfrac{3}{2}x + y + 2 = 0$ **11.**

13. $y = 4x^2 - 20x + 21$, Domain: all real numbers   **15.** $0, 2, -1$ **17.**

19. vertex: $(-2, 2)$
axis of aymmetry: $y = 2$
x-intercepts: 6
y-intercepts: 3, 1

21. center: $(1, 3)$
vertices: $(4, 3), (-2, 3)$
asymptotes: $y = \frac{4}{3}x + \frac{5}{3}, y = -\frac{4}{3}x + \frac{13}{3}$

23. $(-3, 1)$

25.   **27.** $(-3, 1)$ **29.** $3500 in savings account, $3500 in stocks

31. (a) 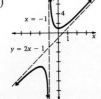   (b)   **33.** (a)   (b)

Chapter 6

Exercise Set 6.1, page 254

1. 2^{-3x} **3.** 5^{2x-4} **5.** 2^{2-x} **7.** 3^{2x-2} **9.** 2^{8-x} **11.** 3^{4x-7} **13.** $2^x(1 + 2^x)$

15.   **17.**   **19.**   **21.**   **23.** 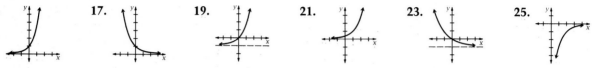 **25.**

27. 3 **29.** $\frac{2}{3}$ **31.** -8 **33.** $\frac{7}{2}$ **35.** No solution **37.** -2 **39.** $\pm\sqrt{2}$ **41.** 0 **43.** 2, 0

45. (a) ≈ 2.7 (b) ≈ 8.8 (c) ≈ 0.3 **47.** Q is squared. Q is cubed. Q is raised to the tenth power.

49.   **51.**   **53.** 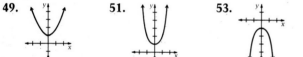 **55.** no, no **57.** yes, no

Exercise Set 6.2, page 261

1. $f^{-1}(x) = 3 - \frac{x}{2}$ **3.** $f^{-1}(x) = \sqrt{x}$ **5.** $f^{-1}(x) = 4 - x^2$ **7.** $f^{-1}(x) = \sqrt[3]{3x} + 1$

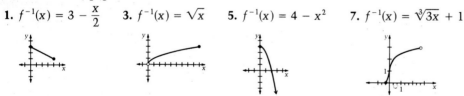

9. [0, 4], [1, 3]   **11.** (0, 9], (0, 3]   **13.** [0, +∞), (−∞, 4]   **15.** $\left[-\frac{1}{3}, \frac{8}{3}\right)$, [0, 3)   **17.** $\log_x 25 = 2$

19. $\log_{16} 4 = x$ **21.** $\log_5 x = 3$ **23.** 8   **25.** $2\sqrt{2}$ **27.** 2   **29.** $\frac{2}{3}$ **31.** $\frac{\sqrt{3}}{3}$

33. **35.** **37.** **39.** **41.** **43.** 2

45. 13.6   **47.** −0.4375   **49.** $2\sqrt{2}$ **51.** −2   **53.** 81   **55.** $2\sqrt{2}$ **57.** All real numbers $x > 0$

59.   **61.**   **63.**   **65.** 2   **67.** 8, 1   **69.** 2, 1

71. By Rule 2 of Exponents $b^1 = b$. By Log/Exp Principle $\log_b b = 1$.   **73.** By Rule 5 of Logarithms $\log_b\left(\frac{1}{x}\right) = \log_b 1 - \log_b x$. By Rule 1 of Logarithms $\log_b\left(\frac{1}{x}\right) = -\log_b x$.

Exercise Set 6.3, page 266

1. 0.6839   **3.** 0.9450   **5.** 0.2718   **7.** 0.5132   **9.** 0.3674   **11.** 0.2041   **13.** 4.724×10^1

15. 3.502×10^{-1} **17.** 1.5×10^{-2} **19.** 9.08×10^{-2} **21.** 3.2941×10^3 **23.** 3.0×10^{-4}

25. 1.5172   **27.** −0.1675   **29.** −1.2441   **31.** 2.5428   **33.** 3.7126   **35.** 4.1644   **37.** 3.54

39. 8.52   **41.** 5.49   **43.** 565   **45.** 302   **47.** 46,100   **49.** 0.177   **51.** 0.00346   **53.** 0.101

55. False, $\log 1 + \log 1 \neq \log(1 + 1)$ **57.** False, $(\log 10)(\log 10) \neq \log[(10)(10)]$ **59.** True
61. False, $\log[100(1)] \neq 100 \log 1$ **63.** False, $\log(10^1) \neq \log(1^{10})$ **65.** 2.0960   **67.** 2.3223
69. Q is increased by 1. Q is increased by $\log_2 3$. Q is increased by 2.

Exercise Set 6.4, page 269

1. (a) 27.0 (b) 3.0 (c) 13.2 (d) 10.4   **3.** (a) -3.6 (b) 5.9 (c) 0.8 (d) 3.9   **5.** (a) -6.4 (b) 2.0 (c) 3.3 (d) -0.5
7. (a) 3.2 (b) -3.8 (c) -3.5 (d) -1.3 **9.** 0.8758   **11.** 0.6757   **13.** -0.5164 **15.** 1.4469
17. 2.6796   **19.** -1.2863 **21.** -1.2921 **23.** -0.0392 **25.** 8.915   **27.** 2.134   **29.** 9,648
31. 0.2202   **33.** 0.5759   **35.** 0.02632   **37.** 5.193×10^{-4} **39.** 1.41   **41.** 0.33   **43.** 0.65
45. 2.62   **47.** -1.59 **49.** 2.59   **51.** 10.54   **53.** (a) 3 (b) 7 (c) 19

Exercise Set 6.5, page 275

1. 172,800   **3.** 1.6 hours   **5.** $2208   **7.** $1774   **9.** 15   **11.** $2226   **13.** 5 years   **15.** 45.2 grams
17. 16,451 years ago   **19.** $13.50   **21.** at least $16,172   **23.** 6.14%   **25.** 7.7%

Chapter Test, page 278

1. 2^{4-7x} **3.**   **5.** No solution   **7.** $f^{-1}(x) = \dfrac{7 - x}{2}$ **9.** $\log_x 125 = 3$ **11.** $\log_7 x = 4$

Domain: $[-1, 5]$
Range: $[1, 4]$

13. $\frac{1}{4}$ **15.** [graph] **17.** 0.6 **19.** 2 **21.** All real numbers $x > 0$ **23.** 0.5977 **25.** 4.51×10^{-3}
27. -0.0237 **29.** 28.9 **31.** 1.0 **33.** -0.5792 **35.** 2.7321 **37.** 4.054×10^{-3} **39.** 0.6 hours
41. $808 **43.** No solution **45.** [graph] **47.** (a) 1.73 (b) 2.59

Chapter 7

Exercise Set 7.1, page 287

1. 67° **3.** 30° **5.** 18 **7.** 10 **9.** 36 **11.** 12 **13.** $\sqrt{13}$ **15.** $\frac{3}{4}$ **17.** 5 **19.** $\frac{32}{5}$ **21.** 10
23. 24 **25.** $\frac{36\sqrt{5}}{5}$ **27.** $\frac{18\sqrt{5}}{5}$ **29.** 42 ft **31.** 17 mi **33.** $\frac{19\sqrt{2}}{2}$ in by $\frac{19\sqrt{2}}{2}$ in, $\frac{361}{2}$ in² **35.** 30°
37. $\frac{5\sqrt{3}}{2}$ **39.** $\frac{5}{2}$ **41.** $\frac{25\sqrt{3}}{8}$ **43.** 240 **45.** $12\sqrt{2}$

Exercise Set 7.2, page 294

1. $\sin \alpha = \frac{3}{5}$ $\csc \alpha = \frac{5}{3}$ **3.** $\sin \alpha = \frac{8}{17}$ $\csc \alpha = \frac{17}{8}$ **5.** $\sin \alpha = \frac{12}{37}$ $\csc \alpha = \frac{37}{12}$
$\cos \alpha = \frac{4}{5}$ $\sec \alpha = \frac{5}{4}$ $\cos \alpha = \frac{15}{17}$ $\sec \alpha = \frac{17}{15}$ $\cos \alpha = \frac{35}{37}$ $\sec \alpha = \frac{37}{35}$
$\tan \alpha = \frac{3}{4}$ $\cot \alpha = \frac{4}{3}$ $\tan \alpha = \frac{8}{15}$ $\cot \alpha = \frac{15}{8}$ $\tan \alpha = \frac{12}{35}$ $\cot \alpha = \frac{35}{12}$

7. $\sin \alpha = \frac{3}{5}$ $\csc \alpha = \frac{5}{3}$ **9.** $\sin \alpha = \frac{\sqrt{2}}{2}$ $\csc \alpha = \sqrt{2}$ **11.** $\sin \alpha = \frac{1}{2}$ $\csc \alpha = 2$
  $\cos \alpha = \frac{4}{5}$ $\sec \alpha = \frac{5}{4}$ $\cos \alpha = \frac{\sqrt{2}}{2}$ $\sec \alpha = \sqrt{2}$ $\cos \alpha = \frac{\sqrt{3}}{2}$ $\sec \alpha = \frac{2\sqrt{3}}{3}$
  $\tan \alpha = \frac{3}{4}$ $\cot \alpha = \frac{4}{3}$ $\tan \alpha = 1$ $\cot \alpha = 1$ $\tan \alpha = \frac{\sqrt{3}}{3}$ $\cot \alpha = \sqrt{3}$

13. $\sin \alpha = \frac{\sqrt{3}}{2}$ $\csc \alpha = \frac{2\sqrt{3}}{3}$ **15.** $\sin \alpha = \frac{1}{2}$ $\csc \alpha = 2$ **17.** $\tan \alpha = \frac{a}{b} = \cot \beta$
  $\cos \alpha = \frac{1}{2}$ $\sec \alpha = 2$ $\cos \alpha = \frac{\sqrt{3}}{2}$ $\sec \alpha = \frac{2\sqrt{3}}{3}$
  $\tan \alpha = \sqrt{3}$ $\cot \alpha = \frac{\sqrt{3}}{3}$ $\tan \alpha = \frac{\sqrt{3}}{3}$ $\cot \alpha = \sqrt{3}$

19. $\sin \beta = \frac{b}{c} = \cos \alpha$ **21.** $3\sqrt{3}$ **23.** $4\sqrt{3}$ **25.** 30   **27.** $14\sqrt{3}$ **29.** $24\sqrt{2}$ **31.** 15

33. 45°   **35.** 60°   **37.** $\frac{4}{5}, \frac{3}{4}$ **39.** $\frac{15}{8}, \frac{15}{17}$ **41.** $\frac{4}{5}, \frac{4}{5}$ **43.** $\frac{24}{25}, \frac{7}{24}$ **45.** $\frac{\sqrt{19}}{10}, \frac{9}{10}$

49. 6   **51.** $30\sqrt{3}$ **53.** $36\sqrt{3}$ ft, 36 ft

Exercise Set 7.3, page 300

1.   **3.**   **5.**   **7.**   **9.**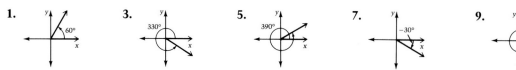

11. [graph showing −725° angle]   **13.** 45°   **15.** 3°   **17.** 322°   **19.** 60°   **21.** 20°, −340°   **23.** 225°, −135°

25. 310°, −50°   **27.** 89°, −271°   **29.** 360°, −360°   **31.** 135°, −225°   **33.** $\frac{1}{2}°$, $-359\frac{1}{2}°$

35. $\sin \alpha = \frac{3}{5}$ $\csc \alpha = \frac{5}{3}$
$\cos \alpha = -\frac{4}{5}$ $\sec \alpha = -\frac{5}{4}$
$\tan \alpha = -\frac{3}{4}$ $\cot \alpha = -\frac{4}{3}$

37. $\sin \alpha = -\frac{5}{13}$ $\csc \alpha = -\frac{13}{5}$
$\cos \alpha = -\frac{12}{13}$ $\sec \alpha = -\frac{13}{12}$
$\tan \alpha = \frac{5}{12}$ $\cot \alpha = \frac{12}{5}$

39. $\sin \alpha = \frac{3}{5}$ $\csc \alpha = \frac{5}{3}$
$\cos \alpha = -\frac{4}{5}$ $\sec \alpha = -\frac{5}{4}$
$\tan \alpha = -\frac{3}{4}$ $\cot \alpha = -\frac{4}{3}$

41. $\sin \alpha = \frac{\sqrt{2}}{2}$ $\csc \alpha = \sqrt{2}$
$\cos \alpha = \frac{\sqrt{2}}{2}$ $\sec \alpha = \sqrt{2}$
$\tan \alpha = 1$ $\cot \alpha = 1$

43. $\sin \alpha = -\frac{\sqrt{2}}{2}$ $\csc \alpha = -\sqrt{2}$
$\cos \alpha = \frac{\sqrt{2}}{2}$ $\sec \alpha = \sqrt{2}$
$\tan \alpha = -1$ $\cot \alpha = -1$

45. $\sin \alpha = -\frac{2\sqrt{5}}{5}$ $\csc \alpha = -\frac{\sqrt{5}}{2}$
$\cos \alpha = \frac{\sqrt{5}}{5}$ $\sec \alpha = \sqrt{5}$
$\tan \alpha = -2$ $\cot \alpha = -\frac{1}{2}$

47. $\sin \alpha = -\frac{2}{3}$ $\csc \alpha = -\frac{3}{2}$ **49.** 1   **51.** undefined   **53.** undefined   **55.** 0   **57.** 0
$\cos \alpha = \frac{\sqrt{5}}{3}$ $\sec \alpha = \frac{3\sqrt{5}}{5}$
$\tan \alpha = -\frac{2\sqrt{5}}{5}$ $\cot \alpha = -\frac{\sqrt{5}}{2}$

59. undefined   **61.** 180°   **63.** 108°   **65.** $\sin \varphi = \frac{5\sqrt{26}}{26}$ $\csc \varphi = \frac{\sqrt{26}}{5}$ **67.** $\sin \varphi = \frac{3}{5}$ $\sec \varphi = -\frac{5}{4}$
$\cos \varphi = -\frac{\sqrt{26}}{26}$ $\sec \varphi = -\sqrt{26}$ $\cos \varphi = -\frac{4}{5}$ $\cot \varphi = -\frac{4}{3}$
$\cot \varphi = -\frac{1}{5}$ $\tan \varphi = -\frac{4}{3}$

69. $\sin \varphi = \frac{3\sqrt{34}}{34}$ $\csc \varphi = \frac{\sqrt{34}}{3}$ **71.** $\sin \varphi = \frac{4}{5}$ $\csc \varphi = \frac{5}{4}$ **73.** $\cos \varphi = -\frac{\sqrt{21}}{5}$ $\csc \varphi = -\frac{5}{2}$
$\cos \varphi = -\frac{5\sqrt{34}}{34}$ $\sec \varphi = -\frac{\sqrt{34}}{5}$ $\tan \varphi = -\frac{4}{3}$ $\sec \varphi = -\frac{5}{3}$ $\tan \varphi = \frac{2\sqrt{21}}{21}$ $\sec \varphi = -\frac{5\sqrt{21}}{21}$
$\tan \varphi = -\frac{3}{5}$ $\cot \varphi = -\frac{3}{4}$ $\cot \varphi = \frac{\sqrt{21}}{2}$

75. $\sin \varphi = -\frac{4}{5}$ $\csc \varphi = -\frac{5}{4}$ **77.** $\sin \varphi = -\frac{3}{5}$ $\sec \varphi = \frac{5}{4}$ **79.** $\sin \varphi = -\frac{\sqrt{51}}{10}$ $\csc \varphi = -\frac{10\sqrt{51}}{51}$
$\tan \varphi = \frac{4}{3}$ $\sec \varphi = -\frac{5}{3}$ $\cos \varphi = \frac{4}{5}$ $\cot \varphi = -\frac{4}{3}$ $\tan \varphi = -\frac{\sqrt{51}}{7}$ $\sec \varphi = \frac{10}{7}$
$\cot \varphi = \frac{3}{4}$ $\tan \varphi = -\frac{3}{4}$ $\cot \varphi = -\frac{7\sqrt{51}}{51}$

Answers to Exercises

Exercise Set 7.4, page 307

1. $\tan 32° = 0.62$ $\sec 32° = 1.18$
   $\csc 32° = 1.89$ $\cot 32° = 1.60$
3. $\sin 66° = 0.91$ $\csc 66° = 1.09$
   $\cos 66° = 0.41$ $\cot 66° = 0.44$
5. $36°$
7. $52°$
9. $45°$

11. $\sin 63°$
13. $\tan 52°$
15. $\sec 87°$
17. $\cos 71°$
19. 0.85
21. 2.56
23. 2.38

25. $\cos \alpha = \dfrac{\sqrt{3}}{2}$ $\csc \alpha = 2$
    $\tan \alpha = \dfrac{\sqrt{3}}{3}$ $\sec \alpha = \dfrac{2\sqrt{3}}{3}$
    $\cot \alpha = \sqrt{3}$

27. $\sin \alpha = \dfrac{3\sqrt{13}}{13}$ $\csc \alpha = \dfrac{\sqrt{13}}{3}$
    $\cos \alpha = \dfrac{2\sqrt{13}}{13}$ $\sec \alpha = \dfrac{\sqrt{13}}{2}$
    $\cot \alpha = \dfrac{2}{3}$

29. $\sin \alpha = \dfrac{2\sqrt{5}}{5}$ $\csc \alpha = \dfrac{\sqrt{5}}{2}$
    $\cos \alpha = \dfrac{\sqrt{5}}{5}$ $\sec \alpha = \sqrt{5}$
    $\tan \alpha = 2$

31. $\sin \alpha = \dfrac{3}{5}$ $\csc \alpha = \dfrac{5}{3}$
    $\tan \alpha = \dfrac{3}{4}$ $\sec \alpha = \dfrac{5}{4}$
    $\cot \alpha = \dfrac{4}{3}$

33. 0.5175
35. 0.3973
37. 1.828
39. 0.8952
41. 0.6111

43. 1.402
45. $28°$
47. $43°$
49. $39°40'$
51. $23°$
53. 0.6368
55. 0.6100
57. 0.3181
59. 0.1352
61. 25.88
63. $3°27'$
65. $69°34'$
67. $26°34'$
69. yes
71. yes
73. yes
75. yes
77. $\dfrac{\sqrt{3}}{2}, \dfrac{1}{2}$
79. $-\sqrt{3}, \dfrac{\sqrt{3}}{3}$
81. $\sin(60° + 30°) = 1 = \sin 60° \cos 30° + \sin 30° \cos 60°$

83. $\cos 90° = 0 = \cos 45° \cos 45° - \sin 45° \sin 45°$

Exercise Set 7.5, page 312

1. positive
3. positive
5. positive
7. positive
9. positive
11. positive
13. negative

15. $20°$
17. $40°$
19. $55°$
21. $70°$
23. $84°$
25. $5°$
27. $85°35'$
29. $-\dfrac{1}{2}$
31. $-\sqrt{3}$
33. 1

35. $\frac{2\sqrt{3}}{3}$ **37.** $-\frac{2\sqrt{3}}{3}$ **39.** $\frac{\sqrt{3}}{3}$ **41.** $\frac{1}{2}$ **43.** -0.4040 **45.** 0.6691 **47.** -1.192 **49.** -0.9994

51. 5.759 **53.** -0.8090 **55.** -1.494 **57.** 0.9957 **59.** 0.1853 **61.** negative **63.** negative

65. negative **67.** negative **69.** negative **71.** positive **73.** negative **75.** negative

Chapter Test, page 314

1. $63°$ **3.** 20 **5.** 30 **7.** $\sin \alpha = \frac{8}{17}$ $\csc \alpha = \frac{17}{8}$ **9.** 48 **11.** $\frac{3\sqrt{10}}{20}, \frac{2\sqrt{10}}{3}$ **13.**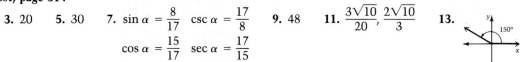

$\cos \alpha = \frac{15}{17}$ $\sec \alpha = \frac{17}{15}$

$\tan \alpha = \frac{8}{15}$ $\cot \alpha = \frac{15}{8}$

15.   **17.** $75°$ **19.** $285°$ **21.** $\frac{12}{13}, -\frac{12}{5}$ **23.** 0 **25.** $13°$ **27.** 1.05

29. $\sin \alpha = \frac{3\sqrt{13}}{13}$ $\csc \alpha = \frac{\sqrt{13}}{3}$ **31.** 0.7386 **33.** negative **35.** positive **37.** $35°$ **39.** -1

$\cos \alpha = \frac{2\sqrt{13}}{13}$ $\sec \alpha = \frac{\sqrt{13}}{2}$

$\cot \alpha = \frac{2}{3}$

41. 0.9793 **43.** $8\sqrt{3}$ **45.** $\sin\varphi = -\dfrac{5}{13}$ $\csc\varphi = -\dfrac{13}{5}$
$\tan\varphi = \dfrac{5}{12}$ $\sec\varphi = -\dfrac{13}{12}$
$\cot\varphi = \dfrac{12}{5}$

47. $\sin(2 \cdot 120°) = -\dfrac{\sqrt{3}}{2} = 2\sin 120° \cos 120°$ **49.** negative

Chapter 8

Exercise Set 8.1, page 324

1. yes   **3.** no   **5.**   **7.**   **9.**

11.   **13.** $\dfrac{\pi}{3}$ **15.** $\dfrac{7\pi}{4}$ **17.** $\dfrac{7\pi}{6}$ **19.** $-\dfrac{\pi}{6}$ **21.** $\dfrac{5\pi}{12}$ **23.** -3π **25.** 270°

27. 150° **29.** −540° **31.** −750° **33.** $\frac{270°}{\pi}$ **35.** 2π in **37.** 8 m **39.** $-\frac{7\pi}{5}$

41. (a) Minute hand: -24π, Hour hand: -2π (b) -2π, $-\frac{\pi}{6}$ (c) $-\pi$, $-\frac{\pi}{12}$ (d) $-\frac{\pi}{6}$, $-\frac{\pi}{72}$

43. 120π, $\frac{3\pi}{5}$ **45.** 5280 radians **47.** (a) $\left(\frac{\sqrt{3}}{2}, \frac{1}{2}\right)$ (b) $\left(\frac{\sqrt{2}}{2}, \frac{\sqrt{2}}{2}\right)$ (c) $\left(\frac{1}{2}, \frac{\sqrt{3}}{2}\right)$

Exercise Set 8.2, page 332

1. undefined **3.** 0 **5.** $\frac{\sqrt{3}}{2}$ **7.** 2 **9.** 1 **11.** $-\frac{1}{2}$ **13.** $\frac{2\sqrt{3}}{3}$ **15.** −1 **17.** $-\frac{\sqrt{2}}{2}$

19. $-\sqrt{3}$ **21.** $\sqrt{2}$ **23.** 0 **25.** $\frac{\sqrt{3}}{3}$ **27.** $-\sqrt{3}$ **29.** $\frac{2\sqrt{3}}{3}$ **31.** −0.53 **33.** −1.65 **35.** 1.05

37. −0.36 **39.** $\sin \frac{13\pi}{6} \sec \frac{5\pi}{3} = 1 = \tan \frac{9\pi}{4}$ **41.** $\tan \frac{\pi}{3} = \sqrt{3} = \dfrac{1 - \cos \frac{4\pi}{3}}{\sin \frac{\pi}{3}}$

43. $\cos \frac{5\pi}{6} = -\frac{\sqrt{3}}{2} = \cos \frac{\pi}{2} \cos \frac{\pi}{3} - \sin \frac{\pi}{2} \sin \frac{\pi}{3}$ **45.** $\frac{3\pi}{4}$ **47.** $\frac{5\pi}{6}$ **49.** $\frac{\pi}{4}$

Exercise Set 8.3, page 343

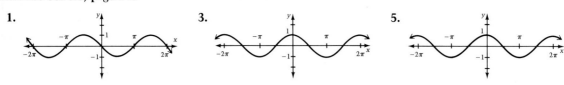

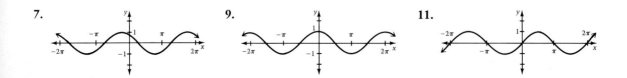

Answers to Exercises **A79**

13. 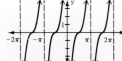   **15.** 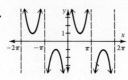   **17.**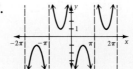

19. 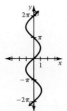   **21.**   **23.** False   **25.** True   **27.** True

29. False   **31.** 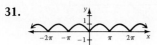   **33.** 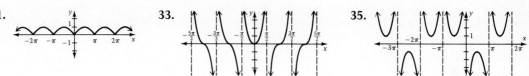   **35.**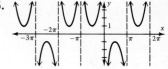

37.    **39.**    **41.**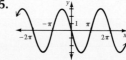

Exercise Set 8.4, page 350

1. 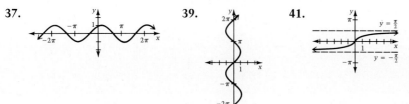   **3.**   **5.**

7. 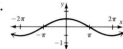 **9.**  **11.** 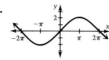 **13.**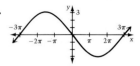

15.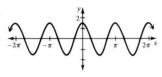

17. amplitude: 2
period: 2π
phase shift: $-\dfrac{\pi}{2}$

19. amplitude: 2
period: 4π
phase shift: $\dfrac{\pi}{3}$
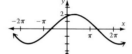

21. amplitude: 2
period: 4π
phase shift $-\dfrac{\pi}{2}$
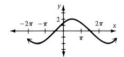

23. amplitude: $\sqrt{3}$
period: 4π
phase shift: $-\pi$
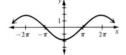

25. amplitude: 3
period: π
phase shift: $\dfrac{\pi}{2}$
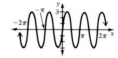

27. amplitude: 3
period: 4π
phase shift: $-\pi$

29. amplitude: 2
period: 2π
phase shift: $\dfrac{\pi}{2}$
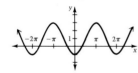

31.

33.

35.

37. **39.** 8, −8 **41.** 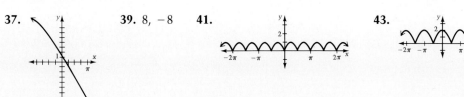 **43.**

45.  **47.** 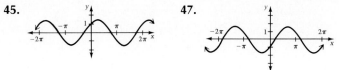 **49.** False **51.** True

Exercise Set 8.5, page 358

1. $\dfrac{\pi}{3}$ **3.** π **5.** $-\dfrac{\pi}{6}$ **7.** $-\dfrac{\pi}{4}$ **9.** 0 **11.** $\dfrac{1}{2}$ **13.** $-\sqrt{3}$ **15.** 0 **17.** $-\dfrac{\sqrt{3}}{2}$ **19.** $-\dfrac{\sqrt{2}}{2}$

21. $-\dfrac{\pi}{4}$ **23.** $\dfrac{\pi}{6}$ **25.** $\dfrac{4}{5}$ **27.** $\dfrac{4}{5}$ **29.** $\dfrac{4}{5}$ **31.** $\dfrac{5}{13}$ **33.** $-\dfrac{2\sqrt{5}}{15}$ **35.** $\dfrac{z^2}{1+z^2}$ **37.** $1 - z^2$

39. $-\dfrac{z\sqrt{1-z^2}}{1-z^2}$ **41.** $\dfrac{2\pi}{3}$ **43.** $-\dfrac{5}{13}$ **45.** $\dfrac{\sqrt{5}}{3}$ **47.** $\dfrac{\sqrt{5}}{2}$ **49.** $-\dfrac{\sqrt{165}}{11}$ **51.** 0.3229 **53.** 1.4312

55. 0.0814 **57.**   **59.**

Chapter Test, page 360

1. yes **3.**   **5.**   **7.** $\dfrac{4\pi}{3}$ **9.** 810° **11.** $\dfrac{3\pi}{2}$, 270° **13.** $\dfrac{9\pi}{4}$

15. $-\dfrac{1}{2}$ **17.** $\dfrac{\sqrt{3}}{3}$ **19.** 0.78 **21.** 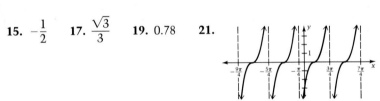 **23.** False **25.** False

27. amplitude: 3
period: π

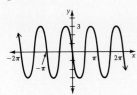

29. amplitude: 4
period: π
phase shift: $-\dfrac{\pi}{2}$

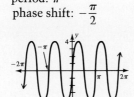

31.

33. $\dfrac{\pi}{6}$ **35.** $\dfrac{4}{5}$

37. True   **39.** False   **41.** 6, −6   **43.** $\dfrac{1}{2}, -\dfrac{1}{2}$ **45.** 4π

Cumulative Test: Chapters 6–8, page 363

1. 2^{2x-5} **3.**   **5.** 2   **7.** $f^{-1}(x) = \dfrac{x+5}{2}$, Domain: $[-5, 1]$, Range: $[0, 3]$ **9.** $\log_{16} 2 = x$

11. **13.** 3   **15.** 0.971   **17.** −0.0855   **19.** 17

21. $\dfrac{40\sqrt{3}}{3}$ **23.** $-\dfrac{5\sqrt{29}}{29}, -\dfrac{2\sqrt{29}}{29}, \dfrac{5}{2}$ **25.** $-\dfrac{\sqrt{2}}{2}$ **27.** $\sqrt{3}$ **29.** $-\dfrac{7\pi}{2}$ **31.** 480°   **33.** −0.7910

35. amplitude: 3
period: π
phase shift: $\dfrac{\pi}{2}$

37. $3h + 14$

39. (a) $\sin \theta = -\dfrac{40}{41}$
$\tan \theta = -\dfrac{40}{9}$
$\csc \theta = -\dfrac{41}{40}$
$\sec \theta = \dfrac{41}{9}$
$\cot \theta = -\dfrac{9}{40}$

(b) $\cos \theta = -\dfrac{24}{25}$
$\tan \theta = \dfrac{7}{24}$
$\csc \theta = -\dfrac{25}{7}$
$\sec \theta = -\dfrac{25}{24}$
$\cot \theta = \dfrac{24}{7}$

Chapter 9

Exercise Set 9.1, page 370

9. $\sin \alpha$ **11.** $\tan \beta$ **13.** $\sin \beta$ **15.** $\tan^2 \varphi$ **17.** $\tan \alpha + 1$ **39.** $\dfrac{\sqrt{1 + \cot^2 \alpha}}{1 + \cot^2 \alpha}$

41. $-\dfrac{\sqrt{1 - \sin^2 \beta}}{1 - \sin^2 \beta}$ **43.** $\dfrac{\sec \gamma \sqrt{\sec^2 \gamma - 1}}{\sec^2 \gamma - 1}$

Exercise Set 9.2, page 377

1. $\dfrac{\sqrt{3}}{2}$ **3.** $\dfrac{\sqrt{6} - \sqrt{2}}{4}$ **5.** $\dfrac{\sqrt{6} - \sqrt{2}}{4}$ **7.** $\dfrac{\sqrt{2} - \sqrt{6}}{4}$ **9.** $\dfrac{\sqrt{2}}{2}$ **11.** $-\dfrac{\sqrt{2}}{2}$ **13.** $\dfrac{\sqrt{6} - \sqrt{2}}{4}$

15. $\dfrac{\sqrt{6} + \sqrt{2}}{4}$ **17.** $\dfrac{\sqrt{6} - \sqrt{2}}{4}$ **19.** $\dfrac{-\sqrt{2} - \sqrt{6}}{4}$ **21.** $\dfrac{\sqrt{3} + 3}{3 - \sqrt{3}}$ **23.** $-2 - \sqrt{3}$ **25.** $\dfrac{\sqrt{3} - 3}{3 + \sqrt{3}}$

27. (a) $-\dfrac{56}{65}$ (b) $\dfrac{33}{65}$ (c) $-\dfrac{56}{33}$ **29.** (a) $-\dfrac{9\sqrt{13}}{169}$ (b) $-\dfrac{46\sqrt{13}}{169}$ (c) $\dfrac{9}{46}$ **31.** (a) $\dfrac{16}{65}$ (b) $-\dfrac{63}{65}$ (c) $-\dfrac{16}{63}$

33. (a) $-\dfrac{3\sqrt{13}}{13}$ (b) $\dfrac{2\sqrt{13}}{13}$ (c) $-\dfrac{3}{2}$ **45.** $\dfrac{\sqrt{3}}{2}\cos\alpha - \dfrac{1}{2}\sin\alpha$ **47.** $\dfrac{\tan\alpha + 1}{1 - \tan\alpha}$ **49.** $\cos\varphi$

51. $\dfrac{\sqrt{2}}{2}(\cos\alpha - \sin\alpha)$ **53.** $\tan(\alpha - \beta) = \dfrac{\tan\alpha - \tan\beta}{1 + \tan\alpha\tan\beta}$ **55.** $\sec(\alpha + \beta) = \dfrac{\sec\alpha\sec\beta}{1 - \tan\alpha\tan\beta}$

57. $\sin 2\alpha = 2\sin\alpha\cos\alpha$

Exercise Set 9.3, page 383

1. $\dfrac{\sqrt{3}}{2}, -\dfrac{1}{2}, -\sqrt{3}$ **3.** $-\dfrac{\sqrt{3}}{2}, -\dfrac{1}{2}, \sqrt{3}$ **5.** $-1, 0$, not defined **7.** $\dfrac{\sqrt{3}}{2}, -\dfrac{1}{2}, -\sqrt{3}$ **9.** $1, 0$, not defined

11. $-\dfrac{\sqrt{3}}{2}, \dfrac{1}{2}, -\sqrt{3}$ **13.** $\dfrac{\sqrt{3}}{2}, -\dfrac{1}{2}, -\sqrt{3}$ **15.** $1, 0$, not defined **17.** $1, 0$, defined

19. (a) $-\dfrac{24}{25}$ (b) $\dfrac{7}{25}$ (c) $-\dfrac{24}{7}$ **21.** (a) $-\dfrac{120}{169}$ (b) $-\dfrac{119}{169}$ (c) $\dfrac{120}{119}$ **23.** (a) $\dfrac{720}{1681}$ (b) $\dfrac{1519}{1681}$ (c) $\dfrac{720}{1519}$

25. (a) $\dfrac{4}{5}$ (b) $\dfrac{3}{5}$ (c) $\dfrac{4}{3}$ **27.** $\dfrac{1}{2}, \dfrac{\sqrt{3}}{2}, \dfrac{\sqrt{3}}{3}$ **29.** $\dfrac{\sqrt{2}}{2}, \dfrac{\sqrt{2}}{2}, 1$

31. $\dfrac{\sqrt{2+\sqrt{3}}}{2}, -\dfrac{\sqrt{2-\sqrt{3}}}{2}, -\dfrac{1}{2-\sqrt{3}}$ **33.** $-\dfrac{\sqrt{2+\sqrt{2}}}{2}, -\dfrac{\sqrt{2-\sqrt{2}}}{2}, \dfrac{\sqrt{2}}{2-\sqrt{2}}$

35. $-\dfrac{\sqrt{2-\sqrt{2}}}{2}, \dfrac{\sqrt{2+\sqrt{2}}}{2}, -\dfrac{\sqrt{2}}{2+\sqrt{2}}$ **37.** $0, 1, 0$ **39.** $\dfrac{\sqrt{3}}{2}, -\dfrac{1}{2}, -\sqrt{3}$

41. $\dfrac{\sqrt{2-\sqrt{3}}}{2}, \dfrac{\sqrt{2+\sqrt{3}}}{2}, \dfrac{1}{2+\sqrt{3}}$ **43.** $\dfrac{\sqrt{2-\sqrt{3}}}{2}, -\dfrac{\sqrt{2+\sqrt{3}}}{2}, -\dfrac{1}{2+\sqrt{3}}$

45. $-\dfrac{\sqrt{2+\sqrt{3}}}{2}, \dfrac{\sqrt{2-\sqrt{3}}}{2}, -\dfrac{1}{2-\sqrt{3}}$ **47.** $\dfrac{3\sqrt{10}}{10}, \dfrac{\sqrt{10}}{10}, 3$ **49.** $\dfrac{2\sqrt{13}}{13}, -\dfrac{3\sqrt{13}}{13}, -\dfrac{2}{3}$

51. $\dfrac{\sqrt{82}}{82}, \dfrac{9\sqrt{82}}{82}, \dfrac{1}{9}$ **53.** $\sqrt{\dfrac{5 + 2\sqrt{5}}{10}}, -\sqrt{\dfrac{5 - 2\sqrt{5}}{10}}, \dfrac{1}{\sqrt{5} - 2}$ **59.** $\cos 3\alpha = 4\cos^3\alpha - 3\cos\alpha$

61. $\sin 4\alpha = 4\sin\alpha\cos\alpha - 8\sin^3\alpha\cos\alpha$ **63.** $2\cos^2 6\alpha - 1$ **65.** $2\sin 4\alpha\cos 4\alpha$

67. $\dfrac{3\tan 2\alpha - \tan^3 2\alpha}{1 - 3\tan^2 2\alpha}$ **69.** $\cot 12\alpha + \csc 12\alpha$ **71.** $\pm 2\sin 4\alpha\sqrt{1 - \sin^2 4\alpha}$ **73.** $\cos\alpha$ **75.** $\sin\alpha$

77. $\cos\alpha$ **79.** 1 **81.** $1 - \sin\alpha$ **83.** $\cos 2\alpha$

Exercise Set 9.4, page 389

1. $\cos 3\alpha$ **3.** $\sin 3\alpha$ **5.** $\tan 4\varphi$ **7.** $1 - \cos\alpha$ **21.** yes   **23.** no   **25.** yes   **27.** $\dfrac{\sqrt{6}}{2}$ **29.** $\dfrac{\sqrt{2}}{2}$

31. $-\dfrac{\sqrt{2}}{2}$ **33.** $-\dfrac{\sqrt{6}}{2}$ **39.** $2\sin 2x\cos x$ **41.** $2\sin 4x\cos 2x$ **43.** $-2\sin 4x\sin 2x$

45. $\dfrac{1}{2}\sin 5x + \dfrac{1}{2}\sin(-x)$ **47.** $\dfrac{1}{2}\cos x - \dfrac{1}{2}\cos 5x$ **49.** $\dfrac{2\sin x\cos x + 1}{\cos^2 x - \sin^2 x}$

Exercise Set 9.5, page 397

1. $\dfrac{4\pi}{3} + 2n\pi, \dfrac{5\pi}{3} + 2n\pi$ **3.** $\dfrac{\pi}{6} + 2n\pi, \dfrac{5\pi}{6} + 2n\pi, 0 + n\pi$ **5.** $\dfrac{\pi}{6} + n\pi, \dfrac{5\pi}{6} + n\pi$

7. $\dfrac{\pi}{3} + 2n\pi, \dfrac{5\pi}{3} + 2n\pi$ **9.** $0 + 2n\pi, \dfrac{3\pi}{2} + 2n\pi$ **11.** $\dfrac{\pi}{6}, \dfrac{5\pi}{6}, \dfrac{3\pi}{2}$ **13.** $\dfrac{\pi}{6}, \dfrac{5\pi}{6}$ **15.** $0, \pi$

17. $\dfrac{\pi}{2}, \dfrac{7\pi}{6}, \dfrac{11\pi}{6}$ **19.** $\dfrac{\pi}{8}, \dfrac{5\pi}{8}, \dfrac{9\pi}{8}, \dfrac{13\pi}{8}$ **21.** $-60°, 60°$ **23.** $0, \dfrac{\pi}{4}, \dfrac{3\pi}{4}, \pi$ **25.** $0°, 60°, 180°$

27. $30°, 90°, 150°, 210°, 270°, 330°, 45°, 135°, 225°, 315°$ **29.** $\dfrac{\pi}{2}, \dfrac{7\pi}{6}, \dfrac{3\pi}{2}, \dfrac{11\pi}{6}$ **31.** $0, \dfrac{\pi}{6}, \dfrac{5\pi}{6}, \pi$

33. $0°, 180°, 48°40', 131°20'$ **35.** $270°, 19°30', 160°30'$ **37.** $210°, 330°, 41°50', 138°10'$

39. 10°30′, 169°30′, 223°, 317°   **41.** no solution

43. 27°20′, 62°40′, 117°20′, 152°40′, 207°20′, 242°40′, 297°20′, 332°40′   **45.** $-\dfrac{\pi}{6}$ **47.** $\dfrac{\pi}{3}$ **49.** $\dfrac{24}{25}$

51. $-\dfrac{7}{25}$ **53.** $-\dfrac{33}{65}$ **55.** 61°16′   **57.** $\dfrac{1}{2}$ **59.** $\theta = \dfrac{\pi}{3}$ or $\theta = \dfrac{5\pi}{3}$, $r = \dfrac{1}{2}$ **61.** 11°19′, 191°19′

Chapter Test, page 400

5. $\dfrac{3 - \sqrt{3}}{3 + \sqrt{3}}$ **9.** $-1, 0,$ not defined   **11.** $-\sqrt{\dfrac{3 - \sqrt{5}}{6}}$ **13.** $-\dfrac{\sqrt{10}}{10}$ **15.** csc x **17.** sin x

19. cos 2φ **25.** $-\dfrac{\sqrt{2}}{2}$ **27.** $\dfrac{1}{2}\sin 7x + \dfrac{1}{2}\sin x$ **29.** $\dfrac{2\pi}{3} + 2n\pi, \dfrac{4\pi}{3} + 2n\pi, 0 + 2n\pi$ **31.** $\dfrac{\pi}{2} + 2n\pi$

33. 90°, 199°30′, 340°30′   **35.** 0°, 98°10′, 261°10′   **37.** $\dfrac{7\pi}{12}$ **39.** 0   **41.** $\dfrac{\sqrt{1 + \cot^2 \alpha}}{1 + \cot^2 \alpha}$

43. (a) $-\sin x$ (b) $-\sin \beta$ **45.** $\dfrac{3 \tan 2x - \tan^3 2x}{1 - 3 \tan^2 2x}$

Chapter 10

Exercise Set 10.1, page 409

1. $\angle A = 45°, a = 32\sqrt{2}, b = 32\sqrt{2}$ **3.** $a = 44\sqrt{3}, \angle B = 30°, c = 88$ **5.** $\angle A = 30°, \angle B = 60°, c = 30$

7. $\angle A = 60°, b = 8\sqrt{3}, c = 16\sqrt{3}$ **9.** $\angle A = 45°, \angle B = 45°, b = 25\sqrt{2}$

11. $\angle B = 54°, c = 120, b = 96$ **13.** $\angle A = 78°, b = 1.9, c = 9.2$

15. $a = 7.6, \angle B = 52°, b = 9.7$ **17.** $a = 0.132, \angle B = 61°40′, c = 0.277$

19. $\angle A = 18°42′, a = 45.69, b = 135.0$ **21.** 21′   **23.** 5800 ft   **25.** 5.2 ft   **27.** 57°

29. (a) 60° (b) 40° (c) 76° (d) 22°   **31.** 2700′   **33.** 17°

Exercise Set 10.2, page 415

1. $b = 20$, $\angle C = 105°$, $c = 27$ **3.** $\angle A = 15°$, $a = 2.9$, $c = 8.0$ **5.** $b = 9.9$, $\angle C = 56°$, $c = 20$
7. $\angle A = 80°$, $a = 46$, $b = 28$ **9.** $\angle A = 41°30'$, $b = 1.043$, $c = 1.011$
11. yes, $\angle B = 90°$, $\angle C = 60°$, $c = 12\sqrt{3}$ **13.** no   **15.** yes, $\angle B = 51°$, $\angle C = 9°$, $c = 9.8$
17. yes, $\angle B = 63°50'$, $\angle C = 63°40'$, $c = 36.4$ or $\angle B = 116°10'$, $\angle C = 11°20'$, $c = 7.97$ **19.** 431 yards
21. N $0°30'$ E   **23.** from first observer: 4.4 mi, from second observer: 6.2 mi   **25.** $x = \dfrac{m \tan \theta \tan \varphi}{\tan \varphi - \tan \theta}$

Exercise Set 10.3, page 420

1. $\angle A = 69°$, $\angle B = 51°$, $c = 11$ **3.** $a = 31$, $\angle B = 34°$, $\angle C = 26°$ **5.** $\angle A = 72°$, $b = 95$, $\angle C = 54°$
7. $a = 27.8$, $\angle B = 13°54'$, $\angle C = 18°30'$ **9.** $\angle A = 37°$, $\angle B = 53°$, $\angle C = 90°$
11. $\angle A = 66°$, $\angle B = 62°$, $\angle C = 52°$ **13.** $\angle A = 29°$, $\angle B = 47°$, $\angle C = 104°$
15. $\angle A = 62°40'$, $\angle B = 36°20'$, $\angle C = 81°$ **17.** 170 m   **19.** 3.9, 7.2   **21.** 47°   **23.** 99.8
25. 52.1   **27.** 672.0   **29.** 7737   **31.** $\angle A = 65°30'$, $\angle B = 69°50'$, $\angle C = 44°40'$, $a = 15.5$ or $\angle A = 114°30'$, $\angle B = 38°10'$, $\angle C = 27°20'$, $a = 23.6$ **33.** 47.0

Exercise Set 10.4, page 426

1. $2\sqrt{34}$, 30.96°   **3.** $2\sqrt{13}$, 326.31°   **5.** 13, 247.38°   **7.** 11, 0°

9. 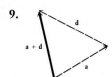   **11.**    **13.** 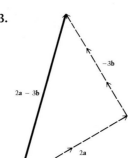   **15.** 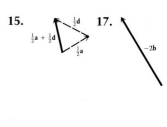   **17.**

19. 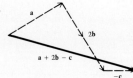 **21.** ⟨5, 12⟩ **23.** ⟨3, −4⟩ **25.** ⟨1, −8⟩ **27.** ⟨2, 2⟩ **29.** $\left(-2\frac{1}{2}, -1\frac{1}{2}\right)$

31. $\left(\frac{5}{2}, \frac{5}{2}\right)$ **33.** ⟨−20, 4⟩ **35.** ⟨5, 19⟩ **37.** ⟨13, 17⟩ **39.** ⟨−21, 2⟩

41. $10, \left\langle -\frac{3}{5}, \frac{4}{5}\right\rangle, \left\langle \frac{6}{5}, -\frac{8}{5}\right\rangle$ **43.** $4, \langle 0, -1\rangle, \langle 0, 2\rangle$ **45.** $2\sqrt{3}, \left\langle \frac{\sqrt{6}}{3}, \frac{\sqrt{3}}{3}\right\rangle, \left\langle -\frac{2\sqrt{6}}{3}, -\frac{2\sqrt{3}}{3}\right\rangle$

47. $\frac{10}{3}, \left\langle \frac{4}{5}, -\frac{3}{5}\right\rangle, \left\langle -\frac{8}{5}, \frac{6}{5}\right\rangle$ **49.** $-\mathbf{i} + 8\mathbf{j}, \sqrt{2}, \frac{1}{\sqrt{2}}\mathbf{i} + \frac{1}{\sqrt{2}}\mathbf{j}$ **51.** $3\mathbf{i} - 16\mathbf{j}, \sqrt{13}, \frac{3}{\sqrt{13}}\mathbf{i} - \frac{2}{\sqrt{13}}\mathbf{j}$

53. $-\frac{19}{2}\mathbf{i} + \frac{13}{2}\mathbf{j}, \frac{\sqrt{26}}{2}, \frac{1}{\sqrt{26}}\mathbf{i} + \frac{5}{\sqrt{26}}\mathbf{j}$ **55.** ⟨−6, 6√3⟩ **57.** ⟨−5√2, 5√2⟩ **59.** ⟨2.7, −7.4⟩

61. $r = 1, s = -2$

Exercise Set 10.5, page 433

1. 569 mph, N 82°6′ E **3.** 427 mph, N 16°19′ W **5.** 391 N = 88 lbs. **7.** 937 N = 211 lbs.

9. −2, 98° **11.** 6, 18° **13.** 0, 90° **15.** −11, 112° **17.** perpendicular **19.** neither **21.** parallel

23. 4827 foot lbs. **25.** 54,067 foot lb.

Exercise Set 10.6, page 438

1. (a) 0, stationary
   (b) $\frac{\sqrt{2}}{4}$, stationary
   (c) $\frac{1}{2}$, stationary
   (d) 0, down
   (e) $-\frac{1}{2}$, stationary
   (f) 0, down
   (g) 0, up

3. (a) -3, stationary
   (b) 3, stationary
   (c) -3, stationary
   (d) -3, stationary
   (e) -3, stationary
   (f) -3, stationary
   (g) -3, stationary

5. (a) 0, stationary
   (b) 0, down
   (c) 0, down
   (d) 0, down
   (e) 0, down
   (f) 0, down
   (g) 0, down

7. amplitude = 6
   period = π
   frequency = $\frac{1}{\pi}$

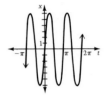

9. amplitude = 3
   period = 4π
   frequency = $\frac{1}{4\pi}$

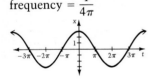

11. $x = \sin\left(t + \frac{\pi}{6}\right)$
    amplitude = 1, period = 2π, frequency = $\frac{1}{2\pi}$

13. $x = \sqrt{2}\sin\left(t - \frac{\pi}{4}\right)$
    amplitude = $\sqrt{2}$, period = 2π, frequency = $\frac{1}{2\pi}$

15. $x = \frac{13}{2}\sin(2t - 1.18)$
    amplitude = $\frac{13}{2}$, period = π, frequency = $\frac{1}{\pi}$

17. $x = 6\cos\frac{2\pi}{3}t$, frequency = $\frac{1}{3}$

19. $x = 6\cos 1.25\pi t$, frequency = 0.625

21. $x = 0.60\cos 2t$, frequency = $\frac{1}{\pi}$

23. $x = 0.02\cos\sqrt{9.8}\,t$, frequency = $\frac{\sqrt{9.8}}{2\pi}$

25. $x = 3 \sin \frac{3}{2}t$ **27.** $x = \frac{3}{2} \sin \frac{2}{3}t$ **29.** multiplied by $\sqrt{2}$ **31.** $x = 0.01 \cos 400\pi t$ **33.** no

Exercise Set 10.7, page 446
1. $A(2, 120°)$ **3.** $C(3, 135°)$ **5.** $E(2, -120°)$
7. $G(-1, -270°)$ **9.** $I(-2, -420°)$ **11.** $K(3, \pi)$
13. $M\left(-2, \frac{5\pi}{6}\right)$ **15.** $Q\left(-4, -\frac{\pi}{2}\right)$

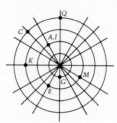

17. $(-1, \sqrt{3})$ **19.** $\left(-\frac{3\sqrt{2}}{2}, \frac{3\sqrt{2}}{2}\right)$ **21.** $(-1, -\sqrt{3})$ **23.** $(0, -1)$ **25.** $(-1, \sqrt{3})$

27. $(-3, 0)$ **29.** $(\sqrt{3}, -1)$ **31.** $(0, 4)$ **33.** $\left(4, \frac{\pi}{2}\right), \left(-4, \frac{3\pi}{2}\right)$ **35.** $\left(\sqrt{2}, \frac{5\pi}{4}\right), \left(-\sqrt{2}, \frac{\pi}{4}\right)$

37. $\left(2, \frac{11\pi}{6}\right), \left(-2, \frac{5\pi}{6}\right)$ **39.** $\left(2, \frac{3\pi}{4}\right), \left(-2, \frac{7\pi}{4}\right)$ **41.** $(5, 2.21), (-5, 5.36)$

43. $(2\sqrt{13}, 0.98), (-2\sqrt{13}, 4.12)$ **45.** $x^2 + y^2 = 9$ **47.** $y = -x$ **49.** $x - 2y = 5$

51. $y^2 + 2x = 1$ **53.** $x^2 - 3y^2 - 8y - 4 = 0$ **55.** $\theta = \frac{\pi}{2}$ **57.** $r(\cos \theta + \sin \theta) = 1$

59. $r - 2 \cos \theta = 0$ **61.** $r^2 \cos^2 \theta = 6r \sin \theta + 9$ **63.** $r^2 \sin 2\theta = 1$ **65.** $r^2 = 16r \sin \theta$

67. **69.** **71.** **73.**

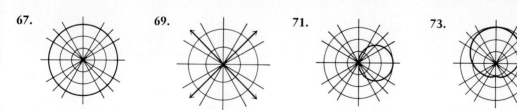

75. $2r \cos \theta - 3r \sin \theta = -4$ **77.** (a) 4.91 (b) 4.91   **79.**   **81.**

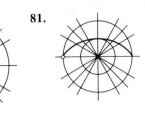

Chapter Test, page 448

1. $\angle A = 30°, a = 4\sqrt{3}, c = 8\sqrt{3}$ **3.** $a = 64.6, \angle B = 56°18', c = 116$

5. $\angle A = 37°40', \angle B = 52°20', c = 59.3$ **7.** $70°$ **9.** $a = 35, b = 53, \angle C = 56°$

11. yes, $\angle B = 117°06', \angle C = 38°36', b = 13.4$ or $\angle B = 14°18', \angle C = 141°24', b = 3.7$

13. 472 yd   **15.** $\angle A = 23°40', \angle B = 37°20', c = 3.32$ **17.** $\angle A = 39°34', \angle B = 54°41', \angle C = 85°45'$

19. 49.1   **21.** (a) magnitude $= 3\sqrt{2}$, direction $135°$ **23.** $\langle 5, 9 \rangle$ **25.** $\dfrac{5\sqrt{2}}{2}, \left\langle -4, 7\dfrac{1}{2} \right\rangle$
(b) $\langle -5, 1 \rangle$

27. $\mathbf{v}_x = -7.3i, \mathbf{v}_y = 4.6j$ **29.** 493 mph, N 74°28' E   **31.** $-4, 117°$

33. amplitude = 4
period = 4π
frequency = $\dfrac{1}{4\pi}$
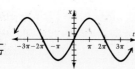

35. $(-2\sqrt{2}, -2\sqrt{2})$

37. $x^2 + \left(y - \dfrac{3}{2}\right)^2 = \dfrac{9}{4}$

39.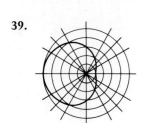

41. $2\sqrt{5}$ **43.** $-2\mathbf{a} - 3\mathbf{b}$ **45.** $54°44'$

Cumulative Test: Chapters 9 and 10, page 451

1. $\dfrac{\sqrt{6} - \sqrt{2}}{4}$ **3.** $\sqrt{3} - 2$ **5.** $\sin 2\alpha = 1$
$\cos 2\alpha = 0$
$\tan 2\alpha$ not defined

7. $\sin 2\alpha = 1$
$\cos 2\alpha = 0$
$\tan 2\alpha$ not defined

9. $\sin \dfrac{\alpha}{2} = \sqrt{\dfrac{15 + \sqrt{161}}{30}}$
$\cos \dfrac{\alpha}{2} = \sqrt{\dfrac{15 - \sqrt{161}}{30}}$

11. 1.58   **17.** $\dfrac{\pi}{2}$ **19.** $\dfrac{\pi}{18}, \dfrac{5\pi}{18}, \dfrac{13\pi}{18}, \dfrac{17\pi}{18}$ **21.** $\dfrac{12}{13}$ **23.** $\dfrac{\sqrt{3}}{2}$ **25.** 24   **27.** 26.8

29. 3705 yd² **31.** $\|\mathbf{c}\| = 5\sqrt{2}$, $\mathbf{u_c} = \left\langle \frac{7\sqrt{2}}{10}, \frac{\sqrt{2}}{10} \right\rangle$, $-\frac{1}{2}\mathbf{c} = \left\langle -\frac{7}{2}, -\frac{1}{2} \right\rangle$ **33.** N 36°52′ E at 25 yd/sec

35. 0, 90° **37.** amplitude = 4, period = 4π, frequency = $\frac{1}{4\pi}$, $x = 2\sqrt{3}$ **39.** $\left(-2, \frac{\pi}{6} \right)$

41. $\sin \theta = 0.8$, $\cos \theta = 0.6$ **43.** 51.4 inches

Chapter 11

Exercise Set 11.1, page 462

1. $i\sqrt{13}$ **3.** $i\sqrt{15}$ **5.** $2i$ **7.** $10i$ **9.** $2i\sqrt{6}$ **11.** $7i\sqrt{2}$ **13.** $-3i$ **15.** $-2i\sqrt{5}$ **17.** $-3i\sqrt{7}$

19. 2, 5 **21.** $-3, 4$ **23.** $-1, -2$ **25.** $\frac{1}{2}, 6$ **27.** 0, 6 **29.** $-\sqrt{3}, 0$ **31.** $\frac{1}{4}, -\sqrt{10}$ **33.** 1

35. $-i$ **37.** -1 **39.** $-i$ **41.** $-8i$ **43.** -81 **45.** $-2\sqrt{14}$ **47.** $4i$ **49.** $-6\sqrt{2}$ **51.** $8 + 9i$

53. $-2 + 8i$ **55.** $10 - 15i$ **57.** $-2 - 10i$ **59.** $-3 - 3i$ **61.** $5 - i$ **63.** $2 + 14i$

65. $-22 + 14i$ **67.** $2 + 15i$ **69.** $-12 + 4i$ **71.** $-24 + 32i$ **73.** $3 + 6i$ **75.** $13 + 11i$

77. $-24 - 31i$ **79.** $-6 - 10i$ **81.** 41 **83.** 2 **85.** $-5 + 12i$ **87.** $24 - 10i$ **89.** $-4 - 3i$

91. $-54 + 54i$ **93.** $28 - 4i$ **95.** $-2, 3$ **97.** $\frac{3}{4}, -2$ **99.** $-2, 5$ **101.** 20, 12 **103.** 0

105. i **107.** True **109.** False **111.** True **113.** True

115. $1^4 = 1$, $i^4 = 1$, $(-1)^4 = 1$, $(-i)^4 = (-1)^4 i^4 = 1$

117. $1^6 = 1$, $\left(\frac{1}{2} + \frac{\sqrt{3}}{2}i\right)^6 = \left[\left(\frac{1}{2} + \frac{\sqrt{3}}{2}i\right)^2\left(\frac{1}{2} + \frac{\sqrt{3}}{2}i\right)\right]^2 = \left[\left(-\frac{1}{2} + \frac{\sqrt{3}}{2}i\right)\left(\frac{1}{2} + \frac{\sqrt{3}}{2}i\right)\right]^2 = (-1)^2 = 1$,
$\left(-\frac{1}{2} + \frac{\sqrt{3}}{2}i\right)^6 = \left[\left(-\frac{1}{2} + \frac{\sqrt{3}}{2}i\right)^2\left(-\frac{1}{2} + \frac{\sqrt{3}}{2}i\right)\right]^2 = \left[\left(-\frac{1}{2} - \frac{\sqrt{3}}{2}i\right)\left(-\frac{1}{2} + \frac{\sqrt{3}}{2}i\right)\right]^2 = (1)^2 = 1$, $(-1)^6 = 1$,
$\left(-\frac{1}{2} - \frac{\sqrt{3}}{2}i\right)^6 = \left[-\left(\frac{1}{2} + \frac{\sqrt{3}}{2}i\right)^6\right] = (-1)^6\left(\frac{1}{2} + \frac{\sqrt{3}}{2}i\right)^6 = (1)(1) = 1$,
$\left(\frac{1}{2} - \frac{\sqrt{3}}{2}i\right)^6 = \left[-\left(-\frac{1}{2} + \frac{\sqrt{3}}{2}i\right)\right]^6 = (-1)^6\left(-\frac{1}{2} + \frac{\sqrt{3}}{2}i\right)^6 = (1)(1) = 1$

119. True   **121.** False, $(i)(i) = -1$ **123.** True

Exercise Set 11.2, page 470

1. $\frac{2}{13} - \frac{3}{13}i$ **3.** $\frac{5}{29} + \frac{2}{29}i$ **5.** $-\frac{1}{2} - \frac{1}{2}i$ **7.** $-\frac{1}{10} + \frac{1}{5}i$ **9.** $0 - \frac{1}{5}i$ **11.** $0 + 7i$

13. $\frac{\sqrt{3}}{7} + \frac{2}{7}i$ **15.** $2 + 2i$ **17.** $\frac{14}{17} - \frac{5}{17}i$ **19.** $\frac{5}{29} - \frac{27}{29}i$ **21.** $\frac{13}{2} - \frac{7}{2}i$ **23.** $-2i$

25. $5 + 3i$ **27.** $-\frac{8}{41} - \frac{10}{41}i$ **29.** $1 + \frac{3}{7}i$ **31.** $-\frac{3}{20} - \frac{1}{20}i$ **33.** $-6i$

35. (a) $\overline{z + w} = 3 + i = \overline{z} + \overline{w}$ (b) $\overline{z - w} = -1 - 7i = \overline{z} - \overline{w}$ (c) $\overline{z \cdot w} = 14 - 2i = \overline{z} \cdot \overline{w}$
(d) $\overline{\left(\frac{z}{w}\right)} = -\frac{1}{2} - \frac{1}{2}i = \frac{\overline{z}}{\overline{w}}$

37. (a) $\overline{z + w} = 17 - 6i = \overline{z} + \overline{w}$ (b) $\overline{z - w} = 7 + 12i = \overline{z} - \overline{w}$ (c) $\overline{z \cdot w} = 87 - 93i = \overline{z} \cdot \overline{w}$
(d) $\overline{\left(\frac{z}{w}\right)} = \frac{33}{106} + \frac{123}{106}i = \frac{\overline{z}}{\overline{w}}$

39. (a) $\overline{z + w} = -7 + 9i = \overline{z} + \overline{w}$ (b) $\overline{z - w} = -5 + i = \overline{z} - \overline{w}$ (c) $\overline{z \cdot w} = -14 - 29i = \overline{z} \cdot \overline{w}$
(d) $\overline{\left(\frac{z}{w}\right)} = \frac{26}{17} + \frac{19}{17}i = \frac{\overline{z}}{\overline{w}}$

41. (a) $\overline{z^2} = 3 + 4i = (\overline{z})^2$ (b) $\overline{z^3} = 2 + 11i = (\overline{z})^3$ **43.** (a) $\overline{z^2} = -32 + 24i = (\overline{z})^2$ (b) $\overline{z^3} = 208 + 144i = (\overline{z})^3$

45. (a) $\overline{z^2} = 24 - 10i = (\overline{z})^2$ (b) $\overline{z^3} = -110 + 74i = (\overline{z})^3$ **47.** $P(2, 5)$ **49.** $P(2, -5)$ **51.** $P(-2, 5)$

53. $P(-2, -5)$ **55.** $P(1, 4)$ **57.** $P(-3, 1)$ **59.** $P(4, 0)$ **61.** $P(-3, 0)$ **63.** $P(0, 6)$ **65.** $P(0, -1)$

67. $P(5, 4)$ **69.** $P(4, -2)$ **71.** $P(-3, 6)$ **73.** $\sqrt{53}$ **75.** $\sqrt{53}$ **77.** 10   **79.** 10   **81.** $3\sqrt{2}$

83. $5\sqrt{2}$ **85.** 2   **87.** 1   **89.** 4   **91.** 4   **93.** $\sqrt{15}$ **95.** $-\frac{7}{25} + \frac{1}{25}i$ **97.** 1   **99.** i

101. $-\frac{23}{26} + \frac{11}{26}i$ **103.** $\frac{1}{10} - \frac{11}{10}i$ **105.** $-\frac{5}{2} - \frac{9}{2}i$ **107.** $\frac{\sqrt{2}}{2}$ **109.** $\frac{\sqrt{2}}{2}$ **111.** $\frac{9}{13} + \frac{7}{13}i$

113. $\frac{4}{17} - \frac{1}{17}i$ **115.** $\frac{15}{7} + 3i$ **117.** $2 + 2i, -2 - 2i$

119. $\overline{z - w} = \overline{(a + bi) - (c + di)} = \overline{(a - c) + (b - d)i} = (a - c) - (b - d)i = (a - bi) - (c - di)$
$= \overline{(a + bi)} - \overline{(c + di)} = \overline{z} - \overline{w}$

121. $\overline{\overline{z}} = \overline{\overline{(a + bi)}} = \overline{a - bi} = a + bi = z$

123. $\frac{1}{2}(z + \overline{z}) = \frac{1}{2}[(a + bi) + \overline{(a + bi)}] = \frac{1}{2}(a + bi + a - bi) = a$

125. $|\overline{z}| = |\overline{a + bi}| = |a - bi| = \sqrt{a^2 + (-b)^2} = \sqrt{a^2 + b^2} = |a + bi| = |z|$

127. False, $|(-1) - 1| \neq |-1| - |1|$ **129.** True   **131.** $z: P(3, 5), \overline{z}: P(3, -5)$

133. $z: P(-3, 5), \overline{z}: P(-3, -5)$ **135.** $z: P(2, -6), \overline{z}: P(2, 6)$ **137.** $z: P(-2, -6), \overline{z}: P(-2, 6)$

139. $z: P(2, 5), w: P(4, -7), z + w: P(6, -2)$ **141.** $z: P(2, 1), w: P(3, -2), zw: P(8, -1)$

Exercise Set 11.3, page 480

1. $\pm 4i$ **3.** $3 \pm 4i$ **5.** $2 \pm i$ **7.** $-\frac{3}{2} \pm \frac{\sqrt{7}}{2}i$ **9.** $1 \pm \frac{1}{2}i$ **11.** $-1 \pm \frac{\sqrt{7}}{2}i$ **13.** $\pm \frac{8}{3}i$

15. $\frac{2}{3} \pm \frac{\sqrt{2}}{3}i$ **17.** $\frac{3 \pm 3\sqrt{5}}{2}$ **19.** $3, 3 \pm i\sqrt{2}$ **21.** $\frac{1}{3}, \pm \frac{\sqrt{5}}{2}i$ **23.** $3, \frac{1}{2} \pm \frac{\sqrt{7}}{2}i$

25. $-2, -1, -1 \pm i\sqrt{2}$ **27.** $2i, -2i, 3$ **29.** $x^2 - 8x + 25 = 0$ **31.** $x^3 - 4x^2 + 25x - 100 = 0$

37. $1, -2, \pm 2i$ **39.** $-1, \frac{3}{2}, -\frac{1}{2} \pm \frac{\sqrt{3}}{2}i$ **41.** $\pm i^{3/2}, \pm i^{1/2}$ **43.** $\pm i\sqrt{5}, \pm\sqrt{5}$ **45.** $\pm\left(\frac{-5 \pm \sqrt{105}}{2}\right)^{1/2}$

Exercise Set 11.4, page 486

1. $3\left[\cos\frac{\pi}{2} + i\sin\frac{\pi}{2}\right]$ **3.** $6[\cos\pi + i\sin\pi]$ **5.** $2\left[\cos\frac{2\pi}{3} + i\sin\frac{2\pi}{3}\right]$ **7.** $2\sqrt{2}\left[\cos\frac{5\pi}{4} + i\sin\frac{5\pi}{4}\right]$

9. $\sqrt{2}\left[\cos\frac{5\pi}{4} + i\sin\frac{5\pi}{4}\right]$ **11.** $2\sqrt{3}\left[\cos\frac{5\pi}{6} + i\sin\frac{5\pi}{6}\right]$ **13.** $4\left[\cos\frac{11\pi}{6} + i\sin\frac{11\pi}{6}\right]$

15. $6\left(\cos\frac{5\pi}{4} + i\sin\frac{5\pi}{4}\right)$ **17.** $13(\cos 67°23' + i\sin 67°23')$ **19.** $25(\cos 196°16' + i\sin 196°16')$

21. 8 **23.** $6 - 2i\sqrt{3}$ **25.** $-4\sqrt{3}$ **27.** $12 + 9i$ **29.** $2(\cos 165° + i\sin 165°)$

31. $\sqrt{5}(\cos(-243°26') + i\sin(-243°26'))$ **33.** $\sqrt{3}\left(\cos\frac{3\pi}{2} + i\sin\frac{3\pi}{2}\right)$

35. $\frac{\sqrt{6}}{3}\left(\cos -\frac{57\pi}{36} + i\sin -\frac{57\pi}{36}\right)$ **37.** -64 **39.** $8i$ **41.** 4^{12}

43. $\sqrt[3]{3}(\cos 30° + i \sin 30°)$
$\sqrt[3]{3}(\cos 150° + i \sin 150°)$
$\sqrt[3]{3}(\cos 270° + i \sin 270°)$

45. $\sqrt[8]{8}(\cos 33°45' + i \sin 33°45')$
$\sqrt[8]{8}(\cos 123°45' + i \sin 123°45')$
$\sqrt[8]{8}(\cos 213°45' + i \sin 213°45')$
$\sqrt[8]{8}(\cos 303°45' + i \sin 303°45')$

47. $\sqrt[10]{2}(\cos 9° + i \sin 9°)$
$\sqrt[10]{2}(\cos 81° + i \sin 81°)$
$\sqrt[10]{2}(\cos 153° + i \sin 153°)$
$\sqrt[10]{2}(\cos 225° + i \sin 225°)$
$\sqrt[10]{2}(\cos 297° + i \sin 297°)$

49. (a) $1(\cos 0° + i \sin 0°)$
$1(\cos 72° + i \sin 72°)$
$1(\cos 144° + i \sin 144°)$
$1(\cos 216° + i \sin 216°)$
$1(\cos 288° + i \sin 288°)$

(b) $1(\cos 0° + i \sin 0°) = 1$
$1(\cos 90° + i \sin 90°) = i$
$1(\cos 180° + i \sin 180°) = -1$
$1(\cos 270° + i \sin 270°) = -i$

(c) $1(\cos 0° + i \sin 0°) = 1$
$1(\cos 45° + i \sin 45°) = \dfrac{\sqrt{2}}{2} + \dfrac{\sqrt{2}}{2}i$
$1(\cos 90° + i \sin 90°) = i$
$1(\cos 135° + i \sin 135°) = -\dfrac{\sqrt{2}}{2} + \dfrac{\sqrt{2}}{2}i$
$1(\cos 180° + i \sin 180°) = -1$
$1(\cos 225° + i \sin 225°) = -\dfrac{\sqrt{2}}{2} - \dfrac{\sqrt{2}}{2}i$
$1(\cos 270° + i \sin 270°) = -i$
$1(\cos 315° + i \sin 315°) = \dfrac{\sqrt{2}}{2} - \dfrac{\sqrt{2}}{2}i$

51. $1 \pm \sqrt{2}i$

53. $\sqrt[3]{2}(\cos 30° + i \sin 30°)$
$\sqrt[3]{2}(\cos 150° + i \sin 150°)$
$\sqrt[3]{2}(\cos 270° + i \sin 270°)$

55. $2(\cos 36° + i \sin 36°)$
$2(\cos 108° + i \sin 108°)$
$2(\cos 180° + i \sin 180°)$
$2(\cos 252° + i \sin 252°)$
$2(\cos 324° + i \sin 324°)$

Chapter Test, page 488

1. $i\sqrt{11}$ **3.** $4i\sqrt{3}$ **5.** $3, -9$ **7.** $0, -3$ **9.** i **11.** $6 - 9i$ **13.** 9 **15.** $-\dfrac{8}{65} - \dfrac{1}{65}i$

17. $\dfrac{15}{13} + \dfrac{29}{13}i$ **19.** (a) $\overline{z + w} = 7 + 2i = \overline{z} + \overline{w}$ (b) $\overline{z - w} = -1 - 4i = \overline{z} - \overline{w}$ (c) $\overline{z \cdot w} = 15 + 5i = \overline{z} \cdot \overline{w}$
(d) $\overline{\left(\dfrac{z}{w}\right)} = \dfrac{9}{25} - \dfrac{13}{25}i = \dfrac{\overline{z}}{\overline{w}}$

21. (a) $\overline{z^2} = 35 - 12i = (\overline{z})^2$ (b) $\overline{z^3} = 198 - 107i = (\overline{z})^3$ **23.** $P(5, -12)$ **25.** $P(0, -7)$ **27.** $\sqrt{41}$

29. 9 **31.** $5 \pm 2i$ **33.** $-\dfrac{1}{2} \pm \dfrac{\sqrt{11}}{2}i$ **35.** $\dfrac{2}{3}, -3, \dfrac{1}{4} \pm \dfrac{\sqrt{31}}{4}i$

37. Degree of polynomial: 4; Roots: $4, -3, \dfrac{5}{2} + \dfrac{\sqrt{19}}{2}i$, and $\dfrac{5}{2} - \dfrac{\sqrt{19}}{2}i$

39. Degree of polynomial: 6; Roots: 0, 2 (multiplicity three), $-2 + 2i\sqrt{3}$, and $-2 - 2i\sqrt{3}$

41. $2, 1 + i$ **43.** $2\sqrt{3}\left(\cos\dfrac{11\pi}{6} + i\sin\dfrac{11\pi}{6}\right)$ **45.** $12\sqrt{3} - 12i, 6i$ **47.** $\dfrac{\sqrt{3}}{2} + \dfrac{1}{2}i$

49. (a) $-5, -2$ (b) $\dfrac{20}{7}, -\dfrac{3}{7}$ **51.** $\dfrac{59}{53} + \dfrac{21}{53}i$ **53.** $\pm 3i^{3/2}, \pm 3i^{1/2}$

Chapter 12

Exercise Set 12.1, page 497

1. $-2, 1, 4, 7, 22, 55$ **3.** $\dfrac{1}{3}, \dfrac{2}{5}, \dfrac{3}{7}, \dfrac{4}{9}, \dfrac{9}{19}, \dfrac{20}{41}$ **5.** $0, \dfrac{1}{2}, \dfrac{2}{3}, \dfrac{3}{4}, \dfrac{8}{9}, \dfrac{19}{20}$ **7.** $4, 7, 12, 19, 84, 403$

9. $6, 2, 0, 0, 30, 272$ **11.** $0, 0, 6, 24, 504, 6840$ **13.** $\dfrac{3}{2}, \dfrac{5}{6}, \dfrac{7}{12}, \dfrac{9}{20}, \dfrac{19}{90}, \dfrac{41}{420}$

15. $\dfrac{1}{2}, \dfrac{1}{4}, \dfrac{1}{8}, \dfrac{1}{16}, \dfrac{1}{512}, \dfrac{1}{1,048,576}$ **17.** $3, 4, 5, 6, 7, 8$ **19.** $1, -1, -5, -13, -29, -61$

21. 3, −4, −3, −10, 3, −4   **23.** 1, 3, 2, −1, −3, −2   **25.** 1, 3, 1, 3, 1, 3   **27.** 4, 7, 10, 13, 16

29. 5, 3, 1, −1, −3   **31.** −2, 3, 8, 13, 18   **33.** $\frac{1}{2}, \frac{3}{4}, 1, \frac{5}{4}, \frac{3}{2}$ **35.** $\frac{1}{4}, \frac{3}{8}, \frac{1}{2}, \frac{5}{8}, \frac{3}{4}$ **37.** 59, 119

39. −19, −49   **41.** 77, 167   **43.** −58, −118   **45.** $\frac{31}{6}, \frac{61}{6}$ **47.** 69, 650   **49.** 0, −280   **51.** 63, 910

53. $\frac{27}{4}, \frac{115}{2}$ **55.** $\frac{27}{8}, \frac{115}{4}$ **57.** 78, 820   **59.** 24, −200   **61.** 48, 1000   **63.** −72, −800   **65.** $8, \frac{220}{3}$

67. n^2 **69.** 2^{n-1} **71.** 1 for $n = 1$, 2^{n-2} for $n \geq 2$ **73.** 3, 7, 21   **75.** 6, 2, −12   **77.** $375   **79.** 666

Exercise Set 12.2, page 501

1. no   **3.** yes   **5.** no   **7.** 32, 256, 2^{n-1} **9.** 64, 512, 2^n **11.** 128, 8192, $\left(\frac{1}{8}\right)4^{n-1}$

13. $\frac{1}{3}, -\frac{1}{81}, -81\left(-\frac{1}{3}\right)^{n-1}$ **15.** 81, 2187, 3^{n-2} **17.** $\frac{243}{2}, -\frac{6561}{16}, -16\left(-\frac{3}{2}\right)^{n-1}$

19. −0.000001, 0.000000001, $0.1(-0.1)^{n-1}$ **21.** $\frac{4}{125}, \frac{32}{15{,}625}, \frac{25}{8}\left(\frac{2}{5}\right)^{n-1}$ **23.** $2^n - 1$, 127

25. $-2(1 - 2^n)$, 254   **27.** $\frac{1 - 4n}{-24}, \frac{5461}{8}$ **29.** $\frac{-243\left[1 - \left(-\frac{1}{3}\right)^n\right]}{4}, -\frac{547}{9}$

31. $\frac{1 - 3^n}{-6}, \frac{1093}{3}$ **33.** $\frac{-32\left[1 - \left(-\frac{3}{2}\right)^n\right]}{5}, -\frac{463}{4}$ **35.** $\frac{[1 - (-0.1)^n]}{11}$, 0.0909091   **37.** $\frac{125\left[1 - \left(\frac{2}{5}\right)^n\right]}{24}, \frac{25{,}999}{5000}$

39. Since $a_1, a_2, a_3, a_4, \ldots$ is a geometric sequence, we know that for $n \geq 1$, $\frac{a_{n+1}}{a_n} = r$. Thus, $\frac{a_{n+2}}{a_n} = r^2$.

41. 9 ft, 3 ft, $\frac{1}{3}$ ft   **43.** 45 ft, 51 ft, $53\frac{2}{3}$ ft   **45.** $80, $86.40, ≈$469.33

47. Since $a \neq b$, we know that $0 < (a-b)^2$ which can be used to show $\sqrt{ab} < \frac{a+b}{2}$.

Exercise Set 12.3, page 508

1. $1 + 4 + 7 + 10 + 13 = 35$ **3.** $13 + 8 + 1 = 22$ **5.** $3 + 3 + 3 + 3 = 12$

7. $0 + 1 - 2 + 3 - 4 = -2$ **9.** $\sum_{n=3}^{6} n$ **11.** $\sum_{n=1}^{5} n^2$ **13.** $\sum_{n=1}^{6} 2n$ **15.** $\sum_{n=1}^{4} (-1)^n$ **17.** 14, 91

19. 21, 1365   **21.** $\frac{7}{9}, \frac{182}{243}$ **23.** $\frac{95}{9}, \frac{3325}{243}$ **25.** 8, 105   **27.** $\frac{125}{7}$ **29.** series has no sum

31. $\frac{20}{7}$ **33.** series has no sum   **35.** $\frac{13}{12}$ **37.** $-\frac{3}{2}$ **39.** $\frac{242}{1125}$ **41.** $\frac{1082}{2475}$ **43.** $\frac{44}{333}$ **45.** $\frac{8}{2475}$

47. $\sum_{n=1}^{\infty} 4\left(-\frac{1}{3}\right)^{n-1}$ **49.** $1, 2, 4, 8, 2^{n-1}$ **51.** $3, 7, 11, 15, 4n - 1$ **53.** $-3, -1, 1, 3, 2n - 5$

55. $\sum_{n=1}^{\infty} (3n + 1)$ **57.** $\sum_{n=1}^{\infty} (n^2 + 1)$ **59.** $\sum_{n=1}^{\infty} (3^n - 1)$ **61.** $\frac{1}{2}$

Exercise Set 12.4, page 513

1. (a) 1 (b) 3 (c) -1 **3.** (a) does not exist (b) does not exist (c) 0   **5.** (a) -3 (b) $\frac{3}{2}$

7. (a) does not exist (b) does not exist   **9.** $\frac{3}{2}$ **11.** 0   **13.** does not exist   **15.** 0

17. $\lim_{n \to +\infty} \frac{2\left[1 - \left(-\frac{1}{2}\right)^n\right]}{3}$ **19.** $\lim_{n \to +\infty} 24\left[1 - \left(\frac{3}{4}\right)^n\right]$ **21.** $\lim_{n \to +\infty} \left(\frac{1}{2} - \frac{1}{n+2}\right)$

23. $\lim_{n \to +\infty} \left(-\frac{13}{12} + \frac{1}{n+2} + \frac{1}{n+3} + \frac{1}{n+4}\right)$ **25.** 0   **27.** 3   **29.** 0

Exercise Set 12.6, page 525

1. $x^6 + 6x^5y + 15x^4y^2 + 20x^3y^3 + 15x^2y^4 + 6xy^5 + y^6$

3. $a^7 - 7a^6b + 21a^5b^2 - 35a^4b^3 + 35a^3b^4 - 21a^2b^5 + 7ab^6 - b^7$ **5.** $81 + 216p + 216p^2 + 96p^3 + 16p^4$

7. $16 + 96p + 216p^2 + 216p^3 + 81p^4$ **9.** $\dfrac{t^3}{27} - \dfrac{t^2}{3} + t - 1$ **11.** $z^5 + 5z^3 + 10z + \dfrac{10}{z} + \dfrac{5}{z^3} + \dfrac{1}{z^5}$

13. $1 - 6x^{1/2} + 15x + 20x^{3/2} + 15x^2 - 6x^{5/2} + x^3$ **15.** $s + 3s^{4/3} + 3s^{5/3} + s^2$ **17.** $1 + 12x + 66x^2 + 220x^3$

19. $256 - 1024s + 1792s^2 - 1792s^3$ **21.** $1 + 22p + 220p^2 + 1320p^3$ **23.** $1140t^3$

25. 70   **27.** $55y^{13/2}$ **29.** $12870p^{16}$ **31.** $i + \dfrac{1}{2}\left(1 - \dfrac{1}{n}\right)i^2 + \dfrac{1}{6}\left(1 - \dfrac{3}{n} + \dfrac{2}{n^2}\right)i^3$ **35.** 3^n

Chapter Test, page 528

1. $3, 0, -1, 0, 3, 35$ **3.** $2, 7, -2, 3, 2, 7$ **5.** $7, 2, -3, -8, -13$ **7.** $63, 123$ **9.** 222   **11.** no

13. $-\dfrac{16}{45}, \dfrac{128}{1215}, \dfrac{27}{10}\left(-\dfrac{2}{3}\right)^{n-1}$ **15.** $6\left[1 - \left(\dfrac{2}{3}\right)^n\right], \dfrac{422}{81}$ **17.** $4 + 9 + 14 + 19 + 24 = 70$ **19.** $\sum_{n=2}^{6} n^2$

21. $7, 63$ **23.** $\dfrac{8}{3}$ **25.** $\dfrac{1}{3}$ **27.** $\dfrac{41}{110}$ **29.** does not exist   **31.** does not exist   **33.** 4   **35.** $-\dfrac{1}{5}$

37. does not exist   **41.** $81 + 540t + 1350t^2 + 1500t^3 + 625t^4$

43. Since $a_1, a_2, a_3, \ldots$ is a geometric sequence, we know that for $n \geq 1$, $\dfrac{a_{n+1}}{a_n} = r.$ Thus, $\dfrac{\frac{1}{a_{n+1}}}{\frac{1}{a_n}} = \dfrac{a_n}{a_{n+1}} = \dfrac{1}{r}.$

Cumulative Test: Chapters 11 and 12, page 531

1. $2 + 26i$ **3.** $-\dfrac{5}{3} + 2i$ **5.** $P(5, -3)$ **7.** 6   **9.** $\dfrac{3}{2} \pm \dfrac{1}{2}i$ **11.** $-1, -4, 4$

13. $4\left(\cos\dfrac{5\pi}{3} + i\sin\dfrac{5\pi}{3}\right)$ **15.** $4\sqrt{2}\left(\cos\dfrac{5\pi}{12} + i\sin\dfrac{5\pi}{12}\right), 2\sqrt{2}\left(\cos\dfrac{11\pi}{12} - i\sin\dfrac{11\pi}{12}\right)$ **17.** $-4 + 0i$

19. $-36, -113$ **21.** $\dfrac{209}{5}$ **23.** $729, -19{,}683, (-3)^{n-2}$ **25.** $\dfrac{16\left[1-\left(\frac{1}{4}\right)^n\right]}{3}, \dfrac{1365}{256}$ **27.** 11, 120

29. series has no sum **31.** (a) 1 (b) 2 **33.** (a) $\dfrac{1}{3} + \dfrac{10}{3}i$ (b) $\dfrac{\sqrt{10}}{5}$

Calculator Exercises

Exercise Set 1.2, page A1
1. 29.2 **3.** -6.2 **5.** 0.3 **7.** 98.4 **9.** $-10{,}590$ **11.** -3

Exercise Set 1.3, page A2
1. $\dfrac{6.8 + 5.7}{10.02} < |8.01 - 9.26|$ **3.** $\dfrac{(6.07)(0.889)}{1 + |1 - 6.19|} > |8.14 - 9.011|$ **5.** -0.0475 **7.** 0.0175 **9.** B

Exercise Set 1.4, page A2
1. 63.87 **3.** 65.01 **5.** $1 - (0.004)^2$ **7.** $(1 + 0.02)^2$ **9.** $<$ **11.** $<$

Exercise Set 1.7, page A3
1. $1 + 0.02$ **3.** $\sqrt{7}$ **5.** $\sqrt[4]{4}$ **7.** Successive roots get closer and closer to 1. **9.** No. The successive squares are getting "arbitrarily" large: larger and larger without any limit.

Exercise Set 1.8, page A4
1. -0.20 **3.** 2.20 **5.** 0.07 **7.** The sum gets closer and closer to 1.

Exercise Set 2.1, page A4
1. -2.9 **3.** 8.2 **5.** 4.6 **7.** 8.255, 8.245 **9.** 167.8 **11.** 21,592.5

Exercise Set 2.2, page A5
1. no real roots **3.** one real root of multiplicity two **5.** no real roots **7.** $-0.4, -1.4$ **9.** $1.2, -2.2$

11. 0.8, 0.5

Exercise Set 2.4, page A5
1. $(-\infty, 4.3)$ **3.** $(2.333, 2.334)$ **5.** $(-\infty, -0.5]$ or $[0.9, +\infty)$ **7.** $(-3.7, 3.7)$

Exercise Set 2.5, page A5
1. 18.25 ft **3.** 53.15 mph, 45.15 mph

Exercise Set 3.2, page A6
1. 1, 2.2, 3.4, 4.6, 5.8, 7 **3.** $-0.34, -0.9, -1.38, -1.78, -2.1, -2.34$

5. 0.678, 0.748, 0.812, 0.872, 0.927, 0.980

Exercise Set 3.4, page A6
1. $y = 0.71x + 5.79$ **3.** $y = -1.25x + 4.12$ **5.** $y = -0.89x + 8.16$ **7.** $y = 5.25x - 3.50$

9. (b) -2 (c) -2 **11.** (b) 1.9 (c) 1.9 **13.** (b) 5.6 (c) 5.6 **15.** (b) -0.4 (c) ≈ -0.39

Exercise Set 3.5, page A7
1. (b) 2.8 (c) 2.75 **3.** (b) 2.8 (c) 2.75 **5.** (b) -0.3 (c) -0.25 **7.** (b) -1.1 (c) -1.055

9. 2.5 sec, 101.3 ft **11.** 1.6 sec, 39.4 ft

Exercise Set 4.2, page A7
1. $[-1.5, -1.25]$ **3.** $[1.75, 2]$ **5.** $[2.25, 2.5]$ **7.** $[2.25, 2.5]$

Exercise Set 4.4, page A8
1. 5.70 **3.** 84.71 **5.** outside **7.** inside **9.** inside **11.** outside

Exercise Set 5.1, page A8
1. (3.04, 0.64) **3.** (3.086, 0.982) **5.** (0.71, 0.71), $(-0.71, -0.71)$ **7.** (0.5, 0.87), $(-0.5, -0.87)$

9. (0.71, 0.71)

Exercise Set 5.3, page A9
1. 9.80 **3.** 32.40 **5.** (a) $(-2.8, 0, 0)$ (b) $(0, -2.1, 0)$ (c) $(0, 0, 17)$

7. (a) (0.2, 0, 0) (b) (0, 0.3, 0) (c) $(0, 0, -0.1)$

Answers to Exercises

Exercise Set 6.1, page A9

1. 8.574, 8.634, 8.694, 8.754, 8.815, 8.877, 8.815   **3.** 2.67   **5.** 31.71   **7.** 1.96   **9.** 0.78   **11.** 1.63

13. -0.17 **15.** 1.5

Exercise Set 6.4, page A10

1. 0.613   **3.** 3.103   **5.** 4.819   **7.** 1.627   **9.** 1.272   **11.** 0.582

Exercise Set 6.5, page A10

1. (a) $27,899.83 (b) 12 years and 12 days   **3.** 231 days

Exercise Set 7.1, page A11

1. $c = 53.19$ **3.** $b = 486.7$ **5.** $c = 0.9568$ **7.** $a = 3267$ **9.** $b = 18.931$

Exercise Set 7.2, page A11

1. $\sin A = 0.8032$ $\csc A = 1.2451$ **3.** $\sin A = 0.5163$ $\csc A = 1.9370$
  $\cos A = 0.5956$ $\sec A = 1.6790$ $\cos A = 0.8564$ $\sec A = 1.1677$
  $\tan A = 1.3485$ $\cot A = 0.7416$ $\tan A = 0.6029$ $\cot A = 1.6588$

5. $\sin A = 0.7508$ $\csc A = 1.3318$ **7.** $\sin A = 0.7986$ $\csc A = 1.2522$
  $\cos A = 0.6604$ $\sec A = 1.5142$ $\cos A = 0.6018$ $\sec A = 1.6617$
  $\tan A = 1.1369$ $\cot A = 0.8796$ $\tan A = 1.3270$ $\cot A = 0.7536$

9. $\sin A = 0.5488$ $\csc A = 1.8222$ **11.** $h = 5.498$ **13.** 90   **15.** $\cos \alpha = 0.9231$, $\cos \beta = 0.3846$
  $\cos A = 0.8360$ $\sec A = 1.1962$
  $\tan A = 0.6565$ $\cot A = 1.5233$

17. $\tan \alpha = 1.8755$, $\cos \beta = 0.8824$ **19.** $\tan \alpha = 0.3430$, $\sin \beta = 0.9459$

Exercise Set 7.3, page A12

1. $\sin \alpha = 0.3656$ $\csc \alpha = 2.7352$ **3.** $\sin \alpha = -0.9442$ $\csc \alpha = -1.0591$
  $\cos \alpha = -0.9308$ $\sec \alpha = -1.0744$ $\cos \alpha = 0.3293$ $\sec \alpha = 3.0372$
  $\tan \alpha = -0.3928$ $\cot \alpha = -2.5459$ $\tan \alpha = -2.8678$ $\cot \alpha = -0.3487$

5. $\sin \alpha = -0.3780 \quad \csc \alpha = -2.6456$
$\cos \alpha = -0.9258 \quad \sec \alpha = -1.0801$
$\tan \alpha = 0.4083 \quad \cot \alpha = 2.4494$

Exercise Set 7.4, page A12

1. $\tan 32° = 0.6249 \quad \sec 32° = 1.1792$ **3.** $\csc 55° = 1.2208 \quad \cos 55° = 0.5738$
$\csc 32° = 1.8871 \quad \cot 32° = 1.6003$ $\sec 55° = 1.7435 \quad \cot 55° = 0.7002$

5. $\cos \alpha = 0.8587 \quad \cot \alpha = 1.6759$ **7.** $\sin \alpha = 0.5672 \quad \csc \alpha = 1.7631$
$\tan \alpha = 0.5967 \quad \sec \alpha = 1.1646$ $\tan \alpha = 0.6886 \quad \sec \alpha = 1.2142$
$\csc \alpha = 1.9516$ $\cot \alpha = 1.4521$

9. 0.5180   **11.** 0.3971   **13.** 0.6122   **15.** 1.2020   **17.** 0.8941   **19.** 1.4267   **21.** 28.00°   **23.** 39.66°

25. 43.01°   **27.** 0.6691   **29.** 0.6249   **31.** -0.8090 **33.** -0.9603 **35.** 1.0457   **37.** 135.60°

39. 275.20°   **41.** 321.60°   **43.** 196.30°

Exercise Set 8.1, page A14

1. IV   **3.** III   **5.** IV   **7.** II   **9.** I   **11.** -0.9301 **13.** -0.8922 **15.** 0.199   **17.** 1.381

19. 720.00°   **21.** $-310.91°$

Exercise Set 8.2, page A14

1. 0.8633   **3.** -0.5000 **5.** -0.8660 **7.** 1.7319   **9.** -0.4755 **11.** -1.6611 **13.** 5.5971

Exercise Set 8.3, page A15

1. $-0.8415, 0.1411, -0.8367, -0.7568, 0, -0.5806$

3. $-0.2130, -0.6002, -0.9789, -0.0730, -0.7071, -0.9863$

5. $-0.6421, -7.0153, 0.6547, -0.8637, 27224.2, 1.4024$

Exercise Set 8.4, page A15

1. $1.9191, -0.8961, 0.2001, 1.0522, -1.3088, 0.5222$ **3.** $-0.8304, 1.6436, -1.3411, 0.3972, -0.1247, 0.8625$

Exercise Set 8.5, page A15
1. 0.1874 **3.** −1.2780 **5.** 0.9533 **7.** 0.6405

Exercise Set 9.2, page A15
1. 0.2588 **3.** 0.2588 **5.** 0.2679 **7.** −0.4799 **9.** 0.3531 **11.** −0.5470
15. 0.9917 **17.** 0.4385 **19.** 2.0494

Exercise Set 9.3, page A16
1. 0.7100 **3.** 0.9917 **5.** 0.4385 **7.** 2.0494

Exercise Set 9.5, page A16
1. 19.27° **3.** 71.57° **5.** 24.09° **7.** 22.5° **9.** 43.95° or 75.52°

Exercise Set 10.1, page A16
1. $b = 36.42$, $c = 43.33$, $\angle B = 57.2°$ **3.** $a = 1422.5$, $c = 1922.3$, $\angle B = 42.27°$
5. $a = 9.353$, $b = 17.451$, $\angle B = 61.81°$ **7.** $c = 3.3780$, $\angle A = 43.52°$, $\angle B = 46.48°$
9. $c = 1.0097$, $\angle A = 12.77°$, $\angle B = 77.22°$ **11.** 48.4 ft

Exercise Set 10.2, page A17
1. $a = 135.4$, $b = 45.7$, $\angle C = 59°$ **3.** $b = 4485.6$, $c = 2535.9$, $\angle A = 71.2°$
5. $c = 10.55$ or 7.62, $\angle B = 84.4°$ or 95.6°, $\angle C = 37°$ or 25.8° **7.** $b = 12{,}864.9$, $\angle A = 23.5°$, $\angle B = 124.8°$
9. $a = 360.9$, $\angle A = 26.2°$, $\angle C = 41.5°$ **11.** 117.4 ft

Exercise Set 10.3, page A17
1. $c = 11264$, $\angle A = 29.1°$, $\angle B = 41.6°$ **3.** $a = 7.558$, $\angle B = 83.3°$, $\angle C = 45.2°$
5. $b = 0.9504$, $\angle A = 63{,}6°$, $\angle C = 54.8°$ **7.** $\angle A = 39.12°$, $\angle B = 24.02°$, $\angle C = 116.86°$
9. $\angle A = 45.01°$, $\angle B = 57.71°$, $\angle C = 77.28°$ **11.** 150.62 in **13.** 505.99 **15.** 6.329 **17.** 236.57 in^2

Exercise Set 10.4, page A18
1. $\langle -5.12, 5.34 \rangle$, magnitude 7.40, direction −46.2° **3.** $\langle 7.53, -1.25 \rangle$, magnitude 7.63, direction −9.43°
5. $\langle 1.34, 12.55 \rangle$ **7.** $\langle -831.82, -80.85 \rangle$

Exercise Set 10.5, page A19

1. 0.52**i**, 9.88**j**   **3.** −883.6**i**, −423.4**j**   **5.** S 38.7 E 6.4 mph   **7.** 1467 ft lb, 2741 ft lb

Exercise Set 10.6, page A19

1. at equilibrium, stationary
0.32 cm above equilibrium
0.55 cm above equilibrium
0.55 cm below equilibrium

3. 0.12 cm below equilibrium
0.24 cm below equilibrium
0.79 cm below equilibrium
0.23 cm below equilibrium

Exercise Set 10.7, page A20

1. (−1.87, 1.15)   **3.** (3.33, 4.22)   **5.** (6.5, 2.2)   **7.** (−6.06, 0.53)

9. 3, 3.84, 4, 3.78, 3, 2.24, 2, 2.08, 3   **11.** 0, 1.4, 2.1, 3.15, 4.4, 5.6, 6.6, 7.17, 8.8

Exercise Set 12.1, page A20

1. 191.8, 395.8, 19,384   **3.** −15,063, −37,163, −1,528,400   **5.** 2.27   **7.** 8,062,500

Exercise Set 12.3, page A20

1. 6.4   **3.** 0.063   **5.** $21,474,831   **7.** $20,807.68

Exercise Set 12.4, page A21

1. 4.01, 4.001, 4.0001, 4   **3.** 32.24, 32.024, 32.0024, 32

5. (a) approx: 1.2496, exact: 1.25, difference: 0.0004
(b) approx: 1.9375, exact: 2, difference: 0.0625
(c) approx: 2.3056, exact: 2.5, difference: 0.1944
approx: 3.0507812, exact: 4, difference: 0.9492188
approx: 3.3616, exact: 5, difference: 1.6384
(d) The five-term approximation is more accurate the closer x is to zero.

7. $0.1\overline{6}$, 0.25, 0.30, $0.3\overline{3}$, 0.357, 0.375, $0.3\overline{8}$, 0.4, 0.409, $0.41\overline{6}$

Exercise Set 12.6, page A22

1. 1.082856, 1.0828567   **3.** 0.81704, 0.8170728   **5.** 128.44867256, 128.44867   **7.** 1.0304237, 1.03

9. 1.2214, 1.2214028; 1.6484375, 1.6487213; 1.8214, 1.8221188; 2.1147461, 2.117; 2.2224, 2.2255409

Index

A-A, *see* Angle-Angle Similarity Theorem
Absolute value
  defined, 11
  distance, 14
  equations involving, 59
  graph of, 102
  inequalities involving, 80
  of complex numbers, 469
Acute angle, 282
  trigonometric ratios of, 289
Acute triangle, 410
Addition
  associative property, 8
  commutative property, 8
  of complex numbers, 460
  of fractional expressions, 46
  of ordinates, 349
  of polynomials, 23
  of radicals, 39
  of vectors, 424
Addition method
  to solve systems of equations, 196
Addition Property
  for Equivalent Equations, 57
  for Equivalent Inequalities, 76
Additive Identity Property, 8
Additive Inverse Property, 8
Air speed, 428
Algebraic equation, 3

Algebraic expression, 3
Altitude of triangle, 284
Ambiguous case, 412
Amplitude, 344
Angle
  acute, 282, 289
  azimuth and, 407
  bearing and, 407
  central, 320
  complementary, 294
  coterminal, 296
  of depression, 406
  of elevation, 406
  initial side of, 296
  obtuse, 282
  quadrantal, 296
  radian measure of, 320
  reference, 309
  right, 282
  of rotation, 296–300
  special, 292, 300, 329
  in standard position, 296
  terminal side of, 296
  vertex of, 282
Angle-Angle Similarity Theorem, 286
Antilogarithm, common, 263
Arc
  and chord, 371
  standard, 319
Arccos, 355

Arcsin, 352
Arctan, 355
Argument of complex number, 480
ASTC memory device, 309
Arithmetic mean, 498
Arithmetic sequence, 494
  sum of n terms, 496
Arithmetic series, 496
Associative Property, 8
Asymptote
  horizontal, 167
  limits, 512
  oblique, 174
  of cosecant function, 341
  of cotangent function, 339
  of exponential function, 251
  of hyperbola, 183–185
  of rational function, 167
  of secant function, 341
  of tangent function, 338
  vertical, 167
Augmented matrix, 217
Axis
  coordinate, 100
  imaginary, 468
  polar, 440
  real, 468
Axis of symmetry of parabola, 131
Azimuth, 407

Index

Base
  e, 273
  of exponential function, 251
  of logarithmic function, 257
  ten, 263
Bearing, 407, 408
Binomial, 22
  coefficient, 520
Binomial Theorem, 521
  proof of, 523–524
Bounded region, 236
Bracketing values, 306

Carbon-14 dating, 274
Center/radius form of a circle, 180
Central angle, 320
Chart, use in problem solving, 88
Chord, 371
Circle
  center, 179
  equation of, 122
  graph of, 178
  radius, 179
  tangent line to, 188
Closed interval, 78
Coefficient
  binomial, 521
  matrix, 217
  numerical, 22
Cofunction identities, 304
Column of a matrix, 216
Common antilogarithm, 263
Common denominator, 46
Common difference, 494
Common factor, 45
  greatest, 28
Common logarithm, 263
Common ratio, 499
Commutative property
  of addition, 8
  of multiplication, 8
Complementary angles, 294
Completing the square, 65
Complex conjugate, 292
Complex fractional expression, 47
Complex numbers
  absolute value, 469
  and radicals, 7
  argument of, 480
  conjugate, 464
  defined, 457
  equality of, 459

Complex numbers (*continued*)
  imaginary part, 457
  modulus of, 480
  operations with, 458
  powers and roots of, 483
  product and quotient of, 481
  properties of, 464
  quotient, 467
  real part, 457
  trigonometric form of, 480
Complex plane, 468
Complex zero, 473
Composite functions, 136
Compound inequality, 79
Compound interest, 273
Conics, 175–186
  circle, 179
  ellipse, 181
  hyperbola, 183
  parabola, 176
Conjugate
  complex, 464
  radical expression, 47
Constant matrix, 217
Constraints, 235
Conversion identities, 387–389
Convex region, 236
Coordinate
  polar, 440
  of point, 12
  plane, 100
  rectangular, 442
Coordinates of an ordered pair, 100
Corner points of a feasible region, 235
Cosecant
  defined, 290
Cosecant function
  domain of, 326
  graph of, 341–342
  period of, 341
  properties of, 341
Cosine
  defined, 290
Cosine function
  domain of, 326
  graph of, 336–337
  inverse of, 354
  period of, 336
  properties of, 336
Cosines, Law of, 416
Cotangent,
  defined, 290

Cotangent function
  domain of, 326
  graph of, 339–340
  period of, 339
  properties of, 339–340
Coterminal angles, 296
Cramer's Rule, 231

Decay
  exponential, 274
  radioactive, 274
Decimal
  repeating, 7
  terminating, 7
Degree (°), 282
  fractional part of, 305
Degree of a polynomial, 22
De Moivre's Theorem, 483
Denominator
  least common, 47
  rationalizing, 41
Dependent variable, 110
Depression, angle of, 406
Determinant
  and Cramer's Rule, 231
  of 2×2 matrix, 225
  of 3×3 matrix, 229
  to solve system of 2 linear equations, 227
Difference
  of cubes, 30
  of squares, 30
Difference identities, 371–376
Digits, significant, 405
Direct variation, 129
Direction, 428
Discriminant, 68
Distance
  defined in one dimension, 14
  formula, 179
Distributive property
  defined, 8
  in simplifying polynomials, 23
  in multiplying polynomials, 24
Division
  of complex numbers, 464
  of polynomials, 161
  of radicals, 38–42
Division algorithm, 479
Division of complex numbers, 483
Domain
  of a function, 108
  of a polynomial function, 151
  of a sequence, 492

Index

Dot product, 432
Double-angle identities, 378, 383

e, 273
Effective rate of interest, 525
Elevation, angle of, 406
Ellipse
  center of, 181
  graph of, 181
  vertices of, 181
Endpoints of an interval, 78
Entry of matrix, 216
Equality
  of complex numbers, 459
  of vectors, 424
Equations
  algebraic, 3
  equivalent, 57
  fractional, 58
  graphing, 101
  linear, 58
  nonlinear, 70
  polar, 443
  polynomial, 70
  properties of, 57
  quadratic, 63
  rectangular form of, 443
  root of, 56
  solution of, 56
  straight line, 101
  systems of linear, 194
  trigonometric, 391–397
Equilateral triangle, 252, 284
Equilibrium, forces in, 430
Equilibrium position, 434
Equivalent equation, 57
Equivalent inequality, 76
Even function, 113
Expanded form, 26
Exponents
  complex numbers, 483
  integer, 16
  negative, 17
  rational, 41
  real, 250
  rules of, 18, 250
Exponential decay, 274
Exponential function, 251
  defined, 251
  equation, 257
  graph of, 251
  natural, 273
  simplifying, 250

Exponential growth, 274
Expression
  algebraic, 3
  simplifying algebraic, 8
Extending the set, 454
Extraneous roots, 73

Factor, 28
Factor Theorem, 168
Factorial notation, 520
Factoring
  by grouping, 31
  patterns, 30
  polynomials, 28
  trinomials, 32
Feasible region, 235
Feasible solutions, 235
First Rule of Polynomial
  Functions, 153
FOIL method, 25
FOIL method for multiplying
  complex numbers, 460
Fractional equation, 58
Fractional expressions, 45
  arithmetic operations with, 46
  complex, 48
  radicals involved in, 47
  reducing, 45
Frequency of harmonic motion,
  435–437
Function
  composite, 136
  defined, 108
  domain of, 108
  evaluate, 110
  even, 113
  exponential, 251
  graph of, 108
  inverse, 255
  linear, 123
  logarithmic, 257
  odd, 113
  one-to-one, 255
  periodic, 333, 344–349
  polynomial, 148
  quadratic, 130
  range of, 108
  rational, 167
  sequence, 316
  trigonometric, 297, 325
  values, 110
  zero of, 157

Geometric mean, 502
Geometric sequence, 499
  sum of n terms, 501
Geometric series, 504
Graphing
  addition of ordinates and, 349
  conic sections, 175
  equations, 100
  functions, 112
  horizontal translations, 116
  inequalities, 77
  parabolas, 101, 176
  polar coordinates and, 440,
    444–445
  polynomial functions, 150, 157
  quadratic functions, 130
  rational functions, 170
  reflection, 114
  sets of ordered pairs, 108
  straight lines, 101
  stretching, 115
  symmetry, 104
  transformations of
    trigonometric functions,
    344–349
  translation, 114
  trigonometric functions,
    333–341
  vertical translation, 116
Greater than ($>$), 12, 76
Greater than or equal to ($\geq$), 76
Greatest common factor, 28
Ground speed, 428
Growth, exponential, 274

Half-angle identities, 378–383
Half-life of radioactive substance,
  275
Half-open interval, 78
Harmonic motion, simple,
  434–438
Head of vector, 423
Horizontal asymptote, 167
Horizontal change in shape, 120
Horizontal line, equation of, 125
Horizontal translations, 116
Hyperbola
  asymptotes of, 183
  graph of, 183
  vertex of, 184
Hypotenuse, 251, 283

i, 283
Identities, 302–305, 366–369
  basic, 367–368
  cofunction, 304
  conversion, 387–389
  difference, 371–376
  double-angle, 378–380, 383
  half-angle, 378–383
  Pythagorean, 303, 367
  quotient, 302, 367
  reciprocal, 302, 367
  sum, 371–376
Identity
  additive, 8
  multiplicative, 8
  properties, 8
Identity statement, 3
Imaginary
  axis, 468
  number, 457
  part of a complex number, 457
  roots of a polynomial, 472
Independent variable, 110
Index of a radical, 36
Index of sigma notation, 503
Induction, mathematical, 515
Inequality, 12
  absolute value and, 80
  compound, 79
  equivalent, 76
  graphing of, 78
  notation, 78
  properties of, 76
  quadratic, 81
  relationship, 12
  systems of, 205
Infinite series, 504
Infinity, 78
Initial point of vector, 423
Initial side of angle, 296
Integer, 6
  exponents, 16
Intercept form of a line, 126
Intercept of a graph, 126
Interest, 272
  effective rate of, 525
Interpolation, 267
Intersection of sets, 85
Interval notation, 78
Inverse
  additive, 8
  multiplicative, 8
  properties, 8
Inverse function
  defined, 255

Inverse sine function, 351
Inverse cosine function, 354
Inverse tangent function, 355
Irrational number, 6–7
Irreducible polynomial, 33
Isosceles triangle, 252, 283, 284

Law of Cosines, 416
Law of Sines, 411
Leading coefficient, 149
Leading term, 149
Least common denominator, 47
Legs of right triangle, 283
Less than ($<$), 12, 76
Less than or equal to ($\leq$), 76
Like radicals, 39
Limit
  evaluate, 502
  of a sequence, 504
Line
  equation of, 100
  general form, 127
  graph of, 101
  point-slope form, 125
  slope of, 123
  slope-intercept form, 126
  two point form, 124
Linear combination of vectors, 426
Linear equation in standard form, 58
Linear equations, systems of 194
  facts about solutions of, 201
Linear function, 123
Linear interpolation, 267
Linear programming, 235
Linear systems of inequality, 205
Logarithmic function, 257
Logarithms
  common, 263
  defined, 257
  equation, 257
  graph, 258
  natural, 273
  rules of, 259
Log/Exp Principle, 257
Lowest terms, 45

Mathematical Induction, 515
Matrix
  and systems of equations, 218
  augmented, 217
  coefficient, 217
  column of, 216

  constant, 217
  determinant of, 225
  entry of, 216
  reduction of, 218
  row of, 216
  square, 216
Maximum function value, 135
Maximum, relative, 158
Mid-point formula, 188
Minimum function value, 135
Minimum, relative, 158
Minor of a determinant, 229
  signed, 229
Minutes, 305
Model, mathematical, 86
Modulus of a complex number, 480
Monomial, 22
Motion, simple harmonic, 434
Multiplication
  associative property, 8
  commutative property, 8
  of complex numbers, 458
  of complex numbers in trigonometric form, 479
  of fractional expressions, 46
  of polynomials, 24–26
  of radicals, 40
  of vectors, 424, 432–433
  scalar, 424
Multiplicative Identity Property, 8
Multiplicative inverse, 8
Multiplicative properties for equivalent equations, 57
  inequalities, 76
  negative numbers, 76
  positive numbers, 76
Multiplicity of a zero, 64

$n!$, 270, 520
n th partial sum, 504
n th root of complex number, 484
n th root of unity, 463, 485
n th Root Theorem, 484
Natural exponential function, 273
Natural logarithm, 273
Negative exponent, 17
Nonlinear equation, 58
  systems of, 201
Norm of vector, 424
Notation
  interval, 78
  set, 78

Index

Number line
  distance, 14
  real, 12

Oblique triangle, 410
Oblique asymptote, 174
Objective function, 235
Obtuse angle, 282
Obtuse triangle, 410
Odd function, 113
One-to-one
  correspondence, 12
  function, 255
Open interval, 78
Order relationship, 12
Ordered pair, 100
Ordered triple, 211
Ordinates, addition of, 349
Origin
  in polar coordinate system, 440
  in a coordinate plane, 100
  on number line, 12
  symmetry, 104

Parabola
  graph, 101, 176
  quadratic function, 130
  vertex form, 131
Parallel lines, slope of, 127
Parallelogram, 252
Partial sum, 504
Pascal's Triangle, 522
Perfect trinomial square, 30
Period
  of cosecant function, 341
  of cosine function, 336
  of cotangent function, 339
  of secant function, 341
  of sine function, 335
  of tangent function, 338
Periodic function, 333
  amplitude of, 344
  using transformations to graph, 344–349
Perpendicular lines, 127
Phase shift, 348
Plotting points, 100
Point-slope form, 125
Polar axis, 440
Polar coordinates
  defined, 440
  graphing and, 444–445
  plotting points using, 440–441

Pole, in polar coordinate system, 440
Polynomial
  arithmetic operations on, 23
  constant, 22
  degree of, 22
  equation, 464–470
  factoring of, 28
  irreducible, 33
  leading coefficient of, 149
  terms of, 149
  zero, 22
Polynomial function, 148
  graphing, 157
  rules of, 153, 157, 158, 161
  standard form, 149
  zero of, 159, 301
Position vector, 424
Power Property for Equations, 73
Prime number, 28
Principal square root, 36
  of a complex number, 456
Problem solving strategies, 86
Product
  dot, 431–432
  scalar, 424
  of vectors, 424, 432–433
Proof by mathematical induction, 515
Properties of real numbers, 8
Pure imaginary number, 457
Pythagorean identities, 303, 367
Pythagorean Theorem, 179, 251, 284

Quadrant, 100
Quadrantal angle, 296
Quadratic equation
  in form, 71
  standard form, 63
Quadratic formula
  derivation of, 66
  statement of, 66
Quadratic function, 130
  defined, 132
  general form, 132
Quadratic inequalities, 81
Quadrilateral, 252
Quantities
  scalar, 421
  vector, 421
Quotient identities, 302, 367

Quotient
  of complex numbers, 464
  of polynomials, 161
  of radicals, 38

Radian measure, 321
Radical sign, 36
Radicals
  expression, 36
  index of, 36
  like, 39
  rules for simplifying, 36, 37
Radicand, 36
Radioactive decay, 274
Radius of a circle, 179
Radius vector, 424
Range of a function, 108
Rational exponents, 41
Rational expression, 45
Rational function, 167
  graphing, 170
Rational number
  decimal representation of, 7
  defined, 6
Rational Root Theorem, 166, 308
Rationalizing denominators, 41
Real exponent, 250
Real number system, 6
Real numbers
  described, 6
  ordering of, 12
  properties of, 8
Real part of a complex number, 457
Reciprocal identities, 302, 367
Reciprocal, negative, 127
Rectangular coordinates, 442
Recursive definition of a sequence, 493
Reference angle
  finding, 309–401
  determining trigonometric function values using, 309
Reflection and graphing, 114, 117
Reflection of point P, 102
Relation, 109
Relative maximum, 158
Relative minimum, 158
Remainder Theorem, 478
Repeating decimal, 7
Resolving into vector components, 429
Right angle, 282

Right triangle, 283, 404
  hypotenuse of, 251, 283
  legs of, 283
  solving, 404
  solving applied problems using, 406
  trigonometric ratios of, 289
Root
  of complex number, 483
  of an equation, 56
  of multiplicity n, 64
  of unity, 469
  square, 36
Rotation, angle of, 296–300
Rounding, 405
Row of a matrix, 216
Row reduction, 218
Rules
  for Horizontal change in shape, 120
  for Vertical change in shape, 119
  of exponents, 18, 250
  of logarithms, 259

S_n
  arithmetic sequence, 496
  geometric sequence, 500
Scalar components of vector, 423
Scalar multiplication, 424
Scalar quantities, 421
Scientific notation, 264
Secant
  defined, 290
Secant function
  domain of, 326
  graph of, 340–341
  period of, 341
  properties of, 341
Sequences
  arithmetic, 494
  described, 492
  as functions, 492
  geometric, 499
  of partial sums, 505
  S_n, 496
  telescoping, 507
  terms of, 493
Series, 504
  arithmetic, 496
  geometric, 504
  infinite, 504
  of partial sums, 504
  telescoping, 507
Set notation, 78

Sets of numbers, 6
Shrinking of graphs, 115, 118
Sigma notation, 503
Significant digits, 405
Similar triangles, 285–287
Sine
  defined, 290
Sine function
  domain of, 326
  graph of, 333–335
  inverse, 351–354
  period of, 335
  properties of, 335
Sines, Law of, 410
Slope
  defined, 123
  of a curve, 514
  parallel lines, 127
  perpendicular lines, 127
Slope-intercept form of a straight line, 126
Solution
  of an equation, 56
  of inequalities, 76
  of systems of equations, 194
Solution set
  equations, 57
  inequalities, 76
  notation, 78
Special angles
  trigonometric ratios of, 292, 300, 329
Speed, 428
  air and ground, 428
Square matrix, 216
Square root
  principal, 36, 460
  property of, 64
Standard arc, 319
Standard form of
  circle, 180
  ellipse, 181–182
  hyperbola, 183–186
  linear equation, 58
  polynomial, 149
  quadratic equation, 63
Standard position of angle, 296
Straight line
  equation of, 100
  general form, 127
  graph of, 101
  point-slope form, 125
  slope of, 123
  slope-intercept form, 126
  two-point form, 124

Stretching of graphs, 115, 118
Substitution method of systems of equations, 199
Subtraction
  of complex numbers, 460
  of fractional expressions, 46
  of polynomials, 23
  of radicals, 39
Sum
  identities, 371–376
  of arithmetic sequence, 496
  of cubes, 30
  of geometric sequence, 506
  partial, 504
Summation notation, 503
Symmetry
  of a graph, 104–105
  with respect to origin, 105
  with respect to x-axis, 104
  with respect to y-axis, 105
Synthetic division, 477
Systems of inequalities, 206
Systems of linear equations
  equivalent, 196
  in three variables, 213
  in two variables, 194
  matrix solution, 218
  solution of, 194
Systems of nonlinear equations, 201

Tail of vector, 423
Tangent
  defined, 290
Tangent function
  domain of, 326
  graph of, 337–339
  inverse, 355
  period of, 338
  properties of, 338
Tangent line to a circle, 188
Telescoping series, 507
Term
  of a polynomial, 22
  of a sequence, 493
Terminal point of vector, 423
Terminal side of angle, 296, 297–299
Terminating decimal, 7
Test value, 162
Tests for symmetry, 104, 105
Theta-ray (θ-ray)
  defined, 440
  opposite of, 440

Index

Transformations, 114–121
Translations
  described, 114
  horizontal, 116
  vertical, 116
Triangle, 282
  acute, 410
  altitude of, 284
  equilateral, 252, 284
  isosceles, 252, 283, 284
  oblique, 410
  obtuse, 410
  right, 283, 289
  similar, 285–287
  solving, 404–408, 410–414, 416–417
Trigonometric equations, 391–397
  inverse, 396
Trigonometric form of complex numbers, 478
Trigonometric ratios
  of acute angle, 289
  finding, 293
Trinomial, 22

u-substitution, 71
Unbounded region, 236
Unit circle, 318
  points and arcs on, 318–320
Unit coordinate vectors, 425
Unit vector, 425
Unity, nth roots of, 485
Union of sets, 85

V-shaped graph, 102
Value of a function, 110

Variable
  dependent, 110
  described, 2
  independent, 110
Variation, direct, 129
Vectors
  addition of, 424
  describing as ordered pairs, 423
  determining magnitude and direction of, 421
  dot product of, 431
  equality of, 424
  forces in equilibrium and, 430
  heads of, 423
  initial points of, 423
  linear combination of, 426
  multiplication of, 424, 432
  norm of, 424
  position, 424
  radius, 424
  resolving into components, 429
  scalar components of, 423
  scalar multiplication of, 424
  solving applied problems using, 428
  tails of, 423
  terminal points of, 423
  unit, 425
  unit coordinate, 425
  vector components of, 429
  zero, 424
Vector components, 429
Vector quantities, 421
Velocity, 428
Vertex
  of angles, 282
  of a hyperbola, 184
  of a parabola, 131
  of triangle, 283
Vertical asymptote, 167
Vertical change in shape, rule for, 119
Vertical line, equation of, 125
Vertical line test, 111

Weighted mean, 50

x-axis, 100
  symmetry about, 104
x-intercept
  of a line, 126
  of a parabola, 132
xy-plane, 100
xyz-coordinate system, 210

y-axis, 100
  symmetry, 105
y-intercept, 126

Zero
  of a function, 157
  of multiplicity m, 64
  of a polynomial, 159, 474
Zero, complex, 473
Zero exponent, 18
Zero factorial, 520
Zero polynomial, 22
Zero vector, 424
Zero Products, Property of, 63